World Scientific Now Publisher Series in Business

GAMES and
DYNAMIC GAMES

World Scientific–Now Publishers Series in Business

ISSN: 2251-3442

Vol. 1   Games and Dynamic Games
by Alain Haurie, Jacek B. Krawczyk and Georges Zaccour

World Scientific – Now Publishers Series in Business: **Vol.1**

GAMES and DYNAMIC GAMES

Alain Haurie
ORDECSYS and University of Geneva, Switzerland

Jacek B Krawczyk
Victoria University of Wellington, New Zealand

Georges Zaccour
HEC Montréal, Canada

Published by

World Scientific Publishing Co. Pte. Ltd.

5 Toh Tuck Link, Singapore 596224

USA office: 27 Warren Street, Suite 401-402, Hackensack, NJ 07601

UK office: 57 Shelton Street, Covent Garden, London WC2H 9HE

British Library Cataloguing-in-Publication Data
A catalogue record for this book is available from the British Library.

Cover art by Zofia Krawczyk-Bernotas

World Scientific–Now Publishers Series in Business — Vol. 1
GAMES AND DYNAMIC GAMES

ISBN-13 978-981-4401-26-5
ISBN-10 981-4401-26-9

In-house Editor: Sandhya Venkatesh

Printed in Singapore by Mainland Press Pte Ltd.

Preface

Game theory can be defined as the study of mathematical models of conflict and cooperation between intelligent, rational decision makers.[1] *Game engineering* can be defined as the
use of game theory to design practical interactive systems.[2] This book deals with games
of competition and cooperation[3] between rational decision makers, the prediction of the
possible results of these games and the design of practical mechanisms to obtain *desired*
outcomes from the interaction between decision makers.

In many games, the strategic interaction between players (individuals, firms, countries,
animals or cells) recurs over time and decisions made during one period affect both current payoffs and future gains. *Dynamic games* provide conceptually rich paradigms and
tools to deal with these situations. The intent of this book is to provide a comprehensive
overview of the theory and applications of equilibrium solution concepts in games and
dynamic games. It contains an introductory chapter followed by ten chapters, which are
divided into three parts. In the first part (Chapters 2–4), we give a general introduction
to the classical noncooperative game theory. In the second part (Chapters 5–7), we cover
deterministic-dynamic games (repeated, multistage and differential games). The third part
(Chapters 8–11) is reserved for stochastic games (games played over event trees, Markov
games, piecewise-deterministic differential games and stochastic-diffusion games).

In an area where many excellent textbooks and monographs are available, convincing
ourselves to offer a new book required a solid argument. We first observed that many of
the existing books do not cover dynamic games at all, or do so very briefly. A second
observation is that textbooks on dynamic games tend to focus on one particular class of
games (e.g., evolutionary, multistage, differential, repeated or stochastic games). Given this
state of affairs and our recent teaching and research experience, where we saw a growing
demand for dynamic game models in economics, management and environmental science,
we concluded that a *general* textbook on dynamic games was very much needed.

Writing a book with such a broad scope provides an exciting opportunity to touch on
different fascinating topics. This comes, however, at the cost of making some tough deci-

---

[1]Definition given by Myerson in [179].

[2]Definition taken from a presentation by R. Adman.

[3]Note that this book does not consider solution concepts for cooperative games that are not based on the equilibrium paradigm. We will however design equilibria in which competitive players can achieve efficient (cooperative)
payoffs.

sions about how far we go in each chapter without going beyond what can be taught in a single course. We thought that, in such a context, we had to keep in mind the following elements while writing this textbook:

This has to be a self-contained book. The book must contain the background in classical game theory that is necessary to fully understand dynamic games. Having everything between the two (front and back) covers simplifies the learning process and conveys a much-needed unified vision for the field. This is relevant at a time when scholars tend to become highly specialized and work in multidisciplinary teams. The first four chapters can be used on their own to teach a short introductory course in classical game theory.

Must-learn material has to be included. As every idea is potentially useful, and given that the definition of what is a must-visit concept, tool, etc., depends on the reader's background and objectives, our task of choosing what to include in each chapter was far from trivial. Our decisions were based on what it takes for a reader to *fully* access the ever-growing applied literature in dynamic games. As a by-product, we hope that the book will arouse the interest of many readers sufficiently to pursue their studies of dynamic games further.

Convincing applications need to be provided. The book must contain strong showcases of the theory at work, not just simple classroom examples. At the end of each chapter we provide a case study we call *game engineering* (GE), to help the reader understand how to model the problem at hand as a game and how to engineer a solution to that game. Taken together, these series of case studies is intended to give the reader a good idea of the variety of problems that can be studied as games or dynamic games. In the GE sections, the reader will also find hints on how to tackle a new class of problems of high social priority, related to global environmental negotiations and the joint exploitation of renewable resources.

The book must be mathematically rigorous yet accessible. A quick survey of the available textbooks and lecture notes on game theory clearly shows that there is no unique writing style used to communicate the main ideas. For example, some books are written in a formal theorem-proof format for readers with advanced mathematical knowledge, others are written in plain English for readers who are not interested in the mathematical aspects of the theory of games. In this book, we try to present all of the concepts and results in a formal way, subject to keeping the mathematics at the level it is usually taught in graduate programs in economics, management science or engineering. The implication of this approach is that some proofs are sketched rather than given in full detail. Students and researchers who are familiar with advanced concepts in, e.g., functional analysis, topology or measure theory, can consult, if they are interested, the original papers referenced in our book where the full proofs first appear. In order to make these difficult sections more accessible, we present numerical illustrations as often as possible and introduce the reader to several methods for numerically solving the games we study.

Our intended readership includes graduate students in economics, management science, engineering and operations research. The book is ideally suited for a two-semester course. Some selected topics can form a one-semester course, e.g., deterministic dynamic games, stochastic games, discrete-time games, continuous-time games, etc. It is hoped that the book will also benefit researchers in social sciences, management and engineering who are interested in using dynamic games in their respective fields.

A. Haurie, Geneva, Switzerland
J. B. Krawczyk, Wellington, New Zealand
G. Zaccour, Montréal, Canada

Acknowledgments

This book uses some results from joint research carried out with many different co-authors. In particular, we would like to thank our colleagues Frédéric Babonneau, Olivier Bahn, Michèle Breton, Dean Carlson, Laurent Drouet, Jerzy Filar, Steffen Jørgensen, Roland Malhamé, Maryse Labriet, Richard Loulou, Francesco Moresino, Odile Pourtallier, Michel Roche, Julien Thénié, Bolek Tolwinski, Jean-Philippe Vial, Marc Vielle and Laurent Viguier for their enriching collaboration.

We also would like to thank Josée Lafrenière who patiently proofread all the chapters, and Francine Benoît who competently produced the final version of the book.

Alain Haurie acknowledges support from the EU-FP6 research program (TOCSIN project), EU-FP7 research program (PLANETS and ERMITAGE projects) and from the GICC research program of France's ministry of ecology, sustainable development, transportation and housing (Ministère de l'Écologie, de l'Énergie, du Développement durable et de la Mer—MEEDDM). Several of the examples and $\mathbb{GE}$ cases given in this book are related to models developed for these research projects.

Jacek B. Krawczyk wishes to thank Paul Calcott and Vlado Petkov, his colleagues from the School of Economics and Finance at Victoria University of Wellington, for their comments, which led to improvements in the clarity of the presentation of certain economic issues.

Georges Zaccour acknowledges the generous financial support of HEC Montréal. He thanks Marilyne Lavoie of GERAD for her help with editing and Gert Jansens for his assistance.

Contents

Stochastic Games   303

Chapter 1

Introduction

1.1 What are dynamic games?

Dynamic Games are mathematical models of strategic interactions between independent agents who control a dynamic system. Such situations occur in military conflicts (e.g., a duel between a bomber and a jet fighter), economic competition (e.g., investments in R&D by computer companies), parlor games (chess, bridge, etc.), international relations (e.g., negotiation of free-trade and environmental agreements) and in many more instances of social interactions.

The above examples concern *dynamic systems*[1] in that the actions of the agents, called players, influence the evolution over time of a system's state, which can mean the aircraft's position and velocity, the high tech firms' stock, of know-how the position of the remaining pieces on a chessboard, the concentrations of pollutants in the atmosphere, etc. The difficulty in deciding what the behavior of these agents should be stems from the fact that each *action* an agent takes at any given time will influence the information received by the other agents, and therefore, their *reaction*, at later times. This difficulty of establishing, and the need for a dynamic-game theory were from the outset, clearly on the minds of von Neumann & Morgenstern, the founding fathers of game theory, when they wrote in their seminal book:

"We repeat most emphatically that our theory is thoroughly static. A dynamic theory would unquestionably be more complete and therefore preferable. But there is ample evidence from other branches of science that it is futile to try to build one as long as the static side is not thoroughly understood" (von Neumann & Morgenstern [182], p. 44).

In von Neumann & Morgenstern's view, dynamic games were bound to be the natural object of the next-step development of a general theory of games. In this sense, dynamic games can be seen as a generalization of classical one-shot games to multi-period games, where strategic interactions carry over in time. From the methodological (and historical) standpoint, we can also note the clear proximity of dynamic games to optimal-control theory, which deals with optimization of dynamic systems. The methods developed for solving optimal-control problems were successfully extended for solving the equilibria of

---

[1]Loosely speaking, a *dynamic system* is one that evolves over time, driven by internal or exogenous forces.

multistage and differential games. Given this, it is fair to state that dynamic games are the offspring of both the classical theory of games and optimal-control theory.

1.2   Objectives and contribution

This textbook has two main objectives. First, it aims to provide a comprehensive introduction to dynamic-game theory; second, it intends to show how this paradigm can be usefully applied to an array of socially important problems in economics, engineering and management science.

Our approach to applying the theory is in the spirit of Robert Aumann's *game engineering*, (see [13]). For Aumann,[2] game engineering comprises the application of game theory either to systems with strict game-play rules or to systems for that these rules must be built. In the second sense, game engineering concerns the design of incentives such that a solution of the game at hand coincides with an outcome desired by a regulator. Aumann mentions final-offer arbitration, auctions, matching, traffic planning in cities, and elections as examples of areas where game engineering has found implementation. We would also include in game engineering the so-called game-theoretic techno-economic models[3] (TEMs), which have been developed for, e.g., the assessment of global policies in energy planning and environmental protection. All of these models combine economic considerations with engineering imperatives. An instructive illustration of such a combination is the model of Nordhaus and Yang [185], where economic reasoning meets the "hard constraints" of engineering, the environment and earth science. Interestingly, this model is a dynamic-game-theoretic offspring of the optimal-control DICE model proposed earlier by Nordhaus [184].

Another point of contact between economics and engineering, which is also mentioned by Aumann in [13], is mathematical programming. Optimization under constraints, in particular linear programming, has been a cornerstone in the development of operations research and has also been linked to the development of activity-analysis models in economics, offering a very useful paradigm for the description of economic-production systems and for the interactions among different economic sectors (see [63], [53]). Recently, as the security of the energy supply and the global environmental threats have become important political issues, techno-economic models have been developed, using mathematical-programming paradigms, to assess policies that can lead to different economic equilibria. Some of the applications and theory developed in this book are based on the ideas and tools of mathematical programming.

---

[2]Robert Aumann is, with Thomas Schelling, the 2005 winner of the Sveriges Riksbank Prize in Economic Sciences in Memory of Alfred Nobel.

[3]This class of models encompasses the economic and the technological aspects of key production sectors such as energy, telecommunication, environmental protection, etc.

1.3 Contribution

A quick Internet search shows the availability of dozens of books and lecture notes on game theory. The reader may therefore legitimately wonder, as the writers originally did, about the need for a new book. First, most of these textbooks are concerned with the *classical* (static) theory of games and do not introduce the reader to the most general *dynamic* games. A nonexhaustive list includes Owen [193], Shubik ([221] and [222]), Aumann [12], and more recently Friedman [87], Fudenberg and Tirole [89], Myerson [179] and Osborne and Rubinstein [192].

Second, books on dynamic games tend to specialize in one particular class of games. In their now-classic monograph, Başar and Olsder ([18]) give an excellent account of multistage and differential games, but, by and large, the book is oriented toward engineering students with a strong mathematical background. Dockner et al. [59] extensively cover the class of differential games and provide numerous applications in economics and management science. Petrosjan [198] concentrates on pursuit-evasion games, whereas Friesz [88] deals with open-loop differential games (and dynamic optimization). Engwerda [68] offers a full coverage of linear-quadratic games, in both discrete and continuous time. In the stochastic games arena, the textbooks by Filar and Vrieze [82] and Neyman and Sorin [183] have clearly marked the field. Aumann and Maschler's book [14] is required reading for repeated games with incomplete information. Sorin [224] provides an extensive treatment of two-player zero-sum repeated games.

The contribution of this book is in its broad coverage of dynamic games and their applications. In our teaching experience, we often felt the need for a textbook that (i) provides, in a single location, comprehensive treatment of all classes of dynamic games,[4] along with their basis in classical game theory; and (ii) helps students, as well as applied researchers and analysts, to use dynamic games in game engineering, including, as previously stated, game-theoretic techno-economic modeling.

1.4 Origins and readership

This book has its origin in a series of lecture notes prepared for several courses on dynamic games taught by the authors at different universities or summer schools. Our intended readership includes graduate students in economics, management science, engineering and operations research who are interested in the analysis of multi-agent optimization problems, with a particular emphasis on the modeling of competitive economic situations. The level of mathematics required does not, for the most part, go beyond that usually expected of such students.

The book is ideally suited for a two-semester course. Some selected topics can form a one-semester course, e.g., deterministic dynamic games, stochastic-dynamic games,

---

[4]An exception is the class of evolutionary games, which has, until now, developed separately from the rest. Excellent books are available in this area, e.g., Hofbauer and Sigmund [125], Weibull [243], Cressman [50], and very recently, Sandholm [217].

discrete-time games, continuous-time games, etc. The book can also benefit researchers in the social sciences, in management or in engineering who are interested in using dynamic games in their fields.

1.5   Organization of the book

The book is organized into three parts: (i) fundamentals of classical game theory; (ii) deterministic dynamic games; and (iii) stochastic games.

The first part covers what a student should know about the description of a game, the various possible ways of defining equilibrium and their characterization, including coupled-constraints equilibria, as well as some extensions and refinements to these equilibrium concepts. This part can be also taught on its own as a rigorous introduction to the classical noncooperative theory of games.

The second part covers repeated games (Chapter 5), multistage games (Chapter 6) and differential games (Chapter 7). Repeated games provide a rich paradigm for incorporating histories of the game into the players' strategies so that they can achieve, at least under some conditions, a cooperative (collectively optimal) outcome. In the multistage-games chapter, we introduce the concept of a dynamic game played by several agents who share the control of a dynamic system observed in discrete time. We first review some fundamental properties of dynamic systems described in a state space. Then, we define equilibrium solutions under different information structures and characterize such solutions through either a coupled-maximum principle or a dynamic-programming equation. We conclude the chapter with an analysis of equilibrium solutions of infinite-horizon systems, where we consider the possibility of including threats in the strategies used by the players as a way to transform some cooperative solutions into equilibria. The differential-games chapter transposes what was done in discrete time in the previous chapter to a continuous-time setting. The starting point is that the system's dynamics is described by differential equations.

The last part deals with stochastic games. In Chapter 8, we deal with a class of stochastic games where the uncertainty is described by an event tree. The formalism developed in this chapter is, in particular, very much relevant to the study of, e.g., dynamic oligopolistic competition in markets characterized by random demand laws, as is the case in many energy and other commodity markets. In the first part of Chapter 9, we present the formalism of stochastic games, as initially proposed by Shapley, by defining Markov games in a two-player zero-sum context. In the second part, we show how to extend the Shapley-Markov game formalism to the case of nonzero-sum stochastic games, which can also be seen as a generalization of the multistage-game paradigm discussed in Chapter 6.

Chapter 10 deals with the class of piecewise deterministic differential games, that is, a class of stochastic games involving the control of a hybrid system by several players. The hybrid nature of the system is due to the fact that its state is composed of a discrete part and a continuous part. A motivation for these games can be found in processes that can "jump" due to, e.g., the discovery of a new technological process. Finally, in Chapter 11, we consider a class of dynamic games called stochastic-diffusion games, where the system is subject to a continuous flow of random shocks. To account for this randomness, the state dynamics are described by stochastic-differential equations. These games are relevant in many areas, e.g., in finance and renewable resources (forests, fisheries), where the evolution of the state (portfolio's wealth, stocks of different species) is naturally stochastic.

We finish each chapter with a brief section that summarizes the chapter's main contents, under the generic title "What have we learned in this chapter?" Then, we propose exercises that require the reader to use the chapter's theory for problem solving. Some exercises are purely mechanical to help the student learn the techniques, whereas others are case studies based on applied research papers.

Finally, at the end of each chapter, we provide a detailed application of dynamic games of the game-engineering variety, in a section called "GE Illustration." These applications deal with a variety of topics, namely:

Environmental issues. Here, we discuss assessment of international emissions agreements, strategic allocation of emissions permits in the European Union, a river-basin pollution problem, and competitive timing of climate policies.

Cartel formation. The vitamin C market in Europe in analyzed.

Macroeconomics. We analyze growth-rate divergences among countries with the same fundamentals.

Adoption of a new technology. We design a subsidy scheme with guaranteed buys by government to accelerate the dissemination of a desirable technology.

Energy issues. We discuss and model competition in the European natural gas market.

Renewable resources. We engineer a cooperative agreement to manage an open-access fishery.

Finance. We solve a game of debt-contract valuation.

PART 1
Classical Theory of Games

Chapter 2

Description of a Game

Dynamic games constitute a subclass of mathematical models studied in what is usually called game theory. It is therefore proper to start our exposition with those basic concepts of *classical* game theory[1] that provide the fundamentals for the theory of dynamic games. At the end of the chapter, to illustrate *game engineering* ($\mathbb{GE}$), we provide an example of how the basic concepts of game theory are used in a techno-economic model of climate change negotiations.

2.1 Basic concepts of game theory

In a *game* we deal with many concepts that relate to the interactions between agents. Below we provide a preliminary list of those concepts that will be further discussed and explained in this chapter.

- *Players*: The agents (humans or automata) interacting and competing in the game. A player[2] can be an agent acting on his sole behalf, e.g., a chess player, an aircraft pilot) or he can represent a set of individuals (e.g., a nation, a corporation, a political party).
- *Actions*: Each player has at his disposal a set of possible actions, called also moves or decisions.
- *Information structure*: A reference to what the players know about the game and its history when they choose an action.
- *Pure strategy*: A rule that associates a player's move with the information available to him at the time that he selects his move.
- *Mixed strategy*: A probability distribution (or measure) on the space of a player's pure strategies. We can thus view a mixed strategy as a random draw from a set of pure strategies.

---

[1]For an exhaustive treatment of most of the definitions of classical game theory see, e.g., Refs. [87, 89, 179, 192, 193, 199] and [221].

[2]Political correctness promotes the usage of gender-inclusive pronouns "they" and "their." However, in games, we will frequently have to address an individual player's action and distinguish it from a collective action taken by a set of several players. As far as we know, in English, this distinction is only possible through traditional, gender-exclusive pronouns: possessive "his", "her" in the possessive case, and personal "he", "she" in the nominative case. In this book, to avoid confusion, we will use "he" and "his" to refer to a single genderless agent.

- *Behavioral strategy*: A rule that defines a random draw of the admissible move as function of the information available.[3] These strategies are intimately linked with mixed strategies. (It was proved early, in [146], that there is a natural correspondence between these two concepts for many games.)
- *Payoffs*: Real numbers measuring desirability of the possible outcomes of the game, e.g., the amounts of money the players may win or loose. Other names for payoffs are *rewards, performance indices* or *criteria, utility measures*, etc.

In the rest of this chapter we give more precise definitions for these concepts. For this purpose we use the formalism of *game trees* to give a representation of the dependence of *outcomes* on players' *actions* and *uncertainties*. We say that the game is thus defined in its *extensive form*.

2.2 Games in extensive form

2.2.1 *Game tree*

A game in extensive form is defined on a *graph*. A graph is a set of nodes connected by arcs. An arc is defined by a pair of nodes, an origin node or *parent* and a termination node or *descendant*. In Figure 2.1 we show a node (called node "1") and 3 arcs, L, M, R originating from this node.

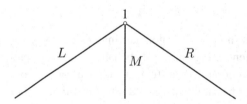

Fig. 2.1   Node and arcs

In a game tree, the nodes indicate game positions. Each position is reached by a succession of moves that define the history of play. The arcs correspond to the possible actions of the player who *has the move* in a given position.[4] To be meaningful, the graph representing a game must have the structure of a tree, that is, all nodes must be connected and there are no cycles.[5] A simple game tree is represented in Figure 2.2. The tree represents

---

[3]A similar concept has been introduced in control theory under the name of *relaxed controls*.

[4]As in parlor games, we use the expression "to have the move" to designate the situation where a player has to select one move.

[5]Two nodes are said to be connected if we can go from one node to the other by following a sequence of (not necessarily directed) arcs forming a path. A cycle is a path where the origin node is the same as the destination node.

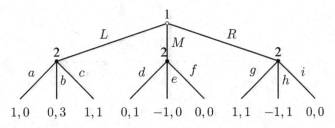

Fig. 2.2 A game tree

a sequence of actions that influence the outcome of a game played by a set of players. In a tree there is a single node without a "parent," called the "root" and a set of nodes without descendants, the "leaves." There is always a single path from the root to any leaf. In a game tree the leaves are marked with numbers that indicate the payoffs, or gains, received by the players as an outcome of their play which is represented by a path from the root to a leaf (see Figure 2.2).

2.2.2   *Description of moves, information and randomness*

A game in extensive form is described by a set of *players*, which can include a particular player called *Nature*, which always plays randomly, and a set of *positions*, which correspond to the nodes on the tree. There is a unique move sequence, called *history*, leading from the root node to each game position represented by a node. At each node one particular player *has the move,* i.e., he has to select a possible action from an admissible set represented by the arcs emanating from the node (see Figure 2.2). When a player selects a move, which corresponds to selecting an arc of the graph, a transition to a new node is defined, where another player has to select his move, etc. *Nature* is a player that selects its moves randomly. The game has a stopping rule described by the terminal nodes of the tree (the "leaves"). This is when the players are paid their rewards or, as we say, their payoffs.

The information that each player is in possession of at each node defines the *information structure* of the game. In full generality the player who has the move may not know exactly at which node of the tree the game is currently located. More exactly, he has the following information: *he knows that the current position of the game is within a given subset of nodes; however, he does not know which specific node it is.* This situation is represented in a game tree by connecting all the nodes that belong to the same *information set* with a dotted line.

In Figure 2.3 we see a set of nodes associated with Player 2 that are linked by a dotted line. Notice that the same number of arcs emanate from each node of the set. They must correspond to the same possible action choice; thus the names given to the arcs are now identical for the three connected nodes. Player 2 selects an arc (a move), knowing that the node is in the information set but not knowing which particular node in this set has

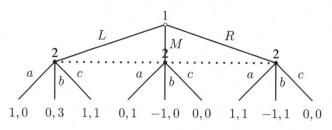

Fig. 2.3   Information set

been reached. To rephrase it, the player does not distinguish between the game positions represented by the connected nodes, but knows that he has to select one of the three actions, which are symbolized by the three arcs (a, b, c) originating from each connected node. We also say that this game has a *simultaneous move information structure*.

Figure 2.4 shows the extensive form of a two-player one-stage game with simultaneous moves and a random intervention of Nature, which plays at the nodes that are marked with n.

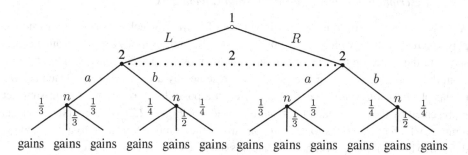

Fig. 2.4   A game in extensive form

This structure means that Player 2 does not know which action has been selected by Player 1 and vice versa. In this figure, Player 1 has to move at the node marked 1 and Player 2 has to move at the nodes marked 2. The information set of the second player is represented by the dotted line connecting the nodes. It says that Player 2 does not know what action Player 1 selects. The nodes marked n are positions where Nature has the move. In this particular case we assume that Nature's move is a random selection among three possible elementary events, with their corresponding probabilities. At the terminal nodes the game stops and the gains (payoffs) are collected.

The extensive form representation of games is inspired by parlor games like *chess*, *poker*, *bridge*, etc., which can be, at least theoretically, correctly described using this framework. In this context, the randomness of Nature's moves is the representation of a chance result of a card draw or a dice toss, realized in the course of the game. The extensive form indeed provides a very detailed description of a game. If the player knows at which node the game is now being played, he knows not only the current node of the game but may also remember the game's entire history (i.e., the sequence of moves by all players) that has led to this node in the game tree. However, the extensive form is rather impractical for analyzing even simple games because the size of the tree increases very fast with the number of steps.[6] An attempt to provide a complete description of a complex game like bridge using extensive form would lead to a huge game tree, a consequence of a combinatorial explosion.

There is another drawback to the extensive form description. To be represented as nodes and arcs, the game histories and actions have to be finite or enumerable. Yet, in many models we want to deal with actions and histories that are described by continuous variables. For such models, we need different methods to describe the rules of the games. For example, in Chapter 7 when we introduce the class of *differential games* we will use differential equations to represent consequences of the players' actions and in Chapter 9 when we deal with *stochastic games* we will use Markovian stochastic processes to describe the dynamics. Nevertheless, the extensive form is useful to conceptualize the structure of a game, using the most elementary mathematical apparatus.

2.2.3 *Von Neumann-Morgenstern utility*

Due to the randomness brought about by Nature's moves, the players will have to compare, and choose from, different *random outcomes* in their decision making. We use Figure 2.5 to illustrate a situation that may confront a player under uncertainty. At node D a decision maker has to choose between actions a_1 and a_2. If he chooses a_1 he faces a random experiment and the game outcome is random, with an expected value of 100. If, however, he chooses action a_2 the outcome is a sure gain of 100.

If the player is *risk neutral* he will be indifferent between these two actions. If he is *risk averse* he will choose action a_2, but if he is a *risk lover* he will choose action a_1.

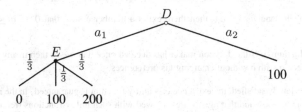

Fig. 2.5 Decision under uncertainty

---

[6]To appreciate the tree's expansion compare the one-step game extensive form in Figure 2.4 with the two-round matching pennies game extensive form that will be shown further on in this chapter, on Figure 2.6.

In order to represent the attitude toward risk of a decision maker, von Neumann and Morgenstern [182] introduced a concept of *cardinal utility* called VNM utility. If we accept the axioms[7] of VNM utility theory then a *rational player* should take the action that leads toward the random outcome with the highest *expected utility* . This behavior is a manifestation of the *principle of maximization of expected utility*, which directs the actions of rational agents.

Assume for example that our player is a risk averter whose utility function assigns the following "util" values to the possible outcomes,[8] as shown in Table 2.1.

Table 2.1   Utility function of a risk averter

Outcomes	Utils
0	0
100	0.75
200	1

Then the expected *utility* of choosing action a_1 is $\frac{1}{3}(0 + 0.75 + 1) = 0.5833$, whereas the expected *utility* of choosing action a_2 is 0.75. So, the *expected utility* of choosing a_2 is higher than the expected utility of action a_1 (obviously, $0.75 > 0.5833$) and the player will select the second action.

If the player is a *risk lover*, his utility function may be defined[9] as in Table 2.2. Here, the expected utility of choosing action a_1 is $\frac{1}{3}(0 + 0.35 + 1) = 0.45$, whereas the utility of choosing action a_2 is 0.35. In this case the player will select action a_1 because it generates 0.45 *utils* while choosing a_2 would have brought him 0.35 *utils* only.

---

[7]There are several *classical* axioms (see, e.g., [179]) formalizing the properties of a *rational agent*'s preferences. These axioms refer to the comparison of lotteries by an agent. The formula $a \succeq b$ denotes "a is preferred to b;" symmetrically, $a \preceq b$ denotes "b is preferred to a." The formula $a \sim b$ denotes "a or b are indifferent;" which means $a \succeq b$ and $a \preceq b$. Using this notation, the axioms that we need to define a rational agent are:

(1) Completeness. Given two lotteries u_1 and u_2 either $u_1 \succeq u_2$ or $u_1 \preceq u_2$.

(2) Transitivity. Between three lotteries u_1, u_2 and u_3 if $u_1 \succeq u_2$ and $u_2 \succeq u_3$ then $u_1 \succeq u_3$.

(3) Monotonicity. If $u_1 \succeq u_2$ and $0 \leq \beta < \alpha \leq 1$, then $\alpha u_1 + (1 - \alpha)u_2 \succeq \beta u_1 + (1 - \beta)u_2$.

(4) Continuity. If $u_1 \succeq u_2$ and $u_2 \succeq u_3$, then there exists a number δ such that $0 \leq \delta \leq 1$, and $u_2 \sim \delta u_1 + (1 - \delta)u_3$.

(5) Substitution (several axioms). If the decision maker has decided once that $a \sim b$ then in any situation where a occurs he can replace a with b without changing his preferences.

If the above axioms are jointly satisfied, the existence of a utility function is guaranteed. In the rest of this book we will assume that this is the case and that agents are endowed with such utility functions (referred to as VNM utility functions) and that their goal is to maximize the expected value of their utility functions. It is in this sense that we refer to *rational agents* in this book.

[8]Risk aversion is represented by the fact that the utility increases less between 100 and 200, than it does between 0 and 100.

[9]Risk loving is represented by the fact that the utility increases more between 100 and 200, than it does between 0 and 100.

Table 2.2   Utility function of a risk lover

Outcomes	Utils
0	0
100	0.35
200	1

The above example illustrates how adoption of VNM utility theory makes it possible to compare different random outcomes and to compare random outcomes with deterministic ones. A player may therefore decide to design a random experiment in order to choose an action when he has the move, and use the principle of expected utility to value this randomized decision. This is the idea behind the proposed use of *behavioral strategies*. Another possibility is for each player to randomly select a strategy to play the game. This is the concept of *mixed strategy*, which we will introduce shortly. Notice that, in this way of playing, one discrete choice among the possible actions or strategies is replaced by a "continuous" one, namely the probability distribution given to each action or strategy in the lottery (random experiment), which will determine the action that is actually implemented.

As a final remark on the foundations of utility theory let us recall that the von Neumann-Morgenstern utility function is defined up to a positive affine transformation[10] of rewards. This says that the player choices will not be affected if the utilities are modified through such a transformation.

2.3   Additional concepts about information

What is known by the players who interact in a game is of paramount importance for the game's outcome. Here, we briefly describe a few typical situations that the players may face.

2.3.1   *Complete and perfect information*

The *information structure* of a game indicates what is known by each player at the time the game starts and at each of their moves.

Complete or incomplete information refers to the information available to the players when they start a game. A player has *complete information* if he knows who the players are, which set of actions is available to each player, what each player's information structure is and what the players' possible outcomes can be. Otherwise, the player has *incomplete information*.

---

[10]A positive affine transformation $x \rightarrow y = f(x)$ is of the form $y = a + bx$ with $b > 0$.

Perfect or imperfect information refers to the information available to a player when he makes a decision about a specific move. In a game defined in extensive form, if each information set consists of a single node, then we say that the players have *perfect information*. If this is not the case the game is one of *imperfect information*.

Remark 2.1. A game with simultaneous moves (e.g., the one shown in Figure 2.4) is one with complete but imperfect information.

2.3.2 Common knowledge

An information structure in a game is *common knowledge* if all of the players know it (it is mutual knowledge) and all of the players know that all other players know it and all other players know that all other players know that all other players know it, and so on.

2.3.3 Perfect recall

When the information structure is such that a player can always remember all of the past moves he has selected along with all the information sets he has attained, we say that the game is of *perfect recall*. Otherwise the game is of *imperfect recall*.

2.3.4 Commitment

A *commitment* is an action to be taken by a player that binds him. This biding and the action itself are known to the other players. In making a commitment a player can persuade the other players to take actions that are favorable to him. To be effective, commitments have to be *credible*. A particular class of commitments are *threats*.

2.3.5 Binding agreement

Binding agreements are restrictions on the possible actions that can be decided by two or more players, under a contract that forces the implementation of the agreement. Usually, a binding agreement requires an outside authority that can monitor the agreement at no cost and impose sanctions on violators that are so severe that cheating is prevented.

2.4   Games in normal form

2.4.1 Playing games through strategies

A strategy is a way of transforming the information into action. The following defines it more precisely:

Definition 2.1. Let $M = \{1, \ldots, m\}$ be the set of players. A pure strategy γ_j for Player j is a mapping that associates an admissible move (action) to every information set available to Player j at any node where he has the move. We call the m-tuple $\underline{\gamma} = (\gamma_j)_{j=1,\ldots m}$ a strategy vector.

Once a strategy is chosen by each player, the strategy vector γ is well defined and the game is played as if it were controlled by automata.[11]

An outcome expressed in terms of its expected utility to Player $j, j \in M$ is associated with the all-player strategy vector $\underline{\gamma}$. If we denote by Γ_j the set of strategies for Player j then the game can be represented by the *payoff mappings*

$$V_j : \Gamma_1 \times \cdots \Gamma_j \times \cdots \Gamma_m \to \mathbf{R}, \quad j \in M \qquad (2.1)$$

that associate a unique (expected utility) payoff $V_j(\underline{\gamma})$ for each player $j \in M$ to a given strategy vector $\underline{\gamma} \in \Gamma_1 \times \cdots \Gamma_j \times \cdots \Gamma_m$. We then say that the game is defined in its *normal* or *strategic* form.

2.4.2 *From extensive form to strategic or normal form*

We will use an example to explain how we can pass from the extensive to the normal form in the description of a game.

Example 2.1. *We consider a simple two-player game, called "matching pennies."[12] The rules of the game are as follows: The game is played in two stages. There are two players and each has a coin. At the first stage each player chooses either head (H) or tail (T) without knowing the other player's choice. Then they reveal their choices to each other. If the coins do not match, Player 1 wins \$5 and Payer 2 wins -\$5 (loses \$5). If the coins match, Player 2 wins \$5 and Payer 1 wins -\$5 (loses \$5). At the second stage, the player who lost at stage 1 has the choice of either stopping the game (Q - "quit") or playing another matching penny subgame with the same type of payoffs as in the first stage. So, his second stage choices are (Q, H, T).*

This game is represented in the extensive form in Figure 2.6. A dotted line connects the different nodes that form an information set for a player. We use numbers '1' and '2' to indicate the player that *has the move* at each node or information set.

In Table 2.3 we have identified the 12 different strategies that can be used by each of the two players in the game of matching pennies. Each player moves twice. In the first move the players have no information; in the second move they know what moves were made at the first stage.

---

[11]The idea of playing games through the use of automata, like e.g., a computer program deciding which action to choose, will be discussed in more detail in chapter 5, when we present the *folk theorem* for repeated games.
[12]This example is borrowed from [87].

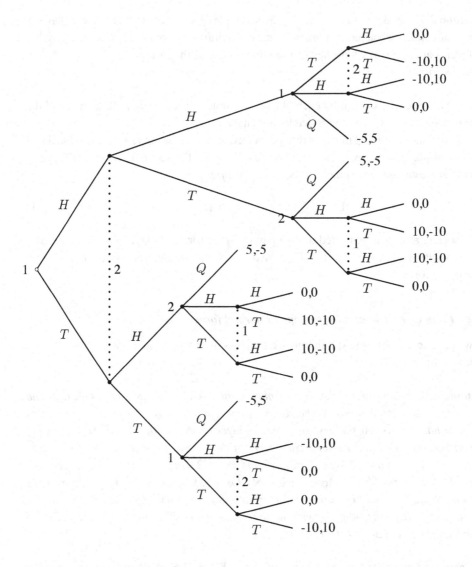

Fig. 2.6   The extensive form tree of the matching pennies game

In this table, each row describes each player's possible action, given the information available to that player. For example, row 1 shows that Player 1 would play Q at round 2 if he played H in round 1 and Player 2 played H in round 1. If Player 1 played H in round 1 and Player 2 played T in round 1 then Player 1, knowing this, would play H in round 2. Clearly this describes a possible course of action for Player 1. There are 12 possible courses of action listed in Table 2.3. The situation is symmetrical for the other player. The 12 possible courses of actions are listed in Table 2.3. The situation is similar (actually, symmetrical) for Player 2.

Table 2.3   List of strategies

	Strategies of Player 1			Strategies of Player 2		
	1st move	2nd move if Player 2 has played		1st move	2nd move if Player 1 has played	
		H	T		H	T
1	H	Q	H	H	H	Q
2	H	Q	T	H	T	Q
3	H	H	H	H	H	H
4	H	H	T	H	T	H
5	H	T	H	H	H	T
6	H	T	T	H	T	T
7	T	H	Q	T	Q	H
8	T	T	Q	T	Q	T
9	T	H	H	T	H	H
10	T	T	H	T	H	T
11	T	H	T	T	T	H
12	T	T	T	T	T	T

Table 2.4 represents the payoff matrix obtained by Player 1 when both players choose one of the 12 possible strategies. The strategies of Player 1 correspond to the rows and those of Player 2 correspond to the columns.

Table 2.4   Payoff matrix for Player 1

	1	2	3	4	5	6	7	8	9	10	11	12
1	-5	-5	-5	-5	-5	-5	5	5	0	0	10	10
2	-5	-5	-5	-5	-5	-5	5	5	10	10	0	0
3	-10	0	-10	0	-10	0	5	5	0	0	10	10
4	-10	0	-10	0	-10	0	5	5	10	10	0	0
5	0	-10	0	-10	0	-10	5	5	0	0	10	10
6	0	-10	0	-10	0	-10	5	5	10	10	0	0
7	5	5	0	0	10	10	-5	-5	-5	-5	-5	-5
8	5	5	10	10	0	0	-5	-5	-5	-5	-5	-5
9	5	5	0	0	10	10	-10	0	-10	0	-10	0
10	5	5	10	10	0	0	-10	0	-10	0	-10	0
11	5	5	0	0	10	10	0	-10	0	-10	0	-10
12	5	5	10	10	0	0	0	-10	0	-10	0	-10

So, the extensive form game defined in Figure 2.6 and played with the strategies indicated above is now represented through a 12×12 payoff matrix for Player 1.

Since what is gained by Player 1 is lost by Player 2 we say that this is a *zero-sum game*. Hence, there is no need to repeat the payoff matrix construction for Player 2, since this matrix is the exact negative of the previous one.[13] The matrix given by Table 2.4 defines the matching pennies game in its normal form since it is a description for this game of the payoff mapping defined in Eq. (2.1).   ◇

2.4.3 *Mixed and behavioral strategies*

The players can introduce randomness in their strategy definition in two possible ways.

Mixed strategies. Since players evaluate game outcomes according to their VNM-utility functions, which permit them to compare random outcomes they can consider selecting their strategy randomly, according to a random experiment or a lottery that they will define. For example, if Player j has p pure strategies, $\gamma_{jk}, k = 1, \ldots, p_j$, he can select the strategy he will play through a lottery, which gives a probability x_{jk} to the pure strategy $\gamma_{jk}, k = 1, \ldots, p_j$. His strategy choice is now one vector of probabilities selected among all the possible probability distributions

$$\mathcal{X}_j = \left\{ \mathbf{x}_j = (x_{jk})_{k=1,\ldots,p_j} \mid x_{jk} \geq 0, \quad \sum_{k=1}^{p_j} x_{jk} = 1. \right\} \tag{2.2}$$

We note that the set $\mathcal{X}_j$ is compact and convex[14] in $\mathbf{R}^p$. This will be important for proving existence of solutions to these games (see Chapter 3).

Behavioral strategies. There is another way of randomizing players' actions.

Definition 2.2. A *behavioral strategy* of Player j is a mapping, which assigns a probability distribution over his set of actions to the information set available to him at a decision node where he has the move.

The difference between *mixed* and *behavioral* strategies is subtle. In a mixed strategy, the player considers the set of possible strategies and picks one at random, according to an appropriately designed lottery. In a behavioral strategy, the player's strategy defines a random experiment in the form of an appropriately designed lottery to select an action at any information set where this player has the move.[15]

A theorem of Harold Kuhn [146], establishes that these two ways of introducing randomness into the choice of actions can be equivalent for games of perfect recall (see Exercise 2.5). We will not develop this aspect of classical game theory here; instead, we refer

---

[13]In a more general situation of a *nonzero-sum* game, a specific payoff matrix will be constructed for each player. The two payoff matrices will have the same dimensions, where the number of rows and columns will correspond to the number of strategies available to Player 1 and 2 respectively.

[14]The reader will notice that this set is the smallest convex set that contains the p extreme points $(x_{jk} = 1,$ $x_{jk'} = 0, k' \neq k)_{k=1,\ldots,p}$ in $\mathbf{R}^p$. This set is called a *simplex* in $\mathbf{R}^p$.

[15]The reader who already has a good knowledge of control theory will certainly recognize the conceptual proximity between behavioral strategies and chattering controls [252].

the interested reader to the book by Myerson [179], which provides an extensive treatment of these topics.

2.5 Looking for a solution to a game in normal form

2.5.1 *Elimination of dominated strategies*

The method of strategy deletion by *iterated strict dominance*, which we describe in this section, provides a first systematic or algorithmic way to find a possible solution to the game. The method is easily applicable to games in normal form, where players strive to maximize their expected utilities.

Consider a game defined in normal form played by m players indexed by $j = 1, \ldots, m$. The finite set of pure strategies for Player j is S_j. The set of mixed strategies for Player j is Σ_j. The reward, or payoff, for Player j when the pure strategy vector $\underline{s} = (s_1, \ldots, s_i, \ldots, s_m)$ has been selected by all players is $r_j(\underline{s})$. With a slight abuse of notation we will also denote $r_j(\underline{\sigma})$ the expected payoff for Player j when the players use the vector of mixed strategies $\underline{\sigma} = (\sigma_1, \ldots, \sigma_i, \ldots, \sigma_m)$. We use $\sigma_j(s_j)$ to denote the probability of using the pure strategy s_j in the mixed strategy σ_j.

Definition 2.3. A pure strategy $s_j \in S_j$ is said to be *strictly dominated* if there exists a mixed strategy $\sigma_j \in \Sigma_j$ for which

$$r_j(\sigma_j, \mathbf{s}_{-j}) > r_j(s_j, \mathbf{s}_{-j}), \quad \forall \mathbf{s}_{-j} \in \mathbf{S}_{-j}, \tag{2.3}$$

where $\mathbf{s}_{-j} \in \mathbf{S}_{-j} = \prod_{i \neq j} S_i$ denotes the strategy vector for all players $i \neq j$ and $\mathbf{S}_j$ is the combined strategy set of players $i \neq j$.

A rational player should not consider a strategy that is strictly dominated as a possible way to play the game, since there is another strategic choice that will yield better rewards under all circumstances.

So let us consider the set of pure strategies of Player j that are not strictly dominated:

$$S_j^1 = \{s_j \in S_j : \text{there is no } \sigma_j \in \Sigma_j \text{ s.t. } r_j(\sigma_j, \mathbf{s}_{-j}) > r_j(s_j, \mathbf{s}_{-j}), \forall \mathbf{s}_{-j} \in \mathbf{S}_{-j}\}.$$

This is the set of pure strategies for Player j that have resisted the strict-dominance test. We do this for each player and we redefine the game by restricting the strategy choices to these undominated pure strategies. For this new game we can repeat the procedure and eliminate the pure strategies that are strictly dominated, and so on according to the recurrence formula

$$S_j^n = \{s_j \in S_j^{n-1} : \text{there is no } \sigma_j \in \Sigma_j^{n-1} \text{ s.t. } r_j(\sigma_j, \mathbf{s}_{-j}) > r_j(s_j, \mathbf{s}_{-j}), \forall \mathbf{s}_{-j} \in \mathbf{S}_{-j}^{n-1}\}$$

where

$$\Sigma_j^{n-1} = \{\sigma_j \in \Sigma_j : \sigma_j(s_j) > 0 \text{ only if } s_j \in S_j^{n-1}\}.$$

Consider the set $S_j^\infty = \cap_1^n S_j^n$, i.e., the set of strategies for Player j that have survived iterated deletion of strictly dominated strategies. Since the game has a finite set of pure strategies for each player, the deletion process should stop after a finite number of iterations.

Definition 2.4. A game is solvable by strict iterated dominance if, for each Player j, the set S_j^∞ reduces to a singleton (i.e., has only one element).

Remark 2.2. Each step of this iteration is based on the assumption that each player thinks "I know that you know ... that I know the payoffs" of the reduced game.

Example 2.2. *Consider a two-player game with the following data:*
$S_1 = \{a, b, c\}$, $S_2 = \{d, e\}$ *and rewards*
$$r_1(a, d) = 1, r_2(a, d) = 4 \ r_1(a, e) = 1, r_2(a, e) = 0\,,$$
$$r_1(b, d) = 4, r_2(b, d) = 1 \ r_1(b, e) = 0, r_2(b, e) = 2\,,$$
$$r_1(c, d) = 0, r_2(c, d) = 0 \ r_1(c, e) = 4, r_2(c, e) = 1\,,$$

which can be summarized in the following table, where rows correspond to Player 1's pure strategies, columns to Player 2's pure strategies and each cell indicates the two players' respective rewards.

	d	e
a	1,4	1,0
b	4,1	0,2
c	0,0	4,1

Notice that, for Player 1 the pure strategy a is strictly dominated by the mixed strategy $\sigma_1(a) = 0, \sigma_1(b) = 0.5, \sigma_1(c) = 0.5$. Indeed the pure strategy "a" yields a payoff of 1 for Player 1 whatever be the strategy choice of Player 2, whereas the mixed strategy proposed above yields an expected payoff of 2 whatever be the strategy choice of Player 2. Since the two other pure strategies are not dominated, we get $S_1^1 = \{b, c\}$. For Player 2 the two pure strategies are undominated so $S_2^1 = \{d, e\}$.

For the next iteration we consider the reduced game defined by the following table where the row corresponding to strategy a has been deleted.

	d	e
b	4,1	0,2
c	0,0	4,1

Here we can see that the pure strategy d of Player 2 is strictly dominated by the pure strategy e. So we get $S_2^2 = \{e\}$. For Player 1 we still have $S_1^2 = \{b, c\}$.

For the next iteration we consider the reduced game defined by the following table where the column corresponding to strategy d has been deleted.

	e
b	0,2
c	4,1

For Player 1 the strategy b is strictly dominated; therefore, we end up with the singleton

	e
c	4,1

Now we have $S_1^3 = \{c\}$ and $S_2^3 = \{e\}$. The game is solvable by strict iterated dominance.

We will see in the next chapter that a solution obtained by strict dominance is a Nash equilibrium, which is a fundamental solution concept in the theory of noncooperative games.

Unfortunately many games are not solvable by iterated strict dominance, as we shall see in Exercise 2.9.

2.5.2 *Rationalizability*

The concept of *Rationalizability* was introduced by Bernheim [27] and Pearce [197]. Like iterated strict dominance (ISD), it derives restrictions on the player's strategy choices from the assumptions that the players' payoffs and rationality are *common knowledge*. In ISD, the starting point is the assumption that a rational player will never use a strategy that is strictly dominated. In rationalizability the starting point is a complementary question: what are the strategies that a rational player could play?

The answer is that a rational player will use only those strategies that are best responses to some beliefs he may have about the opponents' strategies. Because payoffs and rationality are common knowledge these beliefs cannot be arbitrary. A player should expect his opponents to use only strategies that are best responses to beliefs they may have. And those beliefs in turn cannot be arbitrary, etc. This leads to an *infinite recess*.

Definition 2.5. Let Σ_j denote the set of mixed strategies for Player j. Set $\tilde{\Sigma}_j^0 = \Sigma_j$. Define, for each j recursively

$$\tilde{\Sigma}_j^n = \left\{ \sigma_j \in \tilde{\Sigma}_j^{n-1} : \exists \boldsymbol{\sigma}_{-j} \in \prod_{i \neq j} \mathrm{Co}[\tilde{\Sigma}_i^{n-1}] \right.$$

$$\left. \text{s.t. } r_j(\sigma_j, \boldsymbol{\sigma}_{-j}) \geq r_j(\sigma_j', \boldsymbol{\sigma}_{-j}), \forall \sigma_j' \in \tilde{\Sigma}_j^{n-1} \right\}, \quad (2.4)$$

where $\mathrm{Co}[\tilde{\Sigma}_i^{n-1}]$ stands for the convex hull of $\tilde{\Sigma}_i^{n-1}$.

The rationalizable strategies for Player j are the elements of the set $R_j = \bigcap_{n=0}^{\infty} \tilde{\Sigma}_j^n$.

A strategy profile $\underline{\sigma} = (\sigma_1, \ldots, \sigma_m)$ is rationalizable if each component σ_j is rationalizable for each Player j.

Remark 2.3. In words, $\tilde{\Sigma}_{-j}^{n-1}$ is the set of strategies for Player j's opponents that have survived through $n-1$ stages of the elimination process. $\tilde{\Sigma}_j^n$ is the set of surviving strategies for Player j that are best response to some strategy in $\tilde{\Sigma}_{-j}^{n-1}$.

Remark 2.4. The convex hull is used in this definition because Player j is uncertain about which of several strategies in $\tilde{\Sigma}_i^{n-1}$ Player i will use. Even if σ_i' and σ_i'' are in $\tilde{\Sigma}_i^{n-1}$, the mixture $(\frac{1}{2}\sigma_i', \frac{1}{2}\sigma_i'')$ may not be in $\tilde{\Sigma}_i^{n-1}$, hence the need to consider the convex hull.

Theorem 2.1. *The set of rationalizable strategies is nonempty and contains at least one pure strategy for each player. Furthermore each $\sigma_j \in R_j$ is a best response (in Σ_j) to an element of $\prod_{i \neq j} \mathrm{Co}[R_i]$.*

Proof.   We can show inductively that the sets $\tilde{\Sigma}_i^n$ are closed, nonempty and nested and that they contain an element that is a pure strategy. Their infinite intersection is thus nonempty and contains a pure strategy. The existence of an element in $\prod_{i \neq j} \mathrm{Co}[R_i]$ to which $\sigma_j \in R_j$ is a best response is obtained by induction on n. See Refs. [27, 197] for details.   □

Pearce [197] has also proved the following result

Theorem 2.2. *Rationalizability and iterated strict dominance coincide in two-player games.*

We will see in the next chapter, where we explore the concept of an equilibrium solution to a game, that a Nash equilibrium solution is rationalizable. Here, we provide an example to show how rationalizability works in a normal-form game.

Example 2.3. *Consider a 3-player game, where the set of strategies are given by*
$$\Gamma_1 = \{s_1, s_2\}, \quad \Gamma_2 = \{t_1, t_2\}, \quad \Gamma_3 = \{z_1, z_2, z_3, z_4\}$$
for each player, respectively. The payoffs are reported in the following table. It can easily be checked that no strategy is dominated for any of the players. However, z_4 is not rationalizable, that is, it is never a best response for player 3 to any strategy profile of players 1 and 2. Indeed, the best responses for player 3 are $R_3(s_1, t_1) = z_1; R_3(s_2, t_1) = R_3(s_1, t_2) = z_2; R_3(s_2, t_2) = z_3$. This shows that z_4 is never a best response to pure-strategy profiles of players 1 and 2.

Table 2.5   Payoffs for the three players

player 3 plays z_1		player 2		player 3 plays z_2		player 2	
		t_1	t_2			t_1	t_2
player 1	s_1	3,3,9	3,3,0	player 1	s_1	3,3,0	3,3,9
	s_2	3,3,0	3,3,0		s_2	3,3,9	3,3,0

player 3 plays z_3		player 2		player 3 plays z_4		player 2	
		t_1	t_2			t_1	t_2
player 1	s_1	3,3,0	3,3,0	player 1	s_1	2,2,6	2,2,0
	s_2	3,3,0	3,3,9		s_2	2,2,0	2,2,6

For our reasoning to be complete, we also need to consider mixed-strategy profiles
$$(x_{11}s_1 + (1 - x_{11})s_2, x_{21}t_1 + (1 - x_{21})t_2),$$
where x_{jk} denotes the probability that player $j = 1, 2$, selects strategy k. Strategy z_4 is a better response than z_1 if
$$9x_{11}x_{21} \le 6x_{11}x_{21} + 6(1 - x_{11})(1 - x_{21}).$$
Strategy z_4 is a better response than z_2 if
$$9x_{11}(1 - x_{21}) + 9x_{21}(1 - x_{11}) \le 6x_{11}x_{21} + 6(1 - x_{11})(1 - x_{21}).$$
Strategy z_4 is a better response than z_3 if
$$9(1 - x_{11})(1 - x_{21}) \le 6x_{11}x_{21} + 6(1 - x_{11})(1 - x_{21}).$$
For rationalizability, these statements must be simultaneously true. We can verify that the system
$$\begin{cases} 9x_{11}x_{21} \le 6x_{11}x_{21} + 6(1 - x_{11})(1 - x_{21}) \\ 9x_{11}(1 - x_{21}) + 9x_{21}(1 - x_{11}) \le 6x_{11}x_{21} + 6(1 - x_{11})(1 - x_{21}) \\ 9(1 - x_{11})(1 - x_{21}) \le 6x_{11}x_{21} + 6(1 - x_{11})(1 - x_{21}) \end{cases}$$
has no solution. Therefore, z_4 is never a best response to any mixed-strategy profile of players 1 and 2. Hence it is not rationalizable.

2.5.3 *Toward more practical solution concepts*

The reader may be puzzled by the difficulty of searching for strictly non-dominated and rationalizable strategy vector. To his comfort it will be good to learn that we shall not proceed further in this direction to find a rational way of playing games. Instead, in the next chapter we will introduce the solution concepts of *saddle point* for a two-player zero-sum game and of *equilibrium* for an m player game. We will be able to prove existence of such solutions for a large class of games and propose efficient computational method to find these solutions.

2.6 What have we learned in this chapter?

Let us summarize the main ideas and concepts presented in this chapter.

- Games are models of interactions among decision makers called players.
- A game is described in extensive form when the rules of the game are presented in terms of a sequence of moves. The definition also includes an information structure and the possible move outcomes.
- The players are assumed to use VNM utility functions to value their moves' outcomes. The players strive to maximize the expected utility of the random outcomes of the game.
- To play the game a player can use a strategy that tells him which action to take (which move to make), given the available information when he has the move. He can also randomize his strategy, either by drawing one strategy randomly, according to a pre-defined lottery, or by randomizing his choice of actions when it is his turn to move.
- A more compact description of a game is obtained when the mapping, which associates the expected utility with a strategy vector defined for all players, is specified. This is the normal-form description of a game.
- Some games can be solved by iterated deletions of strictly dominated strategies. This method relies on a sequence of guesses and counter-guesses on the strategy choices that could be made by rational players.
- Rationalizable strategy profiles can always be defined in a game in normal form obtained from a game in extensive form.
- In the $\mathbb{GE}$ *illustration*, which concludes this chapter we show how the basic elements of game theory can be used to model some negotiations about climate change mitigation involving several groups of countries.
- In the next chapter we introduce the concept of equilibrium as a possible solution to a game, i.e., a way the game will be played by rational players.

2.7 Exercises

Exercise 2.1. Download Gambit[16] and familiarize yourself with its use to define games in extensive form.

Gambit (Software Tools for Game Theory) is a library of game theory software and tools for the construction and analysis of finite extensive and strategic games. Gambit is designed to be portable across platforms and runs on Linux, Mac OS X and Windows. Gambit is free, open Source software, released under the terms of the GNU General Public License. You may download Gambit from the website `http://www.gambit-project.org/doc/intro.html`.

Exercise 2.2. **Hands-on.** Use Gambit to draw the game tree for the following two-player simultaneous-move matrix game:

$$\begin{bmatrix} 200,\ 100 & 100,\ 150 \\ 50,\ 0 & 150,\ 200 \end{bmatrix} \tag{2.5}$$

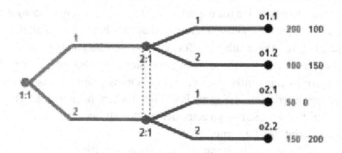

Fig. 2.7   Game tree built with Gambit

Hint: Start Gambit.

- In the interactive window, you will see that two players have by default been defined with colors red and blue, respectively.
- Right click on the black node and then select **Insert move**. A dialogue box appears proposing to assign the move to Player 1 and to define an information set and two possible actions. Click **OK** to accept this definition. The node is now called 1:1. It has two arcs called 1 and 2, respectively and its color is red, since it corresponds to Player 1.
- Now, right click on the upper black node and then select **insert move**. In the dialogue box select **Insert move for Player 2** in the first field and choose **at a new information**

[16]R.D. McKelvey, A.M. McLennan and T.L. Turocy (2010). Gambit: Software Tools for Game Theory, Version 0.2010.09.01. http://www.gambit-project.org.

set in the second field; select **2 actions** and click **OK**. A new pair of arcs appears on the graph. Right click now on the other black node. Select **Insert move for Player 2** in the upper field of the dialogue box and **information set 1** in the second field (there is only one existing information set). Automatically, the number of actions is set to 2. Click **OK**.

- Now you can see the structure of a game with simultaneous moves. What remains is to define the outcomes. Right clicking on the terminal nodes enables you to assign labels to the nodes (for example o1.1, o1.2, o2.1, o2.2). Double clicking on the (u) placed to the right of each terminal node enables you to enter the outcome values; use the tab key to toggle from Player 1 to Player 2. At the end, you should have the game tree as shown in Figure 2.7.

Exercise 2.3. Consider the payoff matrix given in Table 2.4, which represents a payoff matrix in a two player game. Using Gambit, create a game tree and formulate a corresponding extensive form of this game.

Exercise 2.4. A two-player game is given in extensive form in Figure 2.8 shown below. Represent the game in normal form.

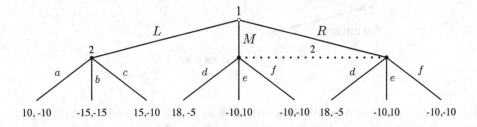

Fig. 2.8 A game in extensive form

Exercise 2.5. Consider the game in extensive form shown in Figure 2.9. The payoffs are not shown because we are only interested in studying the information structure of the game.

(1) Define the set of strategies for every player and present this game in normal form.
(2) Show that for every behavioral strategy there exists an equivalent mixed strategy.
(3) Show that for every mixed strategy, there exists a realization of an equivalent behavioral strategy.

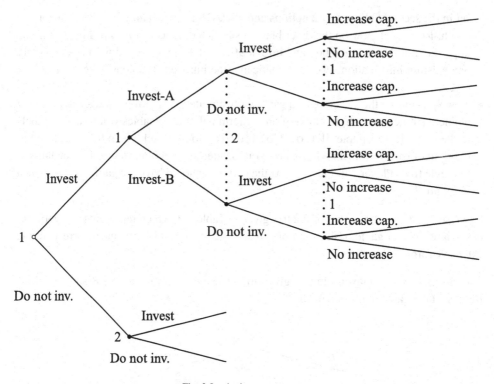

Fig. 2.9   An investment game

Exercise 2.6. You have only $100 and you aspire to be rich. You decide that playing the game of roulette in a casino could put you on the fast track to your first million because you have devised an ingenious betting strategy. A roulette has 37 numbers, of which 18 are black, 18 are red and 1 is green. In stage 1 you bet $1 on black. Whenever you win, you start over at stage 1. If you lose, you double the previous bet and add $1. You remain in this stage whenever you lose. You exit when you have no money left.

(1) Is this system certain to beat the casino, i.e., the other player (nature)?
(2) If not, what is the probability that you will lose and what do these odds depend on?
(3) Describe this game in extensive form.

Exercise 2.7. The winner of a game show (Monty Hall's show on NBC) is presented with three closed doors. Behind two of them are goats and behind the third is a brand new car. The contestant must choose one door. After he has done so, a door is opened to reveal one of the goats. The contestant is now offered two possible actions: (a) randomly select one of the two remaining doors; or (b) stick with the door initially selected. The contestant will take home the prize behind the final door selected.

(1) Is it rational to change the door selection?
(2) What it the probability of winning the car before the first door is opened?
(3) Suppose the player chooses door 1. Write the extensive and the normal forms for this game.
(4) Use Bayes' rule[17] to calculate the probability after the door is opened.
(5) Imagine that there are 1,000 doors and Monty opens 998 of the non-chosen doors; would your answer change, qualitatively?

Exercise 2.8. Use backward induction to solve the Rosenthal's centipede game [214], depicted in Figure 2.10.

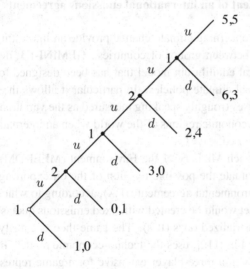

Fig. 2.10 The centipede game

(1) Solve the game using backward induction.
(2) If you were player 2, what would you believe if Player 1 played action u on his first move? Would it make a difference if you believed this was a *typo* or a *thinko*? Does it really make sense to optimize at a node if the conditional probability of your reaching that node is zero?

---

[17] Readers unfamiliar with the Bayes rule are directed to a probability textbook or Wikipedia.

Exercise 2.9. Consider a two-player game with the payoffs given in Table 2.6

Table 2.6   Two-player game

		2		
		L	M	R
	H	4,3	5,1	6,2
1	M	2,1	8,4	3,6
	B	3,0	9,6	7,8

(1) Show that this game is not solvable by the elimination of strictly dominated strategies.
(2) Show that all outcomes that survive the elimination of strictly dominated strategies are rationalizable.

2.8   GE illustration: Assessment of an international emissions agreement

The international negotiations concerning climate change provide an interesting example of interaction and competition between groups of countries. GEMINI-E3, described in Ref. [26] is a computable general equilibrium model that has been designed for the economic assessment of international climate policies. In particular it allows the analyst to evaluate the welfare gains or losses (roughly speaking measured as the variation of household consumption) in different economic regions of the world when an international environmental agreement is implemented.

A research report to the French Ministry of the Environment (MEEDDM) proposed an extensive game model to evaluate the possible reaction of the oil exporting countries (OPEC) to an international environmental agreement (IEA), according to which an international emissions trading market would be created with fixed emissions quotas for developing countries (DCs) and industrialized ones (ICs). The game-theoretic analysis of this problem, which is also presented in [119], uses the techno-economic model GEMINI-E3 to compute the outcome payoffs in a three-player extensive-form game representing the strategic interactions between the parties involved in the negotiation process of an IEA. The players are the following: OPEC, which determines the world price of oil; and, the two groups of countries: the industrialized (ICs) and the developing countries (DCs). These countries negotiate an abatement policy under uncertainty, since the climate sensitivity[18] (CS) is not known yet. The first planning period is 2000–2025, followed by the period 2025-2050. We assume that the actual climate sensitivity will be known by year 2025.

We assume that an international-emissions trading system will be put in place. The international-emissions trading system allows a country to emit *more* than its quota provided that the country buys emissions rights from other countries, which are emitting below

---

[18]Roughly speaking, climate sensitivity is measured using the surface atmospheric temperature change that will occur from a doubling of atmospheric GHG concentrations, compared with the pre-industrial situation. The range of uncertainty for this parameter is between 1.5 and 5°C.

their quotas. Therefore, if a country has a high quota level, it will have the opportunity to supply emission rights to the market. The price of emissions rights is set at the marginal abatement cost for the worldwide economy.

In the IEA, all countries share the common objective of keeping the temperature increase in the 21st century below 2°C, and the increase is understood to be caused by anthropogenic GHG emissions. The CS value, if known, determines a total amount of GHG emissions allowed over the 2000–2050 horizon in order to attain this goal. However, the CS will only become known in 2025, so the decisions for the first period's abatement are made under uncertainty. The environmental agreement is about the sharing of that total amount between the two groups of countries (DCs and ICs). Once a global allowance is given, each group of countries may decide *either* to abate vigorously in the first period in order to leave more room for emissions in the second period, after economic growth has taken place, *or* to be lax in the first period and wait and see how much must be abated in the second period, when more information on the CS is available. Each group of countries thus has to decide if they will use high, average or low quotas in the first period. In the second period the quotas will be defined with the knowledge of the following: the total amount of emissions allowed for the whole period corresponding to the actual CS value and the level of quotas decided in the first period.

Structure of the extensive-form game: The game tree will have three levels of nodes corresponding to different decision stages. (i) At the initial node, OPEC moves. It determines a world price for oil, defined as low, medium or high, for the whole planning horizon. (ii) At the second decision level, DCs and ICs simultaneously chose their respective emissions quotas for the first period. The quotas maybe: high, medium or low. (iii) The third level corresponds to the resolution of uncertainty about climate sensitivity.

When the CS is known in 2025, then the objective to be reached, formulated in terms of the 2050 GHG concentrations is also known. Therefore, the total emissions allowed for the whole horizon are also defined. The abatement that must be achieved in the second period is therefore defined by the climate sensitivity (which determines the concentration goal for 2050) and the abatement already executed in the first period.

Figure 2.11 shows the structure of the extensive game under the restriction that the action choices of each of the players (OPEC, ICs, DCs, nature) are confined to three possible moves: high, medium or low. This graph was produced with Gambit. The first three branches at the initial stage correspond to oil price levels decided by OPEC. The three groups of three branches at the second stage correspond to three possible emissions levels (quotas) agreed by ICs. The nine groups of three branches at the third stage correspond to three possible emissions quotas agreed by DCs. Finally the 27 groups of three branches at the fourth stage correspond to three possible CS outcomes decided upon randomly by nature. The parameters defining this game are detailed below.

A CGE to compute outcomes: The gains of the players are defined as the variations in welfare, associated with the different strategies, for the three groups of countries, as computed with the help of the CGE model GEMINI-E3. Table 2.7 gives the list of countries included in the three groups that are considered as players. Table 2.8 indicates the GDP

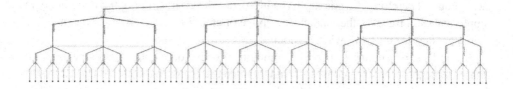

Fig. 2.11   Full game tree

Table 2.7   Geographic zones and players

ICs	DCs	OPEC
USA	Asia	Middle East
Russia	China	
Europe	Brasil	
Canada	Latin America	
Australia	Africa	
New Zealand	Mexico	
Japan		

Table 2.8   GDP and GHG-emissions growth rate in the BAU case

	2010–2020	2020–2030	2030–2040	2040–2050
GDP (AAGR)				
ICs	2.7%	2.4%	1.8%	1.8%
DCs	4.7%	4.1%	3.6%	3.5%
OPEC	3.6%	3.2%	2.6%	2.5%
World	3.2%	2.9%	2.4%	2.4%
GHG (AAGR)				
ICs	1.0%	0.8%	0.5%	0.6%
CDs	2.8%	2.0%	1.9%	1.7%
OPEC	2.0%	0.9%	0.6%	1.0%
World	1.9%	1.4%	1.3%	1.3%

and the GHG emissions growth rates that are assumed in a *business as usual* (BAU) case, where no consideration is given to the climate-change threat. These welfare variations are defined relative to a reference situation where no environmental constraint is imposed on the world economy. It is assumed that a universal agreement exists because it is necessary to maintain the long-term temperature change below 2°C. This can be expressed in terms of concentration targets, in ppmv[19] or total cumulated emissions until 2050, in GTC.[20] Table 2.9 gives these values for the three possible CS values, called high, medium and low, respectively. The medium value has a 0.5 probability and the two other values have a 0.25 probability.

---

[19]Parts per million in volume.
[20]Gigaton of equivalent carbon.

Table 2.9   Concentration targets and associated cumulative quotas

CS	Concentration targets	Cumulative quotas	Probability
High	500 ppmv	450 GTC	0.25
Medium	600 ppmv	500 GTC	0.50
Low	700 ppmv	650 GTC	0.25

The negotiations are about the emissions quotas that will be allocated to ICs and DCs in the first period 2010–2020. These quotas are expressed in cumulative emissions, measured in GTC, over the 20 year period. The three possible choices are:

Quotas in period 1: For each of the two players (ICs and DCS):
high → 120 GTC; medium → 110 GTC; low → 80 GTC.

The IEA also includes a burden sharing agreement where 50% of the total emissions allowed over the whole horizon 2000–2050 will be allocated to ICs and 50% to DCs.

In 2025 CS is known and given the emissions quotas decided in the first period, the cumulative amount of emissions allocated to each player for the 2025-2050 period is determined. There is no more choice to be made by the players. Therefore the game is essentially a one shot simultaneous choice game where OPEC decides a price level for oil and ICs and DCs decide if they start with low, medium or high quotas in an international emissions trading scheme. The consequences of these choices are then computed, using the GEMINI-E3 macro-economic model. They constitute the outcomes of the game.

Now we show the extensive-game representation of the competition between ICs and DCs, which have to choose the first period level of quotas, depending on the world oil price level selected by OPEC. The game has a simultaneous move structure. The outcome payoffs in each of these games were obtained by computing the expected values of the outcomes in the third stage, where, as mentioned before the decision of ICs and DCs are contingent on the quotas selected in stage 2 and the revealed CS.

For example, Figure 2.12 shows the game tree[21] when the price of oil is low and the split of emissions rights is 50%-50%. Below we give all the other extensive game representations of the competition between ICs and DCs, for different levels of oil price and different splits of the total emissions permitted.

In the next chapter, we define the concept of an equilibrium solution. In another GE illustration at the end of the next chapter, we will provide an equilibrium solution to this game.

---

[21]The graphs have been produced using Gambit.

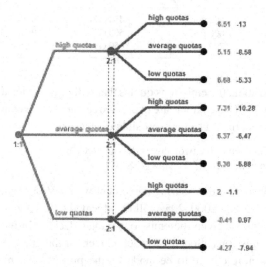

Fig. 2.12   When oil price is low and the split of emissions rights is 50%-50%

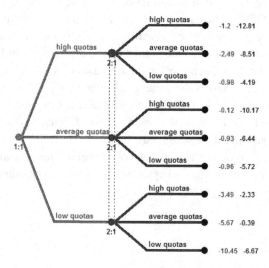

Fig. 2.13   When oil price is medium and the split of emissions rights is 50%-50%

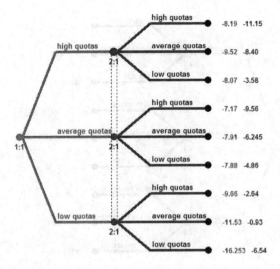

Fig. 2.14 When oil price is high and the split of emissions rights is 50%-50%

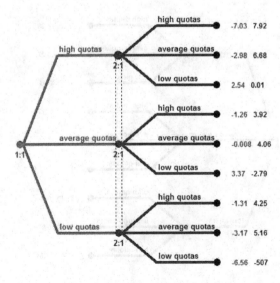

Fig. 2.15 When oil price is low and the split of emissions rights is 60% DC, 40% IC

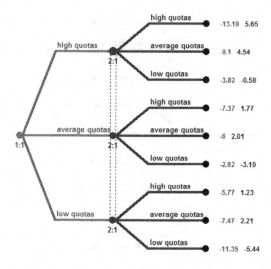

Fig. 2.16   When oil price is medium and the split of emissions rights is 60% DC,40% IC

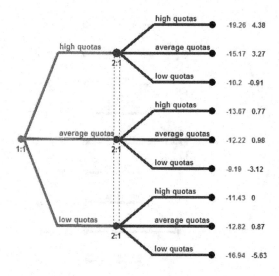

Fig. 2.17   When oil price is high and the split of emissions rights is 60% DC, 40% IC

Chapter 3

Equilibrium Solutions for Noncooperative Games

3.1 Introduction

In this chapter, we deal with *solution concepts* for noncooperative games. In games, a solution concept prescribes the appropriate strategy choices for rational players. Therefore in order for it to have a solution concept, the game needs to be defined in *normal* or *strategic* form. In noncooperative games, players are pursuing maximization of their own utility functions and are unable to form binding commitments. A strategy vector $\underline{\gamma}^*$ that each player finds non-improvable by a unilateral change of his strategy, will be called an *equilibrium solution* to the m-player game.

Recall that an m-person game in normal form is defined by the following data:

$$\{M, (\Gamma_j, V_j : j \in M)\},$$

where $M = \{1, 2, ..., m\}$ is the set of players. For each player $j \in M$, Γ_j is the set of strategies and V_j denotes the payoff function that assigns a real number $V_j(\underline{\gamma})$ to a *strategy vector*

$$\underline{\gamma} \triangleq (\gamma_1, \gamma_2, \ldots, \gamma_m) \in \Gamma_1 \times \Gamma_2 \times \cdots \times \Gamma_m.$$

We will explore the equilibrium solution concept for different classes of games in normal form.

The first class is constituted by so-called *two-player zero-sum matrix games*, which describe conflict situations where there are two players, each using a *mixed strategy* defined over a finite set of *pure strategies*. Moreover, what one player gains the other player loses, which explains why the games are called zero-sum. In this framework, the equilibrium solution is also called a *saddle-point* solution and it can be computed using linear programming.

The second category consists of two-player games, again with a finite pure strategy set for each player but where the payoffs are not zero-sum. These are called *nonzero-sum bimatrix games*, or simply, *bimatrix games*. In this context, we will show that an equilibrium solution can be computed using nonlinear programming techniques like linear complementarity or quadratic programming. Notwithstanding the computational difficulties, the nonzero-sum games can be extended to situations where $m > 2$, in which case the games will be called m-matrix games.[1]

The third category includes m-player *concave games*, which encompasses the previous class of m-matrix games and extends it to problems where strategies are selected from more general convex and compact sets. We will use mathematical programming to prove existence of equilibrium solutions for these more general games and to establish equilibrium uniqueness and computational stability.

3.2 Matrix games

3.2.1 Definitions

Here we consider games as defined in the previous chapter, where the players have a finite set of pure strategies and where the payoffs always sum to zero.

Definition 3.1. A game is zero-sum if the sum of the players' payoffs is always zero. Otherwise the game is nonzero-sum. A two-player zero-sum game is also called a **duel**.

Definition 3.2. A two-player zero-sum game, in which each player has only a finite number of pure strategies to choose from is called a **matrix game**.

3.2.2 Security levels

Let us explore how matrix games can be solved. We number the players 1 and 2 respectively. Conventionally, Player 1 is the maximizer and has k (pure) strategies, say $i = 1, 2, ..., k$, and Player 2 is the minimizer and has n strategies to choose from, say $j = 1, 2, ..., n$. If Player 1 chooses strategy i while Player 2 picks strategy j, then Player 2 pays Player 1 the amount a_{ij}.[2] The set of all possible payoffs that Player 1 can obtain is represented by the $k \times n$ matrix A with entries a_{ij} for $i = 1, 2, ..., k$ and $j = 1, 2, ..., n$. Now, the element in the i-th row and j-th column of the matrix A corresponds to the amount that Player 2 will pay Player 1 if the latter chooses strategy i and the former chooses strategy j. Thus we can say that in the game under consideration, Player 1 (the maximizer) selects rows of A while Player 2 (the minimizer) selects columns of that matrix. As a result of the play, as said above, Player 2 pays Player 1 the amount of money specified by the element in the selected row and column of the matrix.[3]

Example 3.1. *Consider a game defined by the following matrix:*

$$\begin{bmatrix} 3 & 1 & 8 \\ 4 & 10 & 0 \end{bmatrix} \tag{3.1}$$

---

[1] See Section 3.2, below.

[2] Negative payments are allowed. We could also say that Player 1 receives the amount a_{ij} and Player 2 receives the amount $-a_{ij}$.

[3] In a zero-sum game we implicitly assume that the players are risk neutral, and so they strive to maximize their monetary gains.

Which strategy will a rational player select?

The first line of reasoning is to consider the players' *security levels*. It is easy to see that if Player 1 chooses the first row, then, whatever Player 2 does, Player 1 will get a payoff equal to at least 1. By choosing the second row, on the other hand, Player 1 risks getting 0. Similarly, by choosing the first column Player 2 ensures that he will not have to pay more than 4, while selecting the second or third column may cost him 10 or 8, respectively. Thus we say that Player 1's *security level* is 1, which is ensured by choosing the first row, while Player 2's security level is 4, as ensured by the choice of the first column. Hence,

$$1 = \max_i \min_j a_{ij}$$

and

$$4 = \min_j \max_i a_{ij}.$$

◇

The strategy that ensures that Player 1 will get a payoff at least equal to his security level is called his *maximin strategy*. Symmetrically, the strategy that ensures that Player 2 will not have to pay more than his security level is called his *minimax strategy*.

Lemma 3.1. *In any matrix game defined by matrix* $A = [a_{ij}]$*, the following inequality holds:*

$$\max_i \min_j a_{ij} \leq \min_j \max_i a_{ij}. \qquad (3.2)$$

Proof. This result can be proved by the observation that, since both security levels are achievable, they necessarily satisfy the inequality (3.2). Another way to establish (3.2) is to write the following inequalities: For any k and l, we have

$$\min_j a_{kj} \leq a_{kl} \leq \max_i a_{il}$$

and thus

$$\forall k, l, \ \min_j a_{kj} \leq \max_i a_{il}.$$

Therefore we have

$$\max_i (\min_j a_{ij}) \leq \min_j \max_i a_{ij}.$$

□

An important observation is that, if Player 1 has to move first and then Player 2 acts having seen the move made by Player 1, then the maximin strategy is Player 1's best choice, leading to the payoff equal to 1. If the situation is reversed and it is Player 2 who moves first, then his best choice will be the minimax strategy and he will have to pay 4. Now, the question is what happens if the players move simultaneously. Example 3.1 shows that when the players move simultaneously, the minimax and maximin strategies are not satisfactory solutions to this game. In particular, the players may try to improve their payoffs by anticipating one another's strategy. As a result, we will see a process that, in some cases does not converge to any stable solution. For example such an instability will occur in the matrix game described in (3.1).

Now consider another example.

Example 3.2. *Let the matrix game* $A = [a_{ij}]$ *be given as follows*

$$\begin{bmatrix} 10 & -15^* & 20 \\ 20 & -30 & 40 \\ 30 & -45 & 60 \end{bmatrix}. \tag{3.3}$$

Can we find a pure-strategy pair that rational players will play?

It is easy to see that, here,

$$\max_i \min_j a_{ij} = \max\{-15, -30, -45\} = -15$$

and

$$\min_j \max_i a_{ij} = \min\{30, -15, 60\} = -15$$

and so, the pair of maximin and minimax strategies is given by

$$(i, j) = (1, 2).$$

That means that Player 1 chooses the first row while Player 2 selects the second column, which will lead to the payoff of -15. In this situation, we see that a payoff of -15 is guaranteed to Player 1 and a payoff of 15 is guaranteed to Player 2. Furthermore, each player's strategy (row 1 for Player 1 and column 2 for Player 2) is the best reply to his opponent's selected strategy.   ◇

Rational players will play the strategy pair $(1, 2)$ in the above example since it is, obviously, advantageous for them to do so. We can say that, in this example, players' *maximin* and *minimax* strategies "solve" the game.

3.2.3 Saddle points

Let us consider more precisely the class of strategy pairs that have this property of each component being the best reply to the other component, as in Example 3.2.

Definition 3.3. If in a matrix game $A = [a_{ij}]_{i=1,...,k;j=1,...,n}$ there exists a pair (i^*, j^*) such that

$$a_{ij^*} \leq a_{i^*j^*} \leq a_{i^*j} \tag{3.4}$$

for all $i = 1, \ldots, k$ and $j = 1, \ldots, n$, we say that the pair (i^*, j^*) is a **saddle point** in pure strategies of the matrix game defined by A.

We observe that $(i^*, j^*) = (1, 2)$ is a saddle point in the game solved in Example 3.2.

A consequence of the above definition is that both players' security levels are equal at a saddle point in a zero-sum game, i.e.,

$$\max_i \min_j a_{ij} = \min_j \max_i a_{ij} = a_{i^*j^*}.$$

What is less obvious is that a game's saddle point exists if the players' security levels are equal.

Lemma 3.2. *Assume that, in a matrix game,*

$$\max_i \min_j a_{ij} = \min_j \max_i a_{ij} = v.$$

Then, the game admits a saddle point in pure strategies.

Proof. Let i^* and j^* be a strategy pair that yields the security level payoffs v and $-v$ for Player 1 and Player 2, respectively. We thus have that

$$a_{i^*j} \geq \min_j a_{i^*j} = \max_i \min_j a_{ij}, \tag{3.5}$$

$$a_{ij^*} \leq \max_i a_{ij^*} = \min_j \max_i a_{ij} \tag{3.6}$$

for all $i = 1, \ldots, k$ and $j = 1, \ldots, n$. Since

$$\max_i \min_j a_{ij} = \min_j \max_i a_{ij} = a_{i^*j^*} = v$$

we obtain by (3.5)–(3.6)

$$a_{ij^*} \leq a_{i^*j^*} \leq a_{i^*j},$$

which is the saddle-point condition. $\qquad\qquad\qquad\qquad\qquad\qquad\qquad\qquad\qquad\square$

If a saddle-point strategy pair exits, then it provides a solution to the matrix-game problem. Indeed, in Example 3.2, if Player 1 expects Player 2 to choose the second column, then the first row will be his optimal choice. On the other hand, if Player 2 expects Player 1 to choose the first row, then it will be optimal for him to choose the second column. In other words, neither player can gain anything by unilaterally deviating from his saddle point strategy. *In a saddle point each player's strategy constitutes his best reply to the strategy selected by his opponent.* This observation leads to the following remark.

Remark 3.1. Let (i^*, j^*) be a saddle point in a matrix game. Players 1 and 2 cannot improve their payoff by unilaterally deviating from $(i)^*$ or $(j)^*$, respectively. We say that the strategy pair (i^*, j^*) is an equilibrium.

Saddle-point strategies, as shown in Example 3.2, lead to both an equilibrium and a pair of *guaranteed payoffs*. Therefore such a strategy pair, if it exists, provides a solution to a matrix game, which is good in that rational players will adopt it. The problem is that a saddle point, as shown in Example 3.2, may not always exist for a matrix game, if players are restricted to using the discrete (finite) set of strategies that index the rows and the columns of game matrix A. In the next section, we will see that this problem can be alleviated if the players are allowed to mix the pure strategies.

3.2.4   *Mixed strategies*

We indicated in Chapter 2 that a player could mix his pure strategies by using a lottery to decide which strategy to play.

Many matrix games do not possess saddle points in the class of pure strategies; see Example 3.1. However, von Neumann proved in [181] that a saddle-point strategy pair always exists in the class of mixed strategies.

Consider the matrix game defined by an $k \times n$ matrix A. (As before, Player 1 has k pure strategies and Player 2 has n pure strategies.) A mixed strategy for Player 1 is a k-tuple

$$x = (x_1, x_2, \ldots, x_k)$$

where x_i are nonnegative for $i = 1, 2, \ldots, k$, and $x_1 + x_2 + \ldots + x_k = 1$. Similarly, a mixed strategy for Player 2 is an n-tuple

$$y = (y_1, y_2, \ldots, y_n)$$

where y_j are nonnegative for $j = 1, 2, \ldots, n$, and $y_1 + y_2 + \cdots + y_n = 1$. Note that a pure strategy can be considered a particular type of mixed strategy with one coordinate equal to one and all others equal to zero.

Player 1's set of possible mixed strategies constitutes a *simplex*[4] in the space $\mathbf{R}^k$. This is illustrated in Figure 3.1 for $k = 3$. Similarly the set of mixed strategies of Player 2 is a simplex in $\mathbf{R}^n$.

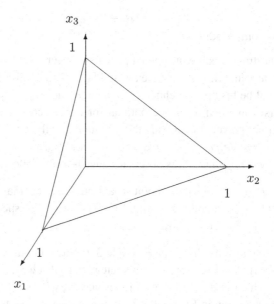

Fig. 3.1   The simplex of mixed strategies

---

[4]A simplex is, by construction, the smallest closed convex set that contains $n + 1$ points in $\mathbf{R}^n$.

The interpretation of a mixed strategy, say x, is that Player 1 chooses his pure strategy i with probability x_i, $i = 1, 2, \ldots, k$. Similarly, Player 2 chooses his pure strategy j with probability y_j, $j = 1, 2, \ldots, n$. Since the two lotteries defining the random draws are independent events, the joint probability that the strategy pair (i, j) is selected is given by $x_i\, y_j$. Therefore, with each pair of mixed strategies (x, y), we can associate an expected payoff given by the quadratic form in x and y (where the superscript $'$ denotes the transposition of a matrix):

$$v \triangleq \sum_{i=1}^{k} \sum_{j=1}^{n} x_i y_j a_{ij} = x' A y.$$

The above-mentioned theorem on the existence of saddle points, which was proved by von Neumann [181], is quoted below.

Theorem 3.1. *Any matrix game has a saddle point in the class of mixed strategies, i.e., there exist probability vectors x and y such that*

$$\max_{x} \min_{y} x' A y = \min_{y} \max_{x} x' A y = (x^*)' A y^* = v^*, \tag{3.7}$$

where v^* is called the value *of the game.*

We will not repeat the complex proof given by von Neumann. Instead, we will formulate a *linear program* (LP) capable of finding a saddle point for the game.

3.2.5 Algorithms for the computation of saddle points

Saddle points in mixed strategies can be obtained as solutions to linear programs. It is easy to show that, for a matrix game, the following two relations hold true[5]:

$$v^* = \max_{x} \min_{y} x' A y = \max_{x} \min_{j} \sum_{i=1}^{k} x_i a_{ij} \tag{3.8}$$

and

$$z^* = \min_{y} \max_{x} x' A y = \min_{y} \max_{i} \sum_{j=1}^{n} y_j a_{ij}. \tag{3.9}$$

These two relations imply that the value of the matrix game can be obtained by solving any of the following two linear programs:

---

[5]Actually this is already a linear-programming result. When the vector x is given, the expression $\min_y x' A y$, with simplex constraints, i.e., $y \geq 0$ and $\sum_j y_j = 1$ defines a linear program. The solution of that LP can always be found at an extreme point of the admissible set. An extreme point in the simplex corresponds to $y_j = 1$ and to the other components equal to 0. Therefore, since Player 1 selects his mixed strategy x expecting the opponent to use his best reply, he can only restrict the search for the best reply to the opponent's set of pure strategies.

(1) Primal problem

$$\max_{x,v} \quad v$$

subject to

$$v \le \sum_{i=1}^{k} x_i a_{ij}, \quad j = 1, 2, \ldots, n,$$

$$1 = \sum_{i=1}^{k} x_i,$$

$$x_i \ge 0, \quad i = 1, 2, \ldots, k.$$

(2) Dual problem

$$\min_{y,z} \quad z$$

subject to

$$z \ge \sum_{j=1}^{n} y_j a_{ij}, \quad i = 1, 2, \ldots, k,$$

$$1 = \sum_{j=1}^{n} y_j,$$

$$y_j \ge 0, \quad j = 1, 2, \ldots, n.$$

The following theorem relates these two programs.

Theorem 3.2. *(Von Neumann* [181]*): Any two-person zero-sum matrix game A has a value.*

Proof. The value v^* of the zero-sum matrix game A is obtained as the common optimal value of the following pair of dual linear programming problems. The respective optimal programs define the saddle-point mixed strategies.

	Primal		Dual
max	v	min	z
subject to	$x'A \ge v1'$	subject to	$A y \le z1$
	$x'1 = 1$		$1'y = 1$
	$x \ge 0$		$y \ge 0$

where

$$1 = \begin{bmatrix} 1 \\ \vdots \\ 1 \\ \vdots \\ 1 \end{bmatrix}$$

denotes a vector of the appropriate dimension with all components equal to 1. It is sufficient to solve only one of the programs.[6] The primal and dual solutions give a saddle-point strategy pair. ☐

Remark 3.2. Simple $n \times n$ games can be solved more easily (see [193]). Suppose A is an $n \times n$ matrix game that does not admit a saddle point in pure strategies. Each player's unique saddle-point mixed strategies and the game value are then given by

$$x = \frac{1'A^D}{1'A^D 1},$$ (3.10)

$$y = \frac{A^D 1}{1'A^D 1},$$ (3.11)

$$v = \frac{\det A}{1'A^D 1},$$ (3.12)

where A^D is the adjoint matrix of A, $\det A$ the determinant of A and 1 the vector of n components equal to 1, as defined before.

Let us illustrate the above formulae using the following example.

Example 3.3. *We want to solve the matrix game*

$$\begin{bmatrix} 1 & 0 \\ -1 & 2 \end{bmatrix}.$$

The game, obviously, has no saddle point (in pure strategies). The adjoint matrix is

$$A^D = \begin{bmatrix} 2 & 0 \\ 1 & 1 \end{bmatrix}.$$

We easily obtain $1'A^D = [3 \ 1]$, $A^D 1 = [2 \ 2]'$, $1'A^D 1 = 4$, $\det A = 2$. Hence the best *mixed strategies* for the players are:

$$x = \begin{bmatrix} \frac{3}{4}, \frac{1}{4} \end{bmatrix}, \quad y = \begin{bmatrix} \frac{1}{2}, \frac{1}{2} \end{bmatrix}$$

and the value of the play is

$$v = \frac{1}{2}.$$

In other words, in the long run, Player 1 is supposed to win 0.5 if he uses the first row 75% of times and the second, 25% of times. Player 2's best strategy will be to use the first and the second column 50% of times, which ensures him a loss of 0.5 (only; using other strategies he would lose more). ◇

---

[6]We refer the reader to any textbook on linear programming to see how to calculate the dual solution once the primal problem has been solved.

3.3 Bimatrix games

3.3.1 *Best reply strategies*

We will now extend the theory we have developed for matrix games to the case of nonzero-sum games. A set of two matrices, or a two-entry matrix called a *bimatrix*, can conveniently represent a two-person nonzero sum game where each player has a finite set of pure strategies. We call such a game a *bimatrix game*.

In a bimatrix game, there are two players, Player 1 and Player 2 who have k and n pure strategies to choose from, respectively. If the players select a pair of pure strategies (i, j), then Player 1 obtains the payoff a_{ij} and Player 2 obtains b_{ij}, where a_{ij} and b_{ij} are entries of $k \times n$ real matrices A and B, correspondingly.

When $a_{ij} + b_{ij} \equiv 0$, then the game is a zero-sum matrix game. Otherwise, the game is nonzero-sum.

In Remark 3.1, page 41, in the context of matrix (i.e., zero-sum) games, we noticed that a saddle point strategy pair constitutes an equilibrium since no player can improve his payoff through a unilateral strategic change. This is because each player has chosen his best reply to the opponent's strategy. We will examine whether the concept of best-reply strategies can also be used to define a solution to a bimatrix game.

Example 3.4. *Consider the bimatrix game defined by the following matrices:*

$$\begin{bmatrix} 52 & 44 & 44 \\ 42 & 46 & 39 \end{bmatrix}$$

and

$$\begin{bmatrix} 50 & 44 & 41 \\ 42 & 49 & 43 \end{bmatrix}.$$

Examine whether there are strategy pairs that constitute a best reply to each other.

It is often convenient to combine the data contained in the two matrices and write them in one matrix whose entries are ordered pairs (a_{ij}, b_{ij}). In this case, we obtain

$$\begin{bmatrix} (52, 50)^* & (44, 44) & (44, 41) \\ (42, 42) & (46, 49)^* & (39, 43) \end{bmatrix}.$$

Consider the two cells indicated by an asterisk $*$. Notice that they correspond to outcomes resulting from best reply strategies. Indeed, at cell "row:1, column:1", if Player 2 sticks to the first column, Player 1 can only worsen his payoff from 52 to 42 if he moves to row 2. Similarly, if Player 1 plays row 1, Player 2 can gain 44 or 41 only, instead of 50, if he abandons column 1. With similar reasoning, we can prove that the payoffs at cell "row:2, column:2" result from another pair of best-reply strategies.   ◇

3.3.2 *Nash equilibria*

We can see from Example 3.4 that a bimatrix game can have many best-reply strategy pairs. However there are other examples where no such pairs can be found among the pure strategies.

Example 3.5. *Consider the following bimatrix game where the set of strategies of Player 1 is* $\{r_1, r_2, r_3\}$*, and the set of strategies of Player 2 is* $\{c_1, c_2, c_3\}$*.*

		Player 2		
		c_1	c_2	c_3
	r_1	$(8,4)$	$(4,0)$	$(2,7)$
Player 1	r_2	$(6,5)$	$(3,6)$	$(5,3)$
	r_3	$(3,9)$	$(4,0)$	$(6,1)$

Identify the best reply strategies.

Let $R_1(c_j)$ and $R_2(r_i)$, $i = 1, 2, 3$, $j = 1, 2, 3$ denote the best reply of, respectively, the first and second player to the opponent's choice of column c_j or row r_i. The best replies are then given as

$$R_1(c_1) = r_1, \ R_1(c_2) = \{r_1, r_3\} \ R_1(c_3) = r_3,$$
$$R_2(r_1) = c_3, \ R_2(r_2) = c_2, \qquad R_2(r_3) = c_1.$$

It is easy to check that there is no pure strategy pair where each strategy component is the best reply to the other.   ◇

Similarly to what we did for the zero-sum games, we will expand the strategy sets for bimatrix games to include mixed strategies. This will allow us to use the Nash equilibrium concept as a solution to bimatrix games. This equilibrium is defined as follows.

Definition 3.4. Let A, B be the players' payoff matrices. A pair of mixed strategies (x^*, y^*) is said to be a Nash equilibrium of the bimatrix game if

$$x^{*\prime} A y^* \geq x' A y^* \quad \text{for every mixed strategy } x, \tag{3.13}$$
$$x^{*\prime} B y^* \geq x^{*\prime} B y \quad \text{or every mixed strategy } y. \tag{3.14}$$

Notice that this definition says that, at equilibrium, no player can improve his payoff by deviating unilaterally from the equilibrium strategy.

Remark 3.3. A Nash equilibrium extends to nonzero-sum games the equilibrium property that holds in the case of the saddle point solution to a zero-sum matrix game. However this solution concept does not guarantee that a player will receive at least the equilibrium payoff. Indeed if his opponent does not use the equilibrium strategy, the game can have any outcome. This is a serious drawback of the equilibrium solution concept.

Remark 3.4. If we were to prescribe a way to play a bimatrix game by rational players, then the prescription should be an equilibrium. Indeed if we prescribe a strategy pair that is not an equilibrium, each player will be tempted to play a different strategy than the one prescribed. The prescription will then be **self-denying**. This is why we consider that a Nash equilibrium is a valid solution concept for a bimatrix game.

The most important contribution of John Nash to the development of game theory is the following theorem[7] [180].

---

[7]Actually the theorem of Nash has been formulated for an m-player game.

Theorem 3.3. *Every finite bimatrix game has at least one Nash equilibrium in mixed strategies.*

We will not give Nash's proof here. It is based on the use of the *Kakutani fixed point theorem*. A slightly different existence theorem will soon be proved in Section 3.4.4 (Theorem 3.6), which will deal with equilibria in concave games. This is a class of games that includes bimatrix games played with mixed strategies. The proof of that theorem will also be based on the Kakutani fixed point theorem.

3.3.3 *Shortcomings of the Nash equilibrium concept*

3.3.3.1 *Multiple equilibria*

As noticed in Example 3.4, a bimatrix game may have several equilibria in pure strategies. There may still be additional equilibria in mixed strategies. The nonuniqueness of Nash equilibria for bimatrix games is a serious theoretical and practical problem.

In Example 3.4, one equilibrium strictly dominates the other equilibrium, i.e., gives both players higher payoffs. Thus, it can be argued that, even without any consultations, the players will naturally pick the strategy pair corresponding to matrix entry $(i, j) = (1, 1)$. However, it is easy to find examples where the situation is not so clear.

Example 3.6. *Consider the following bimatrix game:*

$$\begin{bmatrix} (2, 1)^* & (0, 0) \\ (0, 0) & (1, 2)^* \end{bmatrix}.$$

It is easy to see that this game[8] has two equilibria in pure strategies, marked by the sign *, *neither of which dominates the other. Moreover, Player 1 will obviously prefer the solution in the upper-left corner, while Player 2 will rather choose the bottom-right corner. It is difficult to decide how this game should be played if the players are to arrive at their decisions independently of each another.*

We can further illustrate and interpret the non-uniqueness of equilibrium solution, using one of the classical games.

Example 3.7. *The game of chicken. The expression "game of chicken" is used in everyday language as a metaphor for a situation where two parties engage in a showdown in which they have nothing to gain and only pride stops them from backing down. The game of chicken can be explained by analyzing the behavior of two drivers heading for a single lane bridge from opposite directions. The first to swerve away yields the bridge to the other. If neither player swerves, the result is a potentially fatal head-on collision. It is presumed that the best thing for each driver to do is to stay straight while the other swerves*

---

[8]The example is a classic game and known as the "battle-of-sexes" game. In an American and rather sexist context, the rows represent the woman's choices of going to the theater or the football match while the columns are the man's choices of the same events. In fact, we can easily understand what a mixed-strategy solution means for this example: the couple will be happy if they alternate going to the theater one week and a match the next.

(since the other is then the "chicken" who loses face among his peers, and since a crash is avoided). A crash is presumed to be the worst outcome for both players. This results in a situation where each player, in attempting to secure his best outcome, risks the worst. Possible numerical values for the game outcomes are given in Table 3.1.

Table 3.1   A symmetric game of chicken

	Straight	Swerve
Straight	-100,-100	1,-1
Swerve	-1,1	0,0

Because swerving represents a loss so trivial compared to the crash that will occur if nobody swerves, the reasonable strategy would seem to be to swerve before a crash is likely. Yet, knowing this, if we believe our opponent will act reasonably, we may well decide not to swerve at all, in the belief that the opponent will swerve and that we will then win.

In Exercise 3.2 we will see that this game admits several Nash equilibrium solutions. In fact, the pure strategy equilibria are the two situations wherein one player swerves while the other does not. Obviously, a different pure strategy equilibrium is preferred by each player. There is also one equilibrium where they both use mixed strategies: $(\frac{1}{100}, \frac{99}{100})$. Driving straight ahead with this low probability generates slightly negative payoffs $(-\frac{1}{100}, -\frac{1}{100})$ for each player.   ◇

3.3.3.2   *Inefficiency of Nash equilibrium*

The prisoner's dilemma is a famous example of a bimatrix game, which is used in many contexts to argue that the Nash equilibrium solution can be inefficient.[9]

Example 3.8.

Table 3.2   The prisoner's dilemma

	Suspect II: Refuses to testify	Agrees to testify
Suspect I:		
Refuses	(2, 2)	(10, 1)
Agrees to testify	(1, 10)	(5, 5)*

Suppose that two suspects are held on the suspicion of having committed a serious crime. Each one can be convicted only if the other provides evidence against him; otherwise he will be found guilty on a lesser charge. However, by agreeing to give evidence against the other guy, a suspect can shorten his sentence by half. The prisoners are held

---

[9]An outcome is called *efficient* or *Pareto efficient* if it is impossible to increase the payoff of one player without decreasing the payoff of another player. An outcome that maximizes a weighted sum of the players' payoffs, with strictly positive weights, is efficient. If the outcome is not efficient, it is called *inefficient*.

in separate cells and cannot communicate with each other. The situation is as described in Table 3.2, with the entries giving the length of the prison sentence for each suspect in every possible situation. Notice that, in this case, the players are assumed to minimize rather than maximize the outcome of the play.

The unique Nash equilibrium of this game is given by the pair of pure strategies (agree-to-testify, agree-to-testify), with the outcome that both suspects will spend five years in prison. However, this outcome is strictly dominated by the strategy pair (refuse-to-testify, refuse-to-testify), which is not an equilibrium and thus not a realistic solution to the problem when the players cannot make binding agreements, even though it is efficient. ◇

This example shows that Nash equilibria can sometimes result in outcomes that are far from efficient. There have been a large number of publications in a variety of areas on the prisoner's dilemma. Usually, the publications show that a Nash equilibrium may lead to an outcome that is unsatisfactory from a central planner's point of view. Then, different approaches are offered to circumvent the difficulty. Most of them are ineffective, unless we move into the realm of dynamic games, a topic that will be addressed in the forthcoming chapters.

3.3.4 *Algorithms for the computation of Nash equilibria in bimatrix games*

Linear programming is closely associated with the characterization and computation of saddle points in (zero-sum) matrix games. For bimatrix games, we will rely on algorithms solving either *quadratic-programming* or *complementarity* problems, which we define below. There are also other algorithms (see [12, 193]) that permit us to find an equilibrium of simple bimatrix games. We will first show one such algorithm for 2×2 bimatrix games, then introduce the quadratic programming [168] and the complementarity-problem [156] formulations.

Equilibrium computation in a 2×2 bimatrix game.   For a simple 2×2 bimatrix game, we can easily compute a mixed-strategy equilibrium as shown in the following example.[10]

Example 3.9. *Consider the game with the payoff matrix given below,*

$$\begin{bmatrix} (1,0) & (0,1) \\ (\,1/2,1/3) & (1,0) \end{bmatrix},$$

for which we want to compute a mixed-strategy equilibrium.

We first notice that this game has no pure-strategy equilibrium. Assume Player 2 chooses his equilibrium strategy y (i.e., he uses first column with probability y and uses second column with probability $(1 - y)$) in such a way that Player 1 (in equilibrium) will

---

[10]This approach to the equilibrium computations requires that there be no dominant row (or column) strategy in the game (compare Section 2.5.1).

get as much payoff using the first row as using the second row, i.e.,

$$y + 0(1 - y) = \frac{1}{2}y + 1(1 - y).$$

This is true for $y^* = \frac{2}{3}$.

Symmetrically, assume Player 1 is using a strategy x (i.e., use first row with probability x, use second row with probability $(1 - x)$) such that Player 2 will get as much payoff using the first column as using the second column, i.e.,

$$0x + \frac{1}{3}(1 - x) = 1x + 0(1 - x).$$

This is true for $x^* = \frac{1}{4}$. The players' payoffs will be, respectively, $\frac{2}{3}$ and $\frac{1}{4}$.

Then the pair of mixed strategies

$$(x^*, 1 - x^*),\ (y^*, 1 - y^*)$$

is an equilibrium in mixed strategies.   ◇

Links between quadratic programming and Nash equilibria in bimatrix games.
Mangasarian and Stone (1964) proved the following result, which links quadratic programming with the search for equilibria in bimatrix games.

Consider a bimatrix game (A, B). We introduce scalar variables v^1, $v^2 \in \mathbf{R}$ and associate the following quadratic program with the game:

$$\max \quad [x'Ay + x'By - v^1 - v^2], \tag{3.15}$$

$$\text{s.t.}$$

$$Ay \le v^1 1_k, \tag{3.16}$$

$$B'x \le v^2 1_n, \tag{3.17}$$

$$x, y \ge 0, \tag{3.18}$$

$$x'1_k = 1, \tag{3.19}$$

$$y'1_n = 1. \tag{3.20}$$

We will represent a solution to the game as quadruple (x, y, v^1, v^2) where v^1, v^2 are the payoffs obtained by Player 1 and 2, respectively.

Lemma 3.3. *The following two assertions are equivalent:*

(1) (x, y, v^1, v^2) *is a solution to the quadratic programming problem (3.15)–(3.20);*
(2) (x, y) *is an equilibrium for the bimatrix game.*

Proof. It follows from the constraints that $x'Ay \le v^1$ and $x'By \le v^2$ for any feasible (x, y, v^1, v^2). Hence the maximum of the program is at most 0.

Assume that (x, y) is an equilibrium for the bimatrix game. Then the quadruple

$$(x, y, v^1 = x'Ay, v^2 = x'By)$$

is feasible, i.e., satisfies (3.16)–(3.20); moreover, it gives a 0 value to the objective function (3.15). Hence the equilibrium defines a solution to the quadratic-programming problem (3.15)–(3.20).

Conversely, let (x_*, y_*, v_*^1, v_*^2) be a solution to the quadratic-programming problem (3.15)–(3.20). We know that an equilibrium exists for a bimatrix game (Nash theorem). We know that this equilibrium is a solution to the quadratic-programming problem (3.15)–(3.20) with an optimal value of 0. Hence the optimal program (x_*, y_*, v_*^1, v_*^2) must also give a value of 0 to the objective function and, thus, be such that

$$x_*' A y_* + x_*' B y_* = v_*^1 + v_*^2. \tag{3.21}$$

For any $x \geq 0$ and $y \geq 0$ such that $x' 1_k = 1$ and $y' 1_n = 1$ we have, by (3.19) and (3.20)

$$x' A y_* \leq v_*^1,$$
$$x_*' B y \leq v_*^2.$$

In particular we must have

$$x_*' A y_* \leq v_*^1,$$
$$x_*' B y_* \leq v_*^2.$$

These two conditions with (3.21) imply

$$x_*' A y_* = v_*^1,$$
$$x_*' B y_* = v_*^2.$$

Therefore we can conclude that, for any $x \geq 0$ and $y \geq 0$ such that $x' 1_k = 1$ and $y' 1_n = 1$, we have, by (3.19) and (3.20),

$$x' A y_* \leq x_*' A y_*,$$
$$x_*' B y \leq x_*' B y_*$$

and hence, (x_*, y_*) is a Nash equilibrium for the bimatrix game. $\qquad\square$

A complementarity-problem formulation. We have seen that the search for equilibria could be done through solving a quadratic-programming problem. Here we show that a solution to a bimatrix game can also be obtained as a solution to a complementarity problem.

There is no loss in generality if we assume that the payoff matrices are $k \times n$ and have positive entries ($a_{ij} > 0$ and $b_{ij} > 0$ for all i, j pairs) only. This is not restrictive since the VNM utilities are defined up to an increasing affine transformation. A strategy for Player 1 is defined as a vector $x \in \mathbf{R}^k$ that satisfies

$$x \geq 0, \tag{3.22}$$
$$x' 1_k = 1 \tag{3.23}$$

and similarly for Player 2

$$y \geq 0, \tag{3.24}$$
$$y' 1_n = 1. \tag{3.25}$$

It is easily shown that the pair (x^*, y^*) satisfying (3.22)–(3.25) is an equilibrium if

$$
\begin{aligned}
(x^{*'} A y^*) \mathbf{1}_k &\geq A y^* \quad (A > 0), \\
(x^{*'} B y^*) \mathbf{1}_n &\geq B' x^* \quad (B > 0),
\end{aligned}
\tag{3.26}
$$

i.e., if the equilibrium condition is satisfied for the pure strategies only.

Consider the following set of constraints with $v_1 \in \mathbf{R}$ and $v_2 \in \mathbf{R}$:

$$
\begin{aligned}
v_1 \mathbf{1}_k &\geq A y^* \\
v_2 \mathbf{1}_n &\geq B' x^*
\end{aligned}
\quad \text{and} \quad
\begin{aligned}
x^{*'} (A y^* - v_1 \mathbf{1}_k) &= 0, \\
y^{*'} (B' x^* - v_2 \mathbf{1}_n) &= 0.
\end{aligned}
\tag{3.27}
$$

The relations on the right-hand side are called the *complementarity constraints*. For mixed strategies (x^*, y^*) satisfying (3.22)–(3.25), they simplify to $x^{*'} A y^* = v_1$, $x^{*'} B y^* = v_2$. This shows that the above system (3.27) of constraints is equivalent to the system (3.26).

Define $s_1 = x/v_2$, $s_2 = y/v_1$ and introduce the slack variables u_1 and u_2 so that the system of constraints (3.22)–(3.25) and (3.27) can be rewritten as

$$
\begin{pmatrix} u_1 \\ u_2 \end{pmatrix} = \begin{pmatrix} \mathbf{1}_k \\ \mathbf{1}_n \end{pmatrix} - \begin{pmatrix} 0 & A \\ B' & 0 \end{pmatrix} \begin{pmatrix} s_1 \\ s_2 \end{pmatrix},
\tag{3.28}
$$

$$
0 = \begin{pmatrix} u_1 \\ u_2 \end{pmatrix}' \begin{pmatrix} s_1 \\ s_2 \end{pmatrix},
\tag{3.29}
$$

$$
0 \leq \begin{pmatrix} u_1 \\ u_2 \end{pmatrix}',
\tag{3.30}
$$

$$
0 \leq \begin{pmatrix} s_1 \\ s_2 \end{pmatrix}.
\tag{3.31}
$$

Introducing four obvious new matrices and variables M, q, s, u permits us to rewrite (3.28)–(3.31) in the generic formulation of a complementarity problem:

$$
u = q + M s,
\tag{3.32}
$$

$$
0 = u' s,
\tag{3.33}
$$

$$
u \geq 0,
\tag{3.34}
$$

$$
s \geq 0.
\tag{3.35}
$$

A pivoting algorithm[11] ([156, 155]), called Lemke-Howson algorithm, has been proposed to solve such problems.

Remark 3.5. Once we have computed x and y that solve (3.28)–(3.31), we will reconstruct the original strategies through the formulae

$$
x = \frac{s_1}{s_1' \mathbf{1}_k},
\tag{3.36}
$$

$$
y = \frac{s_2}{s_2' \mathbf{1}_n}.
\tag{3.37}
$$

---

[11]This algorithm could also be used to solve a *quadratic program* [154]. This implies that the task of computing an equilibrium is of the same level of complexity as solving a quadratic-programming problem.

3.3.5 *Computing all equilibria in a bimatrix game*

There are theorems showing that, under fairly weak conditions, the number of Nash equilibria in a bimatrix game is odd. In particular, it is shown in [205] that if y is an odd integer between 1 and $2n - 1$, then there is an $n \times n$ bimatrix game with exactly y Nash equilibria.

The problem of finding all the equilibria in a bimatrix or an m-matrix game is as difficult as finding all the extreme points of a polytope. In [10], Audet and Hansen provide an algorithm to enumerate all Nash equilibria in a bimatrix game.

3.3.6 *Extension to m-player games*

Wilson [246] showed that the Lemke-Howson algorithm [156] can handle the case of an m-player game (i.e., m-matrix game). Interestingly, in the Gambit software, a variety of algorithms is implemented to easily compute all the equilibrium solutions of games defined in extensive form, bimatrix and m-matrix games (see the hands-on exercise).

Judd [133] explains how to reduce the computation of Nash equilibrium to a constrained optimization problem. The proposed method, based on a theorem by McKelvey [170], is only locally convergent but, combined with extensive search, can find all Nash equilibria, assuming there is a finite number of them.

3.3.7 *Solutions obtained by iterated strict dominance or rationalizability*

We saw in Section 2.5.1 that, for some games, a solution can be found by the iterated deletion of strictly dominated strategies. The following lemma shows that the solution thus obtained is necessarily a Nash equilibrium.

Lemma 3.4. *In an m-player game, when the iterated deletion of strictly dominated strategies yields a unique strategy vector, this strategy vector is necessarily a Nash equilibrium.*

Proof. Let $\underline{s}^* = (s_1^*, \ldots, s_i^*, \ldots, s_m^*)$ be the unique *pure* strategy m-tuple obtained in the limit of the procedure of iterated deletion of strictly dominated strategies. Suppose that this does not correspond to a Nash equilibrium. Then there exists[12] a player j and a strategy $s_j \in S_j$ such that the payoff to player j is improved, i.e.,

$$r_j(s_j, \mathbf{s}_{-j}^*) > r_j(\underline{s}^*). \tag{3.38}$$

Then if the singleton $\underline{s}^*$ has been obtained in one round of the iterated-deletion procedure, s_j^* must dominate all other pure strategies in S_j, which contradicts (3.38). More generally, s_j must have been eliminated at some round of the procedure because it is strictly dominated by some other strategy s_j', which is itself eliminated in a later round, being dominated by a strategy s_j'', which is, in turn ultimately eliminated by s_j^* in the last round. By transitivity, s_j^* must then be a better reply to $\mathbf{s}_{-j}^*$ than s_j, which again contradicts (3.38).$\square$

---

[12]Here, we use the same notation as in Section 2.5.1.

Note[13] also that every *pure or mixed-strategy* Nash equilibrium is *rationalizable*, since, if the strategy m-tuple $\underline{\sigma}^* = (\sigma_1, \ldots, \sigma_m)$ is a Nash equilibrium then, for each j, σ_j^* must be element of $\tilde{\Sigma}_j^n$ for each n, according to Definition 2.5.

3.3.8 Rationality and importance of Nash's concept of equilibrium

We want to emphasize the importance of *rationality* in the Nash equilibrium, which was alluded to in Remark 3.4.

A decision maker is *rational* if he makes decisions consistently in pursuit of his objective, which is to maximize the expected value of his own payoff, measured on some utility scale. Nash's concept of equilibrium satisfies a rationality test: *Suppose we want to compute a strategy pair to be used by players in a game that these players understand completely. Then we must specify an equilibrium for them to play; anything else would only be accepted by irrational players. If the specification given were not an equilibrium, then a player could gain by changing his strategy. Thus playing a non-equilibrium strategy would be a self-denying way for rational players to play the game.*

This argument implies that any outcome that is not an equilibrium would be an unreasonable prescription of how players should behave. However, this does not imply that any given equilibrium must be reasonable in all given situations. In brief we cannot recommend a way to play which is not an equilibrium but we cannot say that a particular equilibrium is the only correct way to play, except in a zero-sum two-player game.

3.4 Concave m-person games

3.4.1 Definition

The nonuniqueness of equilibria in bimatrix games and, a fortiori, in m-matrix games poses a problem. If there are many equilibria and if we assume that players cannot communicate or enter into preplay negotiations, we do not know which strategy corresponding to which equilibrium the players may choose. However, if an equilibrium exists and it is unique, then we know what strategies rational agents will play.

In single-agent optimization theory, we know that the strict concavity of the (maximized) objective function and compactness and convexity of the constraint set lead to the existence and uniqueness of a solution. The following question thus arises: *Can we generalize the mathematical programming approach to a situation where the optimization criterion is a Nash-equilibrium? Can we give sufficient conditions for the existence and uniqueness of an equilibrium solution?*

The answers are given by Rosen in his seminal paper [213] dealing with *concave m-person games.*

---

[13]Here, we use the same notation as in Section 2.5.2.

A concave m-person game is characterized by individual strategies represented by vectors in compact[14] subsets of Euclidean spaces ($\mathbf{R}^{m_j}$ for Player j) and by payoffs represented, for each player, by a continuous function that is concave with respect to his own strategic variables. This includes the case of m-matrix games played with mixed strategies, which where studied in the previous sections. Indeed, in a matrix (or m-matrix) game, a player's mixed strategies are represented by elements of a *simplex*, which is a compact convex set, and the payoffs are *bilinear* (or *multilinear*) forms of the strategies. Hence, in matrix games, a player's payoff is a concave function with respect to his own strategic variables. This structure is generalized in concave games in two ways: (i) the strategies are vectors that belong to a general compact convex set rather than remaining restricted to a convex combination of an enumerable set of the elements; and (ii) each player's payoff is represented by a continuous function that is concave with respect to his own strategies, rather than multilinear forms.

Let us thus introduce the following game, defined in normal, or strategic, form.

- $M = \{1, \ldots, m\}$ is the set of players. Each player $j \in M$ controls his strategy $u_j \in U_j$ where U_j is a compact convex subset of $\mathbf{R}^{m_j}$ and m_j is a given integer. Player j receives a payoff $\psi_j(u_1, \ldots, u_j, \ldots, u_m)$ that depends on the actions chosen by all the players. We assume that the reward function $\psi_j : U_1 \times \cdots \times U_j, \cdots \times U_m \mapsto \mathbf{R}$ is continuous in each u_i and concave in u_j.
- A *coupled constraint* set is defined as a proper subset $\mathcal{U}$ of $U_1 \times \cdots \times U_j \times \cdots \times U_m$. The constraint is such that the joint action $\underline{u} = (u_1, \ldots, u_m)$ must be in $\mathcal{U}$, which is also assumed to be a convex set.

Definition 3.5. An equilibrium under the coupled constraint set $\mathcal{U}$ is defined as a decision m-tuple $\underline{u}^* = (u_1^*, \ldots, u_j^*, \ldots, u_m^*) \in \mathcal{U}$ such that for each player $j \in M$

$$\psi_j(\underline{u}^*) \geq \psi_j(u_1^*, \ldots, u_j, \ldots, u_m^*) \tag{3.39}$$
$$\text{for all } u_j \in U_j \text{ s.t. } (u_1^*, \ldots, u_j, \ldots, u_m^*) \in \mathcal{U}.$$

We will occasionally use the abbreviation CCE to refer to this equilibrium, which is *coupled constraint equilibrium*.

Remark 3.6. The consideration of *coupled constraints* is a new feature. Now, each player's strategy space may depend on the strategy of the other players. This may look awkward in the context of nonocooperative games, where the players cannot enter into communication or cannot coordinate their actions. However the concept is mathematically well defined.

Remark 3.7. We find coupled-constraint games well suited to capture some interesting aspects of environmental management. For example, we can think of a global pollution standard as a coupled constraint that is imposed on a finite set of firms, which are competing on the same market. This idea was exploited, e.g., in [109] with the reference to a river valley polluted by three paper mills, and in [64] where coupled constraints were implied by the global emission standards.

---

[14]I.e., closed and bounded.

Remark 3.8. Some authors (see, e.g., [72, 71, 73, 90, 101, 194] or [240]) call a coupled-constraints equilibrium a generalized Nash equilibrium or GNE. We refer to [71] for a comprehensive survey of the GNE and for numerical solutions to this type of equilibrium. Among other topics, the survey includes complementarity formulations of the equilibrium conditions and the solution methods based on variational inequalities.

3.4.2 Existence of equilibria

At a coupled-constraint equilibrium point no player can improve his payoff by a unilateral change in his strategy, which keeps the combined vector in $\mathcal{U}$. We want to show that such a solution exists in a concave game. To do so, it is convenient to reformulate the equilibrium conditions as a *fixed-point* condition for a *point-to-set mapping*.[15] For that purpose, we introduce the *global reaction function* $\theta : \mathcal{U} \times \mathcal{U} \times \mathbf{R}^m_+ \mapsto \mathbf{R}$ defined by

$$\theta(\underline{u}, \underline{v}, \mathbf{r}) = \sum_{j=1}^{m} r_j \psi_j(u_1, \ldots, v_j, \ldots, u_m), \tag{3.40}$$

where the coefficient $r_j > 0$, $j = 1, \ldots, m$ is an arbitrary positive weight given to each player's payoff. The precise role of this weighting scheme will be explained later. At present, we can assume that $r_j \equiv 1$. There, v_j is the j-th component of the combined-action vector

$$\underline{v} = (v_1, \ldots, u_j, \ldots, v_m). \tag{3.41}$$

In equation (3.40), which defines $\theta(\cdot, \cdot, \cdot)$, we treat player j's moves v_j as reactions to the other players' actions $(u_1, \ldots, u_{j-1}, u_{j+1}, \ldots, u_m)$. Consequently, $\underline{v} = (v_1, v_2, \ldots, v_m)$ will be selected as a maximizer of the weighted-response function $\theta(\cdot, \cdot, \cdot)$.

Notice that, even if $\underline{u}$ and $\underline{v}$ are in $\mathcal{U}$, the combined vectors $(u_1, \ldots, v_j, \ldots, u_m)$ are elements of a larger set in $U_1 \times \ldots \times U_m$. As defined in (3.40), the function $\theta(\underline{u}, \underline{v}, \mathbf{r})$ is continuous in $\underline{u}$ and concave in $\underline{v}$ for every fixed $\underline{u}$. This function is helpful to characterize an equilibrium through a fixed-point property, as shown in the following result.

Lemma 3.5. *Let $\underline{u}^* \in \mathcal{U}$ be such that*

$$\theta(\underline{u}^*, \underline{u}^*, \mathbf{r}) = \max_{\underline{u} \in \mathcal{U}} \theta(\underline{u}^*, \underline{u}, \mathbf{r}). \tag{3.42}$$

Then $\underline{u}^*$ is a coupled constraint equilibrium.

Proof. Assume $\underline{u}^*$ satisfies (3.42) but is not a coupled constraint equilibrium, i.e., does not satisfy (3.40). Then, for one player, say ℓ, there would exist a vector

$$\bar{\underline{u}} = (u_1^*, \ldots, u_\ell, \ldots, u_m^*) \in \mathcal{U}$$

such that

$$\psi_\ell(\bar{\underline{u}}) > \psi_\ell(\underline{u}^*).$$

If so, then $\theta(\underline{u}^*, \bar{\underline{u}}) > \theta(\underline{u}^*, \underline{u}^*)$, which is in contradiction to (3.42). $\qquad \square$

---

[15] A point-to-set mapping, or *correspondence*, is a multivalued function that assigns sets to vectors. Similarly to the usual point-to-point mappings, these functions can also be continuous, upper semi-continuous, etc.

This result has two important consequences.

(1) It shows that proving existence of an equilibrium is equivalent to proving that a fixed point exists for an appropriately defined reaction mapping ($\underline{u}^*$ is the best reply, or reaction to $\underline{u}^*$ in (3.42)).
(2) It shows that the equilibrium problem (3.39) and (3.40) defines an associated *implicit maximization problem* (3.42). We say that the latter is implicit because it is defined in terms of the very solution $\underline{u}^*$ that it characterizes.

Let us introduce a coupled reaction mapping to make the fixed point argument more precise.

Definition 3.6. The point-to-set mapping

$$\Gamma(\underline{u}, \mathbf{r}) = \left\{ \underline{v} : \theta(\underline{u}, \underline{v}, \mathbf{r}) = \max_{\underline{w} \in \mathcal{U}} \theta(\underline{u}, \underline{w}, \mathbf{r}) \right\} \tag{3.43}$$

is called the coupled-reaction mapping associated with the positive weighting $\mathbf{r}$. A fixed point of $\Gamma(\cdot, \mathbf{r})$ is a vector $\underline{u}^*$ such that $\underline{u}^* \in \Gamma(\underline{u}^*, \mathbf{r})$.

We can see from Lemma 3.5 that a fixed point of $\Gamma(\cdot, \mathbf{r})$ is a coupled constraint equilibrium.

To prove the following theorem, we need the notion of the upper semi-continuity of a point-to-set mapping and the Kakutani fixed-point theorem.

Definition 3.7. Let $\Phi : \mathbf{R}^m \mapsto 2^{\mathbf{R}^m}$ be a point-to-set mapping.[16] We say that this mapping is upper semi-continuous if, whenever the sequence $\{x_k\}_{k=1,2,\ldots}$ converges in $\mathbf{R}^m$ toward x^0, then any accumulation point y^0 of the sequence $\{y_k\}_{k=1,2,\ldots}$ in $\mathbf{R}^n$, where $y_k \in \Phi(x_k)$, $k = 1, 2, \ldots$, is such that $y^0 \in \Phi(x_0)$.

Theorem 3.4. (Kakutani.) *Let* $\Phi : A \mapsto 2^A$ *be a point-to-set upper semi-continuous mapping, where A is a compact subset (i.e., a closed and bounded subset) of $\mathbf{R}^m$. Then there exists a fixed point for Φ. That is, there exists $x^* \in \Phi(x^*)$ for some $x^* \in A$.*

And here is the existence theorem of coupled-constraint equilibrium.

Theorem 3.5. *Let the mapping $\Gamma(\cdot, \mathbf{r})$ be defined through (3.43). For any positive-weighting $\mathbf{r}$, there exists a fixed point of $\Gamma(\cdot, \mathbf{r})$, i.e., a point $\underline{u}^*$ s.t. $\underline{u}^* \in \Gamma(\underline{u}^*, \mathbf{r})$. Hence a coupled-constraint equilibrium exists.*

Proof.   The proof is based on the Kakutani fixed-point theorem. We are required to show that the point-to-set mapping is upper semicontinuous. This is a consequence of the concavity of the game, the continuity of payoff functions and of compactness of all constraint sets $U_j, j = 1, \ldots, m$ and $\mathcal{U}$.   □

---

[16] By $2^{\mathbf{R}^m}$ we denote the set of all subsets of $\mathbf{R}^m$.

So, in simple terms, we know that coupled-constraint equilibria exist in concave games.

Remark 3.9. This existence theorem is very close, in spirit, to the theorem of Nash [180] proving existence of equilibria in mixed strategies for bimatrix games. It uses a fixed-point result that is topological and not constructive, i.e., it does not provide a computational method. However, the definition of a normalized equilibrium introduced by Rosen [213] establishes a constructive link between mathematical programming and concave games with coupled constraints.

3.4.3 *Normalized equilibria*

3.4.3.1 *Karush-Kuhn-Tucker multipliers*

Suppose that $\mathcal{U}$, the coupled constraint set (3.40), can be defined by a set of inequalities

$$h_k(\underline{u}) \geq 0, \quad k = 1, \ldots, p \tag{3.44}$$

where $h_k : U_1 \times \cdots \times U_m \to \mathbf{R}$, $k = 1, \ldots, p$, are given concave functions. Let us further assume that the payoff functions $\psi_j(\cdot)$ as well as the constraint functions $h_k(\cdot)$ are continuously differentiable and satisfy the constraint-qualification conditions[17] so that the Karush-Kuhn-Tucker multipliers exist for each of the implicit single-agent optimization problems defined below.

Assume that all players, other than Player j, use their strategies u_ℓ^*, $\ell \in M \setminus \{j\}$ while Player j uses u_j. We will denote the corresponding joint-strategy vector by $[u_j, \mathbf{u}_{-j}^*]$.

Then, the equilibrium conditions (3.39)–(3.40) define a single agent optimization problem with a concave objective function and a convex compact admissible set. Under the assumed-constraint qualification, there exists a vector of multipliers

$$\lambda_j = (\lambda_{jk})_{k=1,\ldots,p}$$

such that the Lagrangian

$$L_j(\lambda_j, [u_j, \mathbf{u}_{-j}^*]) = \psi_j([u_j, \mathbf{u}_{-j}^*]) + \sum_{k=1\ldots p} \lambda_{jk} h_k([u_j, \mathbf{u}_{-j}^*]) \tag{3.45}$$

verifies, at the optimum,

$$0 = \frac{\partial}{\partial u_j} L_j((\lambda_j, [[u_j^*, \mathbf{u}_{-j}^*]), \tag{3.46}$$

$$0 \leq \lambda_j, \tag{3.47}$$

$$0 = \lambda_{jk} h_k([u_j^*, \mathbf{u}_{-j}^*]) \quad k = 1, \ldots, p, \tag{3.48}$$

$$0 \leq h_k([u_j^*, \mathbf{u}_{-j}^*]) \quad k = 1, \ldots, p. \tag{3.49}$$

Definition 3.8. We say that the equilibrium is normalized if the different multipliers λ_j for $j \in M$ are colinear with a common vector λ_0, namely,

$$\lambda_j = \frac{1}{r_j} \lambda_0 \tag{3.50}$$

where the coefficient $r_j > 0, j = 1, \ldots, m$ is a weight assigned to Player j.

---

[17]Known from mathematical programming, see, e.g., [167].

We observe that the common multiplier λ_0 is associated with the implicit mathematical programming problem

$$\max_{u \in \mathcal{U}} \theta(\underline{u}^*, \underline{u}, \mathbf{r}). \tag{3.51}$$

This is easy to see when we write the Lagrangian for this problem

$$L_0(\lambda_0, \underline{u}) = \sum_{j \in M} r_j \psi_j([u_j, \mathbf{u}^*_{-j}]) + \sum_{k=1...p} \lambda_{0k} h_k(\underline{u}) \tag{3.52}$$

and the first order necessary conditions

$$0 = \frac{\partial}{\partial u_j} \left\{ r_j \psi_j(\underline{u}^*) + \sum_{k=1,...,p} \lambda_{0k} h_k(\underline{u}^*) \right\}, \quad j \in M, \tag{3.53}$$

$$0 \le \lambda_0, \tag{3.54}$$

$$0 = \lambda_{0k} h_k(\underline{u}^*) \quad k = 1, \dots, p. \tag{3.55}$$

$$0 \le h_k(\underline{u}^*) \quad k = 1, \dots, p. \tag{3.56}$$

Then, the relationship (3.50) is obvious.

3.4.3.2 *An economic interpretation*

In the context of mathematical programming, a Karush-Kuhn-Tucker multiplier can be interpreted as the marginal cost or *shadow price* associated with the right-hand side of a constraint. More precisely, the multiplier measures the sensitivity of the optimum solution to marginal changes in this right-hand-side.

The multiplier also permits a *price decentralization,* in the sense that, through an ad-hoc pricing mechanism,[18] the optimizing agent is induced to satisfy the constraints.

In a normalized equilibrium, the shadow-cost interpretation is not so apparent; however, the *price decomposition principle* is still valid. Once the common multiplier has been defined with the associated weighting $r_j > 0, j = 1, \dots, m$, the coupled constraint will be satisfied by equilibrium-seeking players, playing without the coupled constraint but using the Lagrangians as payoffs

$$L_j(\lambda_j, [u_j, \mathbf{u}^*_{-j}]) = \psi_j([u_j, \mathbf{u}^*_{-j}]) + \frac{1}{r_j} \sum_{k=1...p} \lambda_{0k} h_k([u_j, \mathbf{u}^*_{-j}]),$$

$$j = 1, \dots, m. \tag{3.57}$$

The common multiplier then allows an implicit pricing of the common constraint so that the latter remains compatible with the equilibrium structure. However, to be useful this result necessitates uniqueness. We will provide conditions that guarantee uniqueness of the normalized equilibrium associated with a given weighting scheme $r_j > 0, j = 1, \dots, m$ in the next section.

---

[18] We propose one such mechanism in Section 3.4.5.

3.4.4 *Uniqueness of equilibrium*

In a mathematical-programming framework, the uniqueness of an optimum results from the strict concavity of the objective function to be maximized. In a game structure, uniqueness of the equilibrium will result from a more stringent concavity requirement, which Rosen [213] termed *diagonal strict concavity*.

Consider the following function

$$\Psi(\underline{u}, \mathbf{r}) = \sum_{j=1}^{m} r_j \psi_j(\underline{u}), \tag{3.58}$$

sometimes referred to as the *joint player payoff*. Let us define the *pseudo-gradient* of this function as the row vector

$$g(\underline{u}, \mathbf{r}) = \left(r_1 \frac{\partial}{\partial u_1} \psi_1(\underline{u}) \; r_2 \frac{\partial}{\partial u_2} \psi_2(\underline{u}) \; \cdots \; r_m \frac{\partial}{\partial u_m} \psi_m(\underline{u}) \right). \tag{3.59}$$

We notice that this expression is composed of the partial gradients of the different player payoffs with respect to the decision variables of the corresponding player.

Definition 3.9. The function $\Psi(\underline{u}, \mathbf{r})$ is diagonally strictly concave on $\mathcal{U}$ if the following condition,

$$g(\underline{u}^1, \mathbf{r})(\underline{u}^2 - \underline{u}^1) + g(\underline{u}^2, \mathbf{r})(\underline{u}^1 - \underline{u}^2) > 0 \tag{3.60}$$

holds for every $\underline{u}^1$ and $\underline{u}^2$ in $\mathcal{U}$.

A sufficient condition for $\Psi(\underline{u}, \mathbf{r})$ to be diagonally strictly concave is that the symmetric matrix[19] $[G(\underline{u}, \mathbf{r}) + G(\underline{u}, \mathbf{r})']$ be negative definite for any $\mathbf{u}$ in $\mathcal{U}$, where $G(\underline{u}, \mathbf{r})$ is the Jacobian of $g(\underline{u}, \mathbf{r})$ with respect to $\underline{u}$.

Theorem 3.6. *If $\Psi(\underline{u}, \mathbf{r})$ is diagonally strictly concave on the convex set $\mathcal{U}$, with the assumptions ensuring the existence of the Karush-Kuhn-Tucker multipliers, then there exists a unique normalized equilibrium for the weighting scheme $\mathbf{r} > 0$.*

Proof. Below we sketch the proof given by Rosen [213]. Assume that for some $\mathbf{r} > 0$ we have two equilibria $\underline{u}^1$ and $\underline{u}^2$. Then we must have

$$h(\underline{u}^1) \geq 0, \tag{3.61}$$

$$h(\underline{u}^2) \geq 0 \tag{3.62}$$

and there exist multipliers $\lambda^1 \geq 0, \lambda^2 \geq 0$, such that

$$\lambda^{1'} h(\underline{u}^1) = 0, \tag{3.63}$$

$$\lambda^{2'} h(\underline{u}^2) = 0, \tag{3.64}$$

for which the following equations are true for each player $j \in M$

$$r_j \frac{\partial}{\partial u_j} \psi_j(\underline{u}^1) + \lambda^{1'} \frac{\partial}{\partial u_j} h(\underline{u}^1) = 0, \tag{3.65}$$

$$r_j \frac{\partial}{\partial u_j} \psi_j(\underline{u}^2) + \lambda^{2'} \frac{\partial}{\partial u_j} h(\underline{u}^2) = 0. \tag{3.66}$$

---

[19]The expression in the square brackets is sometimes referred to as the pseudo-Hessian of $\Psi(\underline{u}, \mathbf{r})$.

We multiply (3.65) by $(\underline{u}^2 - \underline{u}^1)'$ and (3.66) by $(\underline{u}^1 - \underline{u}^2)'$ and we sum up to obtain an expression $\beta + \gamma = 0$ where γ represents the sum of the terms that contain the multipliers λ_k^1 and λ_k^2. Due to the concavity of the constraint functions h_k and the nonnegativity of the conditions (3.61)–(3.64) we obtain

$$
\begin{aligned}
\gamma &= \sum_{j \in M} \sum_{k=1}^{p} \{\lambda_k^1 \frac{\partial}{\partial u_j} h_k(\underline{u}1)(\underline{u}^2 - \underline{u}^1) + \lambda_k^2 \frac{\partial}{\partial u_j} h_k(\underline{u}^2)(\underline{u}^1 - \underline{u}^2)\} \\
&\geq \sum_{j \in M} \{\lambda^{1'}[h(\underline{u}^2) - h(\underline{u}^1)] + \lambda^{2'}[h(\underline{u}^1) - h(\underline{u}^2)]\} \\
&= \sum_{j \in M} \{\lambda^{1'} h(\underline{u}^2) + \lambda^{2'} h(\underline{u}^1)\} \geq 0,
\end{aligned}
\tag{3.67}
$$

and

$$
\beta = \sum_{j \in M} r_j \left[\frac{\partial}{\partial u_j} \psi_j(\underline{u}^1)(\underline{u}^2 - \underline{u}^1) + \frac{\partial}{\partial u_j} \psi_j(\underline{u}^2)(\underline{u}^1 - \underline{u}^2) \right].
$$

Since $\Psi(\underline{u}, \mathbf{r})$ is diagonally strictly concave we have $\beta > 0$, which contradicts $\beta + \gamma = 0$. $\qquad\qquad\square$

3.4.5 *Equilibrium enforcement through taxation*

The presence of a coupled constraint in our model suggests that an exogenous body wants to enforce it. We will assume such a body exists and call it "regulator."

After $\underline{u}^*$, a unique normalized CCE, has been computed, a regulator can create an unconstrained (decoupled) game that has $\underline{u}^*$ as its solution. This can be achieved by an *enforcement mechanism*.

Suppose the regulator can modify the players' payoff functions so that they contain some penalty functions triggered by violation of the constraints. The players could then be charged for violating constraint k according to the function

$$
T_{kj}(\lambda_k, r_j, \underline{u}) = \frac{\lambda_k}{r_j} \max(0, -h_k(\underline{u}), \quad j \in M
\tag{3.68}
$$

where λ_k is the Lagrange multiplier associated with the kth constraint.

Here, we can appreciate the role of r_j that is, of player j's weight, which defines his responsibility for the constraints' satisfaction. If the weights are identical $[1, 1, \ldots 1]$ for all players then the penalty term for constraint k is the same for each player j

$$
T_{k,j}(\lambda, 1, \underline{u}) = \lambda_k \max(0, -h_k(\underline{u})).
\tag{3.69}
$$

But, if the weight for player j is, for example, $r_j > 1$ and the weights for the other players are $1, 1, \ldots 1$, then the responsibility of player j for the constraints' satisfaction is lessened. This is reflected in that the privileged player's marginal cost of violating the constraint is smaller than that of the non- or less-privileged players.

The players' payoff functions, so modified, will be

$$
\tilde{\psi}_j(\underline{u}) = \psi_j(\underline{u}) - \sum_k T_{kj}(\lambda_k, r_j, \underline{u}).
\tag{3.70}
$$

Notice that, under this taxation scheme, the penalties remain nominal (i.e., zero) if the constraints are satisfied.

The Nash equilibrium of the new unconstrained (decoupled) game with payoff functions $\tilde{\psi}_j$ is implicitly defined by the equation

$$\tilde{\psi}_j(\underline{u}^{**}) = \max_{u_j \in \mathbf{R}^+} \tilde{\psi}_j([u_j, \mathbf{u}_{-j}^{**}]), \qquad \forall\, j \in M, \tag{3.71}$$

where $\mathbf{u}_{-j}^{**} = (u_i)_{i \neq j}$. It is easy to see that the Karush-Kuhn-Tucker conditions for problem (3.71) are like those for the normalized equilibrium. This implies that the normalized CCE matches the unconstrained equilibrium with penalty functions (3.68) (see [144, 139] and [140] for a discussion and applications).

3.4.6 *A numerical technique*

The strict diagonal concavity property that yields the uniqueness result for Theorem 3.6 also provides an interesting extension of the gradient method for the computation of the unique equilibrium. The basic idea is to project the pseudo-gradient $g(\underline{u}, \mathbf{r})$ on the constraint set $\mathcal{U} = \{\underline{u} : h(\underline{u}) \geq 0\}$ at each step of the algorithm. Denoting the projection $\bar{g}(\underline{u}^\ell, \mathbf{r})$ the algorithm uses the usual steepest ascent step

$$\underline{u}^{\ell+1} = \underline{u}^\ell + \tau^\ell \bar{g}(\underline{u}^\ell, \mathbf{r}').$$

Rosen [213] shows that the step size $\tau^\ell > 0$ can be chosen at each step ℓ sufficiently small to guarantee a reduction of the norm of the projected gradient. This yields a convergence of the procedure toward the unique equilibrium.

3.4.7 *A variational inequality formulation*

The necessary conditions for a coupled-constraint equilibrium can also be given in the form of a variational inequality.

Theorem 3.7. *Under the convexity and differentiability assumptions made above, the vector $\underline{u}^* = (u_1^*, \ldots, u_m^*) \in \mathcal{U}$ is a normalized equilibrium, with weights $r \geq 0$ if and only if it satisfies the following variational inequality (VI):*

$$g(\underline{u}^*, \mathbf{r})(\underline{u} - \underline{u}^*) \leq 0, \qquad \forall \underline{u} \in \mathcal{U}. \tag{3.72}$$

Proof. The variational inequality (3.72) can also be written

$$\sum_{j=1}^{m} r_j \frac{\partial}{\partial u_j} \psi_j(\underline{u}^*)(u_j - u_j^*).$$

Now consider, for each j, the auxiliary function

$$\phi_j(t) = \psi_j(\underline{u}^* + t([u_j, \mathbf{u}_{-j}^*] - \underline{u}^*)).$$

Under the convexity assumptions, $\underline{u}^*$ is a coupled constraint equilibrium if for each j, the function $\phi_j(t)$ achieves a maximum at $t = 0$. Therefore we must have $\phi_j'(0) \leq 0$. It is easy to check that $0 \geq \phi_j'(0) = r_j(u_j - u_j^*)' \frac{\partial}{\partial u_j} \psi_j(\underline{u}^*)$. Hence summing over j we obtain the VI (3.72). □

There is an important theory corpus concerning VIs and the various algorithms that can be used to solve them.[20] Indeed all these algorithms can be utilized to find a normalized equilibrium. An important property required by many VI solution algorithms to guarantee their convergence concerns the monotonicity of VI's operators.

Definition 3.10. The VI

$$F(\underline{u}^*)'(\underline{u} - \underline{u}^*) \leq 0 \quad \forall \underline{u} \in \mathcal{U}$$

has a monotonous (resp. strictly monotonous) operator if the following inequality holds:

$$[F(\underline{u}^1) - F(\underline{u}^2)]'(\underline{u}^1 - \underline{u}^2) \leq (\text{ resp. } <)0 \quad \forall \underline{u}^1, \underline{u}^2 \in \mathcal{U}. \tag{3.73}$$

We can prove that, if the Jacobian matrix $\nabla F(x)$ is such that $[\nabla F(x) + \nabla F(x)']$ is negative definite, then the VI operator $F(\underline{u})$ is strictly monotone.

Remark 3.10. The strict diagonal concavity assumption is then equivalent to the property of strong monotonicity in the operator $g(\underline{u}^*, \mathbf{r})$, in the parlance of variational-inequality theory.

3.4.8 *A nonlinear-complementarity formulation*

Recall the first-order conditions for a normalized equilibrium given in Eqs. (3.53)–(3.55):

$$0 = \frac{\partial}{\partial u_j} \left\{ r_j \psi_j(\underline{u}^*) + \sum_{k=1,\dots,p} \lambda_{0k} h_k(\underline{u}^*) \right\}, \quad j \in M,$$

$$0 \leq \lambda_0,$$

$$0 = \lambda_{0k} h_k(\underline{u}^*) \quad k = 1, \dots, p,$$

$$0 \leq h_k(\underline{u}^*) \quad k = 1, \dots, p.$$

We suppose, in addition to the coupled constraints, that the decision vectors u_j must remain nonnegative. The equilibrium conditions can then be summarized by the following nonlinear complementarity problem

$$0 \leq \begin{pmatrix} u \\ \lambda_0 \end{pmatrix} \perp \begin{pmatrix} -\left(\frac{\partial}{\partial u_j}\{r_j\psi_j(\underline{u}) + \lambda_0' h(\underline{u})\}\right)'_{j=1\dots m} \\ h(\underline{u}) \end{pmatrix} \geq 0. \tag{3.74}$$

The complementarity condition, summarized by the symbol $\perp$, imposes that for each corresponding component in the two vectors, at least one is zero while the other satisfies the inequality constraint. Complementarity problems have been studied extensively in the mathematical programming literature, see, e.g., the book by Ferris and Pang [79]. Ferris and Munson have proposed a very efficient algorithm for solving these problems and this algorithm has been implemented in the GAMS and AMPL modeling languages [77, 78]. Next Section 3.4.9 provides a demonstration of the use of this algorithm to solve an oligopoly model for a Nash equilibrium.

---

[20] See, e.g., the recent book by Konnov [137].

3.4.9 Application to the Cournot oligopoly model

A typical example of a concave game is the oligopoly model, originally formulated by Augustin Cournot in 1838 [49]. In this section, we will apply the results established above to the characterization and computation of the Cournot/Nash equilibrium solution.

3.4.9.1 Assumptions:

Let us recall the basic formulation of the model. We consider a single market on which m firms are competing. The market is characterized by its (inverse) demand law $p = D(Q)$ where p is the market clearing price and $Q = \sum_{j=1,\ldots,m} q_j$ is the total supply of the homogenous good on the market. The firm j faces a cost of production $C_j(q_j)$; hence, letting $q = (q_1, \ldots, q_m)$ represent the production decision vector of the m firms together. The profit of firm j is $\pi_j(q) = q_j D(Q) - C_j(q_j)$. The following assumptions are placed on the model.

Assumption 3.1. The market demand and the firms' cost functions satisfy the following properties: (i) The inverse-demand function is finite valued, nonnegative and defined for all $Q \in [0, \infty)$. It is also twice differentiable, with $D'(Q) < 0$ wherever $D(Q) > 0$. In addition $D(0) > 0$. (ii) $C_j(q_j)$ is defined for all $q_j \in [0, \infty)$, nonnegative, convex, twice continuously differentiable, and $C'(q_j) > 0$. (iii) $Q\, D(Q)$ is bounded and strictly concave for all Q, such that $D(Q) > 0$.

If we assume that each firm j chooses a supply level $q_j \in [0, \bar{q}_j]$, this model satisfies the definition of a concave game à la Rosen. Therefore an equilibrium exists, which we will call "Cournot/Nash" equilibrium. Let us address the uniqueness issue in the duopoly case.

Consider the pseudo-gradient (there are no weights (r_1, r_2) because the constraints are not coupled)

$$g(q_1, q_2)' = \begin{pmatrix} D(Q) + q_1\, D'(Q) - C_1'(q_1) \\ D(Q) + q_2\, D'(Q) - C_2'(q_2) \end{pmatrix}$$

and the Jacobian matrix

$$G(q_1, q_2) = \begin{pmatrix} 2D'(Q) + q_1\, D''(Q) - C_1''(q_1) & D'(Q) + q_1\, D''(Q) \\ D'(Q) + q_2\, D''(Q) & 2D'(Q) + q_2\, D''(Q) - C_1''(q_2) \end{pmatrix}.$$

The negative definiteness of the symmetric matrix

$$\frac{1}{2}[G(q_1, q_2) + G(q_1, q_2)'] =$$

$$\begin{pmatrix} 2D'(Q) + q_1\, D''(Q) - C_1''(q_1) & D'(Q) + \frac{1}{2}Q\, D''(Q) \\ D'(Q) + \frac{1}{2}Q\, D''(Q) & 2D'(Q) + q_2\, D''(Q) - C_1''(q_2) \end{pmatrix}$$

implies the uniqueness of the Cournot/Nash equilibrium. This condition extends easily to the m-firm case.

3.4.9.2 *Computation of a Cournot/Nash equilibrium as a nonlinear complementarity problem*

We have seen that Nash equilibria in m-matrix games could be characterized as the solution to a linear complementarity problem. As shown in Section 3.4.8, a similar formulation holds for a Cournot equilibrium and it will in general lead to a *nonlinear complementarity problem* or NCP. We can rewrite the equilibrium condition as an NCP in canonical form

$$q_j \geq 0 \quad j = 1, \ldots, m, \tag{3.75}$$

$$f_j(q) \geq 0 \quad j = 1, \ldots, m, \tag{3.76}$$

$$q_j \, f_j(q) = 0 \quad j = 1, \ldots, m, \tag{3.77}$$

where

$$f_j(q) = -g_j(q) = -D(Q) - q_j \, D'(Q) + C_j'(q_j) \quad j = 1, \ldots, m. \tag{3.78}$$

The relations (3.75–3.77) express that, at a Cournot equilibrium, either $q_j = 0$ and the gradient $g_j(q)$ is ≤ 0, or $q_j > 0$ and $g_j(q) = 0$. There is now optimization software that solves these types of problems. An extension of the GAMS ([35]) modeling software is provided in [216] as a means to solve this type of complementarity problem. More recently the modeling language AMPL (see [84]) was also adapted to handle a problem of this type and to submit it to an efficient NCP solver like PATH (see [77]).

Example 3.10. *This example comes from the AMPL website*
$\quad$ `http://www.ampl.com/TRYAMPL/index.html`
$\quad$ *Consider an oligopoly with $n = 10$ firms. The production-cost function of firm j is given by*

$$c_i(q_i) = c_i q_i + \frac{\beta_i}{1 + \beta_i} (L_i q_i)^{(1+\beta_i)/\beta_i}. \tag{3.79}$$

The market-demand function is defined by

$$D(Q) = \left(\frac{A}{Q}\right)^{1/\gamma}. \tag{3.80}$$

Therefore

$$D'(Q) = \frac{1}{\gamma}(A/Q)^{1/\gamma - 1}\left(-\frac{A}{Q^2}\right) = -\frac{1}{\gamma}D(Q)/Q. \tag{3.81}$$

The Nash Cournot equilibrium will be the solution of the NLCP (3.75–3.77) where

$$f_i(q) = -D(Q) - q_i \, D'(Q) + C_i'(q_i)$$

$$= -\left(\frac{A}{Q}\right)^{1/\gamma} + \frac{1}{\gamma}q_i \, D(Q)/Q + c_i + (L_i q_i)^{1/\beta_i}.$$

3.4.9.3 *AMPL input*

The following files can be submitted via the web submission tool on the NEOS server site
http://www-neos.mcs.anl.gov/neos/solvers/cp:PATH/AMPL.html

%%%%%%%FILE nash.mod%%%%%%%%%%

```
set Rn := 1 .. 10;
param gamma := 1.2;
param c {i in Rn}; param beta {i in Rn};
param L {i in Rn} := 10;
var q {i in Rn} >= 0;            # production vector
var Q = sum {i in Rn} q[i];
var divQ = (5000/Q)**(1/gamma);
s.t. feas {i in Rn}:
q[i] >= 0 complements
0 <= c[i] + (L[i] * q[i])**(1/beta[i]) - divQ
- q[i] * (-1/gamma) * divQ / Q;
```

%%%%%%%FILE nash.dat%%%%%%%%%%

```
data;
param c :=
 1    5
 2    3
 3    8
 4    5
 5    1
 6    3
 7    7
 8    4
 9    6
10    3 ;
 param beta :=
  1    1.2
  2    1
  3    .9
  4    .6
  5    1.5
  6    1
  7    .7
```

```
  8  1.1
  9    .95
 10    .75 ;
```

```
### %%%%%%%FILE nash.run%%%%%%%%%%
```

```
model nash.mod; data nash.dat;
set initpoint := 1 .. 4; param initval {Rn,initpoint} >= 0;
data;param initval param  initval :   1   2   3   4 :=
 1   1    10   1.0  7
 2   1    10   1.2  4
 3   1    10   1.4  3
 4   1    10   1.6  1
 5   1    10   1.8 18
 6   1    10   2.1  4
 7   1    10   2.3  1
 8   1    10   2.5  6
 9   1    10   2.7  3
10   1    10   2.9  2 ;
for {point in initpoint}
let{i in Rn} q[i] := initval[i,point];
solve;
display max{i in 1.._nccons}
abs(_ccon[i]), min{i in 1.._ncons} _con[i].slack;
```

The result obtained for the different initial points is

```
q [*] :=
 1    7.44155
 2    4.09781
 3    2.59064
 4    0.935386
 5   17.949
 6    4.09781
 7    1.30473
 8    5.59008
 9    3.22218
10   1.67709 ;
```

It shows that firm 5 has a market advantage. It has the highest β value ($\beta_5 = 1.5$), which corresponds to the highest productivity.

3.4.9.4 *Computing the solution of a classical Cournot model*

Many approaches have been proposed to find a solution to a Cournot oligopoly game, or more generally, a concave game. We described above the one based on a solution of an NCP. The following are a few alternatives:

(a) discretize the decision space and use the complementarity algorithm proposed in [156] to compute a solution to the associated bimatrix game;
(b) exploit the fact that the players are uniquely coupled through the condition that $Q = \sum_{j=1,\dots,m} q_j$ and a sort of primal-dual search technique is used [70];
(c) formulate the Nash-Cournot equilibrium problem as a VI. Assumption 3.1 implies Strict Diagonal Concavity (SDC) in the sense of Rosen, i.e., strict monotonicity of the pseudo-gradient operator, hence uniqueness of the solution and convergence of various gradient-like algorithms [137].

3.5 Existence of Nash equilibrium in infinite games with continuous payoffs

In many economic models, like the Cournot model, the set of actions of the agents are continuous. We have seen in the previous section how to obtain a proof of existence of a Nash-equilibrium by using Kakutani theorem applied to reaction correspondences (point-to-set mapping) defined in (3.43). The condition of concavity of payoffs with respect to the player's own strategies can be relaxed to quasi-concavity.[21] The most general existence theorem for pure strategy Nash-equilibria is the following

Theorem 3.8. *In a strategic (or normal) form game whose strategy sets U_j are compact convex subsets of an Euclidean space and payoff functions $\psi_j(\underline{u})$ are continuous in $\underline{u} = (u_1, \dots, u_m)$ and quasi-concave in u_j, there exists a pure-strategy Nash-equilibrium.*

Proof. The proof is very similar to that of Theorem 3.5. Continuous payoffs imply closed-graph reaction correspondences. Quasi-concavity in player's own strategies implies that the correspondences are convex-valued. Then we invoke Kakutani theorem to find a fixed point. □

Now, if the action (pure strategy) set is continuous but nonconvex, it is possible to obtain a Nash-equilibrium in mixed-strategies.

Theorem 3.9. *In a strategic (or normal) form game whose strategy sets U_j are nonempty compact subsets of an Euclidean space and payoff functions $\psi_j(\underline{u})$ are continuous in $\underline{u} = (u_1, \dots, u_m)$, there exists a Nash-equilibrium in mixed-strategy.*

---

[21]A function $f : X \to \mathbf{R}$ is quasiconcave if its upper contour sets $S_\alpha = \{x \in X : f(x) \geq \alpha\}$ are convex, $\forall \alpha \in \mathbf{R}$.

Proof.   Here the mixed strategies are the probability measures over the pure strategy sets U_j, denoted $\mu_j \in \mathcal{P}[U_j]$, endowed with the topology of weak convergence.[22] The introduction of mixed-strategies makes the strategy spaces convex. The payoffs are linear in own strategy and continuous in all strategies. So the reaction correspondences are convex-valued. The space of probability measures on a continuous set is infinite dimensional. We must invoke a more powerful fixed-point theorem than Kakutani, which is provided by Glicksberg [94].   □

3.6   Stackelberg solution

The German economist Heinrich Freiherr von Stackelberg published in 1934 an essay on *Market Structure and Equilibrium*[23] (Marktform und Gleichgewicht [241]), in which he proposed a model of duopoly with asymmetric information structure. The players of this game are a *leader* and a *follower* and they compete in quantity.

Firms engage in Stackelberg competition if one (the leader) has the advantage of moving first. The leader commits himself to use some strategy and the follower adapts: once the leader has made its move, he cannot undo it; it is committed to that action.

3.6.1   *Extensive game representation*

The Stackelberg solution of a noncooperative game can be interpreted as a change from the simultaneous moves to a sequential move information structure. To be more precise, we reproduce below the Figure 2.3 representing the game tree of a simultaneous move "one-shot" game.[24] Player 1 pure strategy set is $\{L, M, R\}$. Player 2 pure strategy set is $\{a, b, c\}$.

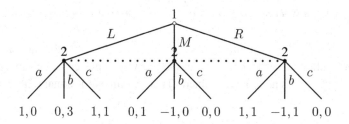

Fig. 3.2   Simultaneous move one shot game

The bimatrix game representation is shown below in Table 3.3. The strategy pair (L, b) is a Nash equilibrium in pure strategies.

---

[22]In a compact metric space X, a sequence of measures μ^n on X converges weakly to a limit μ if $\int_X f d\mu^n \to \int_X f d\mu$ for all continuous function f on X.

[23]Recently reprinted in English, see [242].

[24]We use this term to denote a game where each player moves only once.

Table 3.3 Bimatrix game formulation: simultaneous moves

	a	b	c
L	(1,0)	(0,3)*	(1,1)
M	(0,1)	(-1,0)	(0,0)
R	(1,1)	(-1,1)	(0,0)

If Player 1 has the possibility to move first and Player 2 can observe the move made by Player 1, the game tree is modified as shown in Figure 3.3. Now the game has two stages. At first stage Player 1 moves. At stage 2, Player 2 knows what move has been chosen by Player 1 and selects his own move.

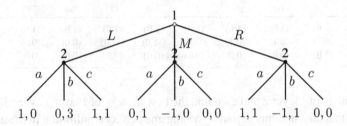

Fig. 3.3 Sequential moves two stage game

The pure strategy set of Player 1 is still $\{L, M, R\}$. The pure strategy set of Player 2 is the set of all mappings $\gamma : \{L, M, R\} \to \{a, b, c\}$. There are 27 such mappings denoted

$$\{aaa, baa, caa, aba, bba, cba, aca, bca, cca,$$
$$aab, bab, cab, abb, bbb, cbb, acb, bcb, ccb,$$
$$aac, bac, cac, abc, bbc, cbc, acc, bcc, ccc, \},$$

where abc represents the mapping γ which sends L onto a, M onto b and R onto c. We will use also $\gamma_1, \ldots, \gamma_{27}$ to refer to these strategies in the bimatrix game representation shown in Table 3.3.

An equilibrium for this bimatrix game is called a *Stackelberg solution* of the original game. We can observe that this new bimatrix game has as Nash equilibria all the 9 pairs $(R, \gamma_1), \ldots, (R, \gamma_9)$, which all give the same pair of payoffs $(1, 1)$. We observe that the leader (Player 1) has a better outcome in the Stackelberg solution, compared with the Nash equilibrium of the original game shown in Table 3.3. This is always true: The Stackelberg solution gives to the leader a payoff which is greater than or equal to the Nash equilibrium payoff in the original game.

There are many equilibrium pairs in the game 3.4. We can considerably reduce the number of candidates if we proceed through sequential optimization, to find first the best

Table 3.4   Bimatrix game formulation: sequential moves

	γ_1	γ_2	γ_3	γ_4	γ_5	γ_6	γ_7	γ_8	γ_9
L	(1,0)	(0,3)	(1,1)	(1,0)	(0,3)	(1,1)	(1,0)	(0,3)	(1,1)
M	(0,1)	(0,1)	(0,1)	(-1,0)	(-1,0)	(-1,0)	(0,0)	(0,0)	(0,0)
R	(1,1)*	(1,1)**	(1,1)*	(1,1)*	(1,1)*	(1,1)*	(1,1)*	(1,1)*	(1,1)*

	γ_{10}	γ_{11}	γ_{12}	γ_{13}	γ_{14}	γ_{15}	γ_{16}	γ_{17}	γ_{18}
L	(1,0)	(1,0)	(1,0)	(1,0)	(1,0)	(1,0)	(1,0)	(1,0)	(1,0)
M	(0,1)	(0,1)	(0,1)	(-1,0)	(-1,0)	(-1,0)	(0,0)	(0,0)	(0,0)
R	(-1,1)	(-1,1)	(-1,1)	(-1,1)	(-1,1)	(-1,1)	(-1,1)	(-1,1)	(-1,1)

	γ_{19}	γ_{20}	γ_{21}	γ_{22}	γ_{23}	γ_{24}	γ_{25}	γ_{26}	γ_{27}
L	(1,0)	(1,0)	(1,0)	(1,0)	(1,0)	(1,0)	(1,0)	(1,0)	(1,0)
M	(0,1)	(0,1)	(0,1)	(-1,0)	(-1,0)	(-1,0)	(0,0)	(0,0)	(0,0)
R	(0,0)	(0,0)	(0,0)	(0,0)	(0,0)	(0,0)	(0,0)	(0,0)	(0,0)

response strategy of Player 2 and then find the best strategy of Player 1. We will see in the next chapter that this amounts to finding a subgame perfect equilibrium in the game 3.4.

The best response of Player 2 to each possible move of Player 2 is shown below:

$$L \to b : (0,3)$$

$$M \to a : (0,1)$$

$$R \to a \text{ or } b : (1,1) \text{ or } (-1,0).$$

We see from the list that the best first move for Player 1 is R and the best reply for Player 2 is the strategy $\gamma^2 = baa$, which is indicated with a double $**$ in the payoff matrix (3.4). We have thus identified a single Stackelberg solution, by using the sequential optimization procedure. The attentive reader will have noticed that the optimal response to R is not uniquely defined.[25]

3.6.2   Duopoly example

Consider now how this solution concept is used in a duopoly game. Let a market for a single good, be described by the (inverse) demand law

$$p = D(q_1 + q_2)$$

$$\pi_j(q_1, q_2) = q_j D(q_1 + q_2) - C_j(q_j),$$

---

[25]The implication is that, even though the "subgame perfect equilibrium" of the game 3.4 is well identified, it is still not straightforward to consider that it is "the" Stackelberg solution, because Player 1, who moves first and chooses R, does not know if Player 2 will implement baa, in which case Player 1 end up winning 1, or bab, in which case he will only win -1.

where p is the clearing price on the market and q_1, q_2 the quantities supplied by the two firms in duopoly; $\pi_j(q_1, q_2)$ is the payoff to firm j, when the duopoly supply is (q_1, q_2). $C_j(q_j)$ is the cost incurred by Firm j when supplying q_j.

Assume Firm 1 commits itself to supply q_1. The optimal response of Firm 2 can be found, under differentiability assumptions, by looking to a 0 of the profit derivative

$$0 = \frac{\partial \pi_2(q_1, q_2)}{\partial q_2}$$
$$= D(q_1 + q_2) + q_2 D'(q_1 + q_2) - C_2'(q_2).$$

This allows us to define a function $\tilde{q}_2(q_1)$ that gives the optimal response function of Player 2 to any feasible q_1.

The best decision of Firm 1 is thus defined as $\max_{q_1} \pi_1(q_1, \tilde{q}_2(q_1))$. Again using the derivative of this function of q_1 only we have the optimality condition

$$0 = \frac{\partial \pi_1(q_1, \tilde{q}_2(q_1))}{\partial q_1} + \frac{\partial \pi_1(q_1, \tilde{q}_2(q_1))}{\partial q_2} \tilde{q}_2'(q_1).$$

To go further, assume that the market has a linear demand law and the firms have quadratic cost functions. The data of the duopoly game becomes:

$$p = D(q_1 + q_2) = a - b(q_1 + q_2),$$
$$C_j(q_j) = c_j q_j + \frac{1}{2} d_j q_j^2,$$
$$\pi_j(q_1, q_2) = q_j (a - b(q_1 + q_2)) - c_j q_j - \frac{1}{2} d_j q_j^2.$$

Now the response function for Firm 2 is a linear (affine) function of q_1:

$$0 = -b q_2 + a - b(q_1 + q_2) - c_2 - d_2 q_2,$$
$$\tilde{q}_2(q_1) = \frac{a - b q_1 - c_2}{2b + d_2}$$

and the optimal supply by Firm 1 is defined by

$$0 = a - 2b q_1 - b \frac{a - b q_1 - c_2}{2b + d_2} + \frac{b^2}{2b + d_2} - c_1 - d_1 q_1,$$
$$q_1^* = \frac{a - b \frac{a + b^2 - c_2}{2b + d_2}}{2b + d_1 - \frac{b^2}{2b + d_2}}.$$

We have thus a well defined Stackelberg solution to the duopoly game.

Remark 3.11. We stated above that the leader secures a higher payoff than what he would obtain in a Nash equilibrium. Note, however, that this does not mean that a player always prefers to be the leader. To illustrate, consider a duopoly competing in prices, with p_j denoting the price of firm $j = 1, 2$. Let the demand of Player j be given by

$$d_j = \frac{p_i - p_j + t}{2t}, \quad i, j = 1, 2, i \neq j,$$

where t is a positive parameter. The payoff of Player j is given by

$$\pi_j = p_j \left(\frac{p_i - p_j + t}{2t} \right), \quad i, j = 1, 2, i \neq j.$$

The unique Nash equilibrium of this game is given by $\left(p_1^N, p_2^N\right) = (t, t)$, and the corresponding payoffs by $\left(\pi_1^N, \pi_2^N\right) = \left(\frac{t}{2}, \frac{t}{2}\right)$. Suppose now that Player 1 is the leader and Player 2 is the follower. It can be easily checked that the unique Stackelberg equilibrium is given by $\left(p_1^S, p_2^S\right) = \left(\frac{3t}{2}, \frac{5t}{4}\right)$, and the profits by $\left(\pi_1^S, \pi_2^S\right) = \left(\frac{9t}{16}, \frac{25t}{32}\right)$. Comparing the payoffs, we get $\pi_1^S > \pi_1^N$ and $\pi_2^S > \pi_1^S$.

3.6.3 *A geometrical illustration.*

The Stackelberg solution is represented geometrically in Figure 3.4 below. The two players are identical. The decision variables of two players are called x_1 and x_2. The iso-profit curves are shown for the two players, and the reaction curves, corresponding to the optimal response functions, for the two players are indicated by the lines $\mathbf{R}_1$ and $\mathbf{R}_2$. The intersection of these two reaction curves is the Cournot equilibrium point, P. The reaction curve $\mathbf{R}_2$ intersect each iso-profit curve of Player 2 at a point where the tangent to this curve is vertical.[26] Player 1 looks for a point located on the reaction curve of Player 2 for which his profit is maximized. It is the point where an iso-profit curve of Player 1 becomes tangent to the reaction curve $\mathbf{R}_2$.

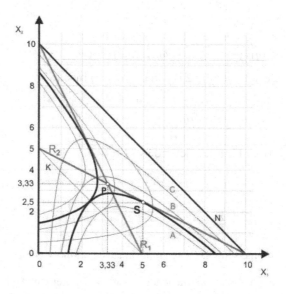

Fig. 3.4   Stackelberg solution vs Cournot solution

3.6.4 *More to come on Stackelberg solution*

We notice that the Stackelberg solution is defined through the introduction of a two-stage game, with an order of play, giving a first move advantage to one player. It will be natural

---

[26]This is because the profit to Player 2 is maximized at this point, when the variable x_1 is fixed.

to visit again this solution concept when we study in more details multistage games or their continuous time counterpart, called differential games.

3.7 What have we learned in this chapter?

Here are the main ideas and concepts presented in this chapter.

- If players' pure-strategy sets are finite, then the normal form of a two-player zero-sum game can be defined by a matrix; respectively, by a bimatrix in nonzero-sum games.
- In a matrix game, a saddle-point solution always exists in the class of mixed strategies (von Neumann theorem). This saddle-point solution provides a guaranteed payoff to each player. In addition, the saddle-point strategies are equilibrium strategies, i.e., each player's strategy is the best reply to the opponent's strategy choice.
- Linear programming can be used to characterize and compute saddle-point solutions.
- In a bimatrix game, an equilibrium always exists in the class of mixed strategies (Nash theorem).
- Quadratic programming can be used to characterize equilibrium solutions. A search for equilibrium can be formulated as a linear complementarity problem; then, a pivoting algorithm provides a computational method to find a Nash equilibrium.
- The equilibrium concept, the existence theorem and the numerical method can be extended to the case of m-player games, where each player has a finite set of pure strategies.
- The equilibrium concept can be extended to a more general class of m-player games where the strategies of each player are elements of a compact and convex set of an Euclidean space and the players' payoffs are concave functions in their own strategies.
- The existence of an equilibrium is proved even when coupled constraints are imposed on the joint strategy space of the m players. Conditions for uniqueness of this equilibrium and for convergence of a pseudo-gradient path toward this unique equilibrium have been proposed by Rosen.
- A coupled constraint equilibrium can also be characterized as a solution to a variational inequality (VI).
- Recently many efficient algorithms have been proposed to solve VIs (see, e.g., [72]). They can be used to find equilibrium solutions in concave m-player games.
- Nash equilibrium solutions in the class of mixed-strategies exist for games with continuous, compact action set and continuous payoffs.
- The Stackelberg solution in a duopoly game is obtained by introducing a two-stage game with a first move advantage given to one player and then looking for a subgame perfect equilibrium in this two-stage game. This solution has already a flavor of dynamic game.
- In the $\mathbb{GE}$ illustration section, we show how a coupled-constraint game model can be useful for the analysis and solution of a river basin pollution problem (RBPP); we also continue the discussion of the IEA model.

- In the next chapter we will discuss some refinements of the Nash equilibrium, to help address some of the concept's shortcomings, which were mentioned in this chapter.

3.8   Exercises

Exercise 3.1. Consider the following battle-of-the-sexes game of perfect but incomplete information where $a \in \{0, 2\}$ and $b \in \{0, 2\}$, each equiprobable. Convert this game into a game of complete but imperfect information using an initial move by nature and describe the game in both normal and extensive forms.

Table 3.5   Battle-of-the-sexes game

		Player 2	
		Opera	Boxing
Player 1	Opera	$3 + a, 1$	$0, 0$
	Boxing	$0, 0$	$1, 3 + b$

Exercise 3.2. Using Gambit software, find all the equilibrium solutions to the Game of chicken defined in Example 3.7 .

Exercise 3.3. A European country X wants to increase its budget spending, but the Maastricht treaty limits the allowable deficit. X's deficit is already at the limit and, in principle, cannot borrow any more money. However, it could borrow via a "detour," in which it purchases currency swaps from a merchant bank at rates lower than the market value, to sell them back at inflated rates at a later date. This technique is not exactly legal and can be detected by Eurostat, a European statistical watchdog. However, it requires that Eurostat makes an effort. We model country X's utility as given in matrix A. That of Eurostat is presented in matrix B. Symbols x and s denote taking action, which for country X means engaging in the currency swaps and, for Eurostat means employing staff to track down these transactions. Symbols $\sim x$ and $\sim s$ correspond to no action taken by either player.

$$A = \begin{array}{c|cc} & \sim x & x \\ \hline \sim s & 2 & 3 \\ s & 4 & 1 \end{array} \qquad B = \begin{array}{c|cc} & \sim x & x \\ \hline \sim s & 4 & 2 \\ s & 1 & 3 \end{array} \tag{3.82}$$

(1) What is the maximin strategy for Eurostat? Is this also a Nash equilibrium?
(2) What is the Nash equilibrium of this game?
(3) Will the Nash equilibrium be affected if country X is punished more harshly by the financial markets if it gets caught?

Exercise 3.4. Consider an industry made up of two identical firms producing a homogeneous good at the unit cost c. Denote by q_i the output of firm $i = 1, 2$. The inverse-demand law is given by

$$p = \max{(1 - q_1 - q_2, 0)}$$

and the profit of player i by

$$\pi_i = (p - c)\, q_i.$$

Firms choose output to maximize profit.

(1) Compute the (Cournot-)Nash equilibrium for the simultaneous game. Is this equilibrium unique?
(2) Suppose that both firms merge. What would their optimal output be? Is this output Pareto-efficient? (See footnote 9, page 49, for the definition of what a Pareto-efficient outcome is.)
(3) Assume (in the initial situation) that firm 1 moves before firm 2. What is the Stackelberg equilibrium of this game?
(4) Draw the extensive and normal forms of these games under the assumption that output choice is discrete and that the only possible output levels are Cournot, monopoly and the two Stackelberg equilibria.
(5) Suppose that firm 1 does not know whether the cost for firm 2 is low (that is c_L) or high (c_H). Find the Bayesian-Nash equilibrium for the Cournot game.

Exercise 3.5. Consider a two-period game.[27] In period A, there is only one firm (called incumbent I) in the market . In period B, another firm (called entrant E) may enter the market and compete against the incumbent. Denote the output of firm $j \in \{I, E\}$ in period $t \in \{A, B\}$ by q_j^t, with $q_E^A \equiv 0$. The inverse-demand laws are given by

$$P^A = \left(1 - q_I^A\right), \quad P^B = \left(1 - q_I^B - q_E^B\right).$$

In period A, the incumbent incurs a unit-production cost C, with $0 < C < 1$. Due to learning-by-doing, the unit cost in period B becomes $C - \lambda_I q_I^A$, where $0 < \lambda_I < 1$. Learning is not entirely firm-specific, that is, the entrant can also benefit from the experience of the incumbent through industrial espionage or by other means. Hence, if player E decides to enter the market, his cost will be $C - \lambda_E q_I^A$, with $0 \leq \lambda_E \leq \lambda_I$. Suppose that the incumbent discounts the profit of period B at the rate δ ($0 \leq \delta < 1$). Both firms are profit maximizers. Their profit functions are given by

$$\pi_I = \pi_I^A + \delta\pi_I^B = q_I^A \left[(1 - q_I^A) - C\right] + \delta q_I^B \left[(1 - q_I^B - q_E^B) - C + \lambda_I q_I^A\right],$$
$$\pi_E = \delta\pi_E^B = \delta q_E^B \left[1 - q_I^B - q_E^B - C + \lambda_E q_I^A\right].$$

(1) What is the myopic output for firm I in period A (i.e., without considering what will happen in period B)?
(2) What is the reaction function for the entrant in period B?
(3) Suppose that firm I ignores the strategic behavior of its rival, what is its optimal output level in this precommitment Nash equilibrium (we will later say "open-loop Nash") in every stage of the game?
(4) Now suppose that firm I anticipates the reaction of the entrant in period B, what is the (subgame-perfect) equilibrium of this game?

---

[27]This exercice is based on Jørgensen and Zaccour [130].

(5) Draw the extensive form representation assuming that the only possible answers are the output levels found. Compare the normal form of this game based on strategies with the normal representation of the game based on actions. What differences do you find?

(6) Weakly dominated actions are actions that never result in a higher payoff for a player, regardless of the other player's action. Delete the weakly dominated actions in the normal-form game of actions. Did you eliminate some Nash equilibria? What Nash equilibrium is left?

(7) What is the effect of an increased output in period A on the incumbent's second-period output?

(8) For what value of learning spillover λ_E will the first period output increase the second period output of the entrant? Hint: look at $\frac{\partial q_I^A}{\partial q_E^B}$.

Exercise 3.6. Denote[28] by $x_i \geq 0$ the strategy of player $i, i = 1, 2$. The objective of the players is to maximize their payoffs given by

$$J_1(x_1, x_2) = -\frac{1}{2}x_1^2 + x_1 x_2,$$
$$J_2(x_1, x_2) = -x_2^2 - x_1 x_2.$$

(1) What is the unique (obvious) Nash equilibrium?

(2) Suppose that a regulator has the power of imposing the following (coupling) constraint on the two players:

$$x_1 + x_2 \geq 1.$$

Determine the optimal solution to the following Pareto-optimization problem:

$$\max J(x_1, x_2) = J_1(x_1, x_2) + J_2(x_1, x_2),$$
$$\text{subject to}: \ x_1 + x_2 \geq 1.$$

(3) Can the regulator implement this solution as a Rosen equilibrium?

Exercise 3.7. Consider the zero-sum matrix game.

$$\begin{bmatrix} 3 & 1 & 8 \\ 4 & 10 & 0 \end{bmatrix}$$

Assume that the players are in the simultaneous-move information structure. Assume that the players try to guess and counterguess the optimal behavior of the opponent, in order to determine their optimal strategy choice. Show that this leads to an unstable process.

Exercise 3.8. Find the value and the saddle-point mixed strategies for the matrix game in Exercise 3.7.

Exercise 3.9. A game was given in extensive form in Figure 2.8 (page 27) and the normal form was obtained in Exercise 2.4 (page 27).

---

[28]The interested reader can consult Krawczyk and Tidball [143] for a general treatment of this context.

(1) What are the players' security levels?
(2) Compute the Nash equilibrium point(s) in pure and/or mixed strategy(-ies) (if they exist).
(3) Indicate the policies that lead to the Pareto-optimal payoffs.
(4) Suppose you would like Player I to use his first-row policy and Player II his first-column policy. How would you redesign the payoff matrix so that element $(1, 1)$ is now preferred by both players (i.e., likely to be played)? Suppose, moreover, that the incentive for the players to change their policy is coming out of your pocket, so that you want to spend as little as possible on motivating the players.

Exercise 3.10. Define the quadratic-programming problem that will find a Nash equilibrium for the bimatrix game

$$\begin{bmatrix} (52, 50)^* & (44, 44) & (44, 41) \\ (42, 42) & (46, 49)^* & (39, 43) \end{bmatrix}.$$

Verify that the entries marked with a $*$ correspond to a solution of the associated quadratic-programming problem.

Exercise 3.11. Do the same as above but instead use the complementarity-problem formulation.

Exercise 3.12. Consider the two-player game where the strategy sets are the intervals $U_1 = [0, 100]$ and $U_2 = [0, 100]$, respectively, and the payoffs $\psi_1(u) = 25u_1 + 15u_1u_2 - 4u_1^2$ and $\psi_2(u) = 100u_2 - 50u_1 - u_1u_2 - 2u_2^2$, respectively. Define the best reply mapping. Find an equilibrium point. Is it unique?

Exercise 3.13. Consider the two player game where the strategy sets are the intervals $U_1 = [10, 20]$ and $U_2 = [0, 15]$, respectively, and the payoffs $\psi_1(u) = 40u_1 + 5u_1u_2 - 2u_1^2$ and $\psi_2(u) = 50u_2 - 3u_1u_2 - 2u_2^2$, respectively. Define the best reply mapping. Find an equilibrium point. Is it unique?

3.9 GE illustration: A river basin pollution problem

We will now present an environmental-management problem that we will model as a game with coupled constraints, to illustrate the theory presented in Section 3.4. These constraints mean that the players' action set is a general convex set $X \subset \mathbf{R}^n$. We follow the analysis presented in [109] and [144].

In this game, we consider three players $j = 1, 2, 3$ located along a river. Each agent is engaged in an economic activity (paper-pulp production, for instance) at a chosen level x_j, but the players must meet environmental conditions set by a local authority.

Pollutants are expelled into the river, where they disperse. Two monitoring stations $\ell = 1, 2$ are located along the river, at which locations the local authority has set maximum pollutant-concentration levels.

The revenue for player j is

$$\mathcal{R}_j(\mathbf{x}) = [d_1 - d_2(x_1 + x_2 + x_3)]x_j \qquad (3.83)$$

with cost

$$\mathcal{F}_j(\mathbf{x}) = (c_{1j} + c_{2j}x_j)x_j. \qquad (3.84)$$

Thus the net profit for player j is

$$\phi_j(\mathbf{x}) = \mathcal{R}_j(\mathbf{x}) - \mathcal{F}_j(\mathbf{x}) = [d_1 - d_2(x_1 + x_2 + x_3) - c_{1j} - c_{2j}x_j]x_j. \qquad (3.85)$$

The constraint on pollution imposed by the local authority at location ℓ is

$$q_\ell(\mathbf{x}) = \sum_{j=1}^{3} \beta_{j\ell}e_j x_j \le K_\ell, \qquad \ell = 1, 2. \qquad (3.86)$$

The economic constants d_1 and d_2 determine the inverse-demand law and are set to 3.0 and 0.01, respectively. The values for constants c_{1j} and c_{2j} are given in Table 3.6, and $K_\ell = 100$, $\ell = 1, 2$.

The $\beta_{j\ell}$ are the decay-and-transportation coefficients from player j to location l, and e_j is the emission coefficient of player j, which are also given in Table 3.6.

Table 3.6   Constants for the river basin pollution game

Player j	c_{1j}	c_{2j}	e_j	β_{j1}	β_{j2}
1	0.10	0.01	0.50	6.5	4.583
2	0.12	0.05	0.25	5.0	6.250
3	0.15	0.01	0.75	5.5	3.750

Solution to the river basin pollution game.   The above coupled-constraint game was solved in [109] using Rosen's projected gradient algorithm; the same game was solved in [144] using the NIRA approach (see [145]).

In this game, the agents maximize profits (3.85), subject to actions satisfying jointly convex constraints (3.86); so this is an example of a coupled-constraint game.

This game was solved for the case of *equal* weights[29] $r_1 = r_2 = r_3 = 1$ and found to have equilibrium $x^* = (21.149, 16.028, 2.722)$, giving net profits $\phi(x^*) = (48.42, 26.92, 6.60)$. The first constraint is active, i.e., $q_1(x^*) = K_1 = 100$ with the Karush-Kuhn-Tucker multiplier $\lambda_1 = 0.574$; the second constraint is inactive ($q_2(x^*) = 81.17$).

---

[29]This corresponds to the case, in which the responsibility for the constraints' satisfaction is shared in solidarity by all players; see Sections 3.4.3.2 and 3.4.5 for interpretations of r_j.

Equilibrium enforcement. Now that the Nash-normalized equilibrium has been found, we can compel the players to obey it by applying Pigouvian taxes in the form of penalty functions (see Section 3.4.5). In this way we create a new, unconstrained (or decoupled) game.

For each constraint, we place a tax on each player in the amount of

$$T_k(\mathbf{x}) = \lambda_k \max(0, q_k(\mathbf{x}) - K_k) , \tag{3.87}$$

where λ_k, $k = 1, 2$ is a penalty coefficient for violating the k-th constraint. Since $T_k(x)$ is a nonsmooth penalty function, there will always exist coefficients λ_k sufficiently large to ensure that agents adhere to the environmental constraints (3.86). In other words, for big λ_k, the waste produced by the agents' optimal solutions will satisfy the environmental standards. We will show that the λ_k obtained in the coupled-constraint equilibrium is of the "right" size and that the equilibrium with the modified payoff functions ϕ_j^*

$$\phi_j^*(\mathbf{x}) = R_j(\mathbf{x}) - F_j(\mathbf{x}) - \sum_k T_k(\mathbf{x})$$

matches x^*.

The new equilibrium problem with payoff functions ϕ_j^* and uncoupled constraints has the Nash equilibrium point x^{**} defined by equation

$$\phi_j^*(\mathbf{x}^{**}) = \max_{x_j \geq 0} \phi_j^*([x_j, \mathbf{x}_{-j}^{**}]), \quad j = 1, \ldots m. \tag{3.88}$$

We make a conjecture based on the general theory of nonsmooth optimization (see, for example [220]) that, for the environmental constraints' satisfaction, the penalty coefficients λ_k should be greater than (or equal to) the Lagrange multipliers corresponding to the constraints (3.86). In our numerical experiments we set λ_k to equal the final Lagrange multipliers for constraint k that were observed during the calculation of the constrained equilibrium with Rosen's algorithm. For this setup, the unconstrained equilibrium x^{**} is equal to the constrained equilibrium x^* (see Figure 3.5).

As stated, for this game only the first constraint $k = 1$ was active. Thus, in the river basin pollution game (with the parameter values given in Table 3.6), the payoff function for player j becomes

$$\phi_j^*(\mathbf{x}) = R_j(\mathbf{x}) - F_j(\mathbf{x}) - T_1(\mathbf{x}) =$$
$$[d_1 - d_2(x_1 + x_2 + x_3) - c_{1j} - c_{2j}x_j]x_j$$
$$- \lambda_1 \max\left(0, \sum_{j=1}^{3} \beta_{j1}e_j x_j - K_1\right). \tag{3.89}$$

In Figure 3.5 cross-sectional graphs of the modified payoff functions illustrate that each payoff function achieves its maximum at the point x^*.

Finally, as we have 3 variables and 2 constraints, if the two constraints were active, any solution satisfying the constraints would be an equilibrium.

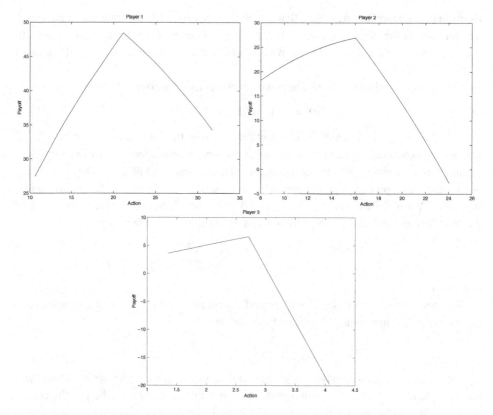

Fig. 3.5   Payoff functions for players 1,2,3 with Pigouvian taxes applied

3.10   GE illustration: Assessment of an IEA (continued)

Scenarios corresponding to equilibrium solutions.   The concept of the equilibrium solution was defined and explained in this chapter. Gambit computes equilibrium solutions
for games defined in their extensive form. Here we show the equilibrium solutions for the
second stage, where ICs and DCs choose their first-period quota, depending on the world
oil-price level selected by OPEC. The outcome payoffs in each of these games were calculated by computing the expected values of the outcomes in the third stage, where, as noted
previously, the decision of the ICs and DCs is contingent on the quotas decided in stage 2
and on the revealed CS.

When the spilt is 50%-50%.

- Figure 3.10 shows the equilibrium quota choices when the price of oil is low.
- Figure 3.10 shows the equilibrium quota choices when the price of oil is medium.
- Figure 3.8 shows the equilibrium quota choices when the price of oil is high.

- Table 3.7 summarizes the payoffs to the three players for the different oil prices.

Table 3.7 Equilibrium payoffs (10^{12})

Player	Low	Average	High
OPEC	-3.23	-0.50	0.77
ICs	6.68	-0.96	-7.88
DCs	-5.33	-5.72	-4.86

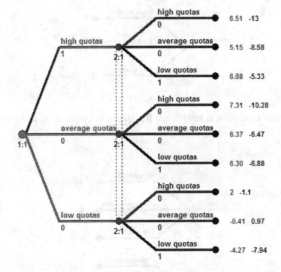

Fig. 3.6   When oil price is *low*, at equilibrium ICs play *high quotas* and DCs play *low quotas*

Based on these simulations, we can make the following observations:

- OPEC will select the high price for world oil which leads to very high cost at equilibrium for DCs and ICs.
- It can be observed that, in this situation, whatever the price of oil, DCs select low quotas in period 1. Indeed, this is because DCs are only allocated limited quotas, in anticipation of future economic growth.
- Considering that it is possible that there will be a high climate sensitivity, which would lead to very stringent emissions constraints in the second period, DCs must hedge this risk by decreasing their emissions in period 1.

When the split is 60% DCs, 40% ICs.   The solution changes sensibly if the agreement gives 40% to ICs and 60% to DCs for the cumulative quotas over the whole 2000–2050 period.

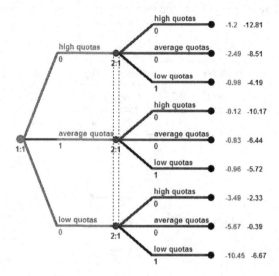

Fig. 3.7   When oil price is *medium*, at equilibrium ICs play *average quotas* and DCs play *low quotas*

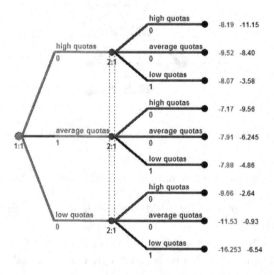

Fig. 3.8   When oil price is *high*, at equilibrium, ICs play *average quotas* and DCs play *low quotas*

Figures 3.9–3.11 and Table 3.8 summarize the equilibrium actions and the payoffs to the three players for the different oil prices.

In that case, OPEC once again selects a high price for oil and ICs and DCs always choose average quotas. A first explanation for this result is related to the optimal timing.

Table 3.8 Equilibrium payoffs (10^{12})

Player	Low	Average	High
OPEC	-3.23	-0.22	1.02
ICs	-0.01	-6.00	-12.22
DCs	4.06	2.00	0.98

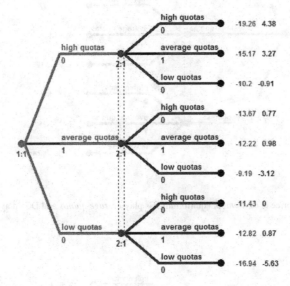

Fig. 3.9 When oil price is *low*, at equilibrium ICs play *average quotas* and DCs play average quotas

Delaying the abatement (low quotas in period 1) can be very costly if, in period 2 countries face a high CS and therefore, very severe reductions of GHG emissions. A contrario early abatement could prove to be costly if the CS is low. Hence the choice of average quotas.

A second element that explains these results is the 60-40 % allocation rule between DCs and ICs. In most cases, this means that DCs will be net sellers of emissions rights. Therefore the choice of a quota level in period 1 is partly dictated by the gains expected in period 2 from selling emissions rights to ICs. Finally a third element of explanation is the influence of the world price of oil. OPEC has a higher profit with high price level.

This imposes a high cost on ICs, which are net importers of oil. This effect is amplified by a devaluation of money for ICs, which triggers a distortion cost in this zone. For DCs, the impact is more limited, as their dependency on oil is weaker and their domestic production is higher.[30]

---

[30]Latin America and African producers are in this zone.

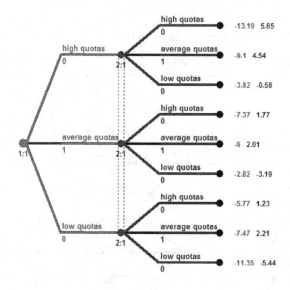

Fig. 3.10   When oil price is *medium*, at equilibrium ICs play *average quotas* and DCs play *average quotas*

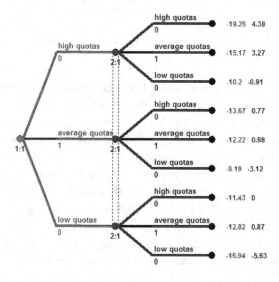

Fig. 3.11   When oil price is *high* , at equilibrium ICs play *average quotas* and DCs play *average quotas*

Chapter 4

Extensions and Refinements of the Equilibrium Concepts

4.1 Introduction

In the previous chapter, we saw that the Nash-equilibrium solution concept has some limitations and drawbacks. For example, in bimatrix or m-matrix games, having multiple equilibria is the rule rather than the exception. Furthermore, the prisoner's dilemma shows that an equilibrium solution can be far from efficient.[1] In general, the Nash-equilibrium solution concept is not robust, in the sense that a slight change in the game data expressing the players' preferences can change the equilibrium solution significantly.

In this chapter, we present a few of the extensions and refinements of the Nash-equilibrium concept, which have been proposed in the game-theory literature to help cope with the drawbacks and shortcomings of the original Nash-equilibrium concept. We first explore the possibilities offered by allowing a correlation between the players' mixed strategies. Then, we look at the possibility of extending the equilibrium solution concept to games with incomplete information, through the use of Bayesian probabilities. The concept of a behavioral strategy is further explored and the possibility of defining a stronger equilibrium concept, which would resist a temporary disruption in the way the players act, is proposed under the name of a *subgame perfect equilibrium*. The design of a robust equilibrium concept, called a *quantal-response equilibrium* is proposed for a new class of games, where the players have random utility functions. Finally, we explore an approach based on the lattice theory, which is capable of establishing conditions for the existence of pure-strategy equilibria in a large class of games used in economic modeling.

4.2 Correlated equilibria

4.2.1 *Relaxing the independence assumption in a Nash equilibrium*

The Nash equilibrium is a solution concept in which the players are expected to act independently and without any inter-agent communication. For many situations, this assumption is too restrictive.[2]

---

[1] See footnote 9, page 49, for an explanation of what we mean by efficiency.

[2] The concept of normalized equilibrium proposed for when the agents have to satisfy "coupled" constraints, which was explored in the previous chapter, was already a breach of this assumption.

Aumann [11] has proposed a possible extension to the Nash equilibrium, to allow for situations where the players can enter into preplay arrangements or can receive recommendations on what to play from an "umpire." If so, their strategy choices could be correlated instead of being independent. In essence, Aumann proposes a solution concept that preserves the equilibrium properties and, at the same time, allows for a degree of communication among the players. He called this type of game solution a *correlated equilibrium* and he showed that, through the correlation mechanism, the players can expect better payoffs in equilibrium than under the usual Nash-equilibrium solution.

4.2.2 Example of a game with correlated equilibria

This example was initially proposed by Aumann [11].

Example 4.1. *Consider a simple bimatrix game defined as follows:*

	c_1	c_2
r_1	5,1	0,0
r_2	4,4	1,5

This game has two pure-strategy equilibria (r_1, c_1) and (r_2, c_2) and a mixed-strategy equilibrium where each player puts the same probability of $\frac{1}{2}$ on each possible pure strategy. Indeed, the equilibrium condition (see Example 3.9, page 50)

$$5p + (1 - p) \times 0 = 4p + (1 - p) \times 1$$

is satisfied for $p = \dfrac{1}{2}$ for either player. So, the respective outcomes are shown in Table 4.1.

Table 4.1   Outcomes of the three equilibria

$(r_1; c_1)$	:	5,1
$(r_2; c_2)$	:	1,5
$(0.5, 0.5; 0.5, 0.5)$	:	2.5,2.5

If the players agree to observe a "coin flip" result,[3] available to both players, and play (r_1, c_1) if the result is "heads," and (r_2, c_2) if it is "tails," then the expected payoff will be

$$\frac{1}{2} \times 5 + \frac{1}{2} \times 1 = 3$$

for either player. This result is a convex combination of the two pure-equilibrium outcomes.

---

[3] Heads and tails with an equal probability of $\dfrac{1}{2}$.

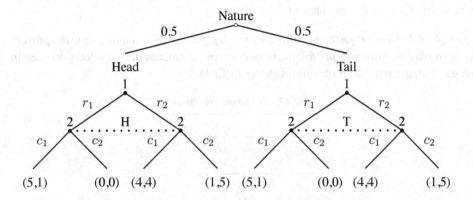

Fig. 4.1 Extensive game formulation with nature playing first

It is easy to see that this way of playing the game defines an equilibrium for the extensive game shown in Figure 4.1. This is an expanded version of the initial game where, at a preliminary stage, nature randomly gives a signal (coin flip or lottery) that will be observed by the players.[4] In this design, the result of the coin flip is "public information," in the sense that it is shared by all the players. We can easily check that the correlated equilibrium is a Nash equilibrium for the expanded game.

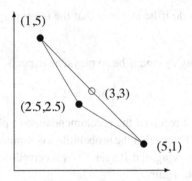

Fig. 4.2 The convex hull of Nash equilibria in Example 4.1

By using a coin flip to determine which equilibrium to play, the players end up in a new type of equilibrium, which yields an outcome located in the convex hull of the set of Nash equilibria of the bimatrix game. We have represented the different outcomes of the game given in Example 4.1 in Figure 4.2. The full circles represent the three Nash-equilibrium outcomes. The triangle defined by these three points is the convex hull of the

---

[4]As before, the dotted lines represent information sets of Player 2.

Nash-equilibrium outcomes. The empty circle represents the outcome obtained by agreeing to play with the coin-flip mechanism.   ◇

Example 4.2. *We now push the example one step further by assuming that the players agree to play according to the following mechanism:* a random device selects one cell in the game matrix with the probabilities shown in Table 4.2.

Table 4.2   Probabilities of signals

	c_1	c_2
r_1	1/3	0
r_2	1/3	1/3

When a cell is selected, then each player is told by an umpire to play the corresponding pure strategy. The trick is that the player is told what to play but is not told what is being recommended to the other player. The information received by each player is no longer public. More precisely, the three possible signal pairs are (r_1, c_1), (r_2, c_1), (r_2, c_2). *When Player 1 receives the signal "play r_2" he knows that there is a $\frac{1}{2}$ probability that Player 2 was told to play c_1, or, also with probability of $\frac{1}{2}$, to play c_2. When Player 1 receives the signal "play r_1," he knows that there is a $\frac{1}{2}$ probability that the other player was told to play c_1.*

We will now prove that the players exercise their best reply strategies when they follow the prescribed play.

Consider what Player 1 can do if he assumes that the other player plays according to the recommendations.

If Player 1 has been told to play r_2 and if he so plays, he expects

$$\frac{1}{2} \times 4 + \frac{1}{2} \times 1 = 2.5$$

since he knows that Player 2 received the recommendation to play c_1 or c_2, each column with probability of $\frac{1}{2}$. (Notice that the probabilities computed here are conditional: we have computed them knowing that Player 1 has received the signal "play r_2.") If instead he played r_1, he would gain

$$\frac{1}{2} \times 5 + \frac{1}{2} \times 0 = 2.5.$$

So, he cannot improve his expected reward by deviating from the signal.

If Player 1 has been told to play r_1 and if he so plays, he knows that there is a probability of 1 that the other player has been told to play c_1 and so he expects 5. If he plays r_2, he expects 4. So, for Player 1, the best reply to Player 2's behavior when the latter plays according to the suggestions of the signalling scheme is to obey the recommendations.

Now we can repeat the computations for Player 2.

If Player 2 has been told to play c_1 and if he plays so, he expects

$$\frac{1}{2} \times 1 + \frac{1}{2} \times 4 = 2.5;$$

if instead he plays c_2 he expects

$$\frac{1}{2} \times 0 + \frac{1}{2} \times 5 = 2.5$$

so, he cannot improve his outcome.

If Player 2 has been told to play c_2 and if he plays so, he expects 5, since there is a probability of 1 that Player 1 has been told to play r_2; if instead he plays c_1, he expects 4; so, he is better off with the suggested play.

Thus we have checked that an equilibrium property holds for this way of playing the game (i.e., when both players follow the recommendations).

All in all, each player expects

$$\frac{1}{3} \times 5 + \frac{1}{3} \times 1 + \frac{1}{3} \times 4 = 3 + \frac{1}{3} = \frac{10}{3}$$

from a game played in this way. This is illustrated in Figure 4.3 where the black spade shows the expected outcome of this mode of play. Aumann called this a **correlated equilibrium**.

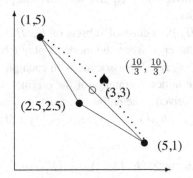

Fig. 4.3 The dominating correlated equilibrium

Indeed, we can now mix these equilibria and still keep the correlated equilibrium property, as indicated by the dotted line in Figure 4.3. Also, we can construct an expanded game in extensive form, for which the correlated equilibrium constructed as above defines a Nash equilibrium (see Exercise 4.1). ◇

In the above example, we see that, by adding a (first) stage to the game, in which nature plays and gives private information to the players, a new class of equilibria can be reached, which dominate some of the original Nash equilibria in the outcome space. If a random device gives all of the players the same information, then the players can reach any payoff that is in the convex hull of the Nash equilibrium outcomes by mixing the different Nash-equilibrium pure strategies. However, if the random device gave each player different information, then the correlated equilibrium could have an outcome that lies outside of the convex hull of Nash-equilibrium outcomes and that dominates some Nash equilibria.

4.2.3   *A general definition of correlated equilibria*

Let us give a general definition of a correlated equilibrium in an m-player normal form game. We will, in fact, give two definitions. The first describes the construct of an *expanded game* with a random device distributing some preplay information to the players. The second definition, which is only valid for m-matrix games, is much simpler, albeit equivalent.

4.2.3.1   *Nash equilibrium in an expanded game.*

Assume that a game is described in normal form, with m players $j = 1, \ldots, m$, their respective strategy sets Γ_j and payoffs $V_j(\gamma_1, \ldots, \gamma_j, \ldots, \gamma_m)$. This will be called the *original normal-form game.* Assume that the players may enter into a phase of *preplay communication,* during which they design a *correlation device* that will randomly provide a signal called the *proposed mode of play.* Let $E = \{1, 2, \ldots, L\}$ be the finite set of the possible modes of play. The correlation device will propose the mode of play ℓ with probability $\lambda(\ell)$. The device will then give the different players some information about the *proposed mode of play.*

More precisely let H_j be a class of subsets of E. H_j will be called the *information structure* of Player j. Player j, when the mode of play ℓ has been selected, receives information denoted $h_j(\ell) \in H_j$. We associate with each player j a *meta-strategy* denoted $\tilde{\gamma}_j : H_j \rightarrow \Gamma_j$ that determines a strategy for the original normal-form game, on the basis of the information received. The entire construct is summarized in the following data quadruple, $(E, \{\lambda(\ell)\}_{\ell \in E}, \{h_j(\ell) \in H_j\}_{j \in M, \ell \in E}, \{\tilde{\gamma}_j : H_j \rightarrow \Gamma_j\}_{j \in M})$, which defines an *expanded game.*

Definition 4.1. The quadruple $(E, \{\lambda(\ell)\}_{\ell \in E}, \{h_j(\ell) \in H_j\}_{j \in M, \ell \in E}, \{\tilde{\gamma}_j^* : H_j \rightarrow \Gamma_j\}_{j \in M})$ defines a correlated equilibrium of the original normal-form game if it is a Nash equilibrium for the expanded game. This means that no player can improve his expected payoff by unilaterally changing his meta-strategy from $\tilde{\gamma}_j^*(h_j(\ell))$ to $\tilde{\gamma}_j(h_j(\ell))$, when he receives the signal $h_j(\ell) \in H_j$.

Therefore, at the correlated equilibrium, the following inequality holds for all $\tilde{\gamma}_j(h_j(\ell))$:

$$\sum_{\ell \in E} \lambda(\ell) V_j([\tilde{\gamma}_j^*(h_j(\ell)), \tilde{\gamma}_{-j}^*(\mathbf{h}_{-j}(\ell))])$$

$$\geq \sum_{\ell \in E} \lambda(\ell) V_j([\tilde{\gamma}_j(h_j(\ell)), \tilde{\gamma}_{-j}^*(\mathbf{h}_{-j}(\ell))]), \tag{4.1}$$

where, as usual, $\tilde{\gamma}^*_{-j}(\mathbf{h}_{-j}(\ell))$ denotes the vector of strategic choices made at equilibrium by the players $i \neq j$.

4.2.3.2 An equivalent definition for m-matrix games

In the case of an m-matrix game, the definition given above can be replaced with the following. This definition is much simpler but, as stated previously, it applies only to this class of games.

Definition 4.2. In an m-matrix game, a correlated equilibrium is a probability distribution $\pi(s)$ over the set of pure strategies $\underline{S} = S_1 \times S_2 \cdots \times S_m$ such that, for every Player j and any mapping $\delta_j : S_j \rightarrow S_j$, the following holds:

$$\sum_{\underline{s} \in \underline{S}} \pi(\underline{s}) V_j([s_j, \mathbf{s}_{-j}]) \geq \sum_{\underline{s} \in \underline{S}} \pi(\underline{s}) V_j([\delta_j(s_j), \mathbf{s}_{-j}]), \quad \forall j \in M \qquad (4.2)$$

where $V_j(\underline{s})$ is the payoff to Player j associated with the pure-strategy m-tuple $\underline{s} \in \underline{S}$ and $\mathbf{s}_{-j} = (s_i)_{i \neq j}$.

In this interpretation of a correlated equilibrium, the m-tuple of pure strategies is recommended as a mode of play with probability $\pi(s)$. Player j may accept the recommendation or may use a meta-strategy, which consists of playing $\delta_j(s_j) \in S_j$ when he receives the recommendation to play s_j. At the correlated equilibrium, there is no such meta-strategy that would improve the expected payoff to Player j when we assume that the other players are playing according to the recommendation they receive.

An equivalent formulation of equation (4.2) is the following

$$\sum_{\mathbf{s}_{-j} \in \mathbf{S}_{-j}} \pi(\mathbf{s}_{-j}|s_j) V_j([s_j, \mathbf{s}_{-j}]) \geq$$

$$\sum_{\mathbf{s}_{-j} \in \mathbf{S}_{-j}} \pi(\mathbf{s}_{-j}|s_j) V_j(s'_j, \mathbf{s}_{-j}]) \quad \forall s'_j \in S_j, \qquad (4.3)$$

where $\pi(\mathbf{s}_{-j}|s_j)$ is the conditional probability of the other players receiving the message "play $\mathbf{s}_{-j}$" knowing that Player j has received the message "play s_j."

4.2.3.3 Linear programming to compute correlated equilibria in m-matrix games

A correlated equilibrium of an m-matrix game is defined by a probability distribution over the set S of pure strategy m-tuples that satisfy the inequalities (4.2). The latter are linear in probabilities $\pi(s)$.

We will apply inequalities (4.2) to the bimatrix game of Example 4.1, page 88. Let us assign the probabilities $\pi_{11}, \pi_{12}, \pi_{21}, \pi_{22}$ to each cell of the bimatrix game. Obviously, the following conditions have to be satisfied by probabilities:

$$\pi_{11}, \pi_{12}, \pi_{21}, \pi_{22} \geq 0,$$

$$\pi_{11} + \pi_{12} + \pi_{21} + \pi_{22} = 1.$$

Now, we write the conditions 4.2 for Player 1 in this particular game. To do this, we have to consider all functions $\delta_1 : \{r_1, r_2\} \to \{r_1, r_2\}$, which are summarized in the following array that contains only the indices of the corresponding elements in the δ_1 mapping:

$$\begin{pmatrix} 1 \; 2 \\ 2 \; 2 \\ 1 \; 1 \\ 2 \; 1 \end{pmatrix}$$

Consider a first possible mapping δ_1, which is not the identity. When Player 1 receives the info "1," he plays "2;" when he receives the info "2," he plays "2." We write the corresponding inequality (4.2) as

$$5\pi_{11} + 0\pi_{12} + 4\pi_{21} + 1\pi_{22} \geq 4\pi_{11} + 1\pi_{12} + 4\pi_{21} + 1\pi_{22}.$$

Now, consider another possible mapping δ_1. When Player 1 receives the info "1," he plays "1;" when he receives the info "2," he plays "1." We write the corresponding inequality (4.2) as

$$5\pi_{11} + 0\pi_{12} + 4\pi_{21} + 1\pi_{22} \geq 5\pi_{11} + 0\pi_{12} + 5\pi_{21} + 0\pi_{22}.$$

Finally, consider the last possible mapping δ_1 which is not the identity. When Player 1 receives the info "1," he plays "2;" when he receives the info "2," he plays "1." We write the corresponding inequality (4.2):

$$5\pi_{11} + 0\pi_{12} + 4\pi_{21} + 1\pi_{22} \geq 4\pi_{11} + 1\pi_{12} + 5\pi_{21} + 0\pi_{22}.$$

We repeat the same procedure for Player 2 and obtain the following inequalities (detailed computations are left to the reader):

$$1\pi_{11} + 0\pi_{12} + 4\pi_{21} + 5\pi_{22} \geq 0\pi_{11} + 0\pi_{12} + 5\pi_{21} + 5\pi_{22},$$

$$1\pi_{11} + 0\pi_{12} + 4\pi_{21} + 5\pi_{22} \geq 1\pi_{11} + 1\pi_{12} + 5\pi_{21} + 4\pi_{22},$$

$$1\pi_{11} + 0\pi_{12} + 4\pi_{21} + 5\pi_{22} \geq 0\pi_{11} + 1\pi_{12} + 5\pi_{21} + 4\pi_{22}.$$

Variables $\pi_{11}, \pi_{12}, \pi_{21}, \pi_{22}$, which satisfy all the above linear constraints, form a polytope (as usual in linear programming). Each element of this polytope defines a correlated equilibrium for the bimatrix game at hand.

To look for different extreme points of the polytope of all correlated equilibria we solve the following linear programming (LP) problem for several different values of the weights ϖ_1, ϖ_2:

$$\max \pi_{11}(\varpi_1 \times 5 + \varpi_2 \times 1) + \pi_{12}(\varpi_1 \times 0 + \varpi_2 \times 0)\pi_{21}(\varpi_1 \times 4 + \varpi_2 \times 4) + \pi_{22}(\varpi_1 \times 1 + \varpi_2 \times 5)$$

subject to

$$\pi_{11}, \pi_{12}, \pi_{21}, \pi_{22} \geq 0,$$

$$\pi_{11} + \pi_{12} + \pi_{21} + \pi_{22} = 1,$$

$$5\pi_{11} + 0\pi_{12} + 4\pi_{21} + 1\pi_{22} \geq 4\pi_{11} + 1\pi_{12} + 4\pi_{21} + 1\pi_{22},$$

$$5\pi_{11} + 0\pi_{12} + 4\pi_{21} + 1\pi_{22} \geq 5\pi_{11} + 0\pi_{12} + 5\pi_{21} + 0\pi_{22},$$

$$5\pi_{11} + 0\pi_{12} + 4\pi_{21} + 1\pi_{22} \geq 4\pi_{11} + 1\pi_{12} + 5\pi_{21} + 0\pi_{22},$$

$$1\pi_{11} + 0\pi_{12} + 4\pi_{21} + 5\pi_{22} \geq 0\pi_{11} + 0\pi_{12} + 5\pi_{21} + 5\pi_{22},$$

$$1\pi_{11} + 0\pi_{12} + 4\pi_{21} + 5\pi_{22} \geq 1\pi_{11} + 1\pi_{12} + 5\pi_{21} + 4\pi_{22},$$

$$1\pi_{11} + 0\pi_{12} + 4\pi_{21} + 5\pi_{22} \geq 0\pi_{11} + 1\pi_{12} + 5\pi_{21} + 4\pi_{22}.$$

It is easy to solve this LP using Excel and its solver (see Exercise 4.2).

If the weights are given by $\varpi_1 = 1, \varpi_2 = 1$, we can easily check that the solution of the LP is $\pi_{11} = \frac{1}{3}, \pi_{12} = 0, \pi_{21} = \frac{1}{3}, \pi_{22} = \frac{1}{3}$. This corresponds to the dominating correlated equilibrium shown in Figure 4.3 and gives an expected payoff of $\frac{10}{3}$ to each player.

If the weights are given by $\varpi_1 = 1, \varpi_2 = 0$, we obtain the solution $\pi_{11} = 1, \pi_{12} = \pi_{21} = \pi_{22} = 0$. This corresponds to a Nash equilibrium and gives an expected payoff of 5 to Player 1, and 1 to Player 2.

If the weights are given by $\varpi_0 = 1, \varpi_2 = 1$, we obtain the solution $\pi_{11} = \pi_{12} = \pi_{21} = 0, \pi_{22} = 1$. This also corresponds to a Nash equilibrium and gives an expected payoff of 1 to Player 1, and 5 to Player 2.

These solutions correspond exactly to the extreme points of the polytope of equilibria already identified for this example in Figure 4.3.

4.3 Bayesian equilibrium with incomplete information

So far, we have considered only games where each player knows everything concerning the game rules, the payoff functions, strategy sets, etc. Hence, we have dealt with *games of complete information*. In this section we will look at a particular class of *games with incomplete information* and explore the class of so-called *Bayesian equilibria*.

4.3.1 *A game with an unknown player type*

As an example of a game of incomplete information, let us consider a situation in which some players do not know the other players' characteristics precisely. For example, in a two-player game, Player 2 may not know exactly what the payoff function of Player 1 is.

Example 4.3. *Consider the case where Player 1 could be one of two types, called θ_1 and θ_2, respectively. We define two matrix games that correspond to these two possible types, respectively:*

θ_1	c_1	c_2
r_1	0,-1	2,0
r_2	2,1	3,0
Game 1		

θ_2	c_1	c_2
r_1	1.5,-1	3.5,0
r_2	2,1	3,0
Game 2		

If Player 1 is of type θ_1, then bimatrix game 1 is played; if Player 1 is type θ_2, then bimatrix game 2 is played; the problem is that Player 2 does not know the type of Player 1.

4.3.2 *A reformulation as a game with imperfect information*

Harsanyi, in [102], proposed the transformation of a game of *incomplete information* into a game with *imperfect information*. (Refer to Section 2.3.1, page 15 for a definition of these concepts.) This transformation introduces a preliminary *chance move*, played by *nature*, which decides randomly the type θ_i of Player 1, as in Example 4.3. The probabilities, denoted p_1 and $p_2 = 1 - p_1$ represent the beliefs held by Player 2 about Player 1's type. So, p_1 and p_2 are prior probabilities about facing a player of type θ_1 or θ_2. It will be assumed that Player 1 knows about these beliefs and also knows that Player 2 knows that he knows, etc. The prior probabilities are thus *common knowledge*.

The information structure[5] in the associated extensive game shown in Figure 4.4, indicates that Player 1 knows his own type when deciding, but that Player 2 observes neither the opponent's type (but does put a probability on it) nor selected action, in this game of simultaneous moves. Call x_i (respectively $1 - x_i$) the probability of Player 1 choosing r_1 (respectively r_2) when he implements a mixed strategy, knowing that he is of type θ_i. Call y (respectively $1 - y$) the probability of Player 2 choosing c_1 (respectively c_2) when he implements a mixed strategy.

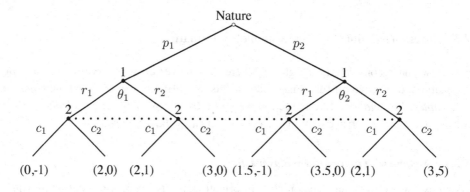

Fig. 4.4   Extensive game formulation of the game of incomplete information

We can define the optimal response of Player 1 to the mixed strategy $(y, 1 - y)$ of Player 2 by solving[6]

$$\max_{i=1,2} a_{i1}^{\theta_1} y + a_{i2}^{\theta_1}(1 - y)$$

if the type is θ_1,

$$\max_{i=1,2} a_{i1}^{\theta_2} y + a_{i2}^{\theta_2}(1 - y)$$

if the type is θ_2.

We can define the optimal response of Player 2 to the pair of mixed strategies $(x_i, 1 - x_i), i = 1, 2$ of Player 1 by solving

$$\max_{j=1,2} p_1(x_1 b_{1j}^{\theta_1} + (1 - x_1)b_{2j}^{\theta_1}) + p_2(x_2 b_{1j}^{\theta_2} + (1 - x_2)b_{2j}^{\theta_2}).$$

---

[5]As usual, the dotted line in Figure 4.4 represents the information set of Player 2.
[6]We call $a_{ij}^{\theta_\ell}$ and $b_{ij}^{\theta_\ell}$ the payoffs of Player 1 and Player 2, respectively, when the type is θ_ℓ.

Let us rewrite these conditions with the data of the game illustrated in Figure 4.4. First consider the reaction function of Player 1:

$$\text{if type} \quad \theta_1 \Rightarrow \max\{0y + 2(1-y), 2y + 3(1-y)\}$$
$$\text{if type} \quad \theta_2 \Rightarrow \max\{1.5y + 3.5(1-y), 2y + 3(1-y)\}.$$

In Figure 4.5, we draw the lines corresponding to these comparisons between two linear functions.

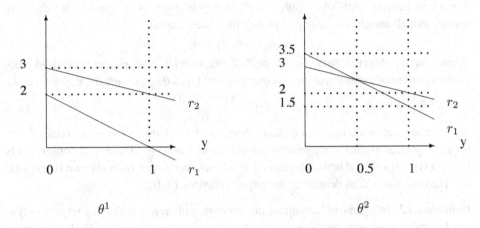

Fig. 4.5   Optimal reaction of Player 1 to Player 2's mixed strategy $(y, 1-y)$

We observe that, if Player 1's type is θ_1, he will always choose r_2, regardless of the value of y, whereas, if his type is θ_2, he will choose r_1 if $y < 0.5$ and switch to r_2 when $y > 0.5$. For $y = 0.5$, the best reply of Player 1 could be any mixed strategy $(x, 1-x)$.

Consider now the optimal reply of Player 2. We know that Player 1, if he is of type θ_1, always chooses action r_2, i.e., $x_1 = 0$. So the best reply conditions for Player 2 can be written as follows:

$$\max\{(p_1(1) + (1-p_1)(x_2(-1) + (1-x_2)1)), (p_1(0) + (1-p_1)(x_2(0) + (1-x_2)0))\},$$

which boils down to

$$\max\{(1 - 2(1-p_1)x_2), 0)\}.$$

We conclude that Player 2 chooses action c_1 if $x_2 < \frac{1}{2(1-p_1)}$, action c_2 if $x_2 > \frac{1}{2(1-p_1)}$ and any mixed action with $y \in [0,1]$ if $x_2 = \frac{1}{2(1-p_1)}$. We can easily conclude from these observations that the equilibria of this game are characterized as follows:

$$x_1 \equiv 0$$
$$\text{if } p_1 \leq 0.5 \quad x_2 = 0, y = 1 \text{ or } x_2 = 1, y = 0 \text{ or } x_2 = \frac{1}{2(1-p_1)}, y = 0.5$$
$$\text{if } p_1 > 0.5 \quad x_2 = 0, y = 1.$$

4.3.3   *A general definition of Bayesian equilibria*

We can generalize the analysis performed with the help of the previous example and introduce the following definitions. Let M be a set of m players. Each player $j \in M$ may be of one type from Θ_j, which is a finite set of types for Player j. Whatever his type, Player j has the same set of pure strategies S_j. Let $\underline{\theta} = (\theta_1, \ldots, \theta_m) \in \underline{\Theta} = \Theta_1 \times \cdots \times \Theta_m$ be a type specification for every player. Then the normal form of the game is specified by the payoff functions

$$V_j(\underline{\theta}; \cdot, \ldots, \cdot) : S_1 \times \cdots \times S_m \to \mathbf{R}, \quad j \in M. \tag{4.4}$$

A prior probability distribution $p(\theta_1, \ldots, \theta_m)$ on $\underline{\Theta}$ is given as *common knowledge*. We assume that all marginal distribution probabilities are nonzero

$$p_j(\theta_j) > 0, \forall j \in M.$$

When Player j observes that he is of type $\theta_j \in \Theta_j$, then he can construct his revised probability distribution conditional on the other players' types $\boldsymbol{\theta}_{-j}$ through the Bayes formula

$$p(\boldsymbol{\theta}_{-j}|\theta_j) = \frac{p([\theta_j, \boldsymbol{\theta}_{-j}])}{p_j(\theta_j)}. \tag{4.5}$$

We can now introduce an *expanded game* where nature randomly draws a type vector $\underline{\theta} \in \underline{\Theta}$ for all players, according to the prior probability distribution $p(\cdot)$. Player j can observe only his own type $\theta_j \in \Theta_j$. Then each player $j \in M$ picks a strategy from his own strategy set S_j. The outcome is then defined by the payoff functions (4.4).

Definition 4.3. In a game of incomplete information, with m players having respective type sets Θ_j and pure strategy sets $S_j = 1, \ldots, m$, a Bayesian equilibrium is a Nash equilibrium in the expanded game, in which each player's pure strategy γ_j is a map from Θ_j to S_j.

The expanded game can be described in normal form as follows. Each player $j \in M$ has a strategy set Γ_j, where a strategy is defined as a mapping $\gamma_j : \Theta_j \to S_j$. Associated with a strategy profile $\underline{\gamma} = (\gamma_1, \ldots, \gamma_m)$, the payoff to Player j is given by

$$V_j(\underline{\gamma}) = \sum_{\theta_j \in \Theta_j} \sum_{\boldsymbol{\theta}_{-j} \in \boldsymbol{\Theta}_{-j}} p_j(\theta_j) \, p(\boldsymbol{\theta}_{-j}|\theta_j)$$

$$V_j([\theta_j, \boldsymbol{\theta}_{-j}]; \gamma_1(\theta_1), \ldots, \gamma_j(\theta_j), \ldots, \gamma_m(\theta_m)). \tag{4.6}$$

As usual, a Nash equilibrium is a strategy profile $\underline{\gamma}^* = (\gamma_1^*, \ldots, \gamma_m^*) \in \Gamma = \Gamma_1 \times \cdots \times \Gamma_m$ such that

$$V_j(\underline{\gamma}^*) \geq V_j(\gamma_j, \underline{\gamma}_{-j}^*) \quad \forall \gamma_j \in \Gamma_j. \tag{4.7}$$

It is easy to see that, since each $p_j(\theta_j)$ is positive, the equilibrium conditions (4.7) lead to the following conditions:

$$\gamma_j^*(\theta_j) = \arg\max_{s_j \in S_j} \sum_{\boldsymbol{\theta}_{-j} \in \boldsymbol{\Theta}_{-j}} p(\boldsymbol{\theta}_{-j}|\theta_j) \, V_j\left([\theta_j, \boldsymbol{\theta}_{-j}]; [s_j, \boldsymbol{\gamma}_{-j}^*(\boldsymbol{\theta}_{-j})]\right) \tag{4.8}$$

$$j \in M.$$

Remark 4.1. Since the sets Θ_j and S_j are finite, the set Γ_j of mappings from Θ_j to S_j is also finite. Therefore, the expanded game is an m-matrix game and according to the Nash theorem, there exists at least one mixed-strategy equilibrium in this game.

4.4 Equilibria in behavioral strategies and subgame perfectness

Let us now explore in more details the concept of behavioral strategy.

4.4.1 *Building up strategies at each information set*

If a game in normal form corresponds to an extensive game where each player will have the move at several successive information sets or nodes, then a natural question arises: *can the players define a mixed strategy by randomizing their choices at each information set instead of randomizing their pure strategies?*

Recall that a mixed strategy for a game in normal form consists of a random sampling from a set of pure strategies. Using a behavioral strategy, i.e., allowing the players to randomly select their moves at each information set, seems to be a more realistic way of playing the game. We will see that, for games with perfect recall, these two ways of playing the game end up with the same set of equilibrium strategies. However, behavioral strategies allow us to introduce a refinement to the equilibrium concept, namely, *subgame perfectness*, which will play an important role in defining the credibility of some equilibrium solutions in dynamic games.

4.4.2 *Multi-agent representation of a game*

Selten [218] proposed a new representation for a game defined in extensive form. He called it the *agent-normal form;* however, this was recast by Myerson [179] as a *multi-agent representation* of a game. The new representation is based on the idea that it should be irrelevant for the game outcome whether Player j is represented by himself at each information set or by a member[7] of his "family" if we assume the same preferences for all family members. In the new representation, the set of players will be indexed by the information set through which the game transits.

Let $s = 1, \ldots, n$ denote a particular information set. If the game has m players, we can define m families of players denoted $(j : s), j = 1, 2 \ldots m; s = 1, 2 \ldots n$. For example, if there are two information sets in a game where Player 1 has the move, then Player 1 can be represented by, in each of these sets, Players $(1 : 1)$ and $(1 : 2)$, respectively. All players $(j : s)$ for $s = 1, 2 \ldots n$ have the same reward and utility function as Player j.

Now, if we consider a mixed strategy for the game played by the extended set of players, we easily notice that it also defines a behavioral strategy for the original game.

As an illustration, consider the following game[8] defined in extensive form in Figure 4.6. We will use U_s (up) and D_s (down) to denote the choices made by the first player and u_s and d_s for the second, where s is the index of the information set (or node, if the information set reduces to a single element).

The initial node, indexed "0," is a chance node, from which the game transits to node $1 : 1$ or $1 : 2$, with probability 0.5. Then, Player 1 observes the node to which the game

---

[7]Myerson [179] calls them *temporary agents*.
[8]This example is borrowed from [179].

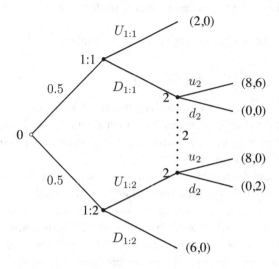

Fig. 4.6 A game in extensive form

has evolved, i.e., he has perfect information at this node. He can either stop the game (with action $U_{1.1}$) or continue (with action $D_{1.1}$) and let Player 2 move. If the game continues, Player 2 does not know the result of the chance move when he has to decide on his move. This is symbolically denoted by the dotted line that joins nodes 2 (as in the game trees used in Chapter 2).

We will compare the normal-form representation and the agent-normal-form representations of this game. Let's first look at the normal-form representation. The set of players is $M = \{1, 2\}$; the set of strategies for Player 2 is $C_2 = \{u_2, d_2\}$. A strategy for Player j, in a normal-form representation of the game, prescribes what he should do in each information set. Consider Player 1. We will use the notation $U_{1:1}/U_{1:2}$ to denote the strategy when he plays "up" in information state (1.1) and "up again" in information state (1.2). Considering all possible choices, the set of strategies of Player 1 contains four elements denoted $C_1 = \{U_{1:1}/U_{1:2}, U_{1:1}/D_{1:2}, D_{1:1}/U_{1:2}, D_{1:1}/D_{1:2}\}$, rather than just $\{U_{1:1}, D_{1:1}, U_{1:2}, D_{1:2}\}$.

The payoffs are specified in the game tree in Figure 4.6 at the ends of the terminal branches (or leaves). The payoffs are shown in Table 4.3, in bimatrix form. We sketch out how the values in the first row are obtained.

$$\text{for } U_{1:1}/U_{1:2} \,\&\, u_2 \quad 5 = 0.5 \cdot 2 + 0.5 \cdot 8, \quad 0 = 0.5 \cdot 0 + 0.5 \cdot 0,$$

$$\text{for } U_{1:1}/U_{1:2} \,\&\, d_2 \quad 1 = 0.5 \cdot 2 + 0.5 \cdot 0, \quad 1 = 0.5 \cdot 0 + 0.5 \cdot 2.$$

Recall that the game starts with a chance node; therefore all payoffs are obtained as expected values.

Table 4.3 The normal-form representation of the game in Figure 4.6

	u_2	d_2
$U_{1:1}/U_{1:2}$	$(5,0)$	$(1,1)$
$U_{1:1}/D_{1:2}$	$(4,0)$	$(4,0)$
$D_{1:1}/U_{1:2}$	$(8,3)$	$(0,1)$
$D_{1:1}/D_{1:2}$	$(7,3)$	$(3,0)$

Clearly, the strategy vectors $(D_{1:1}/U_{1:2}, u_2)$ with payoffs $(8,3)$ is a Nash equilibrium in pure strategies and so is $(U_{1:1}/D_{1:2}, d_2)$ with payoffs $(4,0)$. However, Player 2 can also mix his strategies. Using Gambit,[9] we obtain 3 equilibria, which are given in Table 4.4.

Table 4.4 Nash equilibria for the normal form

	$U_{1:1}U_{1:2}$	$U_{1:1}D_{1:2}$	$D_{1:1}U_{1:2}$	$D_{1:1}D_{1:2}$	u_2	d_2
equilibrium 1	0	1	0	0	1/4	3/4
equilibrium 2	0	1	0	0	0	1
equilibrium 3	0	0	1	0	1	0

Let us now construct the *agent-normal-form representation*. The set of players is $N = \{(1 : 1), (1 : 2), 2\}$, of whom agents $(1 : 1)$ and $(1 : 2)$ are the auxiliary players that represent Player 1 at his two information sets. Agent 2 represents Player 2 at his single information set. The respective strategy sets are $S_{1:1} = \{U_{1:1}, D_{1:1}\}$, $S_{1:2} = \{U_{1:2}, D_{1:2}\}$, $S_2 = \{u_2, d_2\}$. The 3-matrix game corresponding to the agent-normal form is shown in Table 4.5.

Table 4.5 The agent-normal-form representation of game

	u_2		d_2	
	$U_{1.2}$	$D_{1:2}$	$U_{1.2}$	$D_{1.2}$
$U_{1:1}$	$(5,5,0)$	$(4,4,0)$	$(1,1,1)$	$(\mathbf{4,4,0})$
$D_{1:1}$	$(\mathbf{8,8,3})$	$(7,7,3)$	$(0,0,1)$	$(3,3,0)$

The pure-strategy equilibria identified in Table 4.4 are also easily visible in this table (see the bold font).

A *behavioral-strategy profile* for the game in Figure 4.6 is thus defined as a *mixed-strategy profile* of the multiagent representation of the game. We have also computed all the Nash equilibria of the game in agent-normal form shown in Table 4.5, with the following results:

We can easily check that the equilibria called equilibrium 1a, equilibrium 2a and equilibrium 3a in the agent-normal form correspond in the normal form to the equilibria called equilibrium 1, equilibrium 2 and equilibrium 3, respectively.

---
[9]See Exercise 2.1.

Table 4.6   Nash equilibria for the agent normal form

	$U_{1:1}$	$D_{1:1}$	$U_{1:2}$	$D_{1:2}$	u_2	d_2
equilibrium 1a	1	0	0	1	1/4	3/4
equilibrium 2a	1	0	0	1	0	1
equilibrium 3a	0	1	1	0	1	0

This correspondence between equilibria in mixed strategies, and equilibria in behavioral strategies can be established more generally. Let τ_j be a mixed strategy of Player j for the game in normal form. Let $s \in S_j$ be an information state of that player and d_s one possible move at s. Let us call $\sigma_{i,s}(d_s)$ the conditional probability of selecting action d_s at information state s, given that the mixed-strategy realization has picked a pure strategy that is compatible[10] with s. At information states that are not compatible with τ_j, we use any probability distribution on the possible moves. In this way, we can define at least one, and possibly an infinite number of behavioral representation(s) of a mixed strategy.

Reciprocally, let σ_j be a behavioral strategy of Player j. The mixed-strategy representation of σ_j is a probability distribution on pure strategies $c_j \in C_j$ such that

$$\tau_j(c_j) = \Pi_{s\in S_j}\sigma_{i,s}(c_j(s)), \quad \forall c_j \in C_j. \tag{4.9}$$

That is, a mixed strategy, in which Player j's planned move at each information set s has the marginal probability distribution $\sigma_{i,s}$ and is determined independently of his moves at all other information states.

It is straightforward to check that if τ_j is a mixed strategy representation of a behavioral strategy σ_j, then σ_j is a behavioral representation of τ_j.

The following important theorem is taken from Meyerson [179].[11]

Theorem 4.1. *If an extensive-form game is a game of perfect recall, then any equilibrium of its normal-form representation admits a behavioral representation, which is an equilibrium of the multiagent-representation of the game.*

As a consequence of the existence theorem of Nash equilibrium in m-matrix games and of the previous theorem, we also obtain the following result.

Theorem 4.2. *For any extensive-form game with perfect recall, there exists a Nash equilibrium in the behavioral strategies.*

4.4.3   Subgame perfect equilibria

Selten [218] proposed the concept of the *subgame perfect equilibrium* in extensive games, (which will be extended to the context of dynamic games in the forthcoming chapters,)

---

[10] We say a pure strategy c_j is compatible with information state s iff there exists at least one combination of pure strategies for the other players such that a node at which Player j moves with information state s could occur with a positive probability when Player j implements pure strategy c_j.

[11] See [179] for proof.

which will have important implications when discussing the credibility of some strategies involving threats.

In an extensive game Γ, we call a *subroot* any node x where any player who moves at x or thereafter will know that node x has occurred. The subgame Γ_x at subroot x is the game that is derived from the initial extensive game by deleting all nodes and branches that do not follow x.

Definition 4.4. A subgame perfect equilibrium of an extensive game Γ is an equilibrium in behavioral strategies, such that, for any subgame Γ_x of Γ beginning at any possible subroot x, the restriction of these behavioral strategies to this subgame is also an equilibrium in behavioral strategies for the subgame Γ_x.

As an illustration, consider the game in extensive form defined below in Figure 4.7.

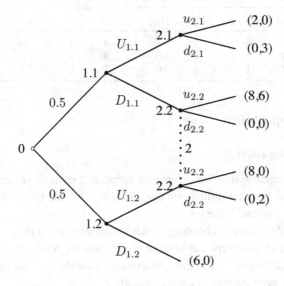

Fig. 4.7   The game in extensive form

The normal form of the game is given by the bimatrix game shown in Table 4.7. This game admits three equilibria, defined in Table 4.8.

These equilibria can be formulated as equilibria in behavioral strategies, as shown in Table 4.9. Among these equilibria, equilibrium 3 is not subgame perfect. Indeed we can consider the subgame starting from the subroot labelled 2.1. For this subgame, there is a single node where Player 2 has the move. His unique equilibrium solution is to select $d_{2.1}$, whereas, in equilibrium 3a, selecting $u_{2.1}$ is recommended.

Table 4.7   The normal form representation of game 4.7

	$u_{2.1}u_{2.2}$	$d_{2.1}u_{2.2}$	$u_{2.1}d_{2.2}$	$d_{2.1}d_{2.2}$
$U_{1.1}U_{1.2}$	$(5,0)$	$(1,1)$	$(4,\frac{3}{2})$	$(0,\frac{5}{2})$
$U_{1.1}D_{1.2}$	$(4,0)$	$(4,0)$	$(3,\frac{3}{2})$	$(3,\frac{3}{2})$
$D_{1.1}U_{1.2}$	$(8,3)$	$(0,1)$	$(8,3)$	$(0,1)$
$D_{1.1}D_{1.2}$	$(7,3)$	$(3,0)$	$(7,3)$	$(3,0)$

Table 4.8   Nash equilibria for the normal form

	$U_{1.1}U_{1.2}$	$U_{1.1}D_{1.2}$	$D_{1.1}U_{1.2}$	$D_{1.1}D_{1.2}$	$u_{2.1}u_{2.2}$	$d_{2.1}u_{2.2}$	$u_{2.1}d_{2.2}$	$d_{2.1}d_{2.2}$
equilibrium 1	0	1	0	0	0	0	0	1
equilibrium 2	0	0	1	0	0	0	1	0
equilibrium 3	0	0	1	0	1	0	0	0

Table 4.9   Nash equilibria for the agent normal form

	$U_{1.1}$	$D_{1.1}$	$U_{1.2}$	$D_{1.2}$	$u_{2.1}$	$d_{2.1}$	$u_{2.2}$	$d_{2.2}$
equilibrium 1a	1	0	0	1	0	1	0	1
equilibrium 2a	0	1	1	0	0	1	1	0
equilibrium 3a	0	1	1	0	1	0	1	0

4.5   Quantal-response equilibrium

4.5.1   *A robust equilibrium concept*

It is assumed in the Nash equilibrium of an m-matrix game, that each player uses a best-response function to define his equilibrium strategy. This is what we shall call *rational expectations* regarding the players' behavior.

Sometimes players make errors in choosing which (pure) strategy to play. Situations like that call for an equilibrium concept that is more *robust* than the Nash equilibrium. In particular the issue of the robustness of an equilibrium stems from the fact that each player may make a mistake in his perception of his utility payoff.

Quantal-response equilibrium (QRE), introduced by McKelvey and Palfrey [171] is a generalization of the Nash equilibrium, which allows for robustness in optimization behavior while maintaining the internal consistency of rational expectations.

4.5.2   *Strategies and payoffs*

Let us first slightly modify the notations used to describe a game.

- $S_i = \{s_{i1}, \ldots, s_{ij}, \ldots, s_{iJ_i}\}$ is the set of pure strategies of Player $i \in I = \{1, \ldots, n\}$. Let $\underline{S} = \prod_i S_i$ be the set of strategy profiles $\underline{s} = (s_1, \ldots, s_n)$ for the n players together.

- For each player i let Σ_i be the simplex of probability measures over S_i. An element $\sigma_i \in \Sigma_i$ is a mixed strategy where $\sigma_i(s_{ij})$ is the probability that Player i chooses pure strategy s_{ij}.
- Payoffs are given by functions $\pi_i(s_i, s_{-i}) : S_i \times \prod_{j \neq i} S_j \to \mathbf{R}$. Given a mixed strategy profile $\underline{\sigma} \in \Sigma = \prod_i \Sigma_i$, Player i's expected payoff is

$$\pi_i(\underline{\sigma}) = \sum_{\underline{s} \in S} p(\underline{s}) \pi_i(\underline{s}),$$

where $p(\underline{s}) = \prod_i \sigma_i(s_i)$ is the probability distribution over pure strategy profiles induced by $\underline{\sigma}$.

4.5.3 *Payoff perturbations*

Let us assume that the players' payoffs are affected by disturbances that are "privately observed." This means that a player observes the disturbance affecting his own utility function, but does not observe the disturbances affecting the utility functions of the other players.

We will use $\pi_{ij}(\underline{\sigma})$ to denote the expected payoff to Player i for adopting the pure strategy s_{ij} when the other players use $\sigma_{-i} = (\sigma_\ell)_{\ell \neq i}$, for each $i \in I$ and $j \in \{1, \ldots, J_i\}$ and for any mixed-strategy profile $\underline{\sigma} \in \Sigma$.

We introduce now "private" errors made by the players in assessing their payoffs. Assume that, for each pure strategy s_{ij}, there is an additional privately observed payoff disturbance, ϵ_{ij}. If so, we can express the *disturbed payoff* as

$$\hat{\pi}_{ij}(\underline{\sigma}) = \pi_{ij}(\underline{\sigma}) + \mu_i \epsilon_{ij}, \tag{4.10}$$

where μ_i is strictly positive. Being private to Player j, the disturbance ϵ_{ij} is not observed by the other players.

4.5.4 *Quantal-response function*

The disturbed payoffs induce choice probabilities on the players' strategy sets. More precisely, assume that the disturbances ϵ_{ij} are unbiased (zero mean) and are independently distributed across players, with absolutely continuous probability measures.

Assume that a player chooses strategy s_{ij} when

$$\hat{\pi}_{ij}(\underline{\sigma}) \geq \hat{\pi}_{ik}(\underline{\sigma}) \quad \forall k \in \{1, \ldots, J_i\}.$$

Then, given this choice behavior, the payoff functions $\pi = (\pi_1, \ldots, \pi_n)$ and the probability-distribution functions (pdfs) $f = (f_1, \ldots, f_n)$, we obtain an induced probability distribution over the actual choices by each player, which is defined as follows:

For a given nominal expected payoff $\pi(\underline{\sigma})$, let $B_{ij}(\pi_i(\underline{\sigma}))$ be the set of realizations of ϵ_{ij} such that strategy s_{ij} has the highest disturbed expected payoff, $\hat{\pi}$. Then

$$P_{ij}(\pi_i(\underline{\sigma})) = \int_{B_{ij}(\pi_i(\underline{\sigma}))} f_i(\epsilon) \, d\epsilon \tag{4.11}$$

is the induced probability that Player i will choose strategy j.

The vector $P_i(\pi_i(\underline{\sigma})) = (P_{i1}(\pi_i(\underline{\sigma})), \ldots, P_{iJ_i}(\pi_i(\underline{\sigma})))$ is a probability distribution over the set of pure strategies of Player i, so it defines a mixed strategy. The function P_i maps $\pi_j(\underline{\sigma})$ into Σ_i. It is called the *quantal-response function* of Player i. It maps a vector of expected payoffs to Player i for each possible pure strategy into a mixed strategy for Player i. That is $P_i : \mathbf{R}^{J_i} \to \Sigma_i$. It can be easily checked that this response function is *single-valued* and *continuous*.

4.5.5 Definition of quantal-response equilibrium

For each given mixed-strategy profile $\underline{\sigma}$, Player i has an expected nominal utility vector $\pi_i(\underline{\sigma})$, which makes him to assign probabilities for each of his pure strategies and thus define a mixed strategy σ_i. Therefore to each $\underline{\sigma}$, $P_i(\pi_i(\underline{\sigma}))$ associates a unique mixed strategy σ_i. Denote $P(\pi) = (P_1(\pi_1), \ldots, P_m(\pi_m))$ the vector quantal response function. Since $P(\pi) \in \Sigma$ and $\pi = \pi(\underline{\sigma})$ is defined as a function of $\underline{\sigma}$, the composite mapping $P \circ \pi(\underline{\sigma}) = P(\pi(\underline{\sigma}))$ maps Σ into itself.

Now we can introduce a concept of equilibrium that is based on the quantal-response functions.

Definition 4.5. Let $f(\epsilon)$ be an admissible vector of pdfs for the disturbances ϵ. A Quantal Response Equilibrium (QRE) is a mixed strategy profile $\underline{\sigma}^*$ such that

$$\underline{\sigma}^* = P(\pi(\underline{\sigma}^*)), \tag{4.12}$$

with P defined in (4.11).

The existence of QRE is easily established through a classical fixed-point argument.

Theorem 4.3. *For any admissible $f(\epsilon)$ there exists a Quantal Response Equilibrium.*

Proof. Use the Brouwer fixed-point theorem[12] as P and π are continuous maps and Σ is a closed convex set. See [171]. □

Remark 4.2. QRE has a number of very interesting features and properties.

- It is based on the idea that choice probabilities are related in a continuous and monotonic way to expected payoffs. In a game-theoretic setting, the choice probability of one player affects the expected utility of the other players and, hence, their choice probabilities.
- It assumes that the players do not necessarily use their best response which may be interpreted as a departure from rationality or as a "bounded rationality" decision framework.
- An alternative interpretation can be made, which makes QREs consistent with the concept of Bayesian equilibrium for games of incomplete information. The game (I, Σ, π) is a complete-information approximation of a game of incomplete information, in

---

[12] Every continuous function f from a convex compact $K \subset \mathbf{R}^n$ to K itself has a fixed point.

which the actual payoffs of the players are private information since the disturbances to Player i's payoffs are known only to i.

4.5.6 *Logit QRE*

The QRE theory will be easier to handle if we assume that the disturbances' probability-distribution functions are exponential.

- Assume the disturbances are independently and identically distributed with pdf $f_i(\epsilon_{ij}) = \lambda e^{-\lambda \epsilon_{ij}}$, where $\lambda > 0$ is the parameter defining this exponential law. For a random variable following this law, $P[X \geq x] = e^{-\lambda x}$.
- Then, it can be shown that the quantal-response function takes the form of a logit[13] distribution, i.e.,

$$P_{ij}(\pi) = \int_{B_{ij}(\pi)} f(\epsilon)\, d\epsilon = \frac{e^{\lambda \pi_{ij}(\sigma)}}{\sum_{k=1}^{J_i} e^{\lambda \pi_{ik}(\sigma)}}. \tag{4.13}$$

This allows us to define a *Logit Quantal Response Equilibrium (LQRE)*:

Definition 4.6. Let the quantal-response function be defined as in (4.13). Then the associated QRE is called a **Logit Quantal Response Equilibrium**. The parameter λ is called the **Sophistication Level.**

4.5.7 *Limit Logit Quantal Response Equilibrium (LLQRE)*

When $\lambda \to 0$ the players tend to choose their strategies totally randomly. When $\lambda \to \infty$ they tend to act rationally and to select the true best response.

Definition 4.7. Let $\underline{\sigma}_t^*$ be the LQRE associated with the sophistication parameter $\lambda = t$. Let $\underline{\sigma}^*$ be a limit of $\{\underline{\sigma}_t^*\}$ when $t \to \infty$. This mixed strategy profile is called a Limit Logit Quantal Response Equilibrium (LLQRE). It is also a Nash equilibrium of the original game in normal form.

Example 4.4. *The asymmetric game of chicken[14] is defined in Table 4.10 (see Example 3.7, page 48, for a symmetric game of chicken).*

Table 4.10   An asymmetric game of chicken

	tough	soft
tough	0,0	6,1 *
soft	1,14 *	2,2

---

[13]The *logit* of a number $p \in (0, 1)$ is defined as: $\mathrm{logit}(p) = \log\left(\frac{p}{1-p}\right) = \ln(p) - \ln(1-p)$.

[14]Taken from [172].

This game has three Nash equilibria: two in pure strategies (indicated by a star in the table cells) and one where they both mix strategies, $\left(\frac{1}{13}, \frac{12}{13}\right)$ for Player 1 (row) and $\left(\frac{4}{5}, \frac{1}{5}\right)$ for Player 2 (column). These equilibria are shown in the Gambit output of Figure 4.8.

👤▦👤 Player 1			tough		soft	
Payoff: 6/5	tough		0	0	6	1
	soft		1	14	2	2
👤▦👤 Player 2						
Payoff: 14/13						

Profiles 1 ▾	All equilibria by enumeration of mixed strategies in strategic game			
#	1: tough	1: soft	2: tough	2: soft
1	$\frac{12}{13}$	$\frac{1}{13}$	$\frac{4}{5}$	$\frac{1}{5}$
2	1	0	0	1
3	0	1	1	0

Fig. 4.8   The equilibria computed by Gambit

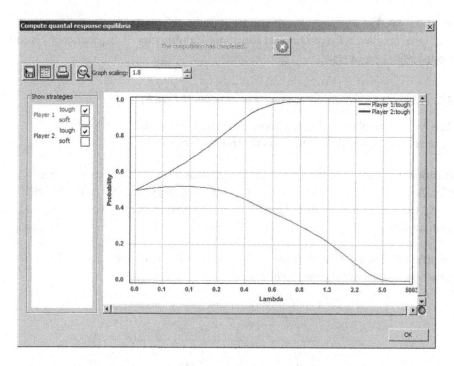

Fig. 4.9   QRE correspondence for the asymmetrical game of chicken

The *logit equilibrium correspondence* maps the evolution of the LQRE when the parameter λ grows from 0 to ∞. Gambit can be used to compute this correspondence, as shown in Figure 4.9. When $\lambda = 0$ both players completely mix their strategies in the LQRE. When $\lambda \to \infty$, the row player plays "soft" with probability of one and the column player almost surely plays "tough." The LQRE solution picks out the equilibrium strategy favoring the player that benefits most from being "tough." ◇

Remark 4.3. Logit equilibria have a number of distinctive properties that make them very different from standard Nash equilibria:

- A Nash equilibrium cannot use a strongly dominated strategy with a positive probability. The refinements of Nash equilibria often eliminate any Nash equilibrium that uses weakly dominated strategies.[15] A logit equilibrium uses every strategy with a positive probability. The probability of adopting a strongly dominated strategy goes to 0 when λ goes to infinity. However, weakly dominated strategies can be used with a positive probability even at the limit.
- In a Nash equilibrium, the mixing probabilities do not depend on the players' own payoffs over the outcomes in the support of the equilibrium; by definition, a player is indifferent between all strategies in the support of the mixed equilibrium. The reason for mixing is to keep the other players at the Nash equilibrium. In an LQRE, by contrast, the mixing probabilities depend on the player's own payoffs.
- The payoff magnitude has an effect on the LQRE. This is indeed different from the Nash equilibrium, where equilibrium solutions are unaffected by changes in the magnitude and scale of the utility function. For an LQRE, the expression of the response function $P_{ij}(\pi_i)$ shows that λ and π_{ij} always enter as scalar multiples of each other. Thus, multiplying all utilities by a constant is equivalent to multiplying λ by a constant. The result will then be that the LQRE will be closer to a Nash equilibrium.

4.6 Supermodular games

In this section, we explore a class of games defined in normal form, for which the existence of equilibria can be proved without invoking a convexity argument for the strategy sets or the payoff functions.

4.6.1 *Tarski's fixed-point theorem*

We have seen that the definition of a Nash equilibrium is related to the construction of a fixed point for a generalized response function (as in equation (3.40), page 57). The existence of such a fixed point and therefore of a Nash equilibrium or, more generally, of a Rosen's coupled-constraints equilibrium, was obtained using arguments from the Kakutani fixed-point theorem (see Theorem 3.4, page 58). To construct the appropriate upper semi-continuous point-to-set mapping, whose fixed point is proved by the Kakutani theorem, the strategy sets have to be, in general, convex and compact subsets of the Euclidean spaces;

---

[15] strategy B weakly dominates strategy A if there is at least one set of opponents' action for which B is superior, and all other sets of opponents' actions give B at least the same payoff as A.

moreover, the payoff of each player has to be concave with respect to his own strategy (see Section 3.4). In particular in m-matrix games, the existence of equilibrium can only be proved when the players use mixed strategies that are elements of a simplex and when the payoffs are the expected utilities, which are linear and hence concave functions of each player's mixed-strategy vector. The question thus arises as to whether it is possible to evoke another general fixed-point theorem using a different topological structure that would not require us to convexify the strategy set via the mixed strategies. The answer is positive and is provided in Tarski's [227] fixed-point theorem, see Theorem 4.4 below.

However, to understand this theorem, we first need to define a mathematical object called a *lattice*. Although lattices are defined in many different domains of mathematics, for our purposes here, we will restrict our considerations to lattices that are subsets of $\mathbf{R}^n$ with a partial order defined by

$$x, y \in \mathbf{R}^n, x \geq y, \text{ iff } x_i \geq y_i, \quad i = 1, n. \tag{4.14}$$

Consider a set X with a partial order[16] defined on its elements. We call such a set a *poset*. Let $H \subseteq X$ and $a \in X$; we say that a is an *upper bound* of H, if $x \leq a$ for all $x \in H$. An upper bound is the *least upper bound* of H or *supremum* of H if, for any upper bound b of H, we have $a \leq b$. We will write $a = \sup_X H$. However, this notation can be justified only if we show the uniqueness of the supremum. Indeed, it is impossible for H to have many, say two, suprema. If a_0 and a_1 were both suprema of H, then $a_0 \leq a_1$ because a_1 is an upper bound and a_0 is a supremum. Similarly, $a_1 \leq a_0$; hence, $a_1 = a_0$ by antisymmetry (see (2) in footnote 16).

The concepts of *lower bound* and *infimum* (or *greatest lower bound*) are similarly defined. That is, a is a *lower bound* of H, if $x \geq a$ for all $x \in H$. A lower bound is the greatest lower bound of H, or infimum of H, if we have $a \geq b$ for any lower bound b of X. We will write $a = \inf_X X$. Here, we also have to show the uniqueness of the infimum: if a_0 and a_1 were both infima of H, then $a_0 \geq a_1$ because a_1 is a lower bound and a_0 is an infimum. Similarly, $a_1 \geq a_0$; hence, $a_1 = a_0$ by the same argument as above.

Definition 4.8. A poset X is a **lattice** if, for any pair $x, y \in X$, both the supremum (also called *join*) $x \vee y \triangleq \sup_X\{x, y\}$ and the infimum (also called *meet*) $x \wedge y \triangleq \inf_X\{x, y\}$ exist and both are elements of X.

Example 4.5. *The space $\mathbf{R}^n$ endowed with the usual partial order, as defined in (4.14) is a lattice.*

Example 4.6. *The set $X = \{(0,0), (0,1), (1,0), (2,2)\} \subset \mathbf{R}^2$ illustrated in Figure 4.10 is a lattice.*

Indeed, it is easy to see that all point pairs have an infimum and a supremum in X. For example $\sup_X\{(0,1), (1,0)\} = (2,2), \inf_X\{(0,1), (1,0)\} = (0,0)$, etc.

---

[16] A relation $\geq$ defined on a set X defines a *partial order* if the following properties hold $\forall x, x^1, x^2 \in X$: (1) $x \geq x$ (reflexivity); (2) if $x^1 \geq x_2$ and $x_2 \geq x^1$ then $x_1 = x_2$ (antisymmetry); (3) if $x \geq x^1$ and $x^1 \geq x^2$ then $x \geq x_2$ (transitivity). If x and y are vectors in $\mathbf{R}^n$, a partial order is defined by the relation $x \geq y$ which means $x_i \geq y_i, i = 1, \ldots n$.

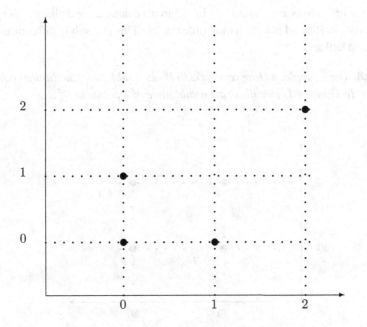

Fig. 4.10   Example of a lattice in $\mathbf{R}^2$

Definition 4.9. A lattice L is a **complete lattice** if for any subset S of L, $\sup_L S$ and $\inf_L S$ exist and belong to L.

It is easy to see that lattice $\mathbf{R}^n$ is complete. The lattice X of example 4.6 illustrated in Figure 4.10 is also complete. If we take $S = \{(0,0), (0,1), (1,0)\}$ then $\inf_X S = (0,0) \in X$ and $\sup_X S = (2,2) \in X$; etc.

We notice that, in particular, a complete lattice L has a largest and a smallest element.

Definition 4.10. A **sublattice** of a lattice L is a nonempty subset of L which is a lattice with the same meet and join operations as L.

Example 4.7. *A subset M of $\mathbf{R}^n$ (where the latter is a lattice), which has the property that $x, y \in M$ imply*

$$\underline{\text{meet}} \quad x \wedge y = \begin{bmatrix} \min\{x_1, y_1\} \\ \min\{x_2, y_2\} \\ \vdots \\ \min\{x_n, y_n\} \end{bmatrix} \in M, \quad (4.15)$$

$$\underline{\text{join}} \quad x \vee y = \begin{bmatrix} \max\{x_1, y_1\} \\ \max\{x_2, y_2\} \\ \vdots \\ \max\{x_n, y_n\} \end{bmatrix} \in M, \quad (4.16)$$

*is therefore a **sublattice**.*

Notice that the operations $x \vee y$ and $x \wedge y$ in the above example are defined with respect to the partial order in $\mathbf{R}^n$ and not the partial order in M. This is a subtle difference between sublattices and lattices.

Example 4.8. *For example, a close rectangle in $\mathbf{R}^n$ is a sublattice. Sublattices can also be discrete sets. In Figure 4.11, we illustrate a sublattice of 4 points in $\mathbf{R}^2$.*

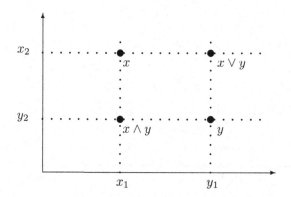

Fig. 4.11   Example of a sublattice in $\mathbf{R}^2$

As another example of a discrete sublattice in $\mathbf{R}^2$, consider the set $K \subset \mathbf{R}^2$ where $K = \{(0,0), (0,1), (1,0), (1,1), (2,2)\}$, illustrated in Figure 4.12. This set is a sublattice of $\mathbf{R}^2$ since it can easily be checked that, for every pair of points x and y, the join $x \vee y$ and the meet $x \wedge y$, defined with respect to the partial order in $\mathbf{R}^n$, are in K. ◇

Notice that the lattice X defined in example 4.6 is not a sublattice of $\mathbf{R}^n$ (could you say why?).

If $x \in L \subset \mathbf{R}^n$ and $y \leq x$ (or, $x \leq y$) for each $y \in L$, then x is the greatest (or, least) element of L. A complete sublattice always has a greatest and a least element. The examples of sublattices of $\mathbf{R}^2$ given above are complete sublattices.

We now quote the Tarski fixed-point theorem, [227].

Theorem 4.4. *Let L be a complete lattice and let $f : L \to L$ be a non-decreasing function (i.e., $x \geq y$ implies $f(x) \geq f(y)$). Then, the set $E = \{x : f(x) = x\}$ of fixed points of f in L is a complete lattice.*

Preliminary remarks.   Since complete lattices cannot be empty, Tarski's theorem guarantees (in particular) the existence of at least one fixed point of f, and even guarantees the existence of the least and of the greatest fixed point. In many practical cases, this property will prove to be an important implication of the theorem. Topkis [232] was the

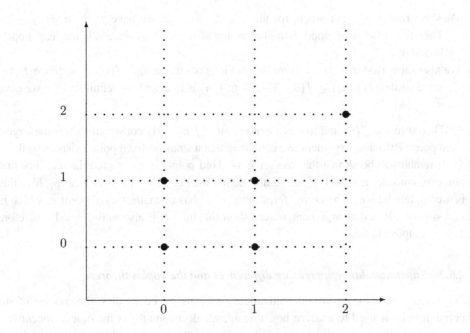

Fig. 4.12   Example of a sublattice in $\mathbf{R}^2$

first to see the implication of this theorem for game theory; he also noticed the proximity of this mathematical background to the concept of *strategic complementarity* in economic theory [233]. We will use this notion later in Section 9.7.3 to discuss the existence of equilibria in stochastic games. Other important contributions in this area can to be found in references [5, 238, 239].

Quoting [239], we can provide some intuition on what Tarski's fixed-point theorem is telling us. It guarantees the existence of a pure strategy Nash equilibrium. It can be visualized by thinking of a function f from [0,1] to [0,1]. Imagine a graph of f and the $45°$ line across the $[0,1] \times [0,1]$ square. If the function is increasing, it will cross the $45°$ line at least once (possibly at the end points (0,0) or (1,1)). The function may make several jumps, but it will nevertheless cross the $45°$ line at some point. We notice, however, that this would not happen if the function were decreasing: discontinuities of the function may lead to it never crossing the $45°$ line. In that case, the equilibrium would not exist.

Now we will prove the Tarski fixed-point theorem (Theorem 4.4).

Proof.   The function f maps L into itself. Consider the set $S = \{x : f(x) \geq x\}$. Because f is non-decreasing (so, order-preserving) this set is nonempty. Indeed, since L is a complete lattice, it has a least element $\underline{x}$ for which, by definition $f(\underline{x}) \geq \underline{x}$.

Since L is complete $u = \sup_L S$ is an element of L. We will show that $u = f(u)$.

We show that $f(u) \geq u$: since, for all $x \in S$, $u \geq x$, we have $f(u) \geq f(x) \geq x$. Therefore $f(u)$ is an upper bound of S and so $f(u) \geq u$, since u is the least upper-bound of S.

We also show that $u \geq f(u)$: from the previous point, we have $f(u) \geq u$. Since f preserves order, $f(f(u)) \geq f(u)$. Therefore $f(u)$ is in S and by definition of u we have $u \geq f(u)$.

Therefore $u = f(u)$ and so u is a fixed point of f in L. By construction, it is the largest fixed point. By a dual argument, we can prove that a smallest fixed point exists as well.

It remains to be shown that the set E of fixed points is a complete lattice. For that purpose, consider a subset $K \subset E$ and define the set $P = \{x : x \geq \sup_L K\}$; this is a complete lattice. Therefore, f restricted to P has a smallest fixed point v, which is $v = \sup_E K$. By a dual argument, we can show that $\inf_E K$ is also well-defined. Therefore E is a complete lattice.   □

4.6.2 *Supermodularity, increasing differences and the Topkis theorem*

Supermodularity is a feature of multivariate functions. In economics, the concept of supermodularity is used to analyze how one agent's decision affects the others' incentives. Supermodularity of the utility (payoff) function is sufficient to yield the monotonicity of optimal actions in the parameter, which is called *parametric* monotonicity.

Definition 4.11. A function $f : L \to R$ on a sublattice l is said to be **supermodular** if, for all x and y in L,

$$f(x) + f(y) \leq f(x \vee y) + f(x \wedge y). \tag{4.17}$$

Example 4.9. *We will show that* $y(k, l) = kl$ *is supermodular where* $k \in R_+$, $l \in R_+$ *and* $y(\cdot, \cdot) : R_+ \times R_+ \mapsto R$. *Consider* (k, l) *and* (k', l') *in* $R_+ \times R_+$ *and assume, without loss of generality, that* $k \geq k'$.

Suppose first that $l \geq l'$. *Then,* $(k, l) \vee (k', l') = (k, l)$ *and* $(k, l) \wedge (k', l') = (k', l')$. *Therefore, obviously,*

$$y(k, l) + y(k', l') \leq y\big((k, l) \vee (k', l')\big) + y\big((k, l) \wedge (k', l')\big).$$

We will now show that the above relationship is also satisfied if $l < l'$ *and hence prove that* $y(k, l)$ *is supermodular, according to Definition 4.11.*

So, let $l < l'$; *then,* $(k, l) \vee (k', l') = (k, l')$ *and* $(k, l) \wedge (k', l') = (k', l)$. *Therefore,*

$$y\big((k, l) \vee (k', l')\big) + y\big((k, l) \wedge (k', l')\big) = kl' + k'l.$$

Let us now observe that

$$kl' + k'l - (kl + k'l') = k(l' - l) - k'(l' - l) = (k - k')(l' - l) \geq 0$$

so we can write that

$$y\big((k, l) \vee (k', l')\big) + y\big((k, l) \wedge (k', l')\big) \geq y(kl) + g(k'l') \geq 0$$

and, finally, that

$$y(k,l) + y(k',l') \leq y\big((k,l) \vee (k',l')\big) + y\big((k,l) \wedge (k',l')\big).$$

Therefore, $y(k,l) = kl$ is supermodular.

If we give an economic interpretation to k and l as capital and labor in a production economy, then we can better appreciate the meaning of supermodularity. In particular, we notice that the parameter l, labor, can amplify a change in k, capital.

Notice that inequality (4.17) can also be written as

$$f(x \vee y) - f(x) \geq f(y) - f(x \wedge y),$$

or

$$f(x \vee y) - f(y) \geq f(x) - f(x \wedge y).$$

The space $\mathbf{R}^n$ and every sublattice are partially ordered; remind that $x \geq y$ iff $x_i \geq y_i$ for $i = 1, \ldots, n$. We call a function $f : L \to \mathbf{R}$ such that $x \geq y$ implies $f(x) \geq f(y)$ *isotone (antitone, $f(y) \geq f(x)$)*.

Increasing differences is an important feature of functions $f : X \times Y \to \mathbf{R}$ that economic agents want to maximize. The feature means that, if f increases in $x \in X$, this growth is amplified by the other variable(s). Formally, we provide this definition.

Definition 4.12. If $f(x,y)$ is a real function on sublattice $L \subset \mathbf{R}^{n+m}$ where $x \in \mathbf{R}^m$ and $y \in \mathbf{R}^n$ and if $f(x,w) - f(x,y)$ is isotone (antitone) in x for all $w \geq y$, then we say that $f(x,y)$ has **increasing (decreasing) differences** in (x,y).

Referring to Figure 4.11, we can see that supermodularity implies increasing differences:

$$f(x \vee y) + f(x \wedge y) \geq f(x) + f(y) \Rightarrow f(x_1, y_2) + f(y_1, x_2) \geq f(x_1, x_2) + f(y_1, y_2)$$

$$\Rightarrow f(y_1, x_2) - f(y_1, y_2) \geq f(x_1, x_2) - f(x_1, y_2)$$

because $x \vee y = (x_1, y_2)$ and $x \wedge y = (y_1, x_2)$.

The concept of increasing differences is linked to the concept of *strategic complementarity* in economic theory. In brief, when the players action space is continuous and, for simplicity, each action is chosen from an interval $[\underline{a}, \overline{a}]$, supermodularity of payoff g_j, where $(j, -j)$ is the player set, implies that an increase in player j's choice a_j increases the marginal payoff $\dfrac{dg_j}{da_{-j}}$ of action a_{-j} for all other players $-j$. That is, if any player j chooses a higher a_j, all other players $-j$ have an incentive to raise their choices a_{-j} too. This is called *strategic complementarity*, because players' strategies are complements to each other. We also mention the opposite case of *submodularity*: it corresponds to the situation of *strategic substitutability* where an increase in a_j lowers the marginal payoff to all other player's choices a_{-j}, so strategies are substitutes. That is, if j chooses a higher a_i, other players have an incentive to pick a lower a_{-j}. In essence, this feature is responsible for existence of equilibria and was captured by Topkis in the following theorem. However, let us first cite a lemma, also due to Topkis, see [231].

Lemma 4.1. *If $f(x)$ is supermodular on the sublattice $L \subset \mathbf{R}^n$, then $f(x)$ has increasing differences on L. If L is a product set and $f(x)$ has increasing differences on L, then $f(x)$ is supermodular on L.*

Finally, let A and B be two subsets of a complete lattice. We can say that lattice A *dominates* lattice B, i.e., $A \geq B$, if $x \in A$ and $y \in B$ implies $x \vee y \in A$ and $x \wedge y \in B$.

We can now prove the following theorem due to Topkis [231].

Theorem 4.5. *Let $f : (x, t) \in L \times T \mapsto f(x, t) \in \mathbf{R}$ where L is a complete lattice and T any ordered set. Suppose that f is supermodular in L and satisfies increasing differences in (x, t). Let $x^*(t) = \operatorname{argmax}_L f(x, t)$ be the set of maximizers of $f(\cdot, t)$ in L. If $x^*(t)$ is nonempty for each t, then $t' \geq t$ implies $x^*(t') \geq x^*(t)$.*

Proof.   Let $x \in x^*(t')$ and $y \in x^*(t)$. Then,

$$0 \geq f(x \vee y, t') - f(x, t') \quad \text{since } x \text{ maximizes } f(\cdot, t')$$
$$\geq f(x \vee y, t) - f(x, t) \quad \text{by increasing differences}$$
$$\geq f(y, t) - f(x \wedge y, t) \quad \text{by supermodularity}$$
$$\geq 0 \quad \text{since } y \text{ maximizes } f(\cdot, t).$$

Therefore, all the inequalities are equalities. We have $x \vee y \in x^*(t')$ and $x \wedge y \in x^*(t)$ and therefore, $x^*(t') \geq x^*(t)$. $\qquad \square$

Remark 4.4. In many applications where the decision variables are continuous, the existence of $x^*(t)$ is guaranteed by the upper-semicontinuity (see Definition 3.7 on page 58) of f and the compactness of L. In fact, if L is a compact set of $\mathbf{R}^n$ and $T = \mathbf{R}^m$ while f is twice differentiable with $\frac{\partial^2 f}{\partial x_i \partial x_j} > 0$, then the theorem applies.

4.6.3   *Equilibrium in a supermodular game*

Consider a game in strategic form. As usual, $M = \{1, \ldots, m\}$ is the set of players. Each player $j \in M$ chooses his action $u_j \in U_j$ where U_j is a subset of $\mathbf{R}^{n_j}$ and n_j is a given integer. Player j receives a payoff

$$\psi_j(u_1, \ldots, u_j, \ldots, u_m)$$

that depends on the actions chosen by all the players. We assume that the product set $\underline{U} = \prod_{j=1}^m U_j$ is a sublattice in $\mathbf{R}^n$, where $n = \sum_{j=1}^m n_j$. The reward function of Player j is denoted $\psi_j : \underline{u} \in \underline{U} \mapsto \psi_j(\underline{u}) \in \mathbf{R}$.

Definition 4.13. The game $(M, (U_j, \psi_j; j \in M))$ defined above is called **supermodular** if $\psi_j(\underline{u})$ is supermodular on $u_j \in U_j$ for each $\mathbf{u}_{-j} \in \Pi_{k \neq j} U_k$ and if, for each $j \in M$, $\psi_j(u_j, \mathbf{u}_{-j})$ has increasing differences in $(u_j, \mathbf{u}_{-j})$.

Example 4.10. *Consider the following bimatrix game, which is a version of the battle of the sexes game (refer to Example 3.6, page 48). Here, each player j can choose a pure strategy $s_j \in \{B, F\}$ and receives a reward $u_j(s_j, s_{-j})$. To show that this game is supermodular, we need to define an order on the strategy sets. Let $B \geq F$ for both players. Then for each player $j = 1, 2$, $u_j(s_j, s_{-i})$ has increasing differences in (s_i, s_{-i}).* ◇

Table 4.11   A bimatrix supermodular game

	B	F
B	2,1	0,0
F	0,0	1,2

Example 4.11. Competitive pricing with substitute products.[17] *There are m firms competing on a market, each supplying a single product. The m products are substitutes for one another. The action for firm j is to select a price p_j for its product. The vector of prices is $p = (p_1, \ldots, p_m)$. Let S_j be the set of possible prices that firm j can select. The set of possible price vectors $\underline{S} = \prod_{j=1}^{m} S_j$ is a sublattice of $\mathbf{R}^m$. The demand for product j depends on the price vector p, and is given by the function $D_j(p)$. There is a unit production cost c_j for product j, so the profit for firm j is defined as $\psi_j(p) = (p_j - c_j)D(p)$. Consider the following two hypotheses:*

H1 *If the price of any product $k \neq j$ increases, then the demand for product j increases; this is the economic definition of substitutability; it also corresponds to the mathematical condition of $D(p)$ being an isotone function of p_k for any $k \neq j$.*

H2 *Cutting the price of product j will result in a greater increase of demand for product j if the price p_k for product k happens to be lower; in other words, the demand for product j is more sensitive to its own price when another product is more competitive because of a lower price. Mathematically this hypothesis corresponds to $D(p)$ having isotone differences in (p_j, p_k) for all $k \neq j$.*

Conditions H1 and H2 imply that $\psi_j(p)$ has isotone differences in (p_j, p_k) for all $k \neq j$. By Lemma 4.1, this is a supermodular game. ◇

Consider a supermodular game as defined at the beginning of this subsection and introduce the combined reaction function[18] $g(\underline{u}, \underline{v}) = \sum_{j=1}^{m} \psi_j(v_j, \mathbf{u}_{-j})$ and the point-to-set map $\Psi(\underline{u}) = \{\underline{v} : g(\underline{u}, \underline{v}) = \sup_{\mathbf{w} \in \underline{U}} g(\underline{u}, \underline{w})\}$ for $\underline{u} \in \underline{U}$. We saw in Section 3.4 that finding an equilibrium point is equivalent to finding a fixed point for the point-to-set mapping $\Psi(\underline{u})$ on $\underline{U}$.

Theorem 4.6. *In a supermodular game $(M, (U_j, \psi_j; j \in M))$, where $\Psi(\underline{u})$ is upper-semicontinuous in $\underline{u}$, the set of equilibrium solutions is nonempty and there always exists a largest and smallest equilibrium. (Topkis 1979, Ref. [232].)*

---

[17]This example is reproduced from reference [232].
[18]Recall that the notation $[v_j, \mathbf{u}_{-j}]$ refers to the vector $\underline{u}$ with u_j replaced by v_j.

Proof.   Using Theorem 4.5, we can show that the point-to-set map $\Psi(\underline{u})$ is ascending in $\underline{u}$. Therefore, for each $\underline{u}$ there exists a least element for $\Psi(\underline{u})$, denoted $\underline{u}(\underline{u})$; a greatest element, denoted $\bar{u}(\underline{u})$; and both these functions are increasing. By Tarski's fixed-point theorem 4.4, the increasing function $\bar{u}(\underline{u})$ admits a greatest fixed point $\bar{v} = \bar{u}(\bar{v})$. Since, by construction, $\bar{u}(\bar{v}) \in \Psi(\bar{v})$, then $\bar{v}$ is an equilibrium point.

Pick any equilibrium point $\bar{v}'$. Then $\bar{v}' \in \Psi(\bar{v}')$ and so $\bar{v}' \leq \bar{u}(\bar{v}')$. Thus $\bar{v}' \leq \sup\{\underline{u} \in \underline{U} : \underline{u} \leq \bar{u}(\underline{u})\} = \bar{v}$.

The existence of a least-equilibrium point is obtained in a similar way. $\square$

So, we have established that pure strategy Nash equilibria exist in supermodular games.

Corollary 4.1. *Topkis has also shown that the above theorem implies the following facts:*

(1) The largest and smallest strategies compatible with iterated strict dominance, rationalizability, correlated equilibrium and Nash equilibrium are the same.

(2) If a supermodular game has a unique Nash equilibrium, then it is dominance solvable, and many learning or adjustment rules will converge to it (e.g., best-response dynamics).

4.7   What have we learned in this chapter?

- We have learned how to define correlated equilibrium solutions and how to use linear programming to find all the correlated equilibria in an m-matrix game.

- We have seen how to use Bayesian probability theory to define equilibrium solution in games of incomplete information.

- We have studied a refinement Nash-equilibrium solution, called subgame perfectness, which can be formulated when we use a multiagent representation of the game.

- We have learned how to define quantal-response equilibrium solution for games where the players' preferences are subject to random shocks. This concept introduces robustness to the equilibrium solution concept.

- We have proved a theorem about the existence of Nash equilibria in games, which do not have to be played on convex strategy sets and with concave payoffs, as the existence theorem in Chapter 3 assumed. Instead, a more general lattice structure of the strategy space and the supermodularity of the payoff functions are sufficient to guarantee the existence of equilibria with a smallest and largest equilibrium solution.

4.8 Exercises

Exercise 4.1. In example 4.1, a correlated equilibrium was constructed for the game

	c_1	c_2
r_1	5,1	0,0
r_2	4,4	1,5

Find the associated extensive-form game for which the proposed correlated equilibrium corresponds to a Nash equilibrium.

Exercise 4.2. Using Excel and the solver tool, solve the LPs defined in subsection 4.2.3.3.

Exercise 4.3. Consider the set $X = \{(0,0), (0,1), (1,0), (2,2)\} \subset \mathbf{R}^2$ illustrated in Figure 4.10. Prove that X is a complete lattice.

Exercise 4.4. Correlated equilibrium

(1) Show that a correlated equilibrium exists in every matrix game.
(2) Show that all convex combinations of mixed Nash equilibria are correlated equilibria.

Exercise 4.5. Building a signaling device

Consider the following two-player game:

		Player 2	
		L	R
Player 1	U	9,9	6,10
	D	10,6	0,0

(1) Find all pure and mixed Nash equilibria and their associated payoffs.
(2) Find a correlated equilibrium based on a coin flip and its associated payoff.
(3) Verify that the strategies used are equilibrium strategies.
(4) Does the perfectly correlated equilibrium have a punch in this example?
(5) Consider the following signaling device:

	L	R
U	α	$(1-\alpha)/2$
D	$(1-\alpha)/2$	0

For what values of α does the use of the signaling device lead to a correlated equilibrium that dominates the mixed-strategy Nash equilibrium?

Exercise 4.6. Entry game

The demand for solar energy panels has increased dramatically due to subsidies for solar-panel installation. Most of these solar panels are produced in China, where production costs are low. A new German firm has the know-how and wants to build a production plant in China, but is afraid that a Chinese incumbent may be contemplating a capacity investment as well. To be profitable, investment needs to be completed before the subsidies end. If both invest, overcapacity will drive down prices. The incumbent can estimate its investment cost very well, but it is well-known that the German entrant only knows that its own investment costs are either low ($c = 1$) or high ($c = 4$), with an equal probability, depending on the cooperation of local party officials. Profits are depicted in the payoff matrix below. Find the Bayesian Nash equilibrium of this game of *asymmetric* information.

		Entrant	
		Enter	Do not enter
Incumbent	Invest	-2,4-c	0,8-c
	Do not invest	2,4	0,6

Exercise 4.7. Somewhat competitive Ph.D. students

Let $M = \{1, \ldots, n\}$ be the set of students enrolled in a Ph.D. course on game theory. Grading is based on an exam and only the student(s) with the best score at this exam will get an A+. It is well-known that not every student derives the same satisfaction from getting an A+, and that the grade is increasing in the amount of effort invested in preparing for the exam. All students are equally intelligent, so that the A+ will be for the student(s) who work(s) the hardest. Student j's utility function is given by

$$u_j = \begin{cases} \theta_j - e_j \text{ if he has A+} \\ -e_j \text{ otherwise} \end{cases} \forall j \in M,$$

where θ_j is student j's type, which is assumed to be distributed over $[0, 1]$ and e_j is student j's effort.

Consider only the symmetric setting, where students of the same type will make the same effort. Assume that all players use a strategy of the form $e_j = b\theta_j^c$.

(1) If Player j uses strategy e_j, what is the probability that he gets an A+? Derive the expected utility function.
(2) Find the Bayesian Nash equilibrium and express the solution in terms of θ_j and n.
(3) Interpret the strategy. Do you think the grading scheme is fair if you know that, in some years, there are 5 students in the class and other years, there are 20?

Exercise 4.8. Political negotiations

Two countries are negotiating a climate agreement. The parties at the table have to make substantial sacrifices to reach a compromise. Environmental agreements are costly in terms of productivity and, if the deal is skewed, then one country may end up carrying the full burden. There is one last dispute and both negotiators must simultaneously decide whether to stand their ground or cave in. If they both cave in, the environmental benefit perceived by public E has to be weighed against the perceived economic loss P of implementing the scheme. If one country stands its ground but the other does not, the deal will be signed and the stubborn player goes home with an agreement that costs him nothing in terms of productivity, while the other player bears the full economic loss. If they both stand ground, they go home with a lesser environmental deal that does not cost as much in economic terms. Negotiations are not so much about the environment or the economics as about politics. Each country knows the political repercussions k that it will face in the next elections if only the lesser deal is signed (it may be seen as a failure or a victory), but is far less certain about the reaction in the other country. The political repercussions have an important impact on the payoffs and are commonly known to be high or low, with equal probability.

The payoffs of the two countries are as follows:

		Country 2	
		Cave in	Stand ground
Country 1	Cave in	$E - P, E - P$	$E - P, E$
	Stand ground	$E, E - P$	$\frac{E}{2} - \frac{P}{2} + k, \frac{E}{2} - \frac{P}{2} + k$

(1) Draw the game tree that represents this game in extensive form.
(2) Assume that $E = 8; P = 8, k_{high} = -12$ and $k_{low} = 4$. Find the pure-strategy Bayesian Nash equilibria of this game.
(3) Argue that there cannot be any mixed-strategy Bayesian Nash equilibria, using dominance and/or rationalizability arguments.
(4) An environmental non-governmental organization says that if, it were known with certainty that k is low for both countries, then a good deal could never be reached. Is that true?

Exercise 4.9. Campesinos in Guatemala

In 2008, the European Union imposed a regulation that all imported animal products had to live up to the HACCP (Hazard Analysis and Critical Control Points) standard. HACPP is a preventive process quality standard. While honey is not a typical animal product, it falls under the HACPP designed for animal products. Even though their product was of excellent quality, this turned out to be a serious problem for local beekeeping cooperatives in Guatemala that export honey to Europe via a fair-trade organization, called *Maya Honey* because it required them to build an expensive processing plant. Such a plant is only cost-efficient if higher volumes of honey than what the cooperatives can individually produce, are processed. Suppose that there are only two cooperatives, *Mieles del Sur* and *Campesinos*. If neither builds the plant, they will have to sell locally, at lower prices. If they both build the plant, overcapacity will result and they will both be in serious trouble. If only one cooperative builds a plant, it becomes a monopolist processor and can charge a steep fee for processing the other cooperative's honey. Both cooperatives have an incentive to invest and both claim to have the means to do so. The payoffs in the table below represent the reduction in profits due to the HACCP regulation.

		Campesinos	
		B	NB
Mieles del Sur	B	-100,-100	-5,-9
	NB	-9,-5	-10,-10

(1) Find the pure and mixed strategies of this game.

(2) Before deciding on the investment, the cooperatives talk to the director of Maya Honey. The director proposes to publicly flip a coin. If the result is heads, then Mieles del Sur will build, if it turns out to be tails, then Campesinos will build. Is it interesting for the cooperatives to accept his proposal? Do they actually need the director to draw up this plan? Draw the extensive-form game tree.

(3) Clearly, the director prefers a quick solution to the problem. Since he is in Europe, he proposes that he flips a coin and communicate the results by telephone but Campesinos suggests that it would be better if the cooperatives got together to flip a coin so that they can both observe the result. Why do you think Campesinos makes this suggestion?

(4) The director replies that the cooperatives can achieve a better result than with a coin flip. He suggests they let him construct an imperfectly correlated device that would send a private signal instead. Is the director right? Can the cooperatives construct such a device without him (for example by programming a computer to construct the signal)?

(5) When the cooperatives get together to flip the coin, Mieles del Sur announces (and proves) that it is able to finance its project a little cheaper than previously thought (see payoff matrix below) and concludes that it should obviously be the one to invest.

		Campesinos	
		B	NB
Mieles del Sur	B	-99,-100	-4,-9
	NB	-9,-5	-10,-10

(a) Suppose that the director favors efficiency. What is the minimal subsidy that the director would have to offer to Campesinos to be sure that Mieles del Sur builds? The subsidy is announced before the decisions are made and may be conditional on the actions of both players.

(b) Suppose that the director favors fairness. What minimal subsidy for the non-builder would make each cooperative equally well-off in the mixed Nash equilibrium?

(c) Is it still possible to construct a signaling device? If yes, what is the effect on the mixed Nash fairness subsidy?

(6) Suppose that a portion of each cooperative's investment cost k is not known with certainty. It is common knowledge that k is either 0 or 1 with equal probability in the matrix game below. Find all the Bayesian Nash equilibria of this game.

		Campesinos	
		B	NB
Mieles del Sur	B	-99-k,-99-k	-4-k,-9
	NB	-9,-4-k	-10,-10

4.9 GE Illustration: A correlated equilibrium model of the strategic allocation of emission allowances in the EU

We now present an illustration of the use of the correlated equilibrium in a series of game-engineering exercises related to the allocation of emission allowances to European Union countries, in the context of the Emission Trading Scheme implemented in these countries. We take our inspiration from [237] where a two-level m-matrix game is designed with countries, or groups of countries as players whose payoffs are the welfare gains of these countries. These players can have a strategic influence on one another's decisions through their allocation of emission allowances, in an international emission trading system. The concept of correlated equilibrium is used to find a solution to this game in the context of EU negotiations.

4.9.1 *The issue of the strategic allocation of allowances*

The two-level game model provides an assessment of the strategic allocation of greenhouse-gas emission allowances in the EU-wide market that was implemented following the Kyoto agreement. It is well-established that a market for emission allowances is an efficient way to implement the abatements recommended by the Kyoto agreement [51, 174, 244]. Economic theory tells us that the way the emission allowances are initially allocated among the different agents in the economy does not matter, in terms of the global welfare effect: Pareto efficiency is achieved irrespective of the initial allocation [47]. However there are market imperfections in the European economies that can challenge the efficiency of trading. In [16], it is shown that some countries can be worse off with trading than without, because of pre-existing tax distortions. Another source of imperfection comes from the limitation of the emission-permits market to some sectors of the economy. These market imperfections may create a situation where some dominant countries strategize their allocation of allowances.

Since the European economies are closely linked, and also open to the rest of the world, the consequences of these strategic choices must be evaluated through the use of a world-wide multi-country computable general equilibrium (CGE) model. The CGE model provides an evaluation of the countries' (or groups') welfare gains when the emission-trading market is implemented. We shall therefore identify a set of m-matrix games, where m represents the possible sets of strategic players and where the payoff matrices are obtained as solutions to the general economic-equilibrium problems, which are associated with the different strategy choices made by the m players. Once an m-matrix game is identified, we look for its solution as defined by a correlated equilibrium [11]. Each correlated equilibrium is obtained as a solution of a linear program.

4.9.2 *Correlated equilibria in a two-level game*

The games are designed in the following way:

The players Countries or groups of countries that may strategize the allocation of allowances.

The strategies Different (contrasted) allocation schemes that can be chosen by these players.

The payoffs The welfare gains (or losses) for each player resulting from the Kyoto emission targets under the EU-wide trading regime.

We say that these games have a two-level structure since the strategies selected by the players have consequences that are calculated from a computable general equilibrium model, as summarized in Figure 4.13 below.

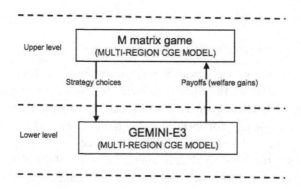

Fig. 4.13   The two-level game structure

The upper-level game will therefore be represented as an m-matrix game, where m is the number of active countries and where the payoffs are obtained by running GEMINI-E3 under the configurations associated with the different possible strategy choices. The upper-level game will be solved through the characterization of the set of *correlated equilibria*.

The game in strategic form is defined by $\Gamma = (M, (S_j)_{j \in M}, (u_j)_{j \in M})$ where M is the set of players, S_j is the set of strategies of Player j and $u_j : \Pi_{i \in M} S_i \to \mathbf{R}$ is the payoff function of Player j. Let us denote $\underline{s} = \Pi_{i \in M} S_i$ and $\mathbf{s} = (s_j)_{j \in M} \in \underline{s}$.

Recall, from Definition 4.2, that a correlated equilibrium is defined by a probability distribution on $\underline{s}$, $\pi(\underline{s}) \geq 0$, $\sum_{\underline{s} \in S} \pi(\underline{s}) = 1$, such that the following inequalities hold

$$\sum_{\underline{s} \in S} \pi(\underline{s}) u_j(\underline{s}) \geq \sum_{\underline{s} \in S} \pi(\underline{s}) u_j(\mathbf{s}_{-j}, \sigma_j), \quad \forall j \in M, \forall \sigma_j \in S_j, \tag{4.18}$$

where $(\mathbf{s}_{-j}, \sigma_j) = (s_i, \ldots, s_{j-1}, \sigma_j, s_{j+1}, \ldots, s_m)$ and $M = \{1, \ldots, m\}$.

The interpretation can be the following: through preplay communication, the agents may send signals to one other, which amounts to each player receiving a recommendation to play a given strategy. The probability $\pi(\underline{s})$ is assigned to the event: *the vector $\underline{s}$ is recommended as a way to play the game*. We have to realize that, in this interpretation, each player knows only the play recommendation that concerns him. The inequalities (4.18) express the fact that a player has no incentive to play other than as recommended. In short, we could view the correlated equilibrium either as the result of playing a game where a mediator sends private information to each player, in the form of a recommendation to play a given strategy, or as the result of the use of communication strategies by the players, where each player would send reports to the other players (and receive reports from the others) before deciding what to do. We can see the relevance of this scheme in the context of EU-wide negotiations for the implementation of a tradable emission-permit scheme.

We saw in Section 4.2 that the set of correlated equilibria in an m-matrix game is a polytope, where the extreme points can be obtained as solutions to linear programs. The set of correlated equilibria contains the set of Nash equilibria. Therefore, if this set of correlated equilibria reduces to a singleton, it is the unique Nash equilibrium for the game.

4.9.3 *The allocation game*

An m-matrix game is defined, where the strategic players are: Germany (DEU), the United Kingdom (UK), Italy (ITA), and the rest of the European Union (REU).[19] Each region has to choose an allocation rule, from among four different rules, in order to allocate the emission allowances across their own economic sectors. As stated by the European Commission, there are three basic approaches, based on (i) historical emissions, (ii) forecast emissions, and (iii) economic efficiency ("least cost" approaches) [69], respectively. We define our four possible allocation rules by considering two "historical emissions" approaches (grandfathering and historical emissions per sector) and two "least cost" approaches (domestic-tax and EU-tax):

- *Grandfathering (GF)*:
  Emission allowances are allocated among sectors according to their historical emissions taking into account a global target of emission reduction at the national level:

$$Q_i^{2010} = E_i^{2001} \times (1 - \text{obj}) \tag{4.19}$$

---

[19]We take Germany, the UK, and Italy as players because they account for around 30%, 17% and 14% of total EU emission quotas respectively, whereas France's emissions, for example, are only 6%.

where Q_i^{2010} are the emission allowances of sector i in 2010, E_i^{2001} represents the emissions of i in the reference year (2001), and `obj` corresponds to a reduction target (25%) applied to eligible sectors in all European countries.[20]

- *Historical emissions (HE)*:

  Under this approach, the total allowance given to a trading sector in a given country is determined by its share of the emissions due to all economic sectors included in the trading scheme, observed in a particular year (e.g., 2001), multiplied by total allowable emissions for the economy; this rule is defined as follows:

$$Q_i^{2010} = \frac{E_i^{2001}}{\sum_j E_j^{2001}} E_{\text{kyoto}}^{2010} \tag{4.20}$$

where E_{kyoto}^{2010} represents emission targets as defined in the Kyoto Protocol.

- *Domestic Tax-based (DT)*: According to this allocation rule, sectoral allowances correspond to those that would occur if a uniform carbon tax were to be implemented at the domestic level; allowances allocated to the trading sectors are defined as follows:

$$Q_i^{2010} = E_i^{\text{DT}} \tag{4.21}$$

where E_i^{DT} stands for the emission allowances for sector i under a uniform national tax that would meet the Kyoto targets.

- *European Tax-based (ET)*:

  According to this allocation rule, sectoral allowances correspond to the ones that would occur if a uniform carbon tax were to be implemented at the European level; allowances allocated to the trading sectors are defined as follows:

$$Q_i^{2010} = E_i^{\text{ET}} \tag{4.22}$$

where E_i^{ET} stands for the emission allowances for sector i under a uniform tax implemented at the European level to reach the aggregated Kyoto emission target.

As explained before, the payoffs are computed by Gemini.E3, a CGE model.[21] The literature has used a variety of possible measurements for the economic impact of climate policies: GDP, change in consumer surplus, discounted present value of consumption, and direct cost [166]. For many economists, the most relevant measurement is surplus, as originally defined by Dupuit. It is expressed in the modern welfare-economics literature by the Compensating Variation of Income (CVI) [25] . In GEMINI-E3, the welfare costs of climate policies are thus measured by the CVI.

Three games are simulated:

- **Game 1**: The 4 players can choose to allocate the emission allowances according to the domestic tax-based approach (DT), or deviate from this rule by giving 10% more (DT+10) or 10% less (DT-10) to the trading sector.

---

[20]The -25% emission target was defined so that the emissions allowances of the trading sectors (recall that all economic sectors are not part of the ETS) are comparable to what is obtained with the other rules (see Figure 4).
[21]GEMINI-E3 is a multi-country, multi-sectorial, dynamic-recursive CGE model that incorporates a highly detailed representation of indirect taxation. A full description of the model is provided in [25, 26].

- **Game 2**: The 4 players can choose to allocate the emission allowances according to the domestic tax-based approach (DT), the grandfathering approach (GF) or the European tax-based approach (ET).
- **Game 3**: The 4 players can choose to allocate the emission allowances according to the domestic tax-based approach (DT), the grandfathering approach (GF) or the historical approach (HE).

In Game 1, we assume that the reference rule for quota allocations corresponds to the one equalizing marginal abatement costs across sectors at the domestic level (DT). In a first-best world, there is no incentive for governments to depart from this allocation rule since welfare is maximized. However, in a second-best world characterized by pre-existing tax distortions, the welfare costs of climate policy might be reduced by reallocating some quotas toward the highly distorted sector [15]. In Game 1, we assess the incentive for EU countries to deviate from the DT allocation by giving quotas of 10% more (or less) to the trading sector. In Game 2, the players can choose among more varied strategies. We allow EU countries to deviate a little more from the DT approach by giving even greater (GF) or lesser (e.g., ET in Germany) quotas to the trading sector. In Game 3, the only way for EU countries to strategize the allocation of quotas is to give more to the trading sector; the GF approach corresponds to a lower deviation from the ET rule than the HE approach.

In order to get the payoffs for an m-matrix game with 4 players and 3 strategies, the GEMINI-E3 model must be run 81 times (3^4). This process is CPU-time-intensive given current software and hardware capabilities. For example, each run takes 30 minutes on a PC Pentium 4 CPU 2.4 GHz with 504 Mo of RAM. Hence, 40.5 hours are needed to solve each game. The computation of correlated equilibria, once the M-matrix game is identified, is very fast.

We can see in Figure 4.14 the emission allowances allocated to the trading sectors under the different allocation rules. We notice that the selected sectors receive greater allowances in all regions with the historical approach than with the other rules. The ranking of the other rules is different from one region to another. For Germany, emission-based approaches (GF and HE) tend to give more emission allowances to the trading sectors than the other rules (DT and ET) do. The European tax-based approach (ET) is the most restrictive for the trading sectors in Germany and the United Kingdom, but not for Italy and the other EU countries.

Table 4.12   Unique correlated equilibria

	Game 1	Game 2	Game 3
Germany	DT	DT	DT
United Kingdom	DT-10	GF	GF
Italy	DT+10	DT	HE
Rest of EU-15	DT+10	GF	GF

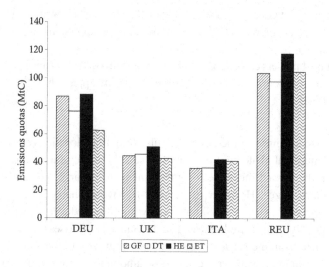

Fig. 4.14   Emission allowances by region under different allocation rules

Table 4.12 gives the unique strategy choices that are obtained when we look for correlated equilibria. It shows that the games have a unique correlated equilibrium, and that the equilibria are always different from the competitive equilibrium solution, where all the countries would play the same DT strategy. Germany, which is the main supplier of emission allowances, tends to rely on the domestic tax-based approach (DT) but the other regions, which are permit buyers, have an incentive to depart from this approach to maximize their own payoffs. In Game 1, Germany plays the domestic tax-based strategy (DT) whereas the United Kingdom decides to give reduced permits to the trading sectors (DT-10), and Italy and the other EU countries are more generous with their eligible sectors (DT+10). In Game 2, Germany and Italy allocate allowances according to the domestic tax-based approach (DT) whereas the United Kingdom and the rest of Europe opt for the grandfathering approach (GF). In Game 3, Germany opts again for the domestic tax-based approach (DT), the United Kingdom and the other EU countries choose the grandfathering approach (GF), and Italy uses the historical approach (HE).

Table 4.13 shows that the cooperative solutions, obtained when the players maximize their joint utility, are different from the correlated equilibria. The table also shows that the cooperative solutions of the games do not correspond to the uniform strategies (e.g., harmonized allocation policy among the regions). Because of market imperfections, the global welfare might be improved by adjusting the sectoral allocation of emission allowances [15, 16]. Table 4.14 shows that the outcomes of the noncooperative equilibria are globally lower than what the players would obtain by playing the cooperative solution. There is no strict dominance of the correlated equilibria by the cooperative solutions with equal weight. Germany and the United Kingdom are always worse off with the noncooperative solution but Italy and the rest of Europe are better off. Italy and the other EU countries thus have an incentive to depart from the cooperative solution.

Table 4.13   Cooperative solutions

	Game 1	Game 2	Game 3
Germany	DT+10	DT	DT
United Kingdom	DT-10	GF	GF
Italy	DT+10	DT	HE
Rest of EU-15	DT	DT	DT

Table 4.14   Payoffs in correlated equilibria versus cooperative solutions

	Cooperative equilibria			Correlated equilibria		
	Game 1	*Game 2*	*Game 3*	*Game 1*	*Game 2*	*Game 3*
Germany	1254	1308	953	599	808	526
United Kingdom	-1646	-1807	-1526	-1880	-2090	-1933
Italy	-2665	-2854	-2726	-2614	-2844	-2709
Rest of EU-15	-4384	-4578	-4482	-4072	-4117	-4095
Total	-7442	-7931	-7781	-7967	-8244	-8211

Figure 4.15 compares the welfare costs by region associated with the correlated equilibria. It is shown that an emission-permit system based on the domestic tax-based approach (Game 1) would be more efficient than other systems, where, e.g., the EU countries would opt for emissions-based approaches (e.g., Game 2 and Game 3). However, even if some countries might be tempted to depart from the domestic tax-based approach and give higher or lower permits to the trading sectors, the equilibrium obtained in Game 1 is better than the equilibria in Game 2 and 3. The gains from emission trading would be reduced in the selling countries (mainly Germany) but the costs would be reduced in the importing countries (UK, Italy, and REU). In other words, this solution might improve the political acceptability of the EU-wide carbon-emission market by limiting the distributive impact of emission trading.

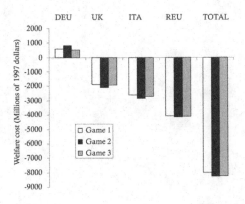

Fig. 4.15   Welfare costs in the three allocation games

PART 2
Deterministic Dynamic Games

Chapter 5

Repeated Games and Memory Strategies

5.1 Introduction

In this chapter, we begin an analysis of *dynamic games*. To define this type of game in an imprecise but intuitive way, we can say that a *dynamic game* is one whose conditions change over time. In a *repeated game*, change is caused by the accumulation of information about the history of the game. As time goes by, the information that the players have at their disposal increases and, since strategies map this (accruing) information into actions, the players' strategic choices are affected. If a game is repeated twice, i.e., the same game is played in two successive stages and a reward is obtained at each stage, it can be easily conceived that, in the second stage, the players can make decisions based on the outcome of the first stage. The situation becomes more complex as the number of stages increases, since the players can base their decisions on the game *histories* represented by sequences of actions and outcomes observed over the increasing number of stages.

We will also consider games that are played over an *infinite* number of stages. These infinitely repeated games are particularly interesting since, at each stage, there are always an infinite number of stages remaining to be played. Therefore, there is no "end-of-horizon effect," which, in finite-horizon games, critically influences the players' behavior. In infinite horizon games, the strategies must be assessed by appraising the infinite streams of rewards. We shall explore different ways of doing this.

In summary, although it is the same game that is considered at each period, the resulting repeated game becomes a fully dynamic system with a complex strategic structure. In this chapter, we will consider repeated bimatrix games and repeated concave m-player games. We will show that the class of equilibria in repeated games expands if we allow for *memory strategies*, i.e., if the players can use information about how the game has been played so far in shaping their current decisions. The class of memory strategies permits players to incorporate *threats* into their strategic choices. In order to be effective, a threat must be *credible*. We will explore the credibility issue and show how the subgame-perfectness property of equilibria introduces a refinement criterion for determining what that makes these threats credible.

We will focus on an interesting result called the *folk theorem*, which shows that the class of equilibria in repeated games can be very large. We will refer to this theorem on various

133

occasions. In the first formulation of this theorem, we will consider an infinitely repeated bimatrix game played with the help of finite automata, which are finite-memory logic machines used to implement strategies. For this type of games, we show that there exists a class of equilibria such that any individually rational outcome of the one-stage bimatrix game can be approximated as closely as desired by the average payoff obtained from one of these equilibria. Further in this chapter, a similar theorem is proved for repeated concave games played over an infinite time horizon, with payoffs consisting of the discounted sum of periodic gains. This result will be even more powerful than the preceding one, since it will show the existence of equilibria that are also Pareto optimal (or *efficient*) and that have the property of being subgame perfect. Both results are based on the definition of memory strategies that include retaliation threats.

In a repeated game, the players' actions have no direct influence on the periodic normal-form games that will be played in forthcoming periods. A more complex dynamic structure is obtained if we assume that the players' actions at one stage influence the type of game to be played in the next stage. This is the case for the class of *multistage* and *Markov games* and, a fortiori concerns the class of *differential games*, where time flows continuously. These fully dynamic games will be discussed in subsequent chapters.

5.2 Repeating a game in normal form

5.2.1 *Representations of the game*

In a repeated game, the inter-temporal (or dynamic) strategy is a way of selecting a stage strategy for each period, on the basis of the knowledge accumulated by the players about the game's history. In fact, repetition of the same game in normal form permits the players to adapt their strategies to the observed history of the game. In particular, they have the opportunity to implement a class of strategies that incorporate threats.

Repeated games have been also called *games in semi-extensive form* by J.W. Friedman [86]. They are called "extensive" because of game's dynamic information structure, and "semi" because a game in normal form is defined at each point in time. The payoff associated with a repeated game is usually defined as the sum of the rewards obtained in each period. However, when the number of periods tends to ∞, the total payoff understood as an infinite sum may not converge to a finite value. There are several ways of dealing with the comparison of infinite streams of payoffs. We will introduce some of them in the subsequent sections.

5.2.2 *Repeated bimatrix games*

Figure 5.1 describes the structure of a repeated bimatrix game. The same bimatrix game is played repeatedly, over T *periods* (or *stages*) that represent the passing of time. In the one-stage game, Player 1 has p pure strategies and Player 2 q pure strategies. The one-stage reward pair associated with the strategy pair (k, ℓ) is given by $(\alpha_{k\ell}^j)_{j=1,2}$. The rewards are accumulated over time.

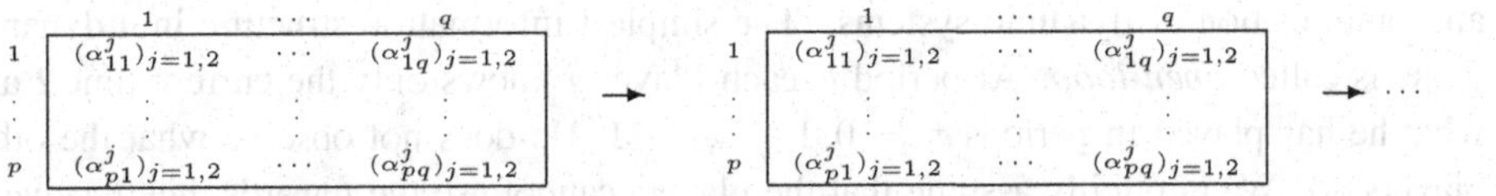

	1	$\cdots$	q
1	$(\alpha_{11}^j)_{j=1,2}$	$\cdots$	$(\alpha_{1q}^j)_{j=1,2}$
$\vdots$	$\vdots$		$\vdots$
p	$(\alpha_{p1}^j)_{j=1,2}$	$\cdots$	$(\alpha_{pq}^j)_{j=1,2}$

$\rightarrow$

	1	$\cdots$	q
1	$(\alpha_{11}^j)_{j=1,2}$	$\cdots$	$(\alpha_{1q}^j)_{j=1,2}$
$\vdots$	$\vdots$		$\vdots$
p	$(\alpha_{p1}^j)_{j=1,2}$	$\cdots$	$(\alpha_{pq}^j)_{j=1,2}$

$\rightarrow \cdots$

Fig. 5.1   A repeated game structure

The information structure corresponds to *perfect information*. So, the players may recall the past history of the game. Notice that the extensive form of those repeated games will result in a complex game tree with a very large or even infinite[1] number of nodes and branches. As stated, the game could be described in a semi-extensive form, i.e., at each period, a normal form (an already aggregated description) will define the consequences of the strategic choices in a one-stage game, as is alluded to in Figure 5.1. Even if the same game seems to be played at each period (it is similar to the others in its normal form description), in fact, the possibility of using the history of the game as a source of information to adapt the strategic choice at each stage makes a repeated game much more complex than an elementary game that is repeated.

5.2.3 *Repeated concave games*

In addition to repeated m-matrix games we will also consider the class of dynamic games where a concave game (as defined in Section 3.4, page 55), is repeated. Let $t \in \{0, 1, \ldots, T-1\}$ be the time index. At each period t the players play a noncooperative game defined by

$$(M; U_j, \psi_j^t(\cdot), j \in M)$$

where (compare Section 3.4), $M = \{1, 2, \ldots, m\}$ is the set of players, U_j is a compact convex set describing the actions available to Player j at each period and $\psi_j^t(\underline{u}(t))$ is the payoff to Player j in period t, when the action vector $\underline{u}(t) = (u_1(t), \ldots, u_m(t)) \in \underline{U} = U_1 \times U_2 \times \cdots \times U_m$ is chosen at period t by the m players. This function is assumed to be concave in $u_j(t)$. We will use $\tilde{\underline{u}} = (\underline{u}(t) : t = 0, 1 \ldots, T-1)$ to denote the action sequence over all T periods. The total payoff of Player j over the T-horizon is then defined by

$$V_j^T(\tilde{\underline{u}}) = \sum_{t=0}^{T-1} \psi_j^t(\underline{u}(t)). \tag{5.1}$$

Open-loop information structure.   We will consider some typical information structures that influence how a dynamic game can be played. Their names, *open-loop* and *closed-loop*[2] are borrowed from control theory, which is the research area dedicated to the design

---

[1]When T becomes infinite.

[2]The *feedback* or *Markov* information structure will be also discussed in Chapter 5.7.5, where we deal with Markov games.

and optimization of dynamic systems. The simplest information structure in a dynamic game is called *open-loop*. At period t, each Player j knows only the current time t and what he has played in periods $\tau = 0, 1, \ldots, t - 1$. He does not observe what the other players do. We implicitly assume that the players cannot use the rewards they receive in each period to infer (e.g., through a filtering procedure) the other players' actions. The open-loop information structure actually eliminates almost every aspect of the dynamic structure of the repeated games. In brief, in an open-loop information structure a player selects an action stream $\tilde{\underline{u}}$ without receiving any information on the other players' action choices. This is indeed the simplest information structure possible.

Closed-loop information structure. At the other end of the continuum of players' knowledge of game history, we have the *closed-loop* information structure, where each Player j knows not only the current time t and what he has played in periods $\tau = 0, 1, \ldots, t - 1$, but also what the other players did in all previous periods. We call *history of the game at period τ* the sequence

$$h(\tau) \equiv (\underline{u}(t) : t = 0, 1 \ldots, \tau - 1).$$

A closed-loop strategy for Player j is thus defined as a sequence of mappings

$$\gamma_j^\tau : h(\tau) \mapsto U_j. \tag{5.2}$$

Due to the repetition of the same game over time, each Player j can adjust his choice of a one-stage action u_j according to the history of the game. This makes it possible to consider memory strategies where threats are included in the announced strategies. For example, a player can declare that some particular histories would trigger a *retaliation* on his part. The description of a *trigger strategy* will therefore include

- a nominal mode of play that contributes to the expected or desired outcomes;
- a retaliatory mode of play that is used as a threat;
- the set of histories that trigger a switch from the nominal mode of play to the retaliatory mode of play.

Notice that the implementation of this sort of strategy requires some form of pre-play communication between players to define the nominal mode of play.

Many authors have studied equilibria obtained in repeated games through the use of *trigger strategies* (see for example, Refs. [87, 206] and [207]). Often, these studies are related to the folk theorem, which we will present shortly.

5.2.4 *Payoffs for infinite-horizon games*

If the game is repeated over an infinite number of periods, then payoffs are represented by infinite streams of rewards

$$\{\psi_j(\underline{u}(t)) : t = 1, 2, \ldots, \infty\}.$$

In this chapter, we will consider two ways to deal with such infinite streams to define a global scalar payoff.

Discounted sum of rewards. We have assumed that the action sets U_j are compact and the functions $\psi_j^t(\cdot)$ are continuous. Therefore the one-stage rewards $\psi_j^t(\underline{u}(t))$ are bounded. If we assume that they are also uniformly bounded,[3] the discounted payoffs

$$V_j(\underline{\tilde{u}}) = \sum_{t=0}^{\infty} (\beta_j)^t \psi_j^t(\underline{u}(t)), \tag{5.3}$$

are well defined, where $\beta_j \in [0, 1)$ is the discount factor of Player j.

Average rewards per period. Assume that the reward functions are stationary, i.e., that they do not change over time,

$$\psi_j^t(\cdot) \equiv \psi_j(\cdot).$$

If the players do not discount their reward streams, then the comparison of strategies implies the consideration of infinite streams of rewards that sum up to infinity. A way to circumvent the difficulty this creates is to consider a *limit average reward per period* in the following way:

$$g_j(\underline{\tilde{u}}) = \liminf_{T \to \infty} \frac{1}{T} \sum_{t=0}^{T-1} \psi_j(\underline{u}(t)). \tag{5.4}$$

This criterion is known in mathematics as the *Cesaro limit* of the sequence of rewards $\{\psi_j(\underline{u}(t)) : t = 1 \ldots \infty\}$.

To link this method of evaluating infinite streams of rewards to the previous one based on discounted sums, we define the *equivalent discounted constant reward*:

$$g_j^{\beta_j}(\underline{\tilde{u}}) = (1 - \beta_j) \sum_{t=0}^{\infty} (\beta_j)^t \psi_j(\underline{u}(t)). \tag{5.5}$$

Since the following holds

$$\sum_{t=0}^{\infty} (\beta_j)^t = \frac{1}{1 - \beta_j}, \tag{5.6}$$

we have the following equality

$$\sum_{t=0}^{\infty} (\beta_j)^t g_j^{\beta_j}(\underline{\tilde{u}}) = \sum_{t=0}^{\infty} (\beta_j)^t \psi_j(\underline{u}(t)); \tag{5.7}$$

hence, the name of *"equivalent discounted constant reward"* for the expression $g_j^{\beta_j}(\underline{\tilde{u}})$.

We expect that, for many instances of these games, the following limiting property will hold true:

$$\lim_{\beta_j \to 1} g_j^{\beta_j}(\underline{\tilde{u}}) = g_j(\underline{\tilde{u}}). \tag{5.8}$$

This limit result is sometime called *Tauberian theorem*.[4]

---

[3]This will be indeed the case if all the reward functions are identical $\psi_j^t(\cdot) \equiv \psi_j(\cdot)$.

[4]In broad terms, Tauberian theorems deal with the sums of a class of non-convergent series. For a recent presentation of the results in this area, which concern a general dynamic programming framework, we refer the reader to Ref. [151].

5.2.5   Threats

In infinite-horizon repeated games each player always has *enough time* to retaliate. He will always have the possibility of implementing an announced threat if the opponents are not playing according to some *nominal sequence* of moves agreed upon by the players before playing (refer to trigger strategies introduced in Section 5.2.3). We will see that the equilibria based on the use of threats constitute a very large class. Therefore the set of closed-loop strategies with a memory information structure is much larger than of open-loop strategies, which often only admits the trivial equilibrium consisting of repeating the static Nash equilibrium at each period.

5.2.6   *Existence of a trivial dynamic equilibrium*

We know, through Rosen's existence theorem (Theorem 3.6, page 61), that a concave static game

$$(M; U_j, \psi_j(\cdot), j \in M)$$

admits an equilibrium $\underline{u}^*$. It should be clear (see Exercise 4.1) that the sequence $\tilde{\underline{u}}^* = (\underline{u}(t) \equiv \underline{u}^* : t = 0, 1 \ldots, T - 1)$ is also an equilibrium in the repeated game in both the open-loop and closed-loop (memory) information structures.

Remark 5.1. Assume that the one-period game is such that a unique equilibrium exists. The open-loop repeated game will also have a unique equilibrium, whereas the closed-loop repeated game will have a plethora of equilibria, as we will see shortly.

5.3   The folk theorem

In this section we present the well-known *folk theorem*, so named because the authorship of the first version, which appeared around 1950, is not well defined.

5.3.1   *Repeated games played by automata*

The idea is that an infinitely repeated game should permit the players to design equilibria, supported by threats, with Pareto-efficient outcomes. Presumably, a well-designed (and credible) threat can entice the other players to cooperate. The equilibrium strategies incorporating threats in such a game can be memory-based because each player can remember the game history.

Because of the infinite horizon, there may arise a need to store an infinite amount of information , which is clearly not feasible. This raises the question of whether we can design strategies that would exploit the history of the game, but with finite information storage capacity. Finite automata offer one way of designing such strategies.

A *finite automaton* is a logical system that, when stimulated by an *input*, generates an *output*. For example an automaton could be used by a player to determine the stage-game action he takes, given the information he has received. The automaton is finite if the length of the logical input (a stream of $0, 1$ bits) it can store is finite. So in our repeated-game context, a finite automaton will not permit a player to base his stage-game action upon an infinite memory of what has happened before. To describe a finite automaton, we need the following:

- A list of all possible stored input configurations; this list is finite and each element is called a *state*. One element of this list is the *initial state*.
- An output function determines the action taken by the automaton, given its current state.
- A transition function tells the automaton how to move from one state to another after it has received a new input element.

In our paradigm of a repeated game played by automata, the input element for automaton a of Player 1 is the stage-game action of Player 2 in the preceding stage; and, those of Player 2 are symmetrical and his actions are determined by automaton b. In this way, each player can choose a way to process the information he receives in the course of the repeated plays, in order to decide on the next action.

An important aspect of a repeated game played by finite automata is that they necessarily generate cycles in the successive visited states, and hence, in the actions taken in the successive stage-games.

Let G be a finite two-player game, i.e., a bimatrix game that defines the one-stage repeated game. Sets U and V contain the pure strategies in game G for Player 1 and 2, respectively. Denote by $g_j(u, v)$ the payoff to Player j in the one-stage game when the pure strategy pair $(u, v) \in U \times V$ has been selected by the two players.

The game is repeated indefinitely. The set of strategies for an infinitely repeated game is enormously rich, but we will restrict the players' strategy choices to the class of finite automata. A pure strategy for Player 1 is an automaton's choice $a \in A$, whose inputs are the actions $v \in V$ of Player 2. Symmetrically, a pure strategy for Player 2 is an automaton's choice $b \in B$, whose inputs are the actions $u \in U$ of Player 1. We associate with the pair of finite automata $(a, b) \in A \times B$ a pair of *cycle average payoffs* per stage, defined as follows:

$$\tilde{g}_j(a, b) = \frac{1}{N} \sum_{n=1}^{N} g_j(u_n, v_n) \quad j = 1, 2, \qquad (5.9)$$

where N is the length of the cycle associated with the pair of automata (a, b) and (u_n, v_n) is the action pair at the n-th stage in this cycle. We notice that the expression (5.9) is also the limit average reward per period due to the cycling behavior of the two automata. We call $\mathcal{G}$ the game defined by the strategy sets A and B and the payoff functions (5.9).

5.3.2 Minimax point and the dominating outcomes

Assume Player 1 wants to threaten Player 2. An effective threat would be to define his action in a one-stage game through the solution of

$$\bar{m}_2 = \min_u \max_v g_2(u, v)$$

where $\bar{m}_2$ is the worst-case payoff for Player 2.

Similarly if Player 2 wants to threaten Player 1, an effective threat would be to define his action in a one-stage game through the solution of

$$\bar{m}_1 = \min_v \max_u g_1(u, v).$$

We call *minimax point* in the payoff space the pair

$$\bar{m} = (\bar{m}_1, \bar{m}_2) \in \mathbf{R}^2. \tag{5.10}$$

Before proceeding any further, we must recall the following definition in topology:

Definition 5.1. Let A and B be subsets of $\mathbf{R}^n$. The set A is *dense* in the set B, if, for any $b \in B$, we can find a sequence $\{a_n\}$ in A, such that

$$\lim_{n \to \infty} a_n = b.$$

Theorem 5.1. *Let $\mathcal{P}_{\bar{m}}$ be the set of outcomes in a matrix game that dominate the minimax point $\bar{m}$. Then, the set of payoffs corresponding to Nash equilibria in pure strategies*[5] *of the game $\mathcal{G}$ is dense in $\mathcal{P}_{\bar{m}}$.*

Proof.   Let $\mathbf{g}_1, \mathbf{g}_2, \ldots, \mathbf{g}_K$ be the pairs of payoffs that appear in the bimatrix game if we write them in vector form, that is,

$$\mathbf{g}_k = (g_1(u_k, v_k), g_2(u_k, v_k)) \tag{5.11}$$

where $(u_k, v_k) \in U \times V$ is a possible pure strategy pair. Let $q_1, q_2, \ldots, q_K$ be nonnegative *rational numbers* that add up to 1. Then the convex combination

$$\mathbf{g}^* = q_1 \mathbf{g}_1 + q_2 \mathbf{g}_2, \ldots, q_K \mathbf{g}_K \tag{5.12}$$

is an element of $\mathcal{P}_{\bar{m}}$ if $g_1^* \geq \bar{m}_1$ and $g_2^* \geq \bar{m}_2$.

First, we notice that the set of these vectors $\mathbf{g}$ is dense in $\mathcal{P}_{\bar{m}}$, since the set of rational numbers is dense in $\mathbf{R}$ and any element of $\mathcal{P}_{\bar{m}}$ is obtained as a convex combination (5.12) with the coefficients in $\mathbf{R}$. Since each q_k is a rational number, it can be written as a ratio of two integers, that is, as a fraction. It is possible to write the K fractions with a common denominator, hence,

$$q_k = \frac{n_k}{N},$$

with

$$n_1 + n_2 + \cdots + n_K = N$$

since the q_k's must sum to 1.

---

[5] A pure strategy in game $\mathcal{G}$ is a deterministic choice made by an automaton; this automaton will process information coming from the repeated use of mixed strategies in the repeated game.

We construct two automata a^* and b^*, such that, together, they play (u_1, v_1) during n_1 stages, (u_2, v_2) during n_2 stages, etc. They complete the sequence by playing (u_K, v_K) during n_K stages and the N-stage cycle begins again. The cycle average payoff of Player j in the $\mathcal{G}$ game played by these automata is given by

$$\tilde{g}_j(a^*, b^*) = \frac{1}{N} \sum_{k=1}^{K} n_k g_j(u_k, v_k) = \sum_{k=1}^{K} q_k g_j(u_k, v_k) = g_j^*.$$

Therefore, these two automata will achieve the payoff pair g.

Then, we refine the structure of these automata as follows:

(1) Let $\tilde{u}(n)$ and $\tilde{v}(n)$ be the pure strategy that Player 1 and Player 2 respectively are supposed to play at stage n, according to the above-defined cycle.
(2) Automaton a^* plays $\tilde{u}(n+1)$ at stage $n+1$ if automaton b has played $\tilde{v}(n)$ at stage n; otherwise, it plays $\bar{u}$, which is the strategy that is min-maxes Player 2's one-shot game payoff, and this stage strategy is played forever.
(3) Symmetrically, automaton b^* plays $\tilde{v}(n+1)$ at stage $n+1$ if automaton a has played $\tilde{u}(n)$ at stage n; otherwise, it plays $\bar{v}$, the strategy that min-maxes Player 1's one-shot game payoff, and this stage strategy is played forever.

Now, it should be clear that the two automata a^* and b^*, defined as above, constitute a Nash equilibrium for game $\mathcal{G}$ whenever g is dominating $\bar{m}$, i.e., $g \in \mathcal{P}_{\bar{m}}$. Indeed, if Player 1 unilaterally changes his automaton to a, then, the automaton b^* will select for all periods except a finite number, the minimax action $\bar{v}$. Consequently, the limit average per period reward obtained by Player 1 is

$$\tilde{g}_1(a, b^*) \leq \bar{m}_1 \leq \tilde{g}_1(a^*, b^*).$$

Similarly for Player 2, a unilateral change to $b \in B$ leads to a payoff

$$\tilde{g}_2(a^*, b) \leq \bar{m}_2 \leq \tilde{g}_2(a^*, b^*).$$

$\square$

Remark 5.2. The attentive reader will have noticed that the threat strategy pair $(\bar{u}, \bar{v})$ is not a Nash equilibrium. Therefore, if and when the threat is used, it is not optimal for each player (it is not the best reply) to continue to use his threat. There is therefore a potential credibility issue in using these threats.

If we imposed a condition that the threats used by the players in their announced strategies had to be credible, then the class of equilibria defined through the above refinements (1)–(3) would be more restricted. We will see that if the threats themselves are Nash equilibria, then, they are credible. These issues will be discussed further on, when we look at the topic of *subgame perfectness*.

5.4   Memory-strategy equilibria in a repeated concave game

In this section we consider repeated m player concave games and we prove for this class of games a similar result to that obtained in Theorem 5.1.

Here, we deal with a situation where m players play, in a succession of periods a concave game as it was defined in Section 3.4. The result will be derived for a problem formulated as an oligopoly game. The theory extends without much change to any concave game.

5.4.1   *Infinite-horizon and Nash-equilibrium threats*

Consider an oligopoly game à la Cournot[6] repeated over an infinite sequence of periods. Let $t = 1, \ldots, \infty$ be the sequence of time periods, and $\underline{q}(t) \in \mathbf{R}^m$, the output-production vector chosen by the m firms at period t. The payoff to Player j in period t corresponding to output $\underline{q}(t)$ is $\psi_j(\underline{q}(t))$.

We denote

$$\tilde{\underline{q}} = \underline{q}(1), \ldots, \underline{q}(t), \ldots$$

the infinite sequence of output decisions of the m firms. Over the infinite time horizon, the players' payoffs are defined as the discounted sums of rewards

$$\tilde{V}_j(\tilde{\underline{q}}) = \sum_{t=0}^{\infty} \beta_j^t \psi_j(\underline{q}(t))$$

where $\beta_j \in [0, 1)$ is the discount factor used by Player j.

We assume that, at each period t, all players have the same information $h(t)$, which is the history of previous outputs decided individually by the competing firms

$$h(t) = (\underline{q}(0), \underline{q}(1), \ldots, \underline{q}(t-1)).$$

Notice that this information vector will become infinite as time goes by. We will deal with this difficulty in the last paragraph of this section.

We consider two possible outcomes for the one-stage game:

(i) The unique[7] Nash (Cournot) equilibrium $\underline{q}^c = (q_j^c)_{j=1,\ldots,m}$; and
(ii) A Pareto-efficient outcome resulting from output levels $\underline{q}^* = (q_j^*)_{j=1,\ldots,m}$, such that $\psi_j(\underline{q}^*) \geq \psi_j(\underline{q}^c)$, $j = 1, \ldots, m$. We say that this Pareto outcome, dominating the Cournot equilibrium, is a *cooperative outcome*.

An obvious equilibrium for the repeated game is defined by the strategy that consists of each Player j repeatedly choosing the output levels $q_j(t) \equiv q_j^c$, $j = 1, 2$, of the one-stage Cournot solution. However, there are many other equilibria that can be obtained through the use of memory strategies.

---

[6]Briefly, the Cournot oligopoly game is a concave game that consists of m players whose payoffs $\psi_j(\cdot)$ depend on the vector of outputs $q = (q_1, \ldots, q_m) \in \mathbf{R}^m$. Readers who are unfamiliar with Cournot games are referred to Exercise 5.3, on page 147.

[7]We assume uniqueness; see Section 3.4.4, page 61.

Let us define for each player

$$\tilde{V}_j^* = \sum_{t=0}^{\infty} \beta_j^t \psi_j(\underline{q}^*), \tag{5.13}$$

$$\Phi_j(\underline{q}^*) = \max_{q_j} \psi_j([q_j, \mathbf{q}_{-j}^*]), \tag{5.14}$$

where, as usual, $\mathbf{q}_{-j}^*$ denotes the vector $(q_i^*)_{i \neq j}$ of output at equilibrium for all players other than j. The following result was proved by J.W. Friedman in [86]:

Lemma 5.1. *There exists an equilibrium for the repeated Cournot game that yields the payoffs $\tilde{V}_j^*$ if the following inequality holds,*

$$\beta_j \geq \frac{\Phi_j(\underline{q}^*) - \psi_j(\underline{q}^*)}{\Phi_j(\underline{q}^*) - \psi_j(\underline{q}^c)}. \tag{5.15}$$

This equilibrium is reached through the use of a so-called trigger strategy, defined as follows:

$$\begin{cases} if \quad (\underline{q}(0), \underline{q}(1), \ldots, \underline{q}(t-1)) = (\underline{q}^*, \underline{q}^*, \ldots, \underline{q}^*) \ then \ q_j(t) = q_j^* \\ otherwise \ \ q_j(t) = q_j^c. \end{cases} \tag{5.16}$$

Proof. If the players use $\underline{q}^*$ then the payoff they obtain is given by[8]

$$\tilde{V}_j^* = \sum_{t=0}^{\infty} \beta_j^t \psi_j(\underline{q}^*) = \frac{1}{1 - \beta_j} \psi_j(\underline{q}^*).$$

Assume Player j decides to deviate unilaterally at period 0. He knows that his deviation will be detected at period 1 and that, thereafter, the Cournot-equilibrium output level $\underline{q}^c$ will be played. So, the best he can hope to obtain from his deviation is the following payoff, which combines the maximal reward he can get in period 0, when the other firms play q_k^*, with the discounted value of an infinite stream of Cournot outcomes:[9]

$$\tilde{V}_j(\tilde{q}) = \Phi_j(\underline{q}^*) + \frac{\beta_j}{1 - \beta_j} \psi_j(\underline{q}^c).$$

This unilateral deviation is not profitable if

$$\tilde{V}_j(\tilde{q}) \leq \tilde{V}_j^* = \sum_{t=0}^{\infty} \beta_j^t \psi_j(\underline{q}^*) = \frac{1}{1 - \beta_j} \psi_j(\underline{q}^*),$$

that is, if the following inequality holds

$$\psi_j(\underline{q}^*) - \Phi_j(\underline{q}^*) + \frac{\beta_j}{1 - \beta_j} (\psi_j(\underline{q}^*) - \psi_j(\underline{q}^c)) \geq 0,$$

which can be rewritten as

$$(1 - \beta_j)(\psi_j(\underline{q}^*) - \Phi_j(\underline{q}^*)) + \beta_j(\psi_j(\underline{q}^*) - \psi_j(\underline{q}^c)) \geq 0,$$

---

[8]Here we use the fact that the $\sum_{k=0}^{\infty} \beta^k = \frac{1}{1-\beta}$ when $\beta < 1$.

[9]Here we use the fact that the $\sum_{k=1}^{\infty} \beta^k = \frac{\beta}{1-\beta}$ when $\beta < 1$.

from which we get

$$\psi_j(\underline{q}^*) - \Phi_j(\underline{q}^*) - \beta_j(\psi_j(\underline{q}^c) - \Phi_j(\underline{q}^*)) \geq 0,$$

and, finally, obtain condition (5.15):

$$\beta_j \geq \frac{\Phi_j(\underline{q}^*) - \psi_j(\underline{q}^*)}{\Phi_j(\underline{q}^*) - \psi_j(\underline{q}^c)}.$$

The same reasoning holds if the deviation occurs at any other period t. $\qquad\square$

Remark 5.3. According to this strategy, the $\underline{q}^*$ output levels correspond to a cooperative mode of play while the $\underline{q}^c$ correspond to a *punitive* mode of play. The punitive mode is a threat. Since the punitive mode of play consists of playing the Cournot solution, which itself is an equilibrium, playing $\underline{q}^c$ is a credible threat. Therefore the players will always choose to play in accordance with the Pareto solution $\underline{q}^*$ and thus, the equilibrium (5.16) defines a nondominated outcome.

Definition 5.2. We will call *collusive equilibrium* the pair of strategies defined through (5.16), which guarantee the nondominated payoffs $\tilde{V}_j^*, j = 1, 2, \ldots m$.

Information storage. The trigger strategy (5.16) requests information on the outputs until $t - 1$, which might imply infinite information storage. However, it is easy to define finite automata to implement this strategy, by simply introducing a *state variable* (call it $\zeta \in \{0, 1\}$) that evolves over time according to the following logic:

$$\zeta(0) = 0$$
$$\zeta(t) = 0 \text{ if } \zeta(t - 1) = 0 \text{ and } \underline{q}(t) = \underline{q}^*$$
$$\zeta(t) = 1 \text{ otherwise.}$$

This is indeed a finite-memory automaton. Now all that is required to implement the trigger strategy is for Player j to use the following automaton, also with finite storage:

$$q_j(t) = q^* \text{ if } \zeta(t) = 0$$
$$q_j(t) = q^c \text{ if } \zeta(t) = 1.$$

The variable ζ is an indicator of the mode of play. The value 0 indicates a cooperative mode and 1, a retaliatory mode of play.

5.4.2 Subgame perfectness

It is important to notice the difference between the threats used in the folk theorem in Section 5.3 and those used in the repeated Cournot game in Section 5.4. In the folk theorem, the threats are the most effective ones in the sense that they aim at maximizing the damage done to the opponent. However, as previously mentioned, these threats lack credibility. This is because implementing them is not the best course of action for the threatening player, given the opponent behavior.

Importantly, in the infinitely repeated Cournot game, threats also constitute an equilibrium for the repeated game. The threats are credible because if they are applied by every player, they constitute the best course of action for all, i.e., they are the optimal reactions to the actions of the opponents. Since that is the case, then we say that the memory (trigger strategy) equilibrium (5.16) is *subgame perfect* in the Selten sense.

The concept of subgame perfectness was proposed and precisely defined by Selten [218], in the realm of the extensive-form representation of a game. Above, we showed its usefulness for refining the equilibrium in the context of a repeated game. In dynamic games, where decisions are represented by continuous variables, the extensive game formalism does not apply. We will later see that the concept of subgame perfectness is also relevant for the dynamic-programming approach and for dynamic games.

5.4.3 *Finite versus infinite-horizon games*

There is an important difference between finitely and infinitely repeated games. If the same game is repeated only a finite number of times (T) then, even with memory strategies, a single game equilibrium will be repeated at each period. This is because retaliation is not possible in the last period.[10] This is then translated backward in time until the initial period.

When the game is repeated over an infinite horizon, there is always enough time to retaliate and possibly, even to resume cooperation. For example, let us consider a repeated Cournot game without discounting, where the players' payoffs are determined by the long-term average of the one-stage rewards. If the players have *perfect information with a delay*,[11] they can observe without error the moves selected by their opponents in previous stages. Actually, there might be a delay of several stages before the observations are realized. In that case, it is easy to show that, to any Pareto-efficient (cooperative) solution dominating a static Nash equilibrium, there corresponds a subgame-perfect equilibrium for the repeated game, based on the use of trigger strategies. These strategies threaten to switch to the noncooperative solution for all remaining periods as soon as a player is observed to depart from the cooperative solution. Since the long-term average only depends on the infinitely repeated one-stage rewards, it is clear that the threat gives rise to the Cournot-equilibrium payoff in the long run. This is the subgame-perfect version of the folk theorem [87].

5.5 What have we learned in this chapter?

In this first chapter that deals with *dynamic games*, we explored a class of strategies called memory- or trigger-strategies. In particular, we studied two classes of repeated games:

(1) A two-player bimatrix game that is played repeatedly over an infinite time horizon;

---

[10]In the oligopoly game played on the finite horizon T, it is impossible for player i to retaliate *after* player j has deviated from an agreed-upon policy in period T, i.e., he did not play $q_j(T) = q_j^*$.

[11]This means that one or several stages or periods pass before the players get the information.

(2) A repeated m-player concave game with payoffs obtained as a discounted sum of rewards over an infinite number of periods.

- We showed how to construct a game played by finite automata, which admits a large class of equilibria—so large, in fact, that any pair of rewards dominating the minmax point (defined in (5.10)) can be approximated, as closely as desired, by the cycle average payoff of an equilibrium of the game played by finite automata.
- We studied the equilibrium property that is based on the integration of a threat into each player's trigger strategy, which corresponds to a search for the minmax reward of his opponent. However, this way of threatening has an important drawback. The pair of threats do not constitute an equilibrium, and therefore, suffer from a lack of credibility. If threats were to be used, it would not be in the best interest of each player to stick to the announced threat strategy.
- We showed for the second class of games, that it is possible to design an equilibrium in the class of trigger strategies, where the threat consists of forever playing the one-stage Nash equilibrium, if the discount factor is close enough to 1.
- The fact that threat strategies constitute equilibrium induces a property called *subgame perfectness*, which is required to make trigger strategies credible.

5.6 Exercises

Exercise 5.1. Prove that the repeated one-stage equilibrium is an equilibrium for the repeated game with additive payoffs and a finite number of periods. Extend the result to infinitely repeated games.

Exercise 5.2. Show that condition (5.15) holds if $\beta_j = 1$. Adapt Lemma 5.1 to the case where the game is evaluated through the long-term average-reward criterion.

Exercise 5.3. Consider a duopoly made up of two symmetric firms. Denote by q_i the quantity produced by firm i at a constant marginal cost c and by Q the total quantity available on the market, i.e., $Q = q_1 + q_2$. Let $P(Q)$ be the inverse-demand function specified as follows:

$$P(Q) = \max\{\alpha - (q_1 + q_2), 0\},$$

where α is a strictly positive parameter. The firms are profit maximizers so that their payoffs are given by

$$\pi_i(q_1, q_2) = (q_i - c) P(Q), \quad i = 1, 2.$$

Suppose that each player has the choice between two actions, namely, collusion (C) or defection (D). When both firms collude, then they produce the monopoly output. If they play D, they produce the Cournot outputs. The resulting payoffs are as follows:[12]

---

[12]For the reader who is not familiar with Cournot duopoly, it is instructive to go through the details of how payoffs are determined in the different cells. Suppose first that the two players collude and maximize their joint profit.

Table 5.1 Payoff matrix

		Firm 2	
		C	D
Firm 1	C	$\frac{(\alpha-c)^2}{8}, \frac{(\alpha-c)^2}{8}$	$\frac{3(\alpha-c)^2}{32}, \frac{9(\alpha-c)^2}{64}$
	D	$\frac{9(\alpha-c)^2}{64}, \frac{3(\alpha-c)^2}{32}$	$\frac{(\alpha-c)^2}{9}, \frac{(\alpha-c)^2}{9}$

It is easy to verify that the two players face a prisoner's dilemma. Indeed, the unique Nash equilibrium is (D, D) and both players are better off cooperating. Denote by r the common discount rate ($0 < r < 1$) and by β the discount factor ($\beta = \frac{1}{1+r}$).

(1) Suppose that the players do not discount their future profits and assume that the stage game described above is played twice. Draw the tree representing the two-period game in extensive form and show that the unique subgame-perfect Nash equilibrium is (D, D) in each stage.
(2) Would the result change if the game were repeated any finite number of times?
(3) Suppose that the players discount their future profits. Would this affect your answers to the two previous questions?
(4) Suppose that the players do not discount their future profits and that there is uncertainty

Assuming that they split the monopoly quantity equally (this makes sense because of the symmetry), then the results are as follows:

$$Q^C = \frac{\alpha - c}{2} \text{ with } q_1^C = q_2^C = \frac{\alpha - c}{4}, \quad P^M = \frac{\alpha + c}{2}, \quad \pi_1^M = \pi_2^M = \frac{(\alpha - c)^2}{8}.$$

In the absence of cooperation, the two firms compete à la Cournot. Assuming an interior solution, first-order equilibrium conditions are

$$\frac{\partial \pi_1(q_1, q_2)}{\partial q_1} = \alpha - 2q_1 - q_2 - c = 0 \Leftrightarrow q_1 = \frac{\alpha - q_2 - c}{2}, \quad (5.17)$$

$$\frac{\partial \pi_2(q_1, q_2)}{\partial q_2} = \alpha - 2q_2 - q_1 - c = 0 \Leftrightarrow q_2 = \frac{\alpha - q_1 - c}{2}. \quad (5.18)$$

It is easy to verify that the Nash equilibrium quantities, price and profits are given by (the superscript D stands for Defection)

$$q_1^D = \frac{\alpha - c}{3}, \quad q_2^D = \frac{\alpha - c}{3}, \quad P^D = \frac{\alpha + 2c}{3}, \quad \pi_1^D = \pi_2^D = \frac{(\alpha - c)^2}{9}.$$

To compute the payoffs when one player, say Player 2, implements his part of the collusion agreement whereas Player 1 deviates, we use the reaction function (5.17) and substitute for $q_2^C = \frac{1-c}{4}$. The resulting equilibrium is given by

$$q_1^D = \frac{3(\alpha - c)}{8}, \quad q_2^C = \frac{\alpha - c}{4}, P = \frac{3\alpha + 5c}{8}, \quad \pi_1^{DC} = \frac{9(\alpha - c)^2}{64}, \quad \pi_2^{DC} = \frac{3(\alpha - c)^2}{32}.$$

Inverting the roles of the two players, leads by symmetry to

$$q_1^C = \frac{\alpha - c}{4}, \quad q_2^D = \frac{3(\alpha - c)}{8}, P = \frac{3\alpha + 5c}{8}, \quad \pi_1^{CD} = \frac{3(\alpha - c)^2}{32}, \quad \pi_2^{CD} = \frac{3(\alpha - c)^2}{64}.$$

We collect these results in the payoff matrix.

about how many times the stage game will be repeated. Assume that both players believe, at the beginning of each period, regardless of how many periods they have played the stage game, that the probability that the game will end at that period is p. Show that cooperation between the two players is now feasible. (Hint: introduce memory strategies where each player chooses his output at each period—assuming that nature has allowed the game to reach this period— as function of the history of the game.)

(5) Redo the same analysis as above but assuming that the players discount their future profits at different rates.

(6) Why does the equilibrium change so drastically when uncertainty is introduced? (Hint: work out your argument thinking in terms of backward induction.)

(7) Suppose again that the game is repeated twice and that there is no discounting. Denote by q_{it} player i's output in period $t = 1, 2$, with

$$q_{it}^C = \frac{\alpha - c}{4}, \quad q_{it}^D = \frac{\alpha - c}{3}, \quad i = 1, 2, \ t = 1, 2.$$

Suppose that each firm announces at the outset of the game its first-period output q_{i1} and its second-period output level as a function of its rival's output in the first period, that is, $q_{i2}(q_{j1}), i, j = 1, 2, i \neq j$. Let $\bar{q}$ be an output level satisfying

$$\frac{\alpha - c}{4} < \bar{q} < \frac{\alpha - c}{3}, \quad i = 1, 2$$

i.e., a quantity lower than the collusion output and higher than the defection output. Consider the following pair of strategies:

$$\left(q_{11} = \bar{q}, \quad q_{12}(q_{21}) = \left\{ \begin{array}{ll} q_{22}^D & \text{if } q_{21} = \bar{q} \\ q^F > q_{12}^D & \text{otherwise} \end{array} \right. \right),$$

$$\left(q_{21} = \bar{q}, \quad q_{22}(q_{11}) = \left\{ \begin{array}{ll} q_{22}^D & \text{if } q_{21} = \bar{q} \\ q^F > q_{22}^D & \text{otherwise} \end{array} \right. \right).$$

For player i, this retaliation strategy reads, "*I will partially cooperate in the first period. If I observe that my rival did the same in the first period, then I will produce the Cournot level in the second period. Otherwise, I will flood the market and produce q^F in the second period.*"

(a) Suppose that $q^F = \frac{2\alpha + c}{3}$ and $\bar{q} = \frac{7(\alpha - c)}{12}$, that is, $\bar{q} = 0.5 q_{it}^C + 0.5 q_{it}^D$. Show that the equilibrium with the above retaliation strategies leads to higher profits than (D, D) in each stage. Is this equilibrium subgame perfect?

(b) Find ranges for $\bar{q}$ and q^F for which this partial cooperation is an equilibrium.

(c) Discuss the results in terms of the folk theorem.

Exercise 5.4. The managers of two neighboring supermarkets must decide weekly on the price of a loss-leader product (i.e., a popular product that, when discounted, attracts a lot (of non-regular) customers to the store, e.g., *Coca Cola*). Both players can choose either to promote the product and offer it at a low price or sell it at the regular price. The associated payoffs are shown in Table 5.2. If both supermarkets select the high price, then each makes

Table 5.2   Payoffs

		Firm 2	
		Low price	Regular price
Firm 1	Low price	10, 10	40, 0
	Regular price	0, 40	20, 20

a profit of 20. Similarly, if both supermarkets choose the low price, then each ends up with a profit of 10. However, if one supermarket offers the product at a low price while the other does not discount it, then the first realizes a profit of 40 and the second does not make any profit. It is easy to verify that the unique Nash equilibrium of the weekly game is both firms choosing the low price, while the Pareto-optimal solution is when both select the regular price. Denote by r the common discount rate ($0 < r < 1$) and by β the discount factor ($\beta = \frac{1}{1+r}$). Suppose that the game is infinitely repeated and that players maximize their discounted stream of profits. Assume that the players use trigger strategies in order to sustain the collusive outcome of 20 each.

(1) Suppose that both firms adopt the *grim-trigger strategy*, that is, that they collude and choose the regular price in the first period and will continue to do so until one of the players cheats on the agreement and defects. After that, they play the Nash-equilibrium strategy for the rest of the game. Establish a condition on the discount factor β for collusion to be sustainable.

(2) Suppose that both firms adopt the *tit-for-tat strategy*, that is, that they collude in the first period and, in future periods, each player selects the action chosen by the competitor in the previous period. Establish a condition on the discount factor β for collusion to be sustainable. Show that *tit-for-tat* is not subgame perfect.

(3) Suppose that both firms adopt the *limited-retaliation strategy*, that is, that they start colluding and then defect for k periods for every defection by any player, followed by reverting to collusion independently of their behavior during the punishment phase. Establish conditions on the discount factor β and on k for collusion to be sustainable.

(4) Suppose that both firms adopt the *win-stay, lose-shift strategy* (also known as Pavlovian strategy), that is, that they collude in the first period and then cooperate at period t if the outcome in period $t - 1$ was either (C, C) or (D, D). Establish a condition on the discount factor β for collusion to be sustainable.

(5) Characterize the Nash equilibrium of this repeated game when one player uses the *tit-for-tat strategy* and the other, the *win-stay, lose-shift strategy*. Is this equilibrium subgame perfect?

5.7   GE illustration: The European vitamin-C cartel

In this GE illustration, we model a collusive agreement between two firms, using the framework of repeated games.

5.7.1 Background

On November 21, 2001, the European Commission sentenced eight firms to pay 855,220,000 euros for the formation of an illegal cartel in the vitamin markets between September 1989 and February 1999. Besides price agreements, there appeared to be agreements on sales quotas, regional segmentation (see [48]) and coordinated price increases. In some markets, formal management structures, the exchange of market data, and common budgeting procedures were used to monitor the agreements.

The stability of the eight-firm cartel can largely be attributed to favorable market conditions. In vitamin markets, homogeneous products are sold to atomistic buyers. Economies of scale for capital- and knowledge-intensive production methods constitute high entry barriers. As such, the cartel was able to supply 70% to 100% of the market.

In this GE Illustration, we focus on the vitamin-C market and the on German export market, for tractability reasons. About 90% of sales worldwide were sealed by only *two* firms (BASF AG and Merck KgaA) (see Ref. [20]); so, we can use the duopoly model, discussed earlier in this chapter, to analyze the game.

Why did the cartel eventually break down? Levinstein and Suslov (Ref. [158]) suggest that the sudden emergence of new Chinese competitors that captured a 30% market share by September 1995, was a major catalyst. The smaller, but state-controlled, Chinese firms overcame entry barriers by using an innovative and less scale-dependent, single-fermentation procedure to produce vitamin C. Since these fringe firms were initially tolerated, they were able to grow very fast by undercutting the collusive price.

5.7.2 The model

To make the game more realistic, we assume that firms can only imperfectly deduce the quantity produced by their cartel partner, from the market price. The production of fringe competitors and imperfect forecasts (because of market volatility) motivate this assumption. Such imperfect monitoring provides an incentive to cheat, but if the volatility is known, firms can still design an optimal trigger strategy that discourages cheating on the agreement. Green and Porter [96] show that every market price below an optimal threshold (even if it is due to sheer chance) must be met with retaliation, in order to deter cheating. Their model proposes that periodic price wars are a device that sustains cooperation over the long-term. If a new entrant can drive down the price enough, the incumbents' collusive agreements will break down.

Denote two rational, expectation-maximizing firms i and j. The firms cannot observe the total supply but they can deduce it from the market price under complete information: in a market equilibrium demand $D(P)$ equals supply Q

$$D(P) = Q = q_i + q_j.$$

We assume that the (inverse) demand function is static and common knowledge. We use the demand and (symmetric) cost functions estimated for US firms in de Roos (Ref. [54]):[13]

$$P = 1805 - 25Q + \epsilon,$$

$$c_i = c_j = 5,$$

where $\epsilon \overset{i.d.d.}{\sim} N(0, \sigma^2)$ reflects an imperfect correlation between price and quantity. This is also common knowledge.

5.7.3 *Complete information*

5.7.3.1 *One-shot game*

For simplicity, let us first assume that $\epsilon = 0$. Each firm can perfectly deduce the quantity produced by its cartel partner from the demand function. In other words, we suppose *complete monitoring* or a closed-loop information structure (with observable actions), in the parlance of Section 5.2.3. In a one-shot game, the Cournot quantity and realized profit for each firm would be

$$q_i^N = q_j^N = 24,$$

$$\pi_i^N = \pi_j^N = 14400.$$

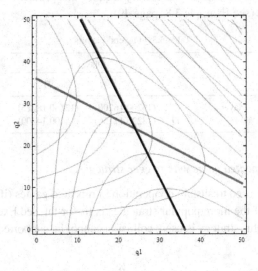

Fig. 5.2 Reaction functions go through the maximum of the value functions and cross at the Cournot-Nash equilibrium

This is depicted in Figure 5.2. In an infinitely repeated game, with a yearly discount factor of $\beta_y = .98$ and monthly price decisions (again as in de Roos (Ref. [54]), the equivalent constant discounted reward is given by

$$\beta = 0.998,$$

$$(1 - \beta)\pi_i^N = (1 - \beta)\pi_j^N = 14400.$$

---

[13]In fact, we simplified their estimate of $Q = 72.41 - 0.039P$.

This compares to the cartel output and its equivalent constant discounted reward profit as follows:

$$q^C = 36,$$
$$(1 - \beta)\pi^C = 32400.$$

As the game is fully symmetric, we equally divide the overall total quantity and the total per-period profit over the firms, i.e.,

$$q_i^C = q_j^C = 18,$$
$$(1 - \beta)\pi_i^C = (1 - \beta)\pi_j^C = 16200.$$

The prices under collusion and in the Cournot equilibrium are given by

$$P^C = 1805 - 25 \cdot 36 = 905 > P^N = 1805 - 25 \cdot 48 = 605.$$

Collusion does not increase the price by 30% as stated in the introduction but by about 50%. This suggests that in reality the firms may not be making the most of collusion.

The table below reports the payoffs in the one-shot game for different strategies, namely, when both firms collude (C), when both firms deviate (D), and when one firm colludes while the other deviates.[14] We clearly identify this problem as another instance of a prisoner's dilemma (see Example 3.8, page 49).

Table 5.3   Players' payoffs

		Player 1	
		C	D
Player 2	C	(16200,16200)	(12150,18225)
	D	(18225,12150)	(14400,14400)

5.7.3.2 *Infinitely repeated game with trigger strategies*

Trigger strategies are used to sustain cooperation in repeated games (if the discount rate is not too high). One of the more popular strategies is tit-for-tat, see Exercise 5.4. It can be proved[15] that, when this strategy is used, collusion cannot be supported if

$$\beta \leq \frac{1}{3}.$$

Otherwise, cooperation is strictly preferred. However, tit-for-tat is *not* subgame perfect. If we evaluated a history where Player 2 has not always cooperated (e.g., because of a glitch),

---

[14]If player i plays (C), the best response of player j is to produce $q_j = 27$. The resulting payoffs are given by

$$\pi_j = (1800 - 25Q)q_i = 18225,$$
$$\pi_i = (1800 - 25Q)q_j = 12150.$$

[15]It is a good exercice for the reader to show these results.

we would see that Player 1 could have done better using another strategy. By contrast, another popular strategy, which is to use a grim trigger

$$s_i(h^t) = \begin{cases} C \text{ if } t = 0 \\ C \text{ if } a_j^\tau = (C, C), \forall \tau = 0, 1, ..., t-1 \\ D \text{ otherwise,} \end{cases}$$

is both a Nash equilibrium and subgame perfect. Both strategies are limits of a more general *limited retaliation* strategy (i.e., cooperate at $t = 0$; when there is a defection, retaliate for T periods; then go back to cooperation), which we use in what follows.

5.7.4 *Incomplete monitoring*

Suppose now that demand fluctuates with $\epsilon \neq 0$. Green and Porter [96] show that, under *incomplete monitoring*, periodic price wars are necessary to stabilize cartels. (Before their work, such a price war was interpreted as the end of a cartel [226].) Intuitively, demand can be lower than expected under incomplete monitoring either because a firm cheated or because demand was temporary lower by chance. If there is no reaction to such a low price, this provides an incentive for future cheating. Therefore, retaliation must always[16] follow low prices, even when they are not caused by deviation. This is similar to complete monitoring where each firm weighs the instant benefit of a deviation against its future loss due to retaliation and compares this to the long-term benefit of cooperation.

We follow Green and Porter [96] and consider only simple trigger strategies.[17] Suppose that players have agreed to play the limited retaliation strategies $(p^R, T, q_i^N, q_i = R(q_j))$ and $(p^R, T, q_j^N, q_j = R(q_i))$. This notation implies that when the price drops below p^R, both firms retaliate by playing (q_i^N, q_j^N) during $T - 1$ retaliation periods; otherwise, they play the cooperative output (q_i, q_j). Both firms begin by cooperating. No rational firm will deviate (if β is sufficiently high) unless a slowdown in demand causes both firms to retaliate at the same time.

Green and Porter [96] identify the expected present value with the following recursive value function

$$V_i(q_i, q_j) = \pi_i(q_i, q_j) + \Pr(p^R \leq P + \epsilon)\beta V_i(q_i, q_j)$$
$$+ \Pr(p^R > P + \epsilon) \left[\beta \pi_i(q_i^N, q_j^N) + ... + \beta^{T-1}\pi_i(q_i^N, q_j^N) + \beta^T V_i(q_i, q_j)\right].$$

The first term renders the payoff when both firms start out cooperating. If $P \geq p^R$ then they will also cooperate in the next period (second term). If $P < p^R$ then the firms retaliate during weeks 1 to T before reverting back to cooperation (last term).

Let us specify additive market uncertainty as $\epsilon \overset{i.d.d.}{\sim} N(0, 100)$. The cumulative density function $F(p^R - P)$ denotes the probability that the market price is lower than p^R or $\Pr(\epsilon < p^R - P)$

$$F(p^R - P) = F(p^R - 1805 + 25q_i + 25q_j) = \frac{1}{2}\left(1 + \text{erf}\left(\frac{p^R - 1805 + 25q_i + 25q_j}{100\sqrt{2}}\right)\right).$$

---

[16]This will be unnecessary if a special monitoring device has been engineered; compare definition (5.2) and equation (6.343) in Chapter 5.7.5.

[17]While they have the benefit of simplicity, these strategies are not subgame perfect. This was shown by Abreu et al. [1] who derived optimal symmetric sequential equilibria.

From the case of complete information, we know that $\beta = .998$, $\pi_i(q_i^N, q_j^N) = 14400$ and $\pi_i(q_i, q_j) = (1800 - 25q_i - 25q_j)q_i$. In fact, if $F(\cdot) = 0$ (market price is always lower than p^R) then we have the cooperative case. After some simplifications, we have

$$V_i(q_i, q_j) = \frac{14400}{1 - .998} + \frac{(1800 - 25q_i - 25q_j)q_i - 14400}{(1 - .998) + (.998 - .998^T)\frac{1}{2}\left(1 + \mathrm{erf}\left(\frac{p^R - 1805 + 25q_i + 25q_j}{100\sqrt{2}}\right)\right)}$$

We depict the reaction correspondences for $p^R = 632$ and $T = 152$ in Figure 5.3.

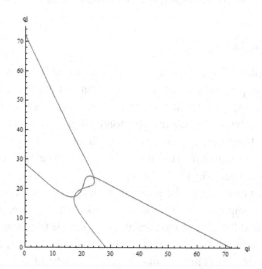

Fig. 5.3 Reaction functions for $p^R = 632$ and $T = 20$ give rise to three Nash equilibria

These reaction functions are not linear and depict three Nash equilibria, expected to occur for many possible parameter setups. One obvious equilibrium is to produce the Cournot output $(24, 24)$ forever. The firms do better if they agree to produce the output leading to a Pareto-optimal Nash equilibrium (*collusive*, see definition 5.2). The obtained equilibria were computed by making use of the symmetry assumptions.

Finally, both firms have to agree on p^R and T. These can be chosen optimally, using a grid search procedure. We calculate the value function in the Pareto-optimal Nash equilibrium for every combination of p^R and for T, i.e., $V^*(p^R, T)$ and simply select the highest one. A fine grid-search indicates[18] that $T^* \to \infty$, $P^* \approx 539$ for $q^{PNE} = 18.9009$ and $V = 15721.2$, which are quite close to the complete-monitoring outcomes.

5.7.5 Conclusion

These results show that, under our parametrization, it is optimal for both players to play a variation of the grim-trigger strategy.[19] As such, the Green and Porter [96] hypothesis

---

[18]Our numerical approximation may not be sensitive enough to discriminate any p^R between 530 and 565, but this has no practical importance.

[19]In fact, we derived a corner solution due to the shape of the distribution (see [203]).

actually reinforces the Stigler hypothesis: since price wars are not temporary, they cannot act as stabalizers; rather, they announce cartel breakup. Agreeing on a very low p^R and a $T \to \infty$ means that the risk of getting caught is low (p^R) but that the punishment is very severe (T). For our parametrization, such an agreement turns out to be more effective than relatively weak punishment with stricter monitoring. Intuitively, severe punishment deters cheating even if the probability of being caught is very low. Higher p^R, but not higher T, increases the probability of *false positives* in detection. Setting p^R very low reduces the probability ($< .001$) that price wars will break out when nobody cheats.

For $q = 18.9009$ the average price is 860, i.e., 42% above the competitive price. This is close to the 30% estimate of de Roos [54]. We have not attempted to accurately estimate price volatility.

We conclude that our model provides an explanation for the cartel breakdown. The entry of Chinese fringe firms (progressively affecting the ability of the dominant cartel to set uncompetitive prices) may not have been enough. Accurate predictions would have led to renegotiation and to somewhat less-stable cartel agreements. We believe that it was the sudden and unpredicted drop in prices that occurred in 1995 that triggered an immediate (and in this case permanent) cartel breakdown.

Chapter 6

Multistage Games

In this chapter, we introduce the concept of a dynamic game played by several agents who share control of a dynamic system observed over stages, in discrete time. Such systems are described by multistage *state equations*[1] and we call such games *multistate*. (In Chapter 7, we will deal with their continuous-time counterparts, called *differential games*, where the systems' dynamics are described by differential equations.) In the present chapter, we will first recall some fundamental properties of dynamic systems described in a state space. Then, we will define equilibrium solutions under different information structures and characterize such solutions through either a coupled maximum principle or dynamic programming. The chapter will conclude with an analysis of equilibrium solutions of infinite-horizon systems, where we consider the possibility of including threats in the strategies used by the players, as a way to transform some cooperative solutions into equilibrium.

6.1 Multistage control systems

6.1.1 *System description in a state space*

Dynamic systems can be defined and studied in a *state space*.[2] A state space contains *state variables*, which provide an exhaustive summary of how different input variable have impacted on the system in the past. By knowing the current "state" and the future time profiles of the input variables, we are able to predict the future behavior of a dynamic system.

Depending on the representation of the time domain, a system's dynamics can be described in discrete or continuous time, which then defines *state equations* as difference equations or differential equations, respectively. The former case is studied in this chapter; the latter will be considered in Chapter 7.

A discrete-time dynamic system described in a *state space* can be formally defined as follows:

---

[1] We explain this notion below in Definition 6.1.

[2] The games will be described and solved using the tools of optimal-control theory. For an easy introduction to optimal control, we refer the interested reader to the classical books of Bryson and Ho [36] and Desoer and Zadeh [250].

Definition 6.1. A dynamic system is described by multistage state equations when (i) the system is observed at discrete time periods $t = 0, 1, \ldots, T$; (ii) the system is characterized at period t by a state variable[3] $\mathbf{x}(t) \in \mathbf{R}^n$ and a control variable $\mathbf{u}(t) \in \mathbf{R}^p$;(iii) the evolution of the state variable is described by a first-order difference equation

$$\mathbf{x}(t+1) = \mathbf{f}(\mathbf{x}(t), \mathbf{u}(t), t) \tag{6.1}$$

where $f : \mathbf{R}^n \times \mathbf{R}^p \times \mathbf{R} \to \mathbf{R}^n$ is a given function; (iv) there is an initial condition for the state variable.

$$\mathbf{x}(0) = \mathbf{x}^0. \tag{6.2}$$

As a consequence of this definition, we know that we can always compute the system's state at time t, $\mathbf{x}(t)$ if we know an initial condition $\mathbf{x}(0) = \mathbf{x}^0$ and control $\mathbf{u}(\tau)$, $\tau = 0, 1, \ldots, t-1$. We call $\tilde{\mathbf{u}}^T = (\mathbf{u}(0), \ldots, \mathbf{u}(T-1))$ the control sequence and $\tilde{\mathbf{x}}^T = (\mathbf{x}(0), \ldots, \mathbf{x}(T))$ the state trajectory generated from $\mathbf{x}(0) = \mathbf{x}^0$ by this control sequence.

6.1.2   *Open-loop and closed-loop control structures*

6.1.2.1   *Definitions*

Consider the case where a single agent is in charge of controlling a dynamic system described by the state equation (6.1). Control can be carried out in two different modes: *open-loop* and *closed-loop*.

Knowing the initial state $\mathbf{x}(0) = \mathbf{x}^0$ and specifying a control sequence $\tilde{\mathbf{u}}^t$, a unique well-defined state trajectory $\tilde{\mathbf{x}}^t$ will result from (6.1). We then say that the system is controlled in open-loop. We sometimes use the following notation to denote an open-loop control:

$$\mathbf{u}(t) = \mu(\mathbf{x}^0, t). \tag{6.3}$$

This emphasizes the fact that the control sequence has been chosen with knowledge of the initial state $\mathbf{x}^0$.

Instead of selecting a whole control sequence, the controller may decide to define the control $\mathbf{u}(t)$ as a function of the information on the system's evolution that is available at time t. This information could be, for example, the current state $\mathbf{x}(t)$ observed at time t. It could also be the past history of the system, meaning all the controls and state values observed in the previous periods, plus the current state. It could be a part of this history, etc. We then say that the system is controlled in closed-loop.

6.1.2.2   *A useful digression toward control engineering*

In control engineering, closed-loop control is related to the concept of *servomechanism*. Historically, the science of control and servomechanisms evolved from the 1788 design

---

[3]In the chapters where we deal with dynamic systems described by multiple state equations, we adopt a notation where vectors and matrices are in boldface style to distinguish them from scalars that are in regular style.

of a regulator device for steam engines by James Watt (a device also called a *centrifugal governor;* see Figure 6.1). The device operates as follows (quoting Wikipedia):

> *The device shown is from a steam engine. Power is supplied to the governor from the engine's output shaft by (in this instance) a belt or chain (not shown) connected to the lower belt wheel. The governor is connected to a throttle valve that regulates the flow of working fluid (steam) supplying the prime mover (prime mover not shown). As the speed of the prime mover increases, the central spindle of the governor rotates at a faster rate and the kinetic energy of the balls increases. This allows the two masses on lever arms to move outward and upward against gravity. If the motion goes far enough, this motion causes the lever arms to pull down on a thrust bearing, which moves a beam linkage, which reduces the aperture of a throttle valve. The rate of working-fluid entering the cylinder is thus reduced and the speed of the prime mover is controlled, preventing over-speeding.*

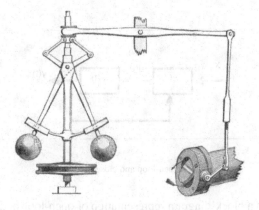

Fig. 6.1   Watt's centrifugal governor

So, in a servomechanism and, in particular, in Watt's centrifugal governor, actions are based on the information about the state fed back to the controller (hence the name *feedback control*).

In other words, feedback control is defined as a function of the current time and state:

$$\mathbf{u}(t) = \sigma(t, \mathbf{x}(t)), \quad t = 0, 1, \ldots \tag{6.4}$$

where the function $\sigma : (t, \mathbf{x}) \to \mathbf{u} \in \mathbf{U}$ is called a *feedback law.*

6.1.2.3   *Discussion*

In general terms, a closed-loop control could determine the control at time t on the basis of *all* information available at time t, which may ultimately be the whole history of the system's evolution up to and including time t. However, the state variable at time t, $\mathbf{x}(t)$, represents an exhaustive summary of the past influence of the inputs on the system and, for practical reasons, closed-loop controls can become feedback controls. This means that instead of relying on the history of the system's evolution, the controller will be content to

use the current state only, to determine the current control. We will see that this simplifica-
tion cannot be extended to the case of noncooperative multistage games (see Section 6.3).

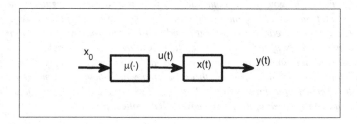

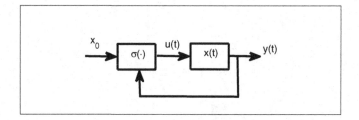

Fig. 6.2   Open-loop and closed-loop control

Figure 6.2 shows a block-diagram representation of open-loop and closed-loop control
structures. In the open-loop configuration (top panel), control $\mathbf{u}(t)$ is determined through
the function $\mu(\cdot)$, which maps the information on the initial condition $\mathbf{x}^0$ and time t into
$u(t)$. The control acts on the state variable $\mathbf{x}(t)$ and this is how output $\mathbf{y}(t)$ is determined.
In the closed-loop case (bottom panel), the output observation is fed back to the function
$\sigma(\cdot)$, which determines the control input.[4]

Both ways of controlling systems are used in practice. The trajectory of interplane-
tary spacecraft is often designed via an open-loop control law. In an airplane or modern
car, many controls are implemented through servomechanisms,[5] and so, in feedback form.
Many of the economic dynamics models, e.g., those related to the theory of economic
growth (see Section 6.1.5) are formulated as open-loop control systems. In industrial-
organization models of competition in markets, "strategic" agents define their actions allowing for the available information; and hence, they are assumed to use closed-loop controls.[6]

---

[4]In particular, if the controller cannot estimate the state from the output, feedback can be based on partial
observations. We will not discuss such feedback in this book.

[5]For example the anti-block braking systems (ABS) used in cars, the automatic landing systems of aircrafts, etc.

[6]We say "are assumed," because commitments (agreements, treaties, schedules, planning processes, etc.) may
force the agents to use the open-loop control even if state observations are available. On the other hand, some

The following example shows the obvious difference between open-loop and feedback controls.

Example 6.1. *If you choose what you wear according to the calendar: "if-it-is-summer-I-wear-a-teeshirt; if-it-is-winter-I-wear-a-sweater," you are using an open-loop control. If, however, you check the actual temperature before you choose a piece of garment, you use a feedback control: "if-it-is-warm-I-wear-a-teeshirt; if-it-is-cold-I-wear-a-sweater." Feedback control is "better" in this case because it adapts to weather uncertainties.*

As in the example given above, dynamic systems may be subjected to random disturbances; we call such systems *stochastic*. If so, the control must be adapted to the changing conditions of the system and closed-loop, feedback or some sort of adaptive control becomes a necessity. We will develop these concepts related to stochastic systems and controls adapted to the history of random disturbances when we study stochastic games in Part 3 of this book.

6.1.3 Linear systems

A rather complete theory has been developed for the control of stationary linear systems, where the state equations take the form:

$$x(t+1) = \mathbf{A}x(t) + \mathbf{B}u(t), \quad x(0) = x^0, \tag{6.5}$$

where $x \in \mathbf{R}^n, u \in \mathbf{R}^m$ and $\mathbf{A}$ and $\mathbf{B}$ are matrices of appropriate dimensions.

For such systems, the properties of the *linear feedback rules* are usually studied. They are defined by:

$$\mathbf{u}(t) = -\mathbf{K}x(t) \tag{6.6}$$

where the matrix $\mathbf{K}$ is called the *feedback gain*. Many "real-life" systems in engineering and physics can be modeled as linear systems.[7]

The theory of linear systems establishes the conditions under which some qualitative properties (e.g., stability or controllability, where the latter is the capacity to reach a target[8]) of the underlying processes hold true. An achievement of control theory for linear systems is that it relates the design of "good" controllers, i.e., that satisfy some imposed goals (either qualitative, like asymptotic stability or target reachability or quantitative, like minimum cost or maximum utility), to the solution of optimal-control problems, as defined in the next section.

---

state variables, like, e.g., the quantity of fish biomass in a management model for an ocean fishery cannot be easily observable. However, the agents may try to establish feedback controls using some proxy variables, e.g., fish prices on a particular market.

[7]For example, the fundamental law of mechanics $f = m\gamma$ which says that the acceleration γ of a body is proportional to the force exerted on it, defines a linear dynamics. In addition, if a system is subjected to small perturbations, its "full," usually nonlinear model can be *linearized*. In this case, the model will also be represented by (6.5). However, we do not intend to delve into the linearization procedure in this book. (Briefly, a linearized model is one where any non-linear function has been replaced by its first-order Taylor-series expansion; for more see, e.g., [165].)

[8]Compare footnotes 11 in this section and 45 on page 201.

6.1.4 *Optimal control*

Assume that the system is controlled over the time horizon $\{0, 1, \ldots, T\}$. We will use the following notation [9]

$$\tilde{\mathbf{u}}^T = \{\mathbf{u}(0), \mathbf{u}(1), \ldots, \mathbf{u}(T-1)\}, \quad \tilde{\mathbf{x}}^T = \{x(0), x(1), \ldots, x(T)\}.$$

A general formulation. Let $g(\mathbf{x}(t), \mathbf{u}(t), t) \in \mathbf{R}$ be a given function that defines a *transition reward*, i.e., a reward the controller obtains when the system receives control input $\mathbf{u}(t)$ and transits from state $\mathbf{x}(t)$ at time t, to state $\mathbf{x}(t+1)$ at time $t+1$. Let $S(x(T)) \in \mathbf{R}$ be a terminal reward or *bequest* function . Then, we can associate a *performance criterion* $J(0, \mathbf{x}^0; \tilde{\mathbf{x}}^T, \tilde{\mathbf{u}}^T)$ with an initial state $\mathbf{x}^0$ at time 0, control input $\tilde{\mathbf{u}}$ and the state trajectory $\tilde{\mathbf{x}}$, which results as a solution of the state equation (like (6.5)). However, to unburden notation, whenever we define a performance-criterion *value*, we will use the plain symbol J to denote the criterion, always properly defined by the symbol $\triangleq$. We will specify the criterion agruments only when we will want to contrast a criterion value obtained for a particular control, or *strategy* (see Section 6.3), to another value obtained for a different control, or strategy.

The performance criterion can be defined as the sum of the terminal reward and transition rewards along the state trajectory,

$$J \triangleq \sum_{t=0}^{T-1} g(\mathbf{x}(t), \mathbf{u}(t), t) + S(\mathbf{x}(T)). \tag{6.7}$$

The *optimal-control problem* consists of finding the best control input $\tilde{\mathbf{u}}^{*T}$, i.e., one that maximizes system performance as measured by criterion (6.7). This problem can be formulated for open-loop control or for closed-loop (or feedback, Markovian) control, depending on the structure that the controller wants to apply it.

When the time horizon tends to infinity $(T \rightarrow \infty)$, we will denote $\tilde{\mathbf{u}} = \{\mathbf{u}(0), \ldots, \mathbf{u}(t), \ldots\}$ and $\tilde{\mathbf{x}} = \{\mathbf{x}(0), \ldots, \mathbf{x}(t), \ldots\}$. The performance criterion takes the form of an infinite sum of transition rewards

$$J \triangleq \sum_{t=0}^{\infty} g(\mathbf{x}(t), \mathbf{u}(t), t). \tag{6.8}$$

This may lead to a definition problem, because the sum may tend to $\pm\infty$. We will see further on in this chapter how to cope with this difficulty.

Linear-quadratic control. Consider the linear system (6.5), repeated below

$$\mathbf{x}(t+1) = \mathbf{A}\mathbf{x}(t) + \mathbf{B}\mathbf{u}(t), \quad \mathbf{x}(0) = \mathbf{x}^0, \tag{6.9}$$

with a quadratic performance criterion

$$J \triangleq \frac{1}{2} \sum_{t=0}^{\infty} (\mathbf{x}(t)'\mathbf{Q}\mathbf{x}(t) + \mathbf{u}(t)'\mathbf{R}\mathbf{u}(t)) \tag{6.10}$$

---

[9]So, the superscript T denotes the length of the action series (T) or states $(T+1)$. Compare the definition formulae (6.21)–(6.23).

where $\mathbf{Q}$ and $\mathbf{R}$ are, respectively, negative definite and negative semi-definite[10] matrices of the appropriate dimensions and $'$ denotes transposition. The combination of a linear model with a quadratic performance index is known in the literature under the name of a *linear-quadratic optimal-control model* .

For this system, it can be proved that, under some conditions concerning matrices $\mathbf{A}$ and $\mathbf{B}$, which we will not develop in this text,[11] there exists an optimal control that maximizes $J(\mathbf{x}^0; \tilde{\mathbf{x}}, \cdot)$ and gives a finite negative value to this performance criterion (6.10). The optimal control is characterized by the following equations:

$$\mathbf{X} = \mathbf{A}'(\mathbf{X} - \mathbf{X}\mathbf{B}(\mathbf{R} + \mathbf{B}'\mathbf{X}\mathbf{B})^{-1}\mathbf{B}'\mathbf{X})\mathbf{A} + \mathbf{Q}, \qquad (6.11)$$

$$\mathbf{K} = (\mathbf{R} + \mathbf{B}'\mathbf{X}\mathbf{B})^{-1}\mathbf{B}'\mathbf{X}, \qquad (6.12)$$

$$\mathbf{u}^*(t) = -\mathbf{K}\mathbf{x}(t). \qquad (6.13)$$

So, to find the optimal control $\tilde{\mathbf{u}}^*$ we first solve the algebraic matrix equation (6.11), called the *algebraic Riccati equation*, whose positive definite matrix solution $\mathbf{X}$ is used to define the *gain matrix* $\mathbf{K}$ in (6.12). Then, we obtain the optimal control by applying the linear feedback law (6.13).

Optimal feedback control. We notice that the solution to the optimal linear-quadratic problem defines the optimal control as a linear feedback law. Obtaining an optimal feedback control directly from the solution procedure is typical of the linear-quadratic model. The optimal feedback law defined by (6.11)–(6.12) is stabilizing the state at zero, i.e., $\mathbf{x}(t) \to 0$, when $t \to \infty$, for any initial state $\mathbf{x}^0$.

In general, it should be possible to establish a link between the optimal control defined in an open-loop structure and the optimal feedback control for the general optimal-control problem defined through (6.1) and (6.7). Indeed, we can expect to solve this problem for all possible initial times t^0 and initial states $\mathbf{x}^0$. Knowing the optimal control, which is implemented at each of these initial states, we can construct, in a pointwise manner, a feedback law $\mathbf{u}^*(t) = \mu^*(t, \mathbf{x})$. We say that, in this manner, we have *synthesized* the optimal control. This technique proved to be effective in the treatment of many *time* optimal-control problems as shown in, e.g., the book by Lee and Markus [150].

Another way to directly obtain an optimal feedback control consists of using dynamic programming [23, 24] to characterize a solution to an optimal-control problem. This approach will be explained in detail in the forthcoming sections of this chapter. Unfortunately, as we will see shortly, the natural correspondence between optimal open-loop and feedback controls does not extend to the case of Nash-equilibrium solutions in noncooperative multistage games.

---

[10]The assumption of matrix negative (semi-)definiteness preserves the interpretation of (6.10) as a concave utility function, which the controller maximizes. Usually, in engineering applications of optimal control, matrices $\mathbf{Q}$ and $\mathbf{R}$ are assumed positive (semi-)definite and then J is a cost that needs be minimized.

[11]These conditions are called controllability of the matrix pair (A, B). This implies that there is a possibility of driving the state of the system to a 0 value through an admissible control, and therefore there are controls that give the quadratic performance index a finite negative value. This makes the optimal-control problem meaningful, even though the performance index is defined as an infinite sum of rewards. Compare footnote 45 on page 201 .

6.1.5 *An optimal economic-growth model*

The capital-accumulation model proposed by Ramsey in [208] is a typical application of the optimal-control methods to economic analysis.

A discrete-time economic-growth model.   Consider an economy producing a malleable output $y(t)$, which can be either invested or consumed, and with two production factors, capital $K(t)$ and labor $L(t)$, where t is time and the horizon is infinite: $t = 0, 1, 2 \ldots, \infty$. The output in period t is determined by the Cobb-Douglas production function:

$$y(t) = A[K(t)]^\alpha [L(t)]^{1-\alpha}, \quad \alpha \in (0, 1). \tag{6.14}$$

The quantity of labor available at each period $L(t) : t = 0, 1, \ldots$ is supposed to be defined exogenously. We may assume that, for example, $L(t)$ is determined by the growth equation

$$L(t+1) = (1 + n(t))L(t), \quad L(0) = L_0, \tag{6.15}$$

where $n(t)$ is the population growth rate in period t, satisfying $n(t) \to 0$ as $t \to \infty$. The accumulation of capital is described by the difference equation

$$K(t+1) = (1 - \delta)K(t) + I(t) \tag{6.16}$$

where $I(t) \geq 0$ is the amount of investment in time t and δ is the capital-depreciation rate in time t. At an initial period $t = 0$, the capital stock is known, $K(0) = K_0$.

The output can be used for consumption $C(t)$ or investment $I(t)$; therefore, we have

$$C(t) = A[K(t)]^\alpha [L(t)]^{1-\alpha} - I(t) \geq 0. \tag{6.17}$$

Assuming the stream of $L(t)$, $t = 0, 1, \ldots$ is known (given exogenously or through (6.15)), the capital-growth equation (6.16) is sufficient to determine any $K(t)$ for a given investment sequence $I(t)$, $t = 0, 1, \ldots$ and therefore to determine $y(t)$ and $C(t)$. We say that equation (6.16) is the state equation for the economy under consideration.[12]

Investment $I(t)$ is the control for this system.[13] We notice that $I(t)$ could be realized as a feedback policy $I(t) = \sigma(t, K(t))$ (see (6.4)), or as an open-loop solution $I(t) = \mu(K_0, t)$ (see (6.3)). The feedback law $\sigma(t, K(t))$ has the interpretation of a rule: "if-*this capital stock at time t*-then-*this investment*." On the other hand, $\mu(K_0, t)$ is a prescription or *schedule* for the investments to be made at each period.

Optimal economic growth.   The social utility of consumption at period t can be measured by the expression

$$\mathcal{U}(L(t), C(t)) = L(t) \ln \left(\frac{C(t)}{L(t)} \right). \tag{6.18}$$

---

[12]If labor were determined by (6.15), then this economy would have two state equations: (6.16) and (6.15).
[13]If constraint (6.17) was non-binding and an investment fund was provided by an exogenous institution, e.g., the IMF, then investment could also be an exogenous forcing factor.

An optimal accumulation path can thus be defined as the feasible time profiles for capital, investment and consumption, which maximize the discounted sum of the social utility of consumption in all periods

$$J \triangleq \sum_{t=0}^{\infty} \beta^t \mathcal{U}(L(t), C(t)), \tag{6.19}$$

where $0 \leq \beta < 1$ is the discount factor and where we have allowed for the state trajectory $\tilde{K} = K(\cdot)$ and control $\tilde{I} = I(\cdot)$ in the performance criterion (note that $K(t)$ and $I(t)$ are implicit in $C(t)$).

With "total" utility defined by (6.19), this economic growth model can be interpreted as an optimal-control model where $\mathbf{x}(t) = K(t)$, $\mathbf{u}(t) = I(t)$, and the right-hand side of the state equation (6.1) is

$$\mathbf{f}(\mathbf{x}(t), \mathbf{u}(t), t) = (1 - \delta)K(t) + I(t).$$

The performance criterion (6.8) becomes

$$J \triangleq \sum_{t=0}^{\infty} \beta^t L(t) \ln \left(\frac{A[K(t)]^\alpha [L(t)]^{1-\alpha} - I(t)}{L(t)} \right).$$

Historically, this deterministic optimal-control problem was analyzed in continuous time[14] and solved using an open-loop structure; see reference [173] for a review of this type of discrete-time economic growth models. There is also a growing literature on feedback solutions to economic problems of growth, environmental protection, fishery management, etc., in particular when the problem context is stochastic, see e.g., the review by Amir [3].

6.2 Description of multistage games in a state-space

Optimal control theory deals with problems of selecting the best controlled input, called optimal control, that optimizes a single performance criterion and that depends on the output of the system. If the controls are distributed among a set of independent actors who strive to optimize their individual performance criteria, we face the situation of a *multistage game in a state space*.[15]

The system's dynamics of a game played in a state space by m players can be represented in discrete time, $t = 0, 1, 2, \ldots, T$, where T is a positive integer called the *horizon* of the game (infinite-horizon games are obtained when T tends to ∞), by the following state equation:

$$\mathbf{x}(t + 1) = \mathbf{f}(\mathbf{x}(t); \mathbf{u}_1(t), \ldots, \mathbf{u}_m(t), t). \tag{6.20}$$

Here, $\mathbf{x}(t) = (x_1(t), \ldots, x_n(t)) \in \mathbf{R}^n$ is the state vector at t, $\mathbf{u}_j(t) \in \mathbf{R}^{p_j}, j = 1, 2 \ldots m$ is the control of Player j at t, and

$$\mathbf{f}(\cdot, \cdot, \cdot) = (f_i(\cdot, \cdot, \cdot))_{j=1,\ldots,n} : \mathbf{R}^n \times \mathbf{R}^{p_1 + \cdots + p_m} \times \mathbf{R}_+ \mapsto \mathbf{R}^n$$

---

[14]In a continuous time setting the state equations are defined as differential equations.
[15]Note that the choice of state variables used to describe a process might not be unique. However, the number of state variables needed to describe that process is unique; this number can be called the *order* of the system.

is the transition function from t to $t + 1$. Hence, equation (6.20) determines the values of the state vector at time $t + 1$, for a given $\mathbf{x}(t) = (x_1(t), \ldots, x_n(t))$ and controls[16] of all players $u_1(t), \ldots, u_m(t)$.

We will use the following notation regarding the players' combined controls and state trajectory:

$$\underline{u}(t) = (\mathbf{u}_1(t), \ldots, \mathbf{u}_m(t)), \tag{6.21}$$

$$\tilde{\mathbf{u}}_j^t = \{\mathbf{u}_j(0), \mathbf{u}_j(1), \mathbf{u}_j(2), \ldots, \mathbf{u}_j(t-1)\}, \tag{6.22}$$

$$\tilde{\underline{u}}^t = \{\underline{u}(0), \underline{u}(1), \underline{u}(2), \ldots, \underline{u}(t-1)\}, \tag{6.23}$$

$$\tilde{\mathbf{x}}^t = \{\mathbf{x}(0), \mathbf{x}(1), \mathbf{x}(2), \ldots, \mathbf{x}(t)\}, \tag{6.24}$$

and we will drop the time exponent t when referring to an infinite stream of controls or states.

We call *history of the game* at period t the vector $h^t = \{t, \tilde{\underline{u}}^t, \tilde{\mathbf{x}}^t\}$, i.e., the sequence of control- and state-variable values that have driven the system, up to period t. The information available to the players when they choose their controls at period t is a part of this history.

In summary, to define a multistage state-space game[17] we need to specify the following:

- The set of players $M = \{1, 2, \ldots m\}$,
- The state equation:

$$\mathbf{x}(t + 1) = \mathbf{f}(\mathbf{x}(t), \underline{u}(t), t), \quad \mathbf{x}(t_0) = \mathbf{x}^0$$

  where $\mathbf{x}(t) \in \mathbf{R}^n$ (possibly $\mathbf{X} \subset \mathbf{R}^n$) is the state variable vector; $\underline{u}(t) = (\mathbf{u}_j(t))_{j=1\ldots m}$, where $\mathbf{u}_j(t) \in \mathbf{U}_j \subset \mathbf{R}^{p_j}$, $j = 1, \ldots, m$ are players' controls, all of appropriate dimensions; and $\mathbf{f}(\cdot, \cdot, \cdot, \cdot)$ is a vector whose components are defined in (6.20);
- The information structure defined by the part of the history vectors $h^t, t = 0, 1 \ldots T$, that is utilized by the players to compute their controls at period t;
- Any other relevant restrictions on the system's variables;
- Player j's payoff function (also called utility function and performance criterion) that each player seeks to maximize. We need to distinguish between the cases when the time horizon is finite and infinite;
- For the finite-horizon case, $T < \infty$, the payoff to Player j is defined by

$$J_j \triangleq \sum_{t=t_0}^{T-1} g_j(\mathbf{x}(t), \underline{u}(t), t) + S_j(\mathbf{x}(T)), \tag{6.25}$$

  where $g_j(\mathbf{x}(t), \underline{u}(t), t) \in \mathbf{R}$ is the *transition reward* of Player j and $S_j(\mathbf{x}(T)) \in \mathbf{R}$ is the *terminal reward* for that player, also called *bequest function*;

---

[16]When we will deal with stochastic systems later in this book, we will see that some "controls" may come from nature and be independent of the players' actions.

[17]Compare the game descriptions in *normal form* in Sections 2.4, 3.1, 4.2 and 5.2.

- For the infinite-horizon case, when $T \to \infty$, the infinite sum of transition rewards may tend to infinity, hence it is not obvious how a performance criterion should be defined or how two performances could be compared if the criterion is not convergent. Several approaches have been proposed for such situations; here we list only the most often used. Here and below we assume that the system is *stationary*, which means that the reward and state-transition functions do not depend explicitly on time t.

(1) *Discounted sum of rewards*

$$J_j \triangleq \sum_{t=0}^{\infty} \beta_j^t g_j(\mathbf{x}(t), \underline{\mathbf{u}}(t)), \tag{6.26}$$

where $0 \leq \beta_j < 1$ is the discount factor for Player j. This is the *discounted* criterion. If the transition reward is a uniformly bounded function, this infinite discounted sum converges to a finite value and payoffs obtained in this manner can be compared. However, we must notice that this criterion gives a diminishing weight to the rewards that occur in the distant future; so, this performance is mostly influenced by what happens in the early periods, and hence, it discriminates against the future generations.

(2) Another criterion, which puts more weight on the rewards obtained in a distant future, is the *limit of average reward* [18]

$$J_j \triangleq \liminf_{T \to \infty} \frac{1}{T} \sum_{t=0}^{T-1} g_j(\mathbf{x}(t), \underline{\mathbf{u}}(t)). \tag{6.27}$$

This criterion is based on the limit of the average reward per period. Notice that, as T tends to infinity, all rewards gained over a finite number of periods will tend to have a negligible influence on the criterion; only what happens in the very distant future, actually at infinity, matters. In essence, this criterion is not affected by what happens in the early periods.

(3) Another approach consists of developing methods to compare streams of rewards even when their sums do not converge. This possibility of comparing agent payoffs is provided by the *overtaking optimality criterion*. We say that $\tilde{\mathbf{u}}^*$ is *overtaking optimal* at $\mathbf{x}_0$, if for any other admissible control $\tilde{\mathbf{u}}$ the following inequality holds:

$$\liminf_{T \to \infty} \left(\sum_{t=0}^{T} g_j(\mathbf{x}^*(t), \underline{\mathbf{u}}^*(t)) - \sum_{t=0}^{T} g_j(\mathbf{x}(t), \underline{\mathbf{u}}(t)) \right) \geq 0, \tag{6.28}$$

where $\tilde{\mathbf{x}}$ and $\tilde{\mathbf{x}}^*$ are the two state trajectories emanating from the same initial state $\mathbf{x}_0$ and generated by these two controls respectively. This condition could be also written as follows:

$$\forall \epsilon > 0, \exists T(\epsilon) \text{ s.t. } \forall T > T(\epsilon)$$

$$\sum_{t=0}^{T} g_j(\mathbf{x}^*(t), \underline{\mathbf{u}}^*(t)) - \sum_{t=0}^{T} g_j(\mathbf{x}(t), \underline{\mathbf{u}}(t)) \geq \epsilon. \tag{6.29}$$

---

[18]This limit is known under the name of *Cesaro limit*.

We obtain a weaker concept of optimality if we replace lim inf by lim sup in expression (6.28). We say that $\tilde{\mathbf{u}}^*$ is *weakly overtaking optimal* at $\mathbf{x}_0$ if for any other admissible control $\tilde{\mathbf{u}}$, the following inequality holds:

$$\limsup_{T \to \infty} \left(\sum_{t=0}^{T} g_j(\mathbf{x}^*(t), \underline{\mathbf{u}}^*(t)) - \sum_{t=0}^{T} g_j(\mathbf{x}(t), \underline{\mathbf{u}}(t)) \right) \geq 0. \tag{6.30}$$

This condition can also be written as follows

$$\forall \epsilon > 0, \forall T > 0, \quad \exists \Theta > T \text{ s.t.}$$

$$\sum_{t=0}^{\Theta} g_j(\mathbf{x}^*(t), \underline{\mathbf{u}}^*(t)) - \sum_{t=0}^{\Theta} g_j(\mathbf{x}(t), \underline{\mathbf{u}}(t)) \geq \epsilon. \tag{6.31}$$

Remark 6.1. When we use the overtaking or weakly overtaking optimality concept we cannot say that a player maximizes this criterion; however, we can use it to determine whether one stream of rewards is better than another one. Finally, we notice that in the overtaking or weakly overtaking optimality criterion, **all** rewards influence the outcome equally, which is in sharp contrast to the two previous criteria.

A stylized model of a plausible conflict situation in fishery management possesses all the ingredients of a dynamic game played in a state space. This was recognized early by several authors, i.e., in references [44, 162, 159]. We will use such a model to illustrate the equilibrium calculation in the forthcoming sections.

Example 6.2. *The Great Fish War. Levhari and Mirman coined this name for the fishery-management model they proposed in* [159]. *Let $x(t)$ be a measure of the quantity of fish at time t. The biological rule of growth in the absence of human intervention is given by*

$$x(t+1) = x(t)^{\alpha}, \tag{6.32}$$

where $0 < \alpha < 1$. Thus a stable steady state of the fish population is $x(t) = x(t+1) = 1$.[19] Two players exploit the fishery. The catch[20] in period t by Player j is $u_j(t)$ and the reward from the consumption of this catch is $g_j(u_j(t)) = \log(u_j(t))$. Over an infinite time horizon the payoff to Player j is

$$J_j \triangleq \sum_{t=0}^{\infty} \beta_j^t \log(u_j(t)), \quad j = 1, 2, \tag{6.33}$$

where $0 < \beta_j < 1$ is the discount factor of Player j. The interdependence between the two players is due to their joint exploitation of the common resource. The state equation for the evolution of the quantity of fish when there is exploitation is

$$x(t+1) = (x(t) - u_1(t) - u_2(t))^{\alpha}. \tag{6.34}$$

This model will be further studied in Exercise 6.7.

---

[19] A normalization of the fish population measure is employed here.
[20] Also measured in normalized units.

In the forthcoming sections we will consider a slightly different fishery-management model where the fish biomass is $x(t)$ in year t. The coefficient $a > 1$ denotes the annual net growth rate of the biomass which evolves according to the equation

$$x(t+1) = ax(t). \tag{6.35}$$

Therefore this population has no stable steady state. The origin $x = 0$ is an unstable steady state. Suppose that there are two players $j = 1, 2$ (e.g., countries) exploiting the fishery and let $u_j(t)$ denote the harvest of Player j in period t. The system's dynamics can then be described by the state equation

$$x(t+1) = a(x(t) - u_1(t) - u_2(t)), \tag{6.36}$$

$u_j(t) \geq 0$, $j = 1, 2$; $0 \leq u_1(t) + u_2(t) \leq x(t)$. We notice that the latter constraint couples the actions of both players and the state of the system. In general, such constraints will be difficult to handle. However, given the players' payoff functions, defined below, this constraint will always be satisfied in equilibrium (see Sections 6.5 and 6.6.2). The initial condition for the problem $x(0) = x^0$ is given.

The state equation (6.36) represents a growth model of fish biomass $x(t)$ where both players have an impact on the growth process through their harvests.

Player j strives to maximize the payoff function of the form

$$J_j \triangleq \sum_{t=t_0}^{T-1} \beta_j^t \sqrt{u_j(t)} + K_j \beta_j^T \sqrt{x(T)} \quad j = 1, 2 \tag{6.37}$$

where $\beta_j \in [0, 1)$ is a discount factor and $K_j > 0$ is a scaling parameter of the fishery's scrap-value. The square root of harvest ($\sqrt{u_j(t)}$) and of the fishery's scrap value ($\sqrt{x(T)}$) reflect the players' diminishing returns from the catch and bequest.

Should the players perceive that the fishery is to be used in perpetuity, then their payoff functions can be modeled as

$$J_j \triangleq \sum_{t=t_0}^{\infty} \beta_j^t \sqrt{u_j(t)} \quad j = 1, 2. \tag{6.38}$$

This model will be further studied in Sections 6.5 and 6.6.2.

In both versions of this fishery-management model the equations (6.34) or (6.36) provide a state-space description for the fishing process at hand and $x(t)$ is the state variable.

6.3   Information structure

The different ways of designing a control of a dynamic system, open-loop versus closed-loop, described in Section 6.1.2, can be associated with different information structures in a multistage game. We will here generalize the developments of Chapter 5 where we studied the open-loop and closed-loop information structures for repeated games.[21]

---

[21] See Section 5.2.3, page 135.

Without loss of generality and for the sake of simplifying the exposition, we will develop the theory for the case of a two-player multistage game defined by the state equation

$$\mathbf{x}(t+1) = \mathbf{f}(\mathbf{x}(t), \mathbf{u}_1(t), \mathbf{u}_2(t), t), \quad \mathbf{x}(t_0) = \mathbf{x}^0 \tag{6.39}$$

where $\mathbf{x}(t) \in \mathbf{R}^n, \mathbf{u}_j \in \mathbf{R}^{m_j}$ and two payoff functions $J_j(t_0, \mathbf{x}^0; \tilde{\mathbf{u}}_1, \tilde{\mathbf{u}}_2)$, $j = 1, 2$, which may have one of the forms (6.25)–(6.26). We will also assume that the players select their controls from a constraint set $U_j \subset \mathbf{R}^{m_j}$.[22] A control is admissible if it satisfies these constraints.

In the game paradigm, the information is transformed into actions by players' strategies. We will have different multistage games depending on the assumptions regarding information type that is used by the players to compute their controls. In general, the most complete information available to a player throughout the play is the game history h^t. Since the total amount of information available at time t tends to increase with t, h^t increases with time. However, in most cases of interest, not all information accumulated up to time t turns out to be relevant for the decision-making process at that point. This can help reduce the vector h^t to another vector of a fixed and finite dimension. This fact is of great importance from an implementation point of view. The possibility of utilizing the game history to generate strategies was discussed in the context of repeated games in Section 5.2, and more specifically in Section 5.2.3. Also, recall the use of finite automata when we proved the folk theorem in Section 5.3.

We will consider the following information structures:

(1) *Open-loop information structure.* Both players use only the knowledge of the initial state x^0 and the time t to determine their controls throughout the play. A *strategy* for Player j is thus determined by a mapping $\mu_j(\cdot, \cdot) : \mathbf{R}^n \times \mathbf{R} \to U_j$ where

$$\mathbf{u}_j(t) = \mu_j(\mathbf{x}^0, t) \quad \text{for all } t,$$

which assigns a control value at each time t, knowing the initial state $\mathbf{x}^0$. The players may have to use this information structure because they do not have access to more information to base their decisions upon. They may also have to use it if the game implies a commitment for all the forthcoming periods.

(2) *Feedback information structure,* frequently referred to as *Markovian information structure.* Both players have access to the information about the current state of the system and "forget," or do not retain, the information from the previous stages. A strategy for Player j is thus determined by a mapping $\sigma_j(\cdot, \cdot) : \mathbf{R} \times \mathbf{R}^n \to U_j$ that determines the control value at period t as a function of t and the current state variable value $\mathbf{x}(t)$

$$\mathbf{u}_j(t) = \sigma_j(t, \mathbf{x}(t)).$$

We then refer to $\sigma_j(\cdot, \cdot)$ as a *feedback* or *Markov* strategy.

---

[22]In the most general setting, the control-constraint sets may also depend on time and current state, i.e., $U_j(t, x) \subset \mathbf{R}^{m_j}$. Also, sometimes the state may be constrained to remain in a subset $\mathbf{X} \subset \mathbf{R}^n$. We will avoid these complications in this textbook.

(3) *Closed-loop information structure.* Both players have access to the information about the history of the game h^t at each period t. This history can concern only the state trajectory, i.e., for all t

$$h^t = \{\mathbf{x}(0), \mathbf{x}(1), \ldots, \mathbf{x}(t)\}$$

or concern the history of all players' actions, i.e., for all t

$$h^t = \{\mathbf{x}(0), \mathbf{u}_1(0), \mathbf{u}_2(0), \ldots, \mathbf{u}_1(t-1), \mathbf{u}_2(t-1)\}.$$

Notice that in the deterministic context of these multistage games, the knowledge of $\mathbf{x}(t)$ for $t > 0$ is implicit because the state trajectory can be reconstructed from the control inputs and the initial state value. We call $\mathcal{H}^t$ the set of all possible game histories at period t.

In both cases, the strategy of Player j is defined by a sequence of mappings $\gamma_{jt}(\cdot)$: $\mathcal{H}^t \to U_j$, which associate a control value at time t with the observed history h^t,

$$\mathbf{u}_j(t) = \gamma_{jt}(h^t).$$

A solution to a game is fundamentally affected by the underlying information structure, as we saw in the introductory chapters on the classical theory of games. This is also true for multistage games as we will see in Section 6.4.

Remark 6.2. The open-loop information structure defined in (1) is often considered as "least interesting" in the context of dynamic games, mainly because it does not lead to **subgame-perfect** equilibrium solutions. This point will be discussed later in this chapter (see Section 6.4.1.2).

Remark 6.3. The difference between information structures (2) and (3) can influence the outcome of a play in an essential manner if the players have some ability to utilize the game history, or "memory," to formulate strategies based on threats (recall Definition 5.2 on collusive equilibrium). In structure (2), players "forget" the opponents' actions and may not be able to implement an adequate punishment. In structure (3), it is obvious "who did what," and penalties may be allocated appropriately. This point will be discussed later in this chapter (see Section 6.9).

We also have two observations regarding the computation of equilibria.

(1) The procedure for computing equilibrium solutions under structures (2), (3) can be substantially more difficult than for structure (1).
(2) Unless otherwise stated, we always assume that the rules of the games, i.e., the dynamics, the control sets and the information structure are common knowledge. As a result, the players' strategies cannot depend explicitly on any private information they may have.

Finally we will illustrate the notion of the information structure using Example 6.2.

Example 6.3. *In Example 6.2, if the fishermen do not know the fish biomass, they are bound to use open-loop strategies, based on some initial x^0 assessed or measured at some point* *in time. If the fishermen could measure the fish biomass reasonably frequently and exactly (e.g., by some advanced process to assess the catchability coefficient or a breakthrough procedure involving satellite photos), then, most certainly, they would strive to compute feedback-equilibrium strategies. But, presumably, measuring the biomass is an expensive process, and thus, an intermediate control structure is practical: update the "initial" value x^0 from time to time and use an open-loop control in between. An information structure of this type is known as* **piecewise open-loop control** *and will be applied to stochastic games in Chapter 10.*

6.4   Strategies and equilibrium solutions

We are now ready to define an equilibrium solution in a multistage game, under various information structures. As with any new game, we must first define a *normal-form representation of the game*, and then introduce the equilibrium conditions.

6.4.1   *Open-loop Nash equilibria in multistage games*

We first define and characterize Nash equilibrium in an *open-loop multistage game*, i.e., in which the players are using open-loop strategies.

6.4.1.1   *Definition*

The multistage game is defined, for the finite-horizon case, by the following utility functions (or performance criterions) and state equations:[23]

$$J_j \triangleq \sum_{t=0}^{T-1} g_j(\mathbf{x}(t), \mathbf{u}_1(t), \mathbf{u}_2(t), t) + S_j(\mathbf{x}(T)), \text{ for } j = 1, 2, \quad (6.40)$$

$$\mathbf{u}_j(t) \;\in\; U_j, \quad (6.41)$$

$$\mathbf{x}(t+1) \;=\; \mathbf{f}(\mathbf{x}(t), \mathbf{u}_1(t), \mathbf{u}_2(t), t), \; t = 0, 1, \ldots T-1, \quad (6.42)$$

$$\mathbf{x}(0) \;=\; \mathbf{x}^0. \quad (6.43)$$

If the game is played in open loop, each player, having observed the initial state x_0 chooses an admissible[24] control sequence $\tilde{\mathbf{u}}_j^T = (\mathbf{u}_j(0), \ldots, \mathbf{u}_j(T-1)), j = 1, 2$. This generates, from the initial position $(0, \mathbf{x}^0)$, a state trajectory $\tilde{\mathbf{x}}^T$ solution to (6.42)–(6.43)

$$\mathbf{x}(1) = \mathbf{f}(\mathbf{x}^0, \mathbf{u}_1(0), \mathbf{u}_2(0), 0),$$

$$\mathbf{x}(2) = \mathbf{f}(\mathbf{x}(1), \mathbf{u}_1(1), \mathbf{u}_2(1), 1),$$

$$\cdots$$

$$\mathbf{x}(T) = \mathbf{f}(\mathbf{x}(T-1), \mathbf{u}_1(T-1), \mathbf{u}_2(T-1), T-1),$$

and the corresponding payoffs according to (6.40). The expression (6.40), subject to (6.41)–(6.43), yields to the normal form of the open-loop multistage game at initial point

---

[23] Also see equations (6.60)–(6.61) page 178.

[24] I.e., which satisfies (6.41) at each $t = 0, 1, \ldots, T-1$.

$(0, \mathbf{x}^0)$. We will, therefore, denote $J_j(0, \mathbf{x}_0; \tilde{\mathbf{u}}_1^T, \tilde{\mathbf{u}}_2^T)$ $\quad j = 1, 2$, this normal form description of the open-loop multistage game, which defines the players' payoffs as functions of their strategies.

We are now ready to proceed with a definition of a Nash equilibrium for this game.

Definition 6.2. A pair of admissible control sequences[25] $\underline{\tilde{\mathbf{u}}}^* = (\tilde{\mathbf{u}}_1^*, \tilde{\mathbf{u}}_2^*)$ is an **open-loop Nash equilibrium** at $(0, \mathbf{x}^0)$ if it satisfies the following equilibrium conditions

$$J_1(0, \mathbf{x}^0; \underline{\tilde{\mathbf{u}}}^*) \geq J_1(0, \mathbf{x}^0; \tilde{\mathbf{u}}_1, \tilde{\mathbf{u}}_2^*) \quad \forall \text{ admissible } \tilde{\mathbf{u}}_1,$$
$$J_2(0, \mathbf{x}^0; \underline{\tilde{\mathbf{u}}}^*) \geq J_2(0, \mathbf{x}^0; \tilde{\mathbf{u}}_1^*, \tilde{\mathbf{u}}_2) \quad \forall \text{ admissible } \tilde{\mathbf{u}}_2.$$

6.4.1.2 *Time consistency and subgame perfectness*

Time consistency is the property, established in the following lemma, which says that the open-loop Nash equilibrium control pair remains an equilibrium throughout the equilibrium trajectory.

Lemma 6.1. *Let $(\tilde{\mathbf{u}}_1^*, \tilde{\mathbf{u}}_2^*)$ be an open-loop Nash equilibrium at $(0, \mathbf{x}^0)$ and let $\tilde{\mathbf{x}}^* = (\mathbf{x}^*(0), \mathbf{x}^*(1), \ldots, \mathbf{x}^*(T))$ be the equilibrium trajectory generated by these controls from $(0, \mathbf{x}^0)$. Then the restriction $(\tilde{\mathbf{u}}_1^*, \tilde{\mathbf{u}}_2^*)_{[\tau, T-1]}$ of these control sequences to the periods $\tau, \ldots, T - 1$ is an open-loop Nash equilibrium at the initial point $(\tau, \mathbf{x}^*(\tau))$, where $\tau = 0, 1, \ldots, T - 1$ and $\mathbf{x}^*(\tau)$ is an intermediate state along the equilibrium trajectory.*

Proof. Assume this property is not true at an intermediate state $(\tau, \mathbf{x}^*(\tau))$ of the equilibrium trajectory. Then, for one player, say Player 1, there exists a control sequence $\tilde{\mathbf{u}}_1^\tau = (\mathbf{u}_1(\tau), \ldots, \mathbf{u}_1(T - 1))$ such that

$$J_1(\tau, \mathbf{x}^*(\tau); (\tilde{\mathbf{u}}_1^*, \tilde{\mathbf{u}}_2^*)_{[\tau, T-1]}) < J_1(\tau, \mathbf{x}^*(\tau); (\tilde{\mathbf{u}}_1^\tau, \tilde{\mathbf{u}}_2^*)_{[\tau, T-1]}).$$

Now by concatenating the controls

$$\tilde{\mathbf{u}}_1^\tau = (\mathbf{u}_1^*(0), \ldots, \mathbf{u}_1^*(\tau - 1))$$

and

$$\tilde{\mathbf{u}}_1^{[\tau, T]} = (\mathbf{u}_1(\tau), \ldots, \mathbf{u}_1(T - 1))$$

and forming

$$\tilde{\mathbf{u}}_1 \equiv (\mathbf{u}_1^*(0), \ldots, \mathbf{u}_1^*(\tau - 1), \mathbf{u}_1(\tau), \ldots, \mathbf{u}_1(T - 1)).$$

Player 1 will be able to use a control $\tilde{\mathbf{u}}_1$, which is admissible and which will yield

$$J_1(0, \mathbf{x}^0; \underline{\tilde{\mathbf{u}}}^*) < J_1(0, \mathbf{x}^0; \tilde{\mathbf{u}}_1, \tilde{\mathbf{u}}_2^*).$$

This contradicts the assumption that $(\tilde{\mathbf{u}}_1^*, \tilde{\mathbf{u}}_2^*)$ is an open-loop Nash equilibrium at $(0, \tilde{x}^0)$. Therefore the lemma is proved. $\qquad \square$

---

[25] From now on, to simplify notations, we will omit the superscript T and refer to $\tilde{\mathbf{u}}_j$ instead of $\tilde{\mathbf{u}}_j^T$ or $\tilde{\mathbf{x}}$ instead of $\tilde{\mathbf{x}}^T$.

In general, an open-loop Nash equilibrium is not *subgame perfect* in the sense of Selten, [218].[26] This means that if a player deviates from the equilibrium control for a while and then decides to play again "properly", then the previously defined equilibrium (i.e., control sequence) will not necessarily be an equilibrium. This lack of subgame perfectness has been considered by some authors as a grave drawback of the open-loop Nash-equilibrium solution concept. Usually, some randomness is invoked in the game, like the "trembling hand" of a controller triggering a deviation from the equilibrium trajectory and therefore a collapse of the equilibrium property. But, if we are in a non-stochastic environment, then the game is deterministic and there is no reason to consider deviations, since it would be detrimental for the player who deviates. Some authors conclude that, in a deterministic context subgame perfectness is not an essential property of equilibrium solutions in multistage games.

Notwithstanding the above positive conclusion we need to remind the reader that an open-loop behavior relies on the assumption that the players can commit, simultaneously, at the beginning of the game, to a completely pre-specified list of actions to be played without any possibility of update or revision during the course of the game. So, from that point of view, a feedback (or Markovian) strategy appears more suited to model the behavior of *strategic* agents, i.e., those who react to available information.

We can summarize this discussion by saying that the difficulties in interpreting the equilibria stem from the fact that equilibria are *not* optima. They are fixed points of reaction functions that depend heavily on the way, in which the strategies are described.

6.4.1.3   *Coupled maximum principle*

We can utilize an optimal-control technique based on the *maximum principle* to characterize open-loop Nash-equilibrium solutions. For that purpose, we use (6.40), (6.42) and define the Hamiltonian for each Player j by

$$H_j(\mathbf{p}_j(t+1), \mathbf{x}(t), \mathbf{u}_1(t), \mathbf{u}_2(t), t) \equiv$$
$$g_j(\mathbf{x}(t+1), \mathbf{u}_1(t), \mathbf{u}_2(t), t) + \mathbf{p}_j(t+1)'\mathbf{f}(\mathbf{x}(t), \mathbf{u}_1(t), \mathbf{u}_2(t), t), \quad (6.44)$$

where $\mathbf{p}_j(t)$ is a *costate* vector[27] in $\mathbf{R}^n$ and $'$ indicates the transposition of the vector $p_j(t+1)$ in a scalar product.

Assumption 6.1. Assume that $\mathbf{f}(\mathbf{x}, \underline{u}, t)$ and $g_j(\mathbf{x}, \underline{u}, t)$ are continuously differentiable in state $\mathbf{x}$ and continuous in controls $\underline{u}$ for each $t = 0, \ldots, T-1$ and $S_j(\mathbf{x})$ is continuously differentiable in $\mathbf{x}$. Assume that, for each j, U_j is compact and convex. Assume also that, for each $t, \mathbf{x}$, the function $H_j(\mathbf{p}, \mathbf{x}, \mathbf{u}_j, \mathbf{u}_{-j}, t)$ is concave in $\mathbf{u}_j$.

We can then formulate the following lemma that provides the necessary conditions that the open-loop equilibrium strategies need to satisfy:

---

[26]Compare Definition 4.4 on page 103 and the contents of Section 5.4.2 starting on page 144.
[27]Also called adjoint vector. This terminology is borrowed from optimal control theory.

Lemma 6.2. *Under Assumption 6.1, if $\underline{\tilde{u}}^*$ is an open-loop Nash equilibrium pair of controls, generating the trajectory $\tilde{x}^*$ from initial state x^0 for the game (6.40), (6.42), then there exist functions of time $p_j(\cdot)$, with values in R^n, such that the following relations hold*

$$u_j^*(t) = \arg \max_{u_j(t) \in U_j} H_j(p_j(t+1), x^*(t), u_j(t), u_{-j}^*(t), t), \qquad (6.45)$$

$$p_j(t)' = \frac{\partial}{\partial x} H_j(p_j(t+1), x^*(t), u_1^*(t), u_2^*(t), t), \qquad (6.46)$$

$$p_j(T)' = \frac{\partial}{\partial x(T)} S_j(x^*(T)), \; j = 1, 2 \,. \qquad (6.47)$$

Proof. The proof is relatively straightforward; it relies on the fact that each player j's best reply to the other player's equilibrium controls is an optimal-control problem, for which the maximum principle in discrete time holds true, under Assumption 6.1; see references [18] or [75] for a more complete proof. ☐

6.4.1.4 *Karush-Kuhn-Tucker conditions and equilibrium formulation as a non-linear complementarity problem*

We can also use a mathematical-programming approach to compute open-loop Nash equilibria, when the following assumption holds:

Assumption 6.2. Assume that $f(x, \underline{u}, t)$ and $g_j(x, \underline{u}, t)$ are continuously differentiable in state x and controls $\underline{u}$ for each $t = 0, \ldots, T-1$ and $S_j(x)$ is continuously differentiable in x. Assume that, for each j, U_j is defined by inequalities $h_j(u_j) \leq 0$, where $h_j : R^{m_j} \to R^{p_j}$, $p_j < m_j$ are given continuously differentiable mappings.

Now we consider the Lagrangians

$$L_j(p_j(t+1), \nu_j(t), x(t), u_1(t), u_2(t), t) \equiv g_j(x(t+1), u_1(t), u_2(t), t)$$
$$+ p_j(t+1)'f(x(t), u_1(t), u_2(t), t) + \nu_j(t)' h_j(u_j(t)), \quad (6.48)$$

Lemma 6.3. *Under Assumption (6.1), if $\underline{\tilde{u}}^*$ is an open-loop Nash equilibrium pair of controls, generating the trajectory $\tilde{x}^*$ from initial state x^0 for the game (6.40), (6.42) and if the constraint qualification conditions of Karush-Kuhn-Tucker hold, then there exist functions of time $p_j(t)$, with values in R^n and functions of time $\nu_j(t)$, with values in R^{p_j}, such that, when forming the Lagrangians (6.48), the following holds true*

$$0 = \frac{\partial}{\partial u_j} L_j(p_j(t+1), \nu_j(t), x^*(t), u_j^*(t), u_{-j}^*(t), t), \qquad (6.49)$$

$$p_j(t)' = \frac{\partial}{\partial x} L_j(p_j(t+1), \nu_j(t), x^*(t), u_1^*(t), u_2^*(t), t), \qquad (6.50)$$

$$0 = \nu_j(t)' h_j(u_j^*(t))\rangle, t = 0, \ldots, T-1, \qquad (6.51)$$

$$0 \leq \nu_j(t), t = 0, \ldots, T-1, \qquad (6.52)$$

$$p_j(T)' = \frac{\partial}{\partial x(T)} S_j(x^*(T)), \; j = 1, 2 \,. \qquad (6.53)$$

Proof. The proof is also straightforward; it relies on the fact that each Player j's best reply to the other players' equilibrium controls is an optimal control problem, for which the necessary optimality conditions, under Assumption 6.1 can be written as above; for a more complete proof see [164]. □

6.4.2   *Feedback-Nash equilibria (FNE) in multistage games*

In this section we present an approach for the characterization of Nash equilibrium solutions in the class of feedback (or Markov) strategies. The approach is based on the dynamic-programming method introduced by Bellman for control systems in [23].

6.4.2.1   *Definition of a feedback-Nash equilibrium in a multistage game*

Consider the following game, defined in normal form[28] at the initial data $(\tau, \mathbf{x}_\tau)$ by the payoff functions

$$J_j(\tau, \mathbf{x}^\tau; \sigma_1, \sigma_2) = \sum_{t=\tau}^{T-1} g_j(\mathbf{x}(t), \sigma_1(t, \mathbf{x}(t)), \sigma_2(t, \mathbf{x}(t)), t) + S_j(\mathbf{x}(T)), j = 1, 2 \quad (6.54)$$

where the state variable values are determined by

$$\mathbf{x}(t+1) = \mathbf{f}(\mathbf{x}(t), \sigma_1(t, \mathbf{x}(t)), \sigma_2(t, \mathbf{x}(t)), t), \ t = \tau, \ldots T-1, \quad (6.55)$$

$$\mathbf{x}(\tau) = \mathbf{x}^\tau, \quad \tau \in \{0, \ldots, T-1\}. \quad (6.56)$$

This game is played in feedback (or Markov) strategies. Each player chooses an admissible feedback $\sigma_j(t, \mathbf{x})$, $j = 1, 2$. This generates a state trajectory from any initial point $(\tau, \mathbf{x}^\tau)$ according to (6.55)–(6.56) and a payoff according to (6.54). We call Σ_j the set of all possible feedback strategies for Player j.

Assuming players strive to maximize (6.54), expressions (6.54)–(6.56) define the normal form of the feedback multistage game.

Definition 6.3. A pair of admissible feedback strategies $\underline{\sigma}^* = (\sigma_1^*, \sigma_2^*)$ is a **feedback-Nash equilibrium** if it satisfies the following equilibrium conditions:

$$J_1(\tau, \mathbf{x}^\tau; \underline{\sigma}^*) \geq J_1(\tau, \mathbf{x}^\tau; \sigma_1, \sigma_2^*) \quad \forall \sigma_1 \in \Sigma_1,$$

$$J_2(\tau, \mathbf{x}^\tau; \underline{\sigma}^*) \geq J_2(\tau, \mathbf{x}^\tau; \sigma_1^*, \sigma_2) \quad \forall \sigma_2 \in \Sigma_2,$$

at any admissible initial point $(\tau, \mathbf{x}^\tau)$.

It is important to notice that the definition of an equilibrium property is proposed at *any* admissible initial point and not solely at the initial data $(0, \mathbf{x}^0)$. This differs considerably from the open-loop Nash equilibrium. This is also why the subgame perfectness property, which we shall discuss later on in this chapter, is built into the definition of this equilibrium-solution concept.

---

[28]To save on notations, we still use the same notation J_j to designate the normal form payoff.

6.4.2.2 *A coupled maximum principle for a smooth feedback-Nash equilibrium*

We say that a feedback strategy pair $\underline{\sigma}(t, \mathbf{x})$ is *smooth* if it is continuous in t and its partial derivatives $\frac{\partial}{\partial \mathbf{x}} \underline{\sigma}(t, \mathbf{x})$ exist and are continuous.

Lemma 6.4. *Under Assumption 6.1 if $\underline{\sigma}^*$ is a smooth feedback-Nash equilibrium pair, generating the trajectory $\tilde{\mathbf{x}}^*$ from initial state $\mathbf{x}^0$ for the game (6.40), (6.42), then there exist functions of time $\mathbf{p}_j(\cdot)$, with values in $\mathbf{R}^n$, such that the following relations hold*

$$\mathbf{u}_j^*(t) = \arg \max_{u_j \in U_j} H_j(\mathbf{p}_j(t+1), \mathbf{x}^*(t), \mathbf{u}_j, \sigma_{-j}^*(t, \mathbf{x}^*(t)), t), \tag{6.57}$$

$$\mathbf{p}_j(t)' = \frac{\partial}{\partial \mathbf{x}} H_j(\mathbf{p}_j(t+1), \mathbf{x}^*(t), \underline{\sigma}^*(t, \mathbf{x}^*(t)), t) + \tag{6.58}$$

$$\frac{\partial}{\partial \mathbf{u}_{-j}} H_j(\mathbf{p}_j(t+1), \mathbf{x}^*(t), \underline{\sigma}^*(t, \mathbf{x}^*(t)), t) \frac{\partial}{\partial \mathbf{x}} \sigma_{-j}^*(t, \mathbf{x}^*(t)),$$

$$t = 0, \ldots, T-1, \quad j = 1, 2,$$

$$\mathbf{p}_j(T)' = \frac{\partial}{\partial \mathbf{x}(T)} S_j(\mathbf{x}^*(T)), \quad j = 1, 2. \tag{6.59}$$

Here σ_{-j}^* denote the strategy of the other player than j.

Proof. The proof is straightforward; it relies on the fact that each player j's best reply to the other player's equilibrium strategies is an optimal-control problem, for which the maximum principle in discrete time holds true, under Assumption 6.1; see references [18] or [75] for a more complete proof. ☐

Remark 6.4. The fact that the partial derivative of the strategy of the other player enters the adjoint equation for the costate variable $\mathbf{p}_j(t)$ of Player j makes these necessary conditions quite difficult to use to characterize a feedback equilibrium. This is mainly because we would need to know the very form of the function $\underline{\sigma}^*$ we are looking for, in order to characterize it.

Also, this lemma shows that we cannot expect that feedback-Nash equilibrium will, in general, generate the same state trajectory that does an open-loop equilibrium.

6.4.2.3 *The issue of the existence of feedback-Nash equilibria*

In general, it is difficult to find conditions that guarantee the existence of feedback-Nash equilibria in multistage games. This is in contrast to open-loop Nash equilibria, which are amenable to existence proofs that are very similar to those used in Chapter 3 for concave games. Extending the concave-game approach to multistage games with the open-loop information structure is relatively straightforward. Unfortunately, this is not the case when we are dealing with feedback strategies. This is because it is difficult to obtain the correct framework for invoking the Kakutani fixed-point theorem, within the space of feedback strategies.

Therefore, even if the equilibrium definitions are relatively easy to formulate for a multi-stage feedback or closed-loop game, we cannot be certain that such a solution exists in general. Notwithstanding this uncertainty, we will prove a *verification theorem,* which shows that *if* we can find a solution to the dynamic programming equations, *then* this solution provides a feedback (Markov) equilibrium. Therefore the existence of a feedback-Nash equilibrium can be established for some particular games, for which an explicit solution of the dynamic-programming equations can be obtained.

On the other hand, perhaps counterintuitively, in a stochastic context, certain multistage (supermodular) games defined on lattices, which were considered in Chapter 9 (see Section 9.7.3, page 353), admit feedback equilibria, which can be established via another fixed point theorem due to Tarski.[29] However, we cannot use this method to prove existence in the deterministic case, which is being considered in this chapter.

6.4.3   *Dynamic programming for multistage games*

It is not difficult to extend the dynamic-programming algorithm, evoked in the context of subgame perfectness in Section 5.4.2 (page 145), to the case of dynamic games. Consider the finite-horizon (maximization) dynamic game played by two players denoted j and "not j", i.e., $-j$, respectively. We use this notation because it would be formally the same where there are $m > 2$ players. In that case, $-j$ would refer to the $m - 1$ opponents of Player j. Payoff to Player j, when the initial point is $(\tau, \mathbf{x}^\tau)$ and the control sequences are $\tilde{\mathbf{u}}_j$ and $\tilde{\mathbf{u}}_{-j}$ is defined[30] by

$$J_j \triangleq \sum_{t=\tau}^{T-1} g_j(\mathbf{x}(t), \mathbf{u}_j(t), \mathbf{u}_{-j}(t), t) + S_j(\mathbf{x}(T)), \quad j = 1, 2, \tag{6.60}$$

where the state trajectory is given by

$$\mathbf{x}(t+1) = \mathbf{f}(\mathbf{x}(t), \mathbf{u}_j(t), \mathbf{u}_{-j}(t), t), \; t = \tau, \tau + 1, \ldots T - 1, \tag{6.61}$$

$$\mathbf{x}(\tau) = \mathbf{x}^\tau. \tag{6.62}$$

We assume that each player can observe the current state $(t, \mathbf{x}(t))$. Players j and $-j$ will therefore determine their controls in period t through the use of feedback strategies $\mathbf{u}_{j(t)} = \sigma_j(t, \mathbf{x})$ and $\mathbf{u}_{-j}(t) = \sigma_{-j}(t, \mathbf{x})$.

Let $(\sigma_j^*(t, \mathbf{x}), \sigma_{-j}^*(t, \mathbf{x}))$ be a feedback-equilibrium solution to the multistage game and $\tilde{\mathbf{x}}^* = (\mathbf{x}^*(\tau), \ldots, \mathbf{x}^*(T))$, the associated trajectory emanating from $((\tau, \mathbf{x}_\tau)..$ The *value function* for Player j is defined as

$$W_j^*(\tau, \mathbf{x}_\tau) = \sum_{t=\tau}^{T-1} g_j(\mathbf{x}^*(t), \sigma_j^*(t, \mathbf{x}^*(t)), \sigma_{-j}^*(t, \mathbf{x}^*(t)), t) + S_j(\mathbf{x}(T)), \tag{6.63}$$

$$W_j^*(T, \mathbf{x}) = S_j(\mathbf{x}), \tag{6.64}$$

$$\mathbf{x}^*(t+1) = \mathbf{f}(\mathbf{x}^*(t), \sigma_j^*(t, \mathbf{x}^*(t)), \sigma_{-j}^*(t, \mathbf{x}^*(t)), t), \; t = \tau, \ldots T - 1, \tag{6.65}$$

$$\mathbf{x}^*(\tau) = \mathbf{x}^\tau. \tag{6.66}$$

---

[29] See Theorem 4.4, page 112.

[30] To simplify the writing we do not use the notation $[\mathbf{u}_j(t), \mathbf{u}_{-j}(t)]$, introduced in previous chapter, but more concisely $\mathbf{u}_j(t), \mathbf{u}_{-j}(t)$.

The value function is the payoff that Player j will receive if the feedback-equilibrium strategy is played from initial point (τ, x^τ). The following result, which is sometimes called the *tenet of transition*, establishes a decomposition of the equilibrium conditions over time.

Lemma 6.5. *The value functions $W_j^*(t, \mathbf{x})$ and $W_{-j}^*(t, \mathbf{x})$, defined by (6.63) and (6.64), satisfy the following recurrent equations, backward in time, also known under the name of the **Bellman equations**:*

$$W_j^*(t, \mathbf{x}^*(t)) = \max_{\mathbf{u}_j} g_j(\mathbf{x}^*(t), \mathbf{u}_j, \sigma_{-j}^*(t, \mathbf{x}^*(t)), t) + \tag{6.67}$$

$$W_j^*(t+1, \mathbf{f}(\mathbf{x}^*(t), \mathbf{u}_j, \sigma_{-j}^*(t, \mathbf{x}^*(t)), t))),$$

$$W_{-j}^*(t, \mathbf{x}^*(t)) = \max_{\mathbf{u}_{-j}} g_{-j}(\mathbf{x}^*(t), \sigma_j^*(t, \mathbf{x}^*(t)), \mathbf{u}_{-j}, t) + \tag{6.68}$$

$$W_{-j}^*(t+1, \mathbf{f}(\mathbf{x}^*(t), \sigma_j^*(t, \mathbf{x}^*(t)), \mathbf{u}_{-j}, t))),$$

$$t = T - 1, T - 2, \dots 0, \tag{6.69}$$

as with the boundary condition (6.64).

Proof. We claim that the following holds

$$\sigma_j^*(t, \mathbf{x}^*(t)) = \arg\max_{\mathbf{u}_j} \left\{ g_j(\mathbf{x}(t), \mathbf{u}_j, \sigma_{-j}^*(t, \mathbf{x}^*(t)), t) + W_j^*(t+1, \mathbf{x}^*(t+1))) \right\}.$$

This means that the feedback-equilibrium control of Player j at point $(t, \mathbf{x}^*(t))$ must be the best reply to the feedback control of Player $-j$ at point $(t, \mathbf{x}^*(t))$, combined with the use of a feedback-equilibrium pair at all successive periods. Assume this is not true. Then there exists a control $\mathbf{u}_j^t$ for which

$$W_j^*(T, \mathbf{x}^*(t)) = g_j(\mathbf{x}^*(t), \sigma_j^*(t, \mathbf{x}^*(t)), \sigma_{-j}^*(t, \mathbf{x}^*(t)), t)$$
$$+ W_j^*(t+1, \mathbf{f}(\mathbf{x}^*(t), \mathbf{u}_j, \sigma_j^*(t, \mathbf{x}^*(t)), \sigma_{-j}^*(t, \mathbf{x}^*(t)), t)))$$
$$< g_j(\mathbf{x}^*(t), \mathbf{u}_j^t, \sigma_{-j}^*(t, \mathbf{x}^*(t)), t)$$
$$+ W_j^*(t+1, \mathbf{f}(\mathbf{x}^*(t), \mathbf{u}_j, \sigma_{-j}^*(t, \mathbf{x}^*(t)), t))). \tag{6.70}$$

We define a new strategy σ_j^+ for Player j, such that

$$\sigma_j^+(\tau, \mathbf{x}) = \sigma_j^*(t, \mathbf{x}) \quad \text{if} \quad \tau \neq t$$
$$\sigma_j^+(t, \mathbf{x}) \equiv \mathbf{u}_j^t.$$

It is easy to check that the payoff generated from the initial point $(t, \mathbf{x}^*(t))$ by the strategy pair $(\sigma_j^+, \sigma_{-j}^*)$ is given by the expression in the RHS of (6.70). This will contradict the equilibrium property at point $(t, \mathbf{x}^*(t))$ for $(\sigma_j^*, \sigma_{-j}^*)$. Therefore the claim is true and, as its direct consequence, the recurrence (6.67) holds true. The same reasoning applies to Player $-j$, which establishes the recurrence (6.68). $\square$

The above lemma establishes a necessary condition for a feedback-Nash equilibrium. It also shows that this equilibrium can be decomposed in time. It illustrates the fact that the equilibrium condition must hold in a set of local games, defined at each possible initial point (τ, x^τ).

In the local game defined in $(\tau, \mathbf{x}^\tau)$, the players' actions are $(\mathbf{u}_j, \mathbf{u}_{-j})$ and the payoffs are given by

$$\mathfrak{h}_j(\tau, \mathbf{x}^\tau; \mathbf{u}_j, \mathbf{u}_{-j}) \equiv g_j(\mathbf{x}^\tau, \mathbf{u}_j, \mathbf{u}_{-j}, \tau,) + W_j^*(t+1, \mathbf{f}(\mathbf{x}^\tau, \mathbf{u}_j, \mathbf{u}_{-j}, \tau)). \quad (6.71)$$

Then, the value at $(\tau, \mathbf{x}^\tau)$ of the feedback-equilibrium pair $(\sigma_j^*(\tau, \mathbf{x}^\tau), \sigma_{-j}^*(\tau, \mathbf{x}^\tau))$ is a Nash equilibrium for this local game.

This suggests the following recursive approach for obtaining a feedback-Nash equilibrium:

At time $T - 1$, for any initial point $(T - 1, \mathbf{x}(T - 1))$, solve the local game with payoffs

$$\mathfrak{h}_j(T - 1, \mathbf{x}(T - 1); \mathbf{u}_j, \mathbf{u}_{-j})$$
$$\equiv g_j(\mathbf{x}(T - 1), \mathbf{u}_j, \mathbf{u}_{-j}, T - 1) + S_j(\mathbf{f}(\mathbf{x}(T - 1), \mathbf{u}_j, \mathbf{u}_{-j}, T - 1)), \quad (6.72)$$
$$\mathfrak{h}_{-j}(T - 1, \mathbf{x}(T - 1); \mathbf{u}_j, \mathbf{u}_{-j})$$
$$\equiv g_{-j}(\mathbf{x}(T - 1), \mathbf{u}_j, \mathbf{u}_{-j}, T - 1) + S_{-j}(\mathbf{f}(\mathbf{x}(T - 1), \mathbf{u}_j, \mathbf{u}_{-j}, T - 1)). \quad (6.73)$$

Assume a Nash equilibrium exists for each of these games and call

$$(\sigma_j^*(T - 1, \mathbf{x}(T - 1)), \sigma_{-j}^*(T - 1, \mathbf{x}(T - 1)))$$

the equilibrium strategy vector. This implies that

$$\sigma_j^*(T - 1, \mathbf{x}(T - 1)) = \arg\max_{\mathbf{u}_j} \big\{ g_j(\mathbf{x}(T - 1), \mathbf{u}_j, \sigma_{-j}^*(T - 1, \mathbf{x}(T - 1)), T - 1) +$$
$$S_j(\mathbf{f}(\mathbf{x}(T - 1), \mathbf{u}_j, \sigma_{-j}^*(T - 1, \mathbf{x}(T - 1)), T - 1)) \big\}$$

and similarly for Player $-j$. Then define

$$W_j^*(T-1, \mathbf{x}(T-1)) \equiv g_j(\mathbf{x}(T-1), \sigma_j^*(T-1, \mathbf{x}(T-1)), \sigma_{-j}^*(T - 1, \mathbf{x}(T-1)), T-1) +$$
$$S_j(\mathbf{f}(\mathbf{x}(T-1), \sigma_j^*(T-1, \mathbf{x}^{T-1}), \sigma_{-j}^*(T-1, \mathbf{x}(T-1)), T-1)).$$

At time $T - 2$, for any initial point $(T - 2, x(T - 2))$, solve the local game with payoffs

$$\mathfrak{h}_j(T - 2, \mathbf{x}(T - 2); \mathbf{u}_j, \mathbf{u}_{-j})$$
$$\equiv g_j(\mathbf{x}(T - 2), \mathbf{u}_j, \mathbf{u}_{-j}, T - 2) + W_j^*(T - 1, \mathbf{f}(\mathbf{x}(T - 2), \mathbf{u}_j, \mathbf{u}_{-j}, T - 2)),$$
$$\mathfrak{h}_{-j}(T - 2, \mathbf{x}(T - 2); \mathbf{u}_j, \mathbf{u}_{-j})$$
$$\equiv g_{-j}(\mathbf{x}(T - 2), \mathbf{u}_j, \mathbf{u}_{-j}, T - 2) + W_{-j}^*(T - 1, \mathbf{f}(\mathbf{x}(T - 2), \mathbf{u}_j, \mathbf{u}_{-j}, T - 2)).$$

We also need to assume that the functions $W_j^*(T - 1, \cdot)$ and $W_j^*(T - 1, \cdot)$, identified in the first step of the procedure, are sufficiently smooth for an equilibrium to exist.[31] Assume that an equilibrium exists everywhere and define

$$(\sigma_j^*(T - 2, \mathbf{x}(T - 2)), \sigma_{-j}^*(T - 2, \mathbf{x}(T - 2)))$$

as the equilibrium strategies in the above game. This implies that

$$\sigma_j^*(T - 2, \mathbf{x}(T - 2)) = \arg\max_{\mathbf{u}_j} \big\{ g_j(\mathbf{x}(T - 1), \mathbf{u}_j, \sigma_{-j}^*(T - 2, \mathbf{x}(T - 2)), T - 2) +$$
$$W_j^*(T - 1, \mathbf{f}(\mathbf{x}(T - 2), \mathbf{u}_j, \sigma_{-j}^*(T - 2, \mathbf{x}(T - 2)), T - 2)) \big\}$$

---

[31] Recall that an equilibrium is a fixed-point for a best-reply function, and that this requires some regularity to exist.

and similarly for Player $-j$. Then define

$$W_j^*(T-2, \mathbf{x}(T-2)) \equiv g_j(\mathbf{x}(T-2), \sigma_j^*(T-2, \mathbf{x}(T-2)), \sigma_{-j}^*(T-2, \mathbf{x}(T-2)), T-2)$$
$$+ W_j^*(T-1, \mathbf{f}(\mathbf{x}(T-2), \sigma_j^*(T-2, \mathbf{x}(T-2)), \sigma_{-j}^*(T-2, \mathbf{x}^{T-2}), T-2)).$$

At time $T-3$, etc. proceed recurrently, defining local games and their solutions for all preceding periods, until period 0 is reached.

In the next theorem, we show that this procedure defines a feedback-Nash equilibrium strategy vector.

Theorem 6.1. *If there exist the value functions $W_j^*(t, \mathbf{x})$ and feedback strategies $(\sigma_j^*, \sigma_{-j}^*)$, which satisfy the local-game equilibrium conditions defined in equations (6.64), (6.68) for $t = 0, 1, 2, \ldots, T-1$, $\mathbf{x} \in \mathbf{R}^n$, then the feedback pair $(\sigma_j^*, \sigma_{-j}^*)$ constitutes a (subgame) perfect equilibrium of the dynamic game with the feedback information structure.[32] Moreover, the value function $W_j^*(\tau, \mathbf{x}^\tau)$ represents the equilibrium payoff of Player j for the game starting at point $(\tau, \mathbf{x}^\tau)$.*

Proof. We proceed again by recurrence. Let us start with points $(T-1, \mathbf{x}(T-1))$. At each of these points, the feedback strategies $(\sigma_j^*, \sigma_{-j}^*)$ cannot be improved by a unilateral change by Player j. This is a consequence of the equilibrium condition for the local game (6.72)–(6.73). Therefore the feedback pair $(\sigma_j^*, \sigma_{-j}^*)$ is an equilibrium solution at any initial point at period $T-1$.

Now consider an initial point $(T-2, \mathbf{x})$ at period $T-2$. We compare the outcome for Player j when he uses any feedback σ_j with the payoff he gets when using σ_j^*, while Player $-j$ uses feedback σ_{-j}^*. Player j expects a payoff equal to

$$W_j(T-2, \mathbf{x}) = g_j(\mathbf{x}, \sigma_j(T-2, \mathbf{x}), \sigma_{-j}^*(T-2, \mathbf{x}), T-2) + \mathfrak{h}_j(T-1, \mathbf{x}(T-1);$$
$$\sigma_j(T-1, \tilde{\mathbf{x}}(T-1)), \sigma_{-j}^*(T-1, \tilde{\mathbf{x}}(T-1)), \quad (6.74)$$

where

$$\tilde{\mathbf{x}}(T-1) = \mathbf{f}(\mathbf{x}, \sigma_j(T-2, \mathbf{x}), \sigma_{-j}^*(T-2, \mathbf{x}), T-2). \quad (6.75)$$

Because of the equilibrium property of $(\sigma_j^*, \sigma_{-j}^*)$ at period $T-1$, we have

$$W_j^*(T-1, \tilde{\mathbf{x}}(T-1)) \geq \mathfrak{h}_j(T-1, \mathbf{x}(T-1);$$
$$\sigma_j(T-1, \tilde{\mathbf{x}}(T-1)), \sigma_{-j}^*(T-1, \tilde{\mathbf{x}}(T-1))). \quad (6.76)$$

Allowing for inequality (6.76) in (6.74) we obtain

$$W_j(T-2, \mathbf{x}) \leq g_j(\mathbf{x}, \sigma_j(T-2, \mathbf{x}),$$
$$\sigma_{-j}^*(T-2, \mathbf{x}), T-2) + W_j^*(T-1, \tilde{\mathbf{x}}(T-1)). \quad (6.77)$$

The RHS of (6.77) is, by construction, less than or equal to $W_j^*(T-2, \mathbf{x})$. Therefore we have proved that no other feedback σ_j can improve σ_j^*, for the last two periods.

The verification would go on similarly for all other periods $T-3, \ldots, 0$. $\square$

---

[32] Such an equilibrium is often called feedback-Nash equilibrium or Markov equilibrium.

Remark 6.5. The above result is a **verification theorem**. Jointly with Lemma 6.5, it shows that the dynamic programming and the Bellman equations are both **necessary** and **sufficient** conditions for a feedback-Nash equilibrium to exist. This means that we need to solve the system of Bellman equations (6.69) for Players j and $-j$ to verify that a pair of feedback-equilibrium strategies exists.

Remark 6.6. It would be useful to find conditions insuring that, at each stage of the dynamic-programming iteration, the local games admit a unique solution. If we were able to show when, at every stage, the local games are diagonally strictly concave[33] (recall Definition 3.9 and Theorem 3.6), then we could guarantee that a unique equilibrium exists

$$\underline{\sigma}(t, \mathbf{x}) \equiv (\sigma_j(t, \mathbf{x}(t)), \sigma_{-j}(t, \mathbf{x}(t))) . \tag{6.78}$$

Unfortunately, diagonal strict concavity for a game at t does not generally imply that the game at $t - 1$ possesses this feature.

6.4.4 *Open-loop versus feedback in deterministic games*

Amir [3] notices that an open-loop equilibrium will remain an equilibrium in deterministic dynamic games, when the strategy spaces are expanded to include feedback (Markovian) or memory-based strategies. The reason is that if all of given player's rivals are using open-loop strategies, the player cannot achieve a higher payoff by using more sophisticated strategies than open-loop. This follows directly by invoking the player's best-response problem which, given the open-loop strategies of the rivals, is a deterministic optimal-control problem whose solution (in terms of the implemented actions) does not depend on an information structure. Nonetheless, no general results are known about the comparison of equilibrium *payoffs* under open-loop versus Markovian behavior.

Open-loop equilibria are typically much simpler to analyze than feedback (Markovian) equilibria. In particular, the usually difficult question of existence of pure-strategy equilibrium (refer to Section 6.4.2.3) is most often straightforward in the open-loop case, where it amounts to using a fixed-point theorem with the action set viewed as a subset of the product

$$\underbrace{R^{p_1 + \cdots p_m} \times R^{p_1 + \cdots p_m} \times \ldots R^{p_1 + \cdots p_m}}_{T \text{ times}}, \tag{6.79}$$

under standard conditions on the primitives, see Section 6.4.1.4. The widespread use of open-loop strategies in the early stages of the development of dynamic games can be attributed to this relative simplicity. However, in general, the commitment to a completely specified course of action over the indefinite future is not a realistic behavioral postulate.

Notwithstanding the above reservation, some *real-life* situations make players seek open-loop equilibria. An oligopolistic R&D problem analyzed in [31] (also, see [225]) is such an example. The so-called *patent races* are also naturally modeled as open-loop games. It is evident that the use of open-loop strategies is justified in those models, because it approximates players' real behavior. In those models, an open-loop information

---

[33] See reference [142] for a dynamic game, for which existence and uniqueness of feedback equilibrium was proved via a theorem similar to Theorem 3.6.

structure means that a firm *cannot* observe its rivals' new technology before choosing its own output level.

On the other hand, the possibility of dealing with the problem of existence of equilibria in open-loop games makes them interesting from a theoretical point of view. In particular, Carlson and Haurie provide in [41] (and also in their earlier works) a complete theory of open-loop dynamic games with coupled state constraints.

6.5 A feedback-Nash equilibrium solution to a fishery-management model

Let us now illustrate how dynamic-programming equations can be solved for the case of the fishery-management game from Example 6.2.

6.5.1 *Intuition*

Intuitively, if one player in Example 6.2 catches "a lot of fish" in one period, then the other player's payoff might decrease dramatically. However, so will the first player's payoff in subsequent periods. Hence, depending on the discount factor, a huge harvest in one period might be in neither player's interest. So, there should exist an equilibrium effort that will guarantee reasonable payoffs for every player, over a reasonably long time. The players' equilibrium harvest strategies will be computed below as solutions to the Bellman equations. We deal here with a finite-horizon case and with the infinite horizon in Section 6.6.2.

We will show how the multistage game formulated in Example 6.2 can be solved explicitly for a feedback-Nash equilibrium. The procedure used to identify a feedback-Nash equilibrium solution exploits the particular form of the functions defining the transition rewards and the state equation. Due to this form, it is possible to suggest a "similar" form for the value functions needed in the dynamic programming equations and then, validate this form through the method of undetermined coefficients.[34] This is explained in detail below.

6.5.2 *Identifying a solution to the dynamic-programming equations*

Let us reproduce below the system's dynamics and the payoff formulations of Example 6.2:

$$x(t + 1) = a(x(t) - u_1(t) - u_2(t)) \tag{6.80}$$

and

$$J_j \triangleq \sum_{t=0}^{T-1} \beta_j^t \sqrt{u_j(t)} + K \beta_j^T \sqrt{x(T)} \quad j = 1, 2 \tag{6.81}$$

where the constants are as in Example 6.2. The controls are subject to the constraint

$$x(t) - u_1(t) - u_2(t) \geq 0, \tag{6.82}$$

---

[34]In broad terms, the procedure for solving the Bellman equation can be compared to a method for finding a particular solution to certain non-homogeneous ordinary differential equations (also called *the lucky guess method*). In either case, we must first conjecture a solution form and then verify it.

which could a priori create additional difficulty in finding an equilibrium.[35] However, the particular formulation of this multistage game will make it so that, at equilibrium, the constraint (6.82) is automatically satisfied.

Let us now apply Theorem 6.1 to this dynamic game. The Bellman equations for the two players are written as follows:

$$W_1^*(t, x) = \max_{u_1} \left\{ \beta_1^t \sqrt{u_1} + W_1^*(t + 1, a(x - u_1 - \sigma_2^*(t, x))) \right\}, \tag{6.83}$$

$$W_2^*(t, x) = \max_{u_2} \left\{ \beta_2^t \sqrt{u_2} + W_2^*(t + 1, a(x - \sigma_1^*(t, x) - u_2)) \right\} \tag{6.84}$$

where $(\sigma_1^*(t, x), \sigma_2^*(t, x))$ is the pair of feedback equilibrium strategies such that

$$x - \sigma_1^*(t, x) - \sigma_2^*(t, x) \geq 0$$

and $W_1^*(t, x(t)), W_2^*(t, x(t))$ are the equilibrium value functions, which satisfy the following terminal conditions:

$$W_j^*(T, x) = K \beta_j^T \sqrt{x(T)}, \quad j = 1, 2. \tag{6.85}$$

We notice that using substitution

$$W_j^*(t, x) = \beta_j^t V_j^*(t, x),$$

simplifies equations (6.83)–(6.84) to

$$V_j^*(t, x) = \max_{u_j} \left\{ \sqrt{u_j} + \beta_j V_j(t + 1, a(x - u_j - \sigma_{-j}^*(t, x))) \right\}, \tag{6.86}$$

$$V_j^*(T, x) = K_j \sqrt{x(T)}, \quad j = 1, 2. \tag{6.87}$$

We call $V_j^*(t, x)$ the current-valued equilibrium value function while $W_j^*(t, x)$ can be referred to as the *present-valued value function*.

To solve (6.86), we will use a procedure, sometimes called the *method of undetermined coefficients*, which consists of the five steps below:

- Guess a functional form for $V_j^*(\cdot)$ with a finite number of undetermined coefficients;
- Substitute it back into (6.86);
- Find $\hat{u}_j$ that maximizes the latter;
- Set this $\hat{u}_j$ to (6.86);
- Identify the parameters of the functional form of $V_j(\cdot)$ by comparing the coefficients of the assumed function with those in the left- and right-hand sides of (6.86) (hence, we will determine the coefficients).

Following this procedure, we will conjecture the Bellman value-functions and verify that they satisfy the Bellman equations. Inspired by the form of the terminal or bequest reward and by the form of the transition reward, which all happen to depend on the square

---

[35]The attentive reader will notice that this constraint has the form of a coupled constraint, as in the Rosen games considered in Chapter 3.

root of the control, we propose the following form for the equilibrium current-valued[36] value function

$$V_j^*(t, x) = C_j(t)\sqrt{x}, \tag{6.88}$$
$$V_j^*(T, x) = K_j\sqrt{x} \tag{6.89}$$

where $C_j(t)$ is a function of t that has to be determined, with a known value at T, given by $C_j(T) = K$. We will prove that (6.88) are the value functions if we show that $C_j(t)$ exists, such that the Bellman equations are satisfied.

Substituting $V_j^*(t, x)$ in (6.86) we get

$$\begin{cases} C_1(t)\sqrt{x} = \max_{u_1}\{\sqrt{u_1} + \beta_1 C_1(t+1)\sqrt{a(x - u_1 - \sigma_2^*(t, x))}\}, \\ C_2(t)\sqrt{x} = \max_{u_2}\{\sqrt{u_2} + \beta_2 C_2(t+1)\sqrt{a(x - \sigma_1^*(t, x) - u_2)}\}. \end{cases} \tag{6.90}$$

Let us regroup the parameters and define the constant ϑ as follows:

$$\vartheta_j = a\beta_j^2. \tag{6.91}$$

We can give an interpretation to ϑ_j: the discount factor β_j reflects how much future payoff is worth to Player j, relative to the current ones, and a is the biomass-regrowth coefficient. So, the constant ϑ can be regarded as an aggregate measure of how the player perceives the future and when $\vartheta > 1$, it indicates a rather optimistic outlook.

Using (6.91) enables us to rewrite (6.90) and obtain

$$C_1(t)\sqrt{x} = \max_{u_1}\{\sqrt{u_1} + \sqrt{\vartheta_1}\,C_1(t+1)\sqrt{x - u_1 - \sigma_2^*(t, x)}\}, \tag{6.92}$$

$$C_2(t)\sqrt{x} = \max_{u_2}\{\sqrt{u_2} + \sqrt{\vartheta_2}\,C_2(t+1)\sqrt{x - \sigma_1^*(t, x) - u_1}\} \tag{6.93}$$

with $C_j(T) = K_j$.

The derivative of the right-hand side of (6.92) must be equal to 0 for an interior solution

$$0 = \frac{1}{2\sqrt{u_1}} - \sqrt{\vartheta_1}\,C_1(t+1)\frac{1}{2\sqrt{x - u_1 - \sigma_2(t, x)}}, \tag{6.94}$$

which yields

$$\frac{1}{2\sqrt{u_1}} = \sqrt{\vartheta_1}\,C_1(t+1)\frac{1}{2\sqrt{x - u_1 - \sigma_2(t, x)}},$$

$$u_1 = \frac{x - u_1 - \sigma_2(t, x)}{\vartheta_1\,(C_1(t+1))^2},$$

$$u_1\left(1 + \frac{1}{\vartheta_1\,(C_1(t+1))^2}\right) = \frac{x - \sigma_2(t, x)}{\vartheta_1\,(C_1(t+1))^2},$$

$$u_1 = \frac{x - \sigma_2(t, x)}{\vartheta_1\,(C_1(t+1))^2 + 1}. \tag{6.95}$$

A similar expression holds for the optimum value u_2. This suggests that we can find an equilibrium in the class of linear feedbacks

$$u_j(t) = F_j(t)\,x(t), \tag{6.96}$$

---

[36] As opposed to the *present-valued value function* $e^{-\beta t}V_j^*(t, x)$.

where $F_j(t)$ is a function of $C_j(t)$, and hence of time. If so, we can write using (6.95)

$$F_1(t)\, x = \frac{(1 - F_2(t))}{\vartheta_1\, (C_1(t+1))^2 + 1}\, x\,, \tag{6.97}$$

$$F_2(t)\, x = \frac{(1 - F_1(t))}{\vartheta_2\, (C_2(t+1))^2 + 1}\, x\,, \tag{6.98}$$

which are two equations in $F_1(t)$ and $F_2(t)$ that admit a solution that defines the feedback equilibrium laws.

When the two players are identical ($\vartheta_1 = \vartheta_2 = \vartheta$, and $K_1 = K_2 = K$) the feedback laws of the two players are identical, $u_j = F(t)x, j = 1, 2$. Therefore we can write

$$F(t) = \frac{(1 - F(t))}{\vartheta\, (C(t+1))^2 + 1}\,,$$

which yields

$$F(t)\left(1 + \frac{1}{\vartheta\, (C(t+1))^2 + 1}\right) = \frac{1}{\vartheta\, (C(t+1))^2 + 1}$$

and

$$F(t) = \frac{1}{2 + \vartheta\, (C(t+1))^2}\,. \tag{6.99}$$

From now on, we will keep the assumption of two identical players.

Substituting $F(t)$ in (6.96) and setting it back to (6.92) yields

$$C(t)\sqrt{x} \equiv \frac{1 + \vartheta[C(t+1)]^2}{\sqrt{2 + \vartheta[C(t+1)]^2}}\sqrt{x}\,. \tag{6.100}$$

Identifying the terms adjacent to $\sqrt{x}$ yields a difference equation in $C(t)$,

$$C(t) = \frac{1 + \vartheta[C(t+1)]^2}{\sqrt{2 + \vartheta[C(t+1)]^2}} \quad \text{with} \quad C(0) = K\,, \tag{6.101}$$

which will allow us to compute the full sequence of coefficients

$$C(1), C(2), \ldots, C(T-1),$$

backward in time, starting from $C_j(T) = K_j$. We have thus determined the equilibrium value functions, as suggested in (6.88).

Having computed the coefficients $C(t)$, we can define the feedback strategies (6.99) forward in time[37] and generate the state evolution as follows

$$\sigma_j^*(0, x_0) = \frac{x_0}{2 + \vartheta[C(1)]^2} \quad j = 1, 2, \tag{6.102}$$

$$x(1) = a\, x_0 \left(1 - \frac{2}{2 + \vartheta[C(1)]^2}\right), \tag{6.103}$$

$$\sigma_j^*(1, x(1)) = \frac{x(1)}{2 + \vartheta[C(2)]^2} \quad j = 1, 2, \tag{6.104}$$

$$x(2) = a\, x(1) \left(1 - \frac{2}{2 + \vartheta[C(2)]^2}\right), \tag{6.105}$$

$$\cdots \quad \cdots \tag{6.106}$$

$$\sigma_j^*(T - 1, x(T - 1)) = \frac{x(T - 1)}{2 + \vartheta[C(T)]^2} \quad j = 1, 2, \tag{6.107}$$

$$x(T) = a\, x(T - 1) \left(1 - \frac{2}{2 + \vartheta[C(T)]^2}\right). \tag{6.108}$$

In addition, calculating $C(0)$ enables us to compute the players' payoff at initial point $(0, x^0)$ as

$$J_j(x^0, \underline{\sigma}^*) = W_j^*(0, x^0) = V_j^*(0, x^0) = C(0)\sqrt{x^0}, j = 1, 2. \tag{6.109}$$

Remark 6.7. Because $\vartheta > 0, 2F(t) < 1$ (see (6.99)), we know that $x(t) \geq u_1(t) - u_2(t)$ and that this constraint that we imposed on the strategies is satisfied.

Remark 6.8. If the players are non-symmetric we can also compute the state progression (6.102)–(6.108) and the equilibrium payoffs (6.109). These will be obtained if we solve the simultaneous equations system (6.97)–(6.98) for $F_1(t)$ and $F_2(t)$, substitute them in (6.96) and set back to (6.92). In this way, two recursions that are similar to (6.101) but applied to $C_1(t)$ and $C_2(t)$ will be obtained. Then, using (6.99), the required progression can be calculated.

6.5.3 *Economic interpretation*

There is an interesting economic interpretation of the solution obtained for this multistage game.

Expression (6.101) defines the backward recurrence

$$C(t + 1) \mapsto C(t), \tag{6.110}$$

which can be contracting, i.e., $\dfrac{C(t)}{C(t+1)} < 1$ or expanding, i.e., $\dfrac{C(t)}{C(t+1)} > 1$ (or steady when the ratio is 1). This can be seen in Figure 6.3, obtained for $\vartheta \in [0.9, 1.1]$ and $C(\cdot) \in [0, 3]$. The 45° plane is in blue-grey. We notice that $C(t)$ can have a value below or above this plane. The solid line at level zero is the projection of the intersection between $C(t)$ and the 45° plane.

---

[37]This approach, in which the coefficients determining the controls are computed by using a backward recurrence and then finding the equilibrium trajectory through a forward recurrence is called a *sweep method* in optimal-control literature.

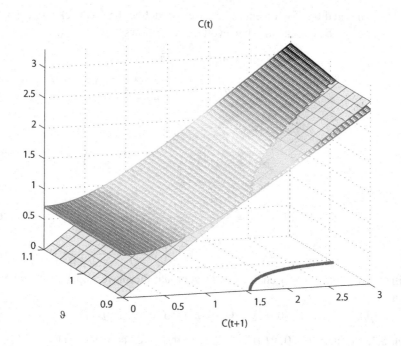

Fig. 6.3   Finite horizon value-function coefficients

If the mapping is expanding backward in time, i.e., $C(t) > C(t + 1)$, then the sequence $C(1), C(2), \ldots, C(T) = K$ is diminishing forward in time; reciprocally, $C(1), C(2), \ldots, C(T) = K$ is an increasing sequence if the mapping is contracting backward in time i.e., $C(t) < C(t + 1)$. Because terms $[C(t + 1)]^2$, $t = 0, \ldots, T$, are in the denominator of (6.99), the harvest intensity increases if mapping (6.101) is expanding and, conversely, harvest intensity decreases if the mapping is contracting.

The former happens if ϑ is sufficiently large, i.e., when the players are optimistic about the future. In such cases, they tend to delay consumption until later. This can take place in fast-growing fisheries, i.e., where $\beta > \dfrac{1}{\sqrt{a}}$. In such fisheries, stock accumulation will be rewarded by future catch (and consumption). Notice, however, that if $T \to \infty$, starvation is optimal for every initial period and all consumption is postponed till infinity. We will see this phenomenon when we discuss the infinite-horizon variant of this game in Section 6.6.2.

If ϑ is small, i.e., when players are not so optimistic about the future, then mapping (6.110) is contracting. So, $C(t)$ increases with t and harvesting intensity decreases. This suggests that when $T \to \infty$, we will obtain a stationary harvesting strategy.[38] This will also be discussed in the infinite-horizon variant of this game in Section 6.6.2.

We can also see that if the bequest-value-function constant $K = C(T)$ is sufficiently large, the mapping is contracting and fishing intensity decreases in t. This can be inter-

---

[38] We say that a strategy is *stationary* if the same decision rule is applied at every time, i.e., $\sigma_j^*(t, x(t)) \equiv \sigma_j^*(\tau, x(\tau)) \equiv \sigma_j^*(x(t))$ for any $t, \tau \in [0, T]$ where $T > 0$ is finite or infinite.

preted as the players' attempt to accumulate fish stocks, to increase the value of the bequest function.

We can analytically characterize the line in plane $\vartheta, C(t+1)$, above which mapping (6.101) is contracting, i.e., the agents' harvesting effort decreases in calendar time.

Contraction backward in time happens when $C(t+1) > C(t)$, i.e.,

$$\frac{1 + \vartheta[C(t)]^2}{\sqrt{2 + \vartheta[C(t)]^2}} > C(t) \Leftrightarrow \left(1 + \vartheta[C(t)]^2\right)^2 > [C(t)]^2(2 + \vartheta[C(t)]^2)$$

$$\Leftrightarrow (1 - \vartheta)\vartheta[C(t)]^4 + (1 - \vartheta)[C(t)]^2 - 1 < 0$$

$$\Leftrightarrow C(t) > \sqrt{\frac{\frac{1}{\sqrt{1-\vartheta}} - 1}{\vartheta}} \tag{6.111}$$

where $t = 0, 1, \ldots, T - 1, T$. So, if $C(T) = K > \sqrt{\dfrac{\frac{1}{\sqrt{1-\vartheta}} - 1}{\vartheta}}$ then mapping (6.101) contracts and the fishermen save the biomass for future generations. The right-hand side (6.111) is represented by the thick line in plane $\vartheta, C(t+1)$ in Figure 6.3. It separates the parameter values, for which the mapping is contracting (above the line) or expanding (below the line).

6.6 Feedback-Nash equilibrium in infinite-horizon multistage games with discounted payoffs

6.6.1 *Verification theorem*

We can extend the method for computing equilibria from Section 6.4.3 to the case of infinite-horizon stationary discounted games. A generic model for such a game is as follows:

$$\mathbf{x}(0) = x^0, \tag{6.112}$$

$$\mathbf{x}(t + 1) = \mathbf{f}(\mathbf{x}(t), \mathbf{u}_1(t), \mathbf{u}_2(t)), \quad t = 0, 1, \ldots, \infty, \tag{6.113}$$

$$J_j \triangleq \sum_{t=0}^{\infty} \beta_j^t g_j(\mathbf{x}(t), \mathbf{u}_j(t), \mathbf{u}_{-j}(t)), \quad j = 1, 2. \tag{6.114}$$

Let $(\sigma_j^*, \sigma_{-j}^*)$ be a pair of stationary feedback-Nash equilibrium strategies. The value function for Player j is defined by

$$\mathbf{x}^*(\tau) = x^\tau, \tag{6.115}$$

$$\mathbf{x}^*(t + 1) = \mathbf{f}(\mathbf{x}^*(t), \sigma_1^*(\mathbf{x}^*(t)), \sigma_2^*(\mathbf{x}^*(t))), \quad t = 0, 1, \ldots, \infty, \tag{6.116}$$

$$W_j^*(\tau, \mathbf{x}^\tau) = \sum_{t=\tau}^{\infty} \beta_j^t g_j(\mathbf{x}^*(t), \sigma_1^*(\mathbf{x}^*(t)), \sigma_2^*(\mathbf{x}^*(t))), \quad j = 1, 2, \tag{6.117}$$

where β_j^t is the power t of the discount factor $\beta_j < 1$ used by Player j. Notice that, due to the geometric discounting, we can always write

$$W_j^*(\tau, \mathbf{x}) = \beta_j^\tau W_j^*(0, \mathbf{x}). \tag{6.118}$$

We will call $V_j^*(\mathbf{x}) \equiv W_j^*(0, \mathbf{x}))$, the current-valued value function. Similar expressions hold for Player $-j$.

We can easily prove an analog to Lemma 6.5 for this infinite-horizon discounted game.

Lemma 6.6. *The current-valued value functions $V_j^*(\mathbf{x})$ and $V_{-j}^*(\mathbf{x})$, defined in (6.118), satisfy the following recurrence equations, backward in time, also called **Bellman equations**:*

$$V_j^*(\mathbf{x}^*(t)) = \max_{\mathbf{u}_j} g_j(\mathbf{x}^*(t), \mathbf{u}_j, \sigma_{-j}^*(\mathbf{x}^*(t))) + \tag{6.119}$$

$$\beta V_j^*(\mathbf{f}(\mathbf{x}^*, \mathbf{u}_j, \sigma_{-j}^*(\mathbf{x}^*(t)), t))),$$

$$V_{-j}^*(\mathbf{x}^*(t)) = \max_{\mathbf{u}_{-j}} g_{-j}(\mathbf{x}^*(t), \sigma_j^*(\mathbf{x}^*(t)), \mathbf{u}_{-j}) + \tag{6.120}$$

$$\beta V_{-j}^*(\mathbf{f}(\mathbf{x}^*(t), \sigma_j^*(\tilde{\mathbf{x}}^*(t)), \mathbf{u}_{-j}))),$$

$$t = 0, 1, \ldots \infty.$$

Proof. The proof is identical to the one proposed for Lemma 6.5, except for obvious changes in the definition of current-valued value functions.   □

Remark 6.9. There is, however, an important difference between Lemma 6.5 and Lemma 6.6. The boundary conditions (6.64), which determine the value function at the final time, do not exist anymore for the infinite-horizon discounted-game case.

This above remark suggests that we can focus on the local current-valued games, at any initial point $\mathbf{x}$, with payoffs defined by

$$\mathfrak{h}_j(\mathbf{x}; \mathbf{u}_j, \mathbf{u}_{-j}) \equiv g_j(\mathbf{x}, \mathbf{u}_j, \mathbf{u}_{-j}) + \beta_j V_j^*(\mathbf{f}(\mathbf{x}, \mathbf{u}_j, \mathbf{u}_{-j})), \tag{6.121}$$

$$\mathfrak{h}_{-j}(\mathbf{x}; \mathbf{u}_j, \mathbf{u}_{-j}) \equiv g_{-j}(\mathbf{x}, \mathbf{u}_j, \mathbf{u}_{-j}) + \beta_j V_{-j}^*(\mathbf{f}(\mathbf{x}, \mathbf{u}_j, \mathbf{u}_{-j})). \tag{6.122}$$

Theorem 6.2. *Let the value functions $V_j^*(\mathbf{x})$ and $V_{-j}^*(\mathbf{x})$ with a stationary feedback strategy vector $(\sigma_j^*, \sigma_{-j}^*)$, such that the following holds true*

$$V_j^*(\mathbf{x}) = \max_{\mathbf{u}_j} \mathfrak{h}_j(\mathbf{x}; \mathbf{u}_j, \sigma_{-j}^*(\mathbf{x})) = \mathfrak{h}_j(\mathbf{x}; \sigma_j^*(\mathbf{x}), \sigma_{-j}^*(\mathbf{x})), \tag{6.123}$$

$$V_{-j}^*(\mathbf{x}) = \max_{\mathbf{u}_{-j}} \mathfrak{h}_{-j}(\mathbf{x}; \sigma_j^*(\mathbf{x}), \mathbf{u}_{-j}) = \mathfrak{h}_{-j}(\mathbf{x}; \sigma_j^*(\mathbf{x}), \sigma_{-j}^*(\mathbf{x})), \tag{6.124}$$

where $\mathfrak{h}_j$ and $\mathfrak{h}_{-j}$ are defined as in (6.121) and (6.122). Then $(\sigma_j^*, \sigma_{-j}^*)$ is a pair of stationary feedback-Nash equilibrium strategies and $V_j^*(\mathbf{x})$ (resp. $V_{-j}^*(\mathbf{x})$) is the current-valued equilibrium value function for Player j (resp. Player $-j$).

Proof. Let us consider the auxiliary finite-horizon game, with periods $0, 1, \ldots, T$, the same transition rewards, and dynamics as in the infinite-horizon game, and with terminal payoff functions $\beta^T V_j^*(\mathbf{x}), \beta^T V_{-j}^*(\mathbf{x})$ at period T. Applying Theorem 6.1, it is easy to check that this finite-horizon game admits the stationary feedback pair $(\sigma_j^*, \sigma_{-j}^*)$ as a feedback-Nash equilibrium. Furthermore the equilibrium payoff for the finite game is the same as the equilibrium payoff for the infinite-horizon game, i.e., $V_j^*(\mathbf{x})$ at point $(0, \mathbf{x})$.

Now consider any feedback law $\sigma_j(t, \mathbf{x})$ for Player j and compare the payoffs obtained by Player j when the feedback[39] pairs $(\sigma_j(t, \cdot), \sigma_{-j}^*(\cdot))$ are used for all t with the equilibrium payoff, starting from any initial point $(\tau, \mathbf{x}^\tau)$. This payoff is given by the infinite sum

$$V_j(\tau, \mathbf{x}^\tau) = \sum_{t=\tau}^{\infty} \beta^t \, g_j(\mathbf{x}(t), \sigma_j(t, \mathbf{x}(t)), \sigma_{-j}^*(\mathbf{x}(t))), \qquad (6.125)$$

where

$$\tilde{\mathbf{x}} = (\mathbf{x}(\tau), \mathbf{x}(\tau + 1), \dots, \mathbf{x}(t), \dots)$$

is the state trajectory generated by $(\sigma_j(t, \cdot), \sigma_{-j}^*(\cdot))$ from the initial point $(\tau, \mathbf{x}^\tau)$. We claim that the following holds true:

$$\beta^\tau V_j^*(\mathbf{x}^\tau) \geq \sum_{t=\tau}^{T-1} \beta^t \, g_j(\mathbf{x}(t), \sigma_j(t, \mathbf{x}(t)), \sigma_{-j}^*(\mathbf{x}(t))) + \beta^T V_j^*(\mathbf{x}(T)). \qquad (6.126)$$

This is a direct consequence of $(\sigma_j^*, \sigma_{-j}^*)$ being a feedback-Nash equilibrium for the finite-horizon game with terminal reward $\beta^T V_j^*(\mathbf{x})$. Let us write the inequality (6.126) as follows

$$\beta^\tau V_j^*(\mathbf{x}^\tau) - \beta^T V_j^*(\mathbf{x}(T)) \geq \sum_{t=\tau}^{T-1} \beta^t \, g_j(\mathbf{x}(t), \sigma_j(t, \mathbf{x}(t)), \sigma_{-j}^*(\mathbf{x}(t))) \qquad (6.127)$$

and let T tend to ∞; we then obtain $\beta^T \to 0$, and so

$$\beta^\tau V_j^*(\mathbf{x}^\tau) \geq \sum_{t=\tau}^{\infty} \beta^t \, g_j(\mathbf{x}(t), \sigma_j(t, \mathbf{x}(t)), \sigma_{-j}^*(\mathbf{x}(t))) = V_j(\tau, \mathbf{x}^\tau), \qquad (6.128)$$

which shows that $\sigma_j(t, \cdot), \sigma_{-j}^*(\mathbf{x}(t))$ does not improve the value function compared to the one obtained for $(\sigma_j^*, \sigma_{-j}^*)$ at point $(\tau, \mathbf{x}^\tau)$. This can be repeated at any feasible initial point, and so, the theorem is proved. $\qquad \square$

Remark 6.10. A feedback equilibrium is, by construction, subgame perfect.

Remark 6.11. Theorems 6.1 and 6.2, the verification theorems, provide sufficient conditions for feedback-Nash (Markovian) equilibria. This means that if we manage to solve the corresponding Bellman equations, then we can claim the equilibrium's existence. A consequence of this fact is that the methods for solving Bellman equations are of prime interest to economists and managers wanting to characterize or implement feedback equilibria.

Remark 6.12. The dynamic-programming algorithm requires that we determine the value functions $W_j(t, \mathbf{x})$, or $W_j(\mathbf{x})$, for every $\mathbf{x} \in \mathbf{X} \subset \mathbf{R}^n$. This can be achieved in practice only if $\mathbf{X}$ is a finite set, **or** the system of Bellman equations has an "easy" analytical solution, e.g., one that could be obtained through the **method of undetermined coefficients**. If $W_j(\cdot, \mathbf{x})$ is affine or quadratic in $\mathbf{x}$ then the method is easily applicable.

---

[39] Recall that $\sigma_{-j}^*(\cdot)$ is stationary and so does not depend on t.

Remark 6.13. The case of quadratic value functions occurs when the dynamic game has linear dynamics and quadratic-payoff functions. These problems are called **linear-quadratic games** and feedback equilibria of such games can be expressed using coupled **Riccati** equations. We dedicate Section 6.7 to linear-quadratic games.

Remark 6.14. In general, however, when $\mathbf{X}$ is a subset of $\mathbf{R}^n$ and W_j cannot be assumed to be linear or quadratic, or of some other pre-determined form (like in Sections 6.5 and 6.6.2, below), the only way to tackle the problem of computation of W_j is to approximate $\mathbf{X}$ by a finite set, say $\mathbf{X}^h$, and to compute an equilibrium for the new game with the discrete state space $\mathbf{X}^h$ (this equilibrium may or may not converge to the original game equilibrium as $h \to 0$). The practical limitation of this approach is that the cardinality of $\mathbf{X}^h$ tends to be high (this is the famous **curse of dimensionality** mentioned by Bellman in his original work on dynamic programming).

6.6.2 An application: An infinite-horizon feedback-Nash equilibrium solution to the fishery's problem

We show here how the dynamic game formulated in Example 6.2 and solved for the case of a finite horizon in Section 6.5, can be solved when the players plan to use the fishery in perpetuity. The method, based on dynamic programming is useful as it provides an example of an infinite-horizon dynamic games, which we can solve with Markovian strategies. Although our model is stylized, we are able to discuss several important economic issues arising in long-term competition.

We repeat the game formulation from Example 6.2. We have the state equation

$$x(t+1) = a(x(t) - u_1(t) - u_2(t)), \qquad (6.129)$$

with the payoffs to the players:

$$J_j \triangleq \sum_{t=0}^{\infty} \beta^t \sqrt{u_j(t)} \quad j = 1, 2 \qquad (6.130)$$

where the constants are as in Example (6.2). We endeavor to compute the pair of feedback stationary equilibrium strategies (σ_1^*, σ_2^*) that satisfy (6.123)–(6.124). The strategies must also be feasible, in that $x - \sigma_1^*(x) - \sigma_2^*(x) \geq 0$.

We apply Theorem 6.2 to this game. The Bellman equations for the players are

$$W_1^*(x) = \max_{u_1} \left\{ \sqrt{u_1} + \beta W_1^*(a(x - u_1 - \sigma_2^*(x))) \right\}, \qquad (6.131)$$

$$W_2^*(x) = \max_{u_2} \left\{ \sqrt{u_2(t)} + \beta W_2^*(a(x - \sigma_1^*(x) - u_2)) \right\} \qquad (6.132)$$

where $W_1^*(x), W_2^*(x)$ are the Bellman value functions and $(\sigma_1^*(x), \sigma_2^*(x))$ is the pair of equilibrium strategies.

Assuming sufficient regularity of the right-hand sides of (6.131) and (6.132), i.e., differentiability, then the feedback-equilibrium Nash harvest strategies have to satisfy

$$
\begin{cases}
\dfrac{1}{2\sqrt{u_1}} = \beta \dfrac{\partial}{\partial u_1} W_1^*(a(x - u_1 - \sigma_2^*(x))), \\[4mm]
\dfrac{1}{2\sqrt{u_2}} = \beta \dfrac{\partial}{\partial u_2} W_2^*(a(x - u_2 - \sigma_1^*(x))).
\end{cases}
\tag{6.133}
$$

We will assume the value functions[40] $W_1^*(x) = C\sqrt{x}$ and $W_2^*(x) = C\sqrt{x}$. Notice that, because of symmetry, we assume that the value functions are the same for each player. Now, we expand the necessary conditions (6.133) and solve them simultaneously; we obtain

$$
\sigma_1^*(x) = \sigma_2^*(x) = \frac{x}{2 + \beta^2 C^2 a}.
\tag{6.134}
$$

We will show later on that these strategies are feasible. Substituting these forms for u_1 and u_2 in the Bellman equations (6.131), (6.132) yields the following identity (after some algebra)

$$
C\sqrt{x} \equiv \frac{1 + \beta^2 C^2 a}{\sqrt{2 + \beta^2 C^2 a}} \sqrt{x}.
\tag{6.135}
$$

Hence,

$$
C = \frac{1 + \beta^2 C^2 a}{\sqrt{2 + \beta^2 C^2 a}}.
\tag{6.136}
$$

If C exists such that (6.136) is satisfied, then $W_1^*(x), W_2^*(x)$ solve the Bellman equations.

Solving for C involves a double-quadratic equation. The only solution, which is real and positive

$$
\bar{C} = \frac{1}{\beta} \sqrt{\frac{\frac{1}{\sqrt{1 - a\beta^2}} - 1}{a}}
\tag{6.137}
$$

requires

$$
a\beta^2 < 1.
\tag{6.138}
$$

Remark 6.15. Here we use the original a and β rather than $\vartheta = a\beta^2$ introduced in Section 6.5. Symbol ϑ was useful at that time, otherwise the three parameters $a, \beta, C(t + 1)$ would have had to be taken into account. However, here, analyzing a, β will be sufficient.

Remark 6.16. A necessary condition for a feedback equilibrium in the fishery's game (6.129)–(6.130) is (6.138). We notice that this condition will **not** be satisfied by economic systems with fast growth and very patient (or forward-looking) players and, as a consequence, games played in such economies will not have equilibrium solutions. Intuitively, the players wait for an infinite amount of time, to catch an infinite amount of fish. (Compare Lemma 6.7 and also Lemma 6.12 in this chapter's GE illustration.)

---

[40]To solve a functional equation like (6.131), we again use the *method of undetermined coefficients*.

Substituting C in the strategies and value functions yields

$$\sigma_1^*(x) = \sigma_2^*(x) = \frac{x}{\frac{1}{\sqrt{1-a\,\beta^2}}+1}\,, \tag{6.139}$$

$$W_1^*(x) = W_2^*(x) = \frac{1}{\beta}\sqrt{\frac{\frac{1}{\sqrt{1-a\,\beta^2}}-1}{a}}\,\sqrt{x}\,. \tag{6.140}$$

In economics, a model of growth represented by equation (6.129) is classified as an "AK" model.[41] A known feature of AK models is that zero is their unique steady state. The zero steady state would predict extinction in our fishery's game. Let us examine under which conditions the equilibrium strategies (6.139) will lead to extinction.

Let us compute a steady-state solution. Substitute the equilibrium strategies in the state equation (6.129).

$$x(t+1) = a\left(x(t) - \frac{x(t)}{\frac{1}{\sqrt{1-a\,\beta^2}}+1} - \frac{x(t)}{\frac{1}{\sqrt{1-a\,\beta^2}}+1}\right)$$

$$= x(t)\cdot a\left(1 - \frac{2}{\frac{1}{\sqrt{1-a\,\beta^2}}+1}\right) = x(t)\cdot a\,\frac{1-\sqrt{1-a\,\beta^2}}{1+\sqrt{1-a\,\beta^2}}\,. \tag{6.141}$$

We can rewrite (6.141) in the following form:

$$x(t+1) = \lambda x(t) \tag{6.142}$$

where

$$\lambda \equiv a\,\frac{1-\sqrt{1-a\,\beta^2}}{1+\sqrt{1-a\,\beta^2}}$$

is the state equation's *eigenvalue*. We first notice that $\lambda > 0$, if (6.138) is satisfied.

We know from the difference-equations analysis (see, e.g., [165]) that (6.142) has a unique steady state at zero, which is asymptotically stable if $\lambda < 1$ and unstable if $\lambda > 1$. We can formulate the following lemmas concerning the long-term exploitation of the fishery in game (6.129)–(6.130).

Lemma 6.7. *Feedback-equilibrium strategies (6.139) lead to the fishery's extinction when the steady state of (6.142) is asymptotically stable, i.e., when*

$$1 < a < \frac{2}{\beta} - 1.$$

Proof. We need to show that

$$\lambda = a\,\frac{1-\sqrt{1-a\,\beta^2}}{1+\sqrt{1-a\,\beta^2}} < 1\,. \tag{6.143}$$

---

[41] Any linear growth model where "capital" x (or 'K') expands proportionally to the growth coefficient a is called an *AK model*.

Then,

$$a \left(1 - \sqrt{1 - a\,\beta^2}\right) < 1 + \sqrt{1 - a\,\beta^2}\,,$$

$$a - 1 < (a + 1)\sqrt{1 - a\,\beta^2}$$

and, after some more algebra

$$\left(-\frac{2}{\beta} - 1\right) < a < \left(\frac{2}{\beta} - 1\right).$$

However, by assumption, the growth coefficient a is bigger than 1. Thus

$$1 < a < \left(\frac{2}{\beta} - 1\right). \tag{6.144}$$
$$\Box$$

We also notice that

$$\left(\frac{2}{\beta} - 1\right) < \frac{1}{\beta^2} \quad \text{for} \quad \beta < 1, \tag{6.145}$$

so (6.144) is compatible with the equilibrium existence condition (6.138).

Lemma 6.8. *Feedback-equilibrium strategies (6.139) do **not** lead to the fishery's extinction if*

$$\frac{2}{\beta} - 1 < a < \frac{1}{\beta^2},$$

which means that the steady state of (6.142) is unstable.

Proof. So, we require

$$\lambda = a \frac{1 - \sqrt{1 - a\,\beta^2}}{1 + \sqrt{1 - a\,\beta^2}} > 1. \tag{6.146}$$

We know from the previous lemma that the unique steady state (zero) is unstable if

$$a > \frac{2}{\beta} - 1.$$

On the other hand, (6.138) has to be satisfied for the equilibrium to exist. We know from (6.145) that there are a, which satisfy both. $\Box$

We can graphically characterize the above value functions and strategies; see Figures 6.4 and 6.5.

In the first of these figures, we see two lines in the plane (a, β): the solid line represents $a = \frac{1}{\beta^2}$ and the thinner (dotted) line shows where $a = \frac{2}{\beta} - 1$. We observe that the value function diverges to infinity when (a, β) are above the solid line and that the value decreases, rather sharply, when they are below the thin line. The decrease in value can be attributed to extensive fishing (actually, overfishing leading to extinction), which is visible in Figure 6.5 when a, β belong to the region below the dotted line.

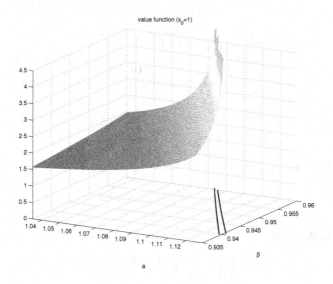

Fig. 6.4   Dependence of value function on β and a

An obvious conclusion drawn from Figures 6.4 and 6.5 is that the larger β, i.e., the more "forward-looking" the players are, the smaller the catch.

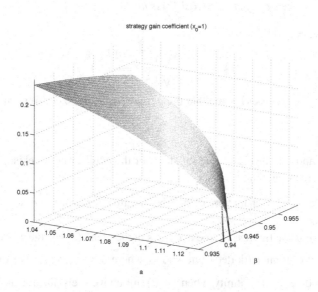

Fig. 6.5   Dependence of strategy on β and a

6.7 Linear-quadratic games

6.7.1 *General presentation*

We consider here an m-person generalization of the optimal-control problem discussed in Section 6.1.4. We have also invoked this class of dynamic games in Remarks 6.12 and 6.13. These are the so-called *linear-quadratic* multistage games.

An important feature of these games is that they admit a solution that can be fully characterized analytically. They can also model an array of economic and managerial problems, see, e.g., [59] (also see Sections 7.6 and 7.7.1 on pages 261 and 265, respectively[42]).

In a general[43] linear-quadratic game, player j's payoff is given by

$$J_j \triangleq \sum_{t=0}^{T-1} \left\{ \frac{1}{2}\mathbf{x}(t)'\mathbf{Q}^j(t)\mathbf{x}(t) + \frac{1}{2}\mathbf{u}_j(t)'\mathbf{R}^j(t)\mathbf{u}_j(t) \right. \tag{6.147}$$

$$\left. + \sum_{i \neq j} \mathbf{u}_i(t)'\mathbf{R}^{ij}(t)\mathbf{u}_j(t) + \mathbf{q}^j(t)\mathbf{x}(t) + \mathbf{p}^j(t)\mathbf{u}_j(t) \right\}$$

$$+ \frac{1}{2}\mathbf{x}(T)'\mathbf{Q}^j(T)\mathbf{x}(T) + \mathbf{q}^j(T)\mathbf{x}(T), \; j = 1, 2, \dots m$$

and the state equation are defined by

$$\mathbf{x}(t+1) = \mathbf{A}(t)\mathbf{x}(t) + \sum_{j \in M} \mathbf{B}_j(t)\mathbf{u}_j(t), \, t = 0, 1, \dots, T-1, \tag{6.148}$$

with initial conditions

$$\mathbf{x}(0) = \mathbf{x}^0. \tag{6.149}$$

This game model can allow for any number of players, state and control variables. Here,[44] $\mathbf{x}(t)$ and $\mathbf{u}_j(t)$ denote, respectively, the state vector (an element of $\mathbf{R}^n$) and Player j's action vector (an element of $\mathbf{R}^{l_j}$), at time t; $\mathbf{A}(t), \mathbf{B}_j(t), \mathbf{Q}^i(t), \mathbf{R}^{ij}(t), \mathbf{q}^j(t), \mathbf{p}^j(t)$ are matrices and vectors of appropriate dimensions, whereas $\mathbf{R}^j(t)$ is negative definite, and $\mathbf{Q}^j(t+1)$ is symmetric and negative semi-definite.

We look for a feedback-Nash equilibrium solution and will solve the DP equations (see Section 6.4.3), where the value functions are assumed to take the form

$$V_j^*(t, \mathbf{x}) = \frac{1}{2}\mathbf{x}'\mathbf{S}^j(t)\mathbf{x} + \mathbf{h}^j(t)\mathbf{x} + s^j(t). \tag{6.150}$$

We will identify affine feedback laws that are the equilibrium strategies, corresponding the to the above value functions (6.150)

$$\sigma_j^*(t, \mathbf{x}) = -\mathbf{P}^j(t)\mathbf{x} + \mathbf{r}^j(t). \tag{6.151}$$

---

[42]Those sections discuss continuous-time games. However, the underlying problems could also be represented in discrete time and hence the reader should see which "real-life" problems can be modeled as linear-quadratic games (discrete *or* continuous time).

[43]Notice that the objective function matrices and vectors $\mathbf{Q}, \mathbf{R}, \mathbf{q}$, *etc* depend on time and could comprise a discount factor.

[44]As before, the transpose operation is denoted by the "prime" $'$ sign.

The local game defined by the DP equation, at stage t and state $\mathbf{x}$, defines payoffs as functions of control actions $\mathbf{u}_1, \ldots, \mathbf{u}_m$,

$$\frac{1}{2}\mathbf{x}'\mathbf{Q}^j(t)\mathbf{x} + \frac{1}{2}\mathbf{u}_j'\mathbf{R}^j(t)\mathbf{u}_j + \sum_{i \neq j}\mathbf{u}_i'\mathbf{R}^{ij}(t)\mathbf{u}_j + \mathbf{q}^j(t)\mathbf{x} + \mathbf{p}^j(t)\mathbf{u}_j$$

$$+\frac{1}{2}[\mathbf{A}(t)\mathbf{x} + \sum_{i \in M}\mathbf{B}_i(t)\mathbf{u}_i]'\mathbf{S}^j(t+1)[\mathbf{A}(t)\mathbf{x} + \sum_{i \in M}\mathbf{B}_i(t)\mathbf{u}_i]$$

$$+\mathbf{h}^j(t+1)[\mathbf{A}(t)\mathbf{x} + \sum_{i \in M}\mathbf{B}_i(t)\mathbf{u}_i]$$

$$j = 1, \ldots, m. \tag{6.152}$$

The local-game payoff of Player j must be maximized with respect to $\mathbf{u}_j$. This implies that the gradient in $\mathbf{u}_j$ must be equal to 0:

$$0 = \mathbf{u}_j'\mathbf{R}^j(t) + \sum_{i \neq j}\mathbf{u}_i'\mathbf{R}^{ij}(t) + \mathbf{p}^j(t) + [\mathbf{A}(t)\mathbf{x} + \sum_{i \in M}\mathbf{B}_i(t)\mathbf{u}_i]'\mathbf{S}^j(t+1)\mathbf{B}_j(t)$$

$$+\mathbf{h}^j(t+1)\mathbf{B}_j(t),$$

$$j = 1, \ldots, m. \tag{6.153}$$

Hence

$$0 = \mathbf{u}_j'[\mathbf{R}^j(t) + \mathbf{B}_j(t)'\mathbf{S}^j(t+1)\mathbf{B}_j(t)] + \mathbf{x}'\mathbf{A}(t)'\mathbf{S}^j(t+1)\mathbf{B}_j(t)$$

$$+\sum_{i \neq j}\mathbf{u}_i'(\mathbf{R}^{ij}(t) + \mathbf{B}_i(t)'\mathbf{S}^j(t+1)\mathbf{B}_j(t)) + \mathbf{p}^j(t+1) + \mathbf{h}^j(t+1)\mathbf{B}_j(t),$$

$$j = 1, \ldots, m. \tag{6.154}$$

Expressing the control according to the feedback law (6.151) we get

$$0 = [-\mathbf{P}^j(t)\mathbf{x} + \mathbf{r}^j(t)]'[\mathbf{R}^j(t) + \mathbf{B}_j(t)'\mathbf{S}^j(t+1)\mathbf{B}_j(t)] + \mathbf{x}'\mathbf{A}(t)'\mathbf{S}^j(t+1)\mathbf{B}_j(t)$$

$$+\sum_{i \neq j}[-\mathbf{P}^i(t)\mathbf{x} + \mathbf{r}^i(t)]'[\mathbf{R}^{ij}(t) + \mathbf{B}_i(t)'\mathbf{S}^j(t+1)\mathbf{B}_j(t)]$$

$$+\mathbf{p}^j(t+1) + \mathbf{h}^j(t+1)\mathbf{B}_j(t),$$

$$j = 1, \ldots, m. \tag{6.155}$$

Identifying the terms in $\mathbf{x}$ gives

$$0 = -\mathbf{P}^j(t)'[\mathbf{R}^j(t) + \mathbf{B}_j(t)'\mathbf{S}^j(t+1)\mathbf{B}_j(t)] + \mathbf{A}(t)'\mathbf{S}^j(t+1)\mathbf{B}_j(t)$$

$$-\sum_{i \neq j}\mathbf{P}^i(t)'[\mathbf{R}^{ij}(t) + \mathbf{B}_i(t)'\mathbf{S}^j(t+1)\mathbf{B}_j(t)], \tag{6.156}$$

while identifying the constant terms gives

$$= \mathbf{r}^j(t)'[\mathbf{R}^j(t) + \mathbf{B}_j(t)'\mathbf{S}^j(t+1)\mathbf{B}_j(t)] + \sum_{i \neq j}\mathbf{r}^i(t)'[\mathbf{R}^{ij}(t) + \mathbf{B}_i(t)'\mathbf{S}^j(t+1)\mathbf{B}_j(t)]$$

$$+\mathbf{p}^j(t+1) + \mathbf{h}^j(t+1)\mathbf{B}_j(t) \tag{6.157}$$

for $j = 1, 2, \ldots, m; t = 0, 1, \ldots, T - 1$.

These conditions are complemented by the recurrence on the value functions

$$\frac{1}{2}\mathbf{x}'\mathbf{S}^j(t)\mathbf{x} + \mathbf{h}^j(t)\mathbf{x} + s^j(t) = \frac{1}{2}\mathbf{x}'\mathbf{Q}^j(t)\mathbf{x} + \frac{1}{2}\mathbf{u}'_j\mathbf{R}^j(t)\mathbf{u}_j$$

$$+ \sum_{i \neq j} \mathbf{u}'_i\mathbf{R}^{ij}\mathbf{u}_j + \mathbf{q}^j(t)\mathbf{x} + \mathbf{p}^j(t)\mathbf{u}_j$$

$$+ \frac{1}{2}[\mathbf{A}(t)\mathbf{x} + \sum_{i \in M}\mathbf{B}_i(t)\mathbf{u}_i]'\mathbf{S}^j(t+1)[\mathbf{A}(t)\mathbf{x} + \sum_{i \in M}\mathbf{B}_i(t)\mathbf{u}_i]$$

$$+ \mathbf{h}^j(t+1)[\mathbf{A}(t)\mathbf{x} + \sum_{i \in M}\mathbf{B}_i(t)\mathbf{u}_i] + s^j(t+1),$$

$$j = 1, \ldots, m, \tag{6.158}$$

where the controls are replaced by the affine feedback laws,

$$\frac{1}{2}\mathbf{x}'\mathbf{S}^j(t)\mathbf{x} + \mathbf{h}^j(t)\mathbf{x} + s^j(t) = \frac{1}{2}\mathbf{x}'\mathbf{Q}^j(t)\mathbf{x}$$

$$+ \frac{1}{2}\left[-\mathbf{P}^j(t)\mathbf{x} + \mathbf{r}^j(t)\right]'\mathbf{R}^j(t)[-\mathbf{P}^j(t)\mathbf{x} + \mathbf{r}^j(t)]$$

$$+ \sum_{i \neq j}[-\mathbf{P}^i(t)\mathbf{x} + \mathbf{r}^i(t)]'\mathbf{R}^{ij}[-\mathbf{P}^j(t)\mathbf{x} + \mathbf{r}^j(t)]$$

$$+ \mathbf{q}^j(t)\mathbf{x} + \mathbf{p}^j(t)[-\mathbf{P}^j(t)\mathbf{x} + \mathbf{r}^j(t)]$$

$$+ \frac{1}{2}[\mathbf{A}(t)\mathbf{x} + \sum_{i \in M}\mathbf{B}_i(t)(-\mathbf{P}^i(t)\mathbf{x} + \mathbf{r}^i(t))]'\mathbf{S}^j(t+1)[\mathbf{A}(t)\mathbf{x}$$

$$+ \sum_{i \in M}\mathbf{B}_i(t)(-\mathbf{P}^i(t)\mathbf{x} + \mathbf{r}^i(t))]$$

$$+ \mathbf{h}^j(t+1)[\mathbf{A}(t)\mathbf{x} + \sum_{i \in M}\mathbf{B}_i(t)(-\mathbf{P}^i(t)\mathbf{x} + \mathbf{r}^i(t))] + s^j(t+1),$$

$$j = 1, \ldots, m. \tag{6.159}$$

Identifying the quadratic terms in both sides we obtain the recurrence

$$\frac{1}{2}\mathbf{S}^j(t) = \frac{1}{2}\,\mathbf{Q}^j(t) + \frac{1}{2}\mathbf{P}^j(t)'\mathbf{R}^j(t)\mathbf{P}^j(t) + \sum_{i \neq j}\mathbf{P}^i(t)'\mathbf{R}^{ij}\mathbf{P}^j(t)$$

$$+ \frac{1}{2}[\mathbf{A}(t) - \sum_{i \in M}\mathbf{B}^i(t)\mathbf{P}^i(t)]'\mathbf{S}^j(t+1)[\mathbf{A}(t) - \sum_{i \in M}\mathbf{B}_i(t)\mathbf{P}^i(t)]. \tag{6.160}$$

for $j = 1, 2, \ldots, m; t = 0, 1, \ldots, T-1$.

Identifying the linear terms in $\mathbf{x}$ we get

$$\mathbf{h}^j(t) = -\mathbf{r}^j(t)'\mathbf{R}^j(t)\mathbf{P}^j(t) - \sum_{i \neq j}\mathbf{r}^j(t)'\mathbf{R}^{ij}(t)'\mathbf{P}^i(t) - \sum_{i \neq j}\mathbf{r}^i(t)'\mathbf{R}^{ij}(t)\mathbf{P}^j(t)$$

$$+ \mathbf{q}^j(t) - \mathbf{p}^j(t)\mathbf{P}^j(t) + [\sum_{i \in M}\mathbf{B}_i(t)\mathbf{r}^i(t)]'\mathbf{S}^j(t+1)[\mathbf{A}(t) - \sum_{i \in M}\mathbf{B}_i(t)\mathbf{P}^i(t)]$$

$$+ \mathbf{h}^j(t+1)[\mathbf{A}(t) - \sum_{i \in M}\mathbf{B}_i(t)\mathbf{P}^i(t)] \tag{6.161}$$

for $j = 1, 2, \ldots, m; t = 0, 1, \ldots, T-1$.

Finally identifying the constant term

$$s^j(t) = \frac{1}{2}\mathbf{r}^j(t)'\mathbf{R}^j(t)\mathbf{r}^j(t) + \sum_{i \neq j}\mathbf{r}^i(t)'\mathbf{R}^{ij}\mathbf{r}^j(t) + \mathbf{p}^j(t)\mathbf{r}^j(t)$$

$$+ \frac{1}{2}\left[\sum_{i \in M}\mathbf{B}_i\mathbf{r}^i(t)\right]'\mathbf{S}^j(t+1)\left[\sum_{i \in M}\mathbf{B}_i\mathbf{r}^i(t)\right]$$

$$+\mathbf{h}^j(t+1)\sum_{i \in M}\mathbf{B}_i(t)\mathbf{r}^i(t) + s^j(t+1) \qquad (6.162)$$

for $j = 1, 2, \ldots, m; t = 0, 1, \ldots, T - 1$.

In stage T the boundary conditions are $\mathbf{S}^j(T) = \mathbf{Q}^j(T), \mathbf{h}^j(T) = q^j(T), s^j(T) = 0$, for $j = 1, 2, \ldots, m$.

Theorem 6.3. *Define the matrices* $\mathbf{S}^j(t)$, $\mathbf{h}^j(T)$, $s^j(t)$, $\mathbf{P}^j(t)$ *and* $\mathbf{r}^j(t)$ *such that the recurrence equations (6.156), (6.157), (6.160, 6.161) and (6.162) are verified for* $j = 1, 2, \ldots, m; t = 0, 1, \ldots, T - 1$, *with the boundary conditions*

$$\mathbf{S}^j(T) = \mathbf{Q}^j(T), \quad \mathbf{h}^j(T) = 0, \quad s^j(T) = 0,$$

for $j = 1, 2, \ldots, m$. *If the matrices*

$$[\mathbf{R}^j(t) + \mathbf{B}_j(t)'\mathbf{S}^j(t+1)\mathbf{B}_j(t)]$$

are definite negative for $j = 1, 2, \ldots, m; t = 0, 1, \ldots, T - 1$, *then (i)*

$$\sigma_j^*(t, \mathbf{x}) = -\mathbf{P}^j(t)\mathbf{x}(t+1) + \mathbf{r}^j(t), \quad j = 1, 2, \ldots, m; t = 0, 1, \ldots, T - 1$$

is a feedback Nash equilibrium strategy m-tuple; (ii)

$$V_j^*(t, \mathbf{x}) = \frac{1}{2}\mathbf{x}'\mathbf{S}^j(t)\mathbf{x} + \mathbf{h}^j(t)\mathbf{x} + s^j(t), \quad j = 1, 2, \ldots, m; t = 0, 1, \ldots, T$$

is the equilibrium value function for each player, each stage and each initial state.

Proof. It suffices to apply the verification theorem. The negative definiteness of $[\mathbf{R}^j(t) + \mathbf{B}_j(t)'\mathbf{S}^j(t+1)\mathbf{B}_j(t)]$ guarantees that the RHS of the DP equation is maximized in the player's own action. Hence the equilibrium is obtained as a solution of the DP equations. □

Remark 6.17. The following remarks summarize the solution properties and extensions.

(1) The game solution defined by equations (6.156), (6.157), (6.160) and (6.162) is valid for any dimensionality of the problem, i.e., these formulae are the same for any number of players, and any dimension of the state and control variables.
(2) Equations (6.156), (6.157), (6.160, 6.161) and (6.162) are often referred to as **coupled Riccati equations**.
(3) The equilibrium strategies (6.151) are affine functions of the state $\mathbf{x}$ and the equilibrium value functions are quadratic forms of state vector $\mathbf{x}(t)$.
(4) A stochastic version of this game is obtained if the state equations include an additive Gaussian vector (i.i.d. across time). A **certainty equivalence** applies to these linear-quadratic-Gaussian games, with no qualitative changes in the solution compared to the deterministic version (see Chapter 9).

6.7.2 Infinite-horizon LQ game

Next, consider the *infinite-horizon* non-discounted stationary version of the game, obtained by letting $T \to \infty$ and $\mathbf{A}, \mathbf{B}_j, \mathbf{Q}^j, \mathbf{R}^j$ be time-invariant. We further assume that $\mathbf{R}^{ij} \equiv 0$ and that the reward function has no linear terms. Sufficient conditions on the primitives that guarantee existence are *not* known at this point, contrary to the optimal-control case. Nonetheless, the following partial answer (involving assumptions on derived objects) is known.

Consider the following matrix equations involving the matrices $\mathbf{S}^j$ and $\mathbf{P}^j$

$$[\mathbf{R}^{jj} + \mathbf{B}'_j \mathbf{S}^j \mathbf{B}_j]\mathbf{P}^j + \mathbf{B}'_j \mathbf{S}^j \sum_{i \neq j} \mathbf{B}_i \mathbf{P}^i = \mathbf{B}'_j \mathbf{S}^j \mathbf{A}, \quad j = 1, 2, \ldots, M, \tag{6.163}$$

where $\mathbf{S}^j$ is defined by

$$\mathbf{S}^j = \mathbf{F}^{j'} \mathbf{S}^j \mathbf{F}^j + \sum_{i \in M} \mathbf{P}^{i'} \mathbf{R}^{ji} \mathbf{P}^i + \tilde{\mathbf{Q}}^j \tag{6.164}$$

with

$$\mathbf{F}^j \triangleq \mathbf{A} - \sum_{j \neq i} \mathbf{B}_i \mathbf{P}^i \tag{6.165}$$

and

$$\tilde{\mathbf{Q}}^j \triangleq \mathbf{Q}^j + \sum_{i \neq j} \mathbf{P}^{i'} \mathbf{R}^{ji} \mathbf{P}^i. \tag{6.166}$$

Proposition 6.1. *Suppose that there exist matrices $\{\mathbf{S}^j, \mathbf{P}^j\}$ satisfying (6.163)–(6.166). If the pair $(\mathbf{F}^j, \mathbf{B}_j)$ is **stabilizable**[45] then*

- *There is a feedback-stationary equilibrium where Player j's strategy is*

$$\sigma^{j*}(\mathbf{x}) = -\mathbf{P}^j x$$

  and his (finite) equilibrium payoff is

$$\frac{1}{2}\mathbf{x}'(0)\mathbf{S}^j \mathbf{x}(0);$$

- *The resulting equilibrium system's dynamics $\mathbf{x}(t+1) = \mathbf{F}\mathbf{x}(t)$ are stable.*

Proof. From the viewpoint of Player j, when the other players use the linear feedbacks $\sigma^{i*}(\mathbf{x}) = -\mathbf{P}^i \mathbf{x}, i \neq j$, the problem reduces to the control over an infinite horizon of a linear system with a quadratic performance criterion. The conditions listed above guarantee that this auxiliary infinite horizon control problem has a solution. This solution drives the state to 0, and the payoff is finite. (See [36] for a proof of this result for the optimal-control problem.) $\qquad\square$

---

[45]The terms *stabilizable* and *observable* are common in control theory. We say that the pair, or the underlying dynamic system, is stabilizable if the matrix $[\mathbf{B}^j, \mathbf{F}^j \mathbf{B}_j, \mathbf{F}^{j^2} \mathbf{B}_j, \ldots, \mathbf{F}^{j^{n-1}} \mathbf{B}_j]$ has full rank. Intuitively, *stabilizability* ensures the existence of a control that "stabilize the system," i.e., drive the state to 0 in infinite time. *Detectability* means that an original state can be "detected" after finite observations. Mathematically, the pair $(\mathbf{F}^j, \overline{\mathbf{Q}}^j)$ is detectable and $(\overline{\mathbf{F}}'_j \overline{\mathbf{Q}}^{j'})$ is observable.

Remark 6.18. The conditions of the preceding proposition are not easy to satisfy. It is difficult to give, a priori, assumptions under which these conditions hold, because they involve the feedback gains $\mathbf{P}^i$ that are to be determined. It has also been shown that infinite horizon LQ games can have solutions that are not related to the limit when $T \to \infty$ of the finite-horizon solution.

6.7.3 *An example of a linear-quadratic game*

Consider the following linear-quadratic game. It corresponds to a dynamic oligopoly model à la Cournot. The state equations describe the accumulation process of the production capacity x_j of firm $j = 1, 2, 3$. The coefficient $a_j \geq 0$ describes the decay (or depreciation) of the capacity x_j over one period. The control variable u_j is the investment (capacity increase) of firm j in one period. The state equations are

$$x_j(t+1) = a_j x_j(t) + u_j(t), \ j = 1, 2, 3. \tag{6.167}$$

We can think of $x_j(t)$ as the production of firm j that is supplied to the market. Then, each player j strives to maximize the payoff

$$J_j \triangleq \sum_0^{T-1} \left((P(t) - c_j)x_j(t) - e_j(u_j(t))^2 \right), j = 1, 2, 3, \tag{6.168}$$

where

$$P(t) = \alpha - \delta(x_1(t) + x_2(t) + x_3(t)) \tag{6.169}$$

is the inverse-demand law ($A > 0, B > 0$) and $e_j(u_j(t))^2$ ($e_j > 0$) describes cost associated with investment u_j.

6.7.3.1 *Formulation in matrix notations*

Using the symbols introduced in the preceding section yields:

$$\mathbf{A} = \begin{bmatrix} a_1 & 0 & 0 \\ 0 & a_2 & 0 \\ 0 & 0 & a_3 \end{bmatrix}, \tag{6.170}$$

$$\mathbf{B}_1 = \begin{bmatrix} 1 \\ 0 \\ 0 \end{bmatrix}, \quad \mathbf{B}_2 = \begin{bmatrix} 0 \\ 1 \\ 0 \end{bmatrix}, \quad \mathbf{B}_3 = \begin{bmatrix} 0 \\ 0 \\ 1 \end{bmatrix}, \tag{6.171}$$

$$\mathbf{Q}^1 = \begin{bmatrix} -2\delta & -\delta & -\delta \\ -\delta & 0 & 0 \\ -\delta & 0 & 0 \end{bmatrix}, \tag{6.172}$$

$$\mathbf{Q}^2 = \begin{bmatrix} 0 & -\delta & 0 \\ -\delta & -2\delta & -\delta \\ 0 & -\delta & 0 \end{bmatrix},$$

$$\mathbf{Q}^3 = \begin{bmatrix} 0 & 0 & -\delta \\ 0 & 0 & -\delta \\ -\delta & -\delta & -2\delta \end{bmatrix},$$

$$\mathbf{R}^{11} = [-2e_1], \quad \mathbf{R}^{22} = [-2e_2], \quad \mathbf{R}^{33} = [-2e_3], \quad \mathbf{R}^{ij} \equiv 0 \text{ if } i \neq j, \quad (6.173)$$

$$\mathbf{q}^1 = [\alpha - c_1 \ 0 \ 0], \quad \mathbf{q}^2 = [0 \ \alpha - c_2 \ 0], \quad \mathbf{q}^3 = [0 \ 0 \ \alpha - c_3], \quad (6.174)$$

$$\mathbf{m}^1 = [0], \quad \mathbf{m}^2 = [0], \quad \mathbf{m}^3 = [0]. \quad (6.175)$$

6.7.3.2   A numerical illustration

We provide a numerical illustration for a case where the players are symmetrical. We assume that $T = 6$ and

$$\mathbf{A} = \begin{bmatrix} 0.9 & 0 & 0 \\ 0 & 0.9 & 0 \\ 0 & 0 & 0.9 \end{bmatrix}, \quad (6.176)$$

$$\mathbf{B}_1 = \begin{bmatrix} 1 \\ 0 \\ 0 \end{bmatrix}, \quad \mathbf{B}_2 = \begin{bmatrix} 0 \\ 1 \\ 0 \end{bmatrix}, \quad \mathbf{B}_3 = \begin{bmatrix} 0 \\ 0 \\ 1 \end{bmatrix}, \quad (6.177)$$

$$\mathbf{Q}^1 = \begin{bmatrix} -0.1 & -0.05 & -0.05 \\ -0.05 & 0 & 0 \\ -0.05 & 0 & 0 \end{bmatrix},$$

$$\mathbf{Q}^2 = \begin{bmatrix} 0 & -0.05 & 0 \\ -0.05 & -0.1 & -0.05 \\ 0 & -0.05 & 0 \end{bmatrix},$$

$$\mathbf{Q}^3 = \begin{bmatrix} 0 & 0 & -0.05 \\ 0 & 0 & -0.05 \\ -0.05 & -0.05 & -0.1 \end{bmatrix}, \quad (6.178)$$

$$\mathbf{R}^{11} = [-0.01], \quad \mathbf{R}^{22} = [-0.01], \quad \mathbf{R}^{33} = [-0.01], \quad \mathbf{R}^{ij} \equiv 0 \text{ if } i \neq j, \quad (6.179)$$

$$\mathbf{q}^1 = [0.5 \ 0 \ 0], \quad \mathbf{q}^2 = [0 \ 0.5 \ 0], \quad \mathbf{q}^3 = [0 \ 0 \ 0.5]. \quad (6.180)$$

Because the payoff does not include terminal (or bequest) rewards we set

$$\mathbf{S}^1(6) = \mathbf{S}^2(6) = \mathbf{S}^3(6) = \mathbf{0}, \quad (6.181)$$

$$\mathbf{n}^1(6) = \mathbf{n}^2(6) = \mathbf{n}^3(6) = \mathbf{0}, \quad (6.182)$$

$$s^1(6) = s^2(6) = s^3(6) = \mathbf{0}. \quad (6.183)$$

Using equations (6.156) for $t = 5$, we get

$$\mathbf{P}^1(5) = [0 \ 0 \ 0], \quad \mathbf{P}^2(5) = [0 \ 0 \ 0], \quad \mathbf{P}^3(5) = [0 \ 0 \ 0], \quad (6.184)$$

and using equations (6.157) we obtain

$$\mathbf{r}^1(5) = [0], \quad \mathbf{r}^2(5) = [0], \quad \mathbf{r}^3(5) = [0]. \quad (6.185)$$

This makes sense, since $t = 5$ is the last period where control is used, and there is no terminal reward; hence, the controls must be equal to 0, whatever the value of $\mathbf{x}(5)$.

Using equations (6.160) for $t = 5$, we get

$$\mathbf{S}^1(5) = \begin{bmatrix} -0.10 & -0.05 & -0.05 \\ -0.05 & 0 & 0 \\ -0.05 & 0 & 0 \end{bmatrix},$$

$$\mathbf{S}^2(5) = \begin{bmatrix} 0 & -0.05 & 0 \\ -0.05 & -0.10 & -0.05 \\ 0 & -0.05 & 0 \end{bmatrix},$$

$$\mathbf{S}^3(5) = \begin{bmatrix} 0 & 0 & -0.05 \\ 0 & 0 & -0.05 \\ -0.05 & -0.05 & -0.10 \end{bmatrix}. \tag{6.186}$$

Using equations (6.161) for $t = 5$, we get

$$\mathbf{h}^1(5) = \begin{bmatrix} 0.5 & 0 & -0 \end{bmatrix},$$
$$\mathbf{h}^2(5) = \begin{bmatrix} 0 & 0.5 & 0 \end{bmatrix},$$
$$\mathbf{h}^3(5) = \begin{bmatrix} 0 & 0 & 0.5 \end{bmatrix}. \tag{6.187}$$

Using equations (6.162) for $t = 5$, we get

$$s^1(5) = s^2(5) = s^3(5) = 0. \tag{6.188}$$

Now we repeat these calculations for $t = 4$. Using equations (6.156) for $t = 4$, we get

$$\mathbf{P}^1(4) = \begin{bmatrix} 0.7857 & 0.03152 & 0.03152 \end{bmatrix},$$
$$\mathbf{P}^2(4) = \begin{bmatrix} 0.03152 & 0.7857 & 0.03152 \end{bmatrix},$$
$$\mathbf{P}^3(4) = \begin{bmatrix} 0.03152 & 0.03152 & 0.7857 \end{bmatrix}, \tag{6.189}$$

and using equations (6.157) we obtain

$$\mathbf{r}^1(4) = \begin{bmatrix} 2.381 \end{bmatrix}, \quad \mathbf{r}^2(4) = \begin{bmatrix} 2.381 \end{bmatrix}, \quad \mathbf{r}^3(4) = \begin{bmatrix} 2.381 \end{bmatrix}. \tag{6.190}$$

Using equations (6.160) for $t = 4$, we get

$$\mathbf{S}^1(4) = \begin{bmatrix} -0.10666 & -0.05045 & -0.05045 \\ -0.05045 & 0.00014 & 0.00014 \\ -0.05045 & 0.00014 & 0.00014 \end{bmatrix},$$

$$\mathbf{S}^2(4) = \begin{bmatrix} 0.00014 & -0.05044 & 0.00014 \\ -0.05045 & -0.10666 & -0.05045 \\ 0.00014 & -0.05045 & 0.00014 \end{bmatrix},$$

$$\mathbf{S}^3(4) = \begin{bmatrix} 0.00014 & 0.00014 & -0.05045 \\ 0.00014 & 0.00014 & -0.05045 \\ -0.05045 & -0.05045 & -0.10666 \end{bmatrix}. \tag{6.191}$$

Using equations (6.161) for $t = 4$, we get

$$\mathbf{h}^1(4) = \begin{bmatrix} 0.52993 & -0.00935 & -0.00935 \end{bmatrix},$$
$$\mathbf{h}^2(4) = \begin{bmatrix} -0.00935 & 0.52993 & -0.00935 \end{bmatrix},$$
$$\mathbf{h}^3(4) = \begin{bmatrix} -0.00935 & -0.00935 & 0.52993 \end{bmatrix}. \tag{6.192}$$

Using equations (6.162) for $t = 5$, we get

$$s^1(4) = s^2(4) = s^3(4) = 0.31179. \tag{6.193}$$

We repeat these calculations for $t = 3, 2, 1$ and finally we obtain for $t = 0$ the following results

$$\mathbf{P}^1(0) = \begin{bmatrix} 0.79752 \ 0.03033 \ 0.03033 \end{bmatrix},$$
$$\mathbf{P}^2(0) = \begin{bmatrix} 0.03033 \ 0.79752 \ 0.03033 \end{bmatrix},$$
$$\mathbf{P}^3(0) = \begin{bmatrix} 0.03033 \ 0.03033 \ 0.79752 \end{bmatrix} \tag{6.194}$$

and

$$\mathbf{r}^1(0) = \begin{bmatrix} 1.17001 \end{bmatrix}, \quad \mathbf{r}^2(0) = \begin{bmatrix} 1.17001 \end{bmatrix}, \quad \mathbf{r}^3(0) = \begin{bmatrix} 1.17001 \end{bmatrix}. \tag{6.195}$$

The matrices of the equilibrium value functions are

$$\mathbf{S}^1(0) = \begin{bmatrix} -0.10683 & -0.05038 & -0.05038 \\ -0.05038 & 0.00012 & 0.00012 \\ -0.05038 & 0.00012 & 0.00012 \end{bmatrix},$$

$$\mathbf{S}^2(0) = \begin{bmatrix} 0.00012 & -0.05038 & 0.00012 \\ -0.05038 & -0.10666 & -0.05038 \\ 0.00012 & -0.05038 & 0.00012 \end{bmatrix},$$

$$\mathbf{S}^3(0) = \begin{bmatrix} 0.00012 & 0.00012 & -0.05038 \\ 0.00012 & 0.00012 & -0.05038 \\ -0.05038 & -0.05038 & -0.10666 \end{bmatrix}, \tag{6.196}$$

and

$$\mathbf{h}^1(0) = \begin{bmatrix} 0.54508 \ -0.014131 \ -0.014131 \end{bmatrix},$$
$$\mathbf{h}^2(0) = \begin{bmatrix} -0.014131 \ 0.54508 \ -0.014131 \end{bmatrix},$$
$$\mathbf{h}^3(0) = \begin{bmatrix} -0.014131 \ -0.014131 \ 0.54508 \end{bmatrix}. \tag{6.197}$$

Finally

$$s^1(0) = s^2(0) = s^3(0) = 1.786. \tag{6.198}$$

In summary the feedback law for Player 1 is given in the following table and symmetrically for the other players.

Assume, the initial state is $\mathbf{x}^0 = \begin{bmatrix} 1 \\ 1 \\ 1 \end{bmatrix}$.

The equilibrium state trajectory is shown below. The controls, as computed from the feedback laws, are given below The transition rewards for each player are then the following: We can check the calculations by comparing the sum of these rewards ($= 2.15$) and

Table 6.1   Equilibrium feedback law

t	0	1	2	3	4	5
$[\mathbf{P}^1(t)]'$						
	0.80	0.80	0.80	0.80	0.79	0.00
	0.03	0.03	0.03	0.03	0.04	0.00
	0.03	0.03	0.03	0.03	0.04	0.00
$r^1(t)$						
	1.03	1.03	1.03	1.03	2.38	0.00

Table 6.2   Equilibrium state trajectory

t	0	1	2	3	4	5	T
$x_1(t)$	1.00	1.21	1.08	1.08	1.08	2.43	2.18
$x_2(t)$	1.00	1.21	1.08	1.08	1.08	2.43	2.18
$x_3(t)$	1.00	1.21	1.08	1.08	1.08	2.43	2.18

Table 6.3   Equilibrium controls

t	0	1	2	3	4	5	T
$u_1(t)$	0.31	-0.01	0.10	0.11	1.46	0.00	
$u_2(t)$	0.31	-0.01	0.10	0.11	1.46	0.00	
$u_3(t)$	0.31	-0.01	0.11	0.11	1.46	0.00	

Table 6.4   Equilibrium transition rewards

t	0	1	2	3	4	5	T
	0.35	0.39	0.37	0.36	0.35	0.33	-

the value function at $\mathbf{x}^0$,

$$
V_1^*(0, \mathbf{x}^0) = \frac{1}{2} \begin{bmatrix} 1 \\ 1 \\ 1 \end{bmatrix}' \begin{bmatrix} -0.10683 & -0.05038 & -0.05038 \\ -0.05038 & 0.00012 & 0.00012 \\ -0.05038 & 0.00012 & 0.00012 \end{bmatrix} \begin{bmatrix} 1 \\ 1 \\ 1 \end{bmatrix} \tag{6.199}
$$

$$
+ \begin{bmatrix} 0.54508 & -0.014131 & -0.014131 \end{bmatrix} \begin{bmatrix} 1 \\ 1 \\ 1 \end{bmatrix}
$$

$$
+ \frac{1}{2}[0.31][-0.01][0.31] = 2.15.
$$

Remark 6.19. The state-trajectory profile shows an interesting behavior. The state converges rapidly (after two periods, to a steady-state value $(1.08, 1.08, 1.08)$, which is quitted only at the last control period ($t = 5$). The feedback law remains the same for periods $t = 0, 1, 2, 3$, and changes only in period $t = 4$. We can infer from this observation that the equilibrium-feedback law for an infinite-horizon problem should be close to the one defined here for $t = 0$. Notice also that the value functions will tend to ∞ when $T \to \infty$.

Unfortunately, this nice asymptotic behavior of the feedback-Nash solution to a particular linear-quadratic game cannot be proved to hold for a general class of these LQ games.

6.8 Two-person zero-sum linear-quadratic game and robust controller design

A subclass of linear-quadratic multistage games deals with two-player *zero-sum* dynamic games. These problems are close (in spirit) to optimal-control problems and provide the backbone theory for the design of *robust controllers*. The study of this class of games will also permit us to show that, for two-person zero-sum multistage games, the feedback and open-loop Nash equilibrium solutions (which become a *saddle-point* for 0-sum games) coincide.

6.8.1 *Problem formulation*

Some economic or managerial situations can be modeled as zero-sum dynamic games. A game of this kind is when an agent strives to maximize his payoff (performance index) *subject* to uncontrolled disturbances that can be counteracting his efforts. In this situation, the agent who wants to design a robust controller assumes that the disturbances will take the worst case form, as if they were the result of a strategy of an opponent aiming at the minimization of the performance index.

Consider the two-person zero-sum linear-quadratic game described by the state equation

$$\mathbf{x}(t+1) = \mathbf{A}(t)\mathbf{x}(t) + \mathbf{B}_1(t)\mathbf{u}_1(t) + \mathbf{B}_2(t)\mathbf{u}_2(t), \tag{6.200}$$
$$t = 0, 1, \ldots, T-1,$$
$$\mathbf{x}(0) = \mathbf{x}^0 \neq 0, \tag{6.201}$$

and the performance index, or payoff

$$J \triangleq \sum_{t=0}^{T-1} \frac{1}{2} \left(\mathbf{x}(t)'\mathbf{Q}(t)\mathbf{x}(t) - \mathbf{u}_1(t)'\mathbf{u}_1(t) + \mathbf{u}_2(t)'\mathbf{u}_2(t) \right)$$
$$+ \frac{1}{2}\mathbf{x}(T)'\mathbf{Q}(T)\mathbf{x}(T), \tag{6.202}$$

where $\mathbf{x}(t), t = 1, 2, \ldots, T$ are the state variables, $\mathbf{u}_1(t), \mathbf{u}_2(t), t = 1, 2, \ldots, T-1$ are the players' control vectors each belonging to an appropriate dimensional Euclidean space; $\mathbf{A}(t), \mathbf{B}_1(t), \mathbf{B}_2(t), \mathbf{D}(t), \mathbf{Q}(t)$ are matrices of compatible dimensions and $\mathbf{Q}(t)$ is negative semidefinite with $\mathbf{Q}(0) = 0$; the game horizon is T.

6.8.2 Open-loop saddle-point solution

An open-loop equilibrium solution to the game defined by (6.200), (6.202), will be a pair of control sequences $(\tilde{\mathbf{u}}_1^*, \tilde{\mathbf{u}}_2^*)$ satisfying the following saddle-point condition: for any admissible control sequences $\tilde{\mathbf{u}}_1$ and $\tilde{\mathbf{u}}_2$

$$J(0, \mathbf{x}^0; \tilde{\mathbf{u}}_1, \tilde{\mathbf{u}}_2^*) \leq J(0, \mathbf{x}^0; \tilde{\mathbf{u}}_1^*, \tilde{\mathbf{u}}_2^*) \leq J(0, \mathbf{x}^0 \tilde{\mathbf{u}}_1^*, \tilde{\mathbf{u}}_2), \tag{6.203}$$

where $J(0, \mathbf{x}^0; \tilde{\mathbf{u}}_1, \tilde{\mathbf{u}}_2)$ is the normal form representation of the open-loop game defined by (6.200)–(6.202).

Theorem 6.4. *For the two-person linear-quadratic multistage game described by (6.200), (6.202), assume the following conditions hold*

$$\mathbf{I} + \mathbf{B}_2(t)'\mathbf{S}(t+1)\mathbf{B}_2(t) < 0 \quad \forall t = 0, \ldots, T-1 \tag{6.204}$$

where $\mathbf{S}(t+1)$ *is given by*

$$\begin{aligned} \mathbf{S}(t) = \mathbf{Q}(t) + \mathbf{A}(t)'\mathbf{S}(t+1)\mathbf{A}(t) + \\ \mathbf{A}(t)'\mathbf{S}(t+1)[\mathbf{I} - \mathbf{B}_2(t)'\mathbf{S}(t+1)\mathbf{B}_2(t)]\mathbf{B}_2(t)S(t+1)\mathbf{A}(t) \end{aligned} \tag{6.205}$$

for $t = 0, \ldots, T-1$ *and*

$$\mathbf{S}(T) = \mathbf{Q}(T). \tag{6.206}$$

Let $\mathbf{\Lambda}(t)$ *and* $\mathbf{M}(t)$ *be appropriate dimensional matrices defined by*

$$\mathbf{\Lambda}(t) = \mathbf{I} - (\mathbf{B}_1(t)\mathbf{B}_1(t)' - \mathbf{B}_2(t)\mathbf{B}_2(t)')\mathbf{M}(t+1) \tag{6.207}$$

and

$$\mathbf{M}(t) = \mathbf{Q}(t) + \mathbf{A}(t)'\mathbf{M}(t+1)\mathbf{\Lambda}(t)^{-1}\mathbf{A}(t); \quad \mathbf{M}(T) = \mathbf{Q}(T). \tag{6.208}$$

Then,
i) $\mathbf{\Lambda}(t)$ *is invertible,* $\forall t = 0, \ldots T-1$,

ii) *the game admits a unique open-loop saddle-point solution given by*

$$\mathbf{u}_1^*(t) = \mathbf{B}_1(t)'\mathbf{M}(t+1)\mathbf{\Lambda}(t)^{-1}\mathbf{A}(t)\mathbf{x}^*(t), \tag{6.209}$$
$$\mathbf{u}_2^*(t) = -\mathbf{B}_2(t)'\mathbf{M}(t+1)\mathbf{\Lambda}(t)^{-1}\mathbf{A}(t)\mathbf{x}^*(t) \tag{6.210}$$

where $\mathbf{x}^*(t)$ *is the saddle-point trajectory determined from*

$$\mathbf{x}^*(t+1) = \mathbf{\Lambda}(t)^{-1}\mathbf{A}(t)\mathbf{x}^*(t); \quad \mathbf{x}^*(0) = \mathbf{x}^0. \tag{6.211}$$

Proof.   It can be shown (see [18], Lemma 1, p. 247) that conditions (6.204), (6.206) hold if and only if $L(\mathbf{x}_0; \tilde{\mathbf{u}}_1, \tilde{\mathbf{u}}_2)$ is strictly convex in $\tilde{\mathbf{u}}_2$ for all $\tilde{\mathbf{u}}_1$. So the game is a strictly concave-convex game, which admits a unique saddle-point solution. We can characterize it, using the coupled maximum principle of Lemma 6.2, which becomes in this case a saddle-point maximum principle.

So, let us form the Hamiltonian

$$H(\mathbf{p}(t+1), \mathbf{x}^*(t), \mathbf{u}_1, \mathbf{u}_2, t) = \frac{1}{2}(\mathbf{x}^*(t)'\mathbf{Q}(t)\mathbf{x}^*(t) - \mathbf{u}_1'\mathbf{u}_1 + \mathbf{u}_2'\mathbf{u}_2) \qquad (6.212)$$
$$+ \mathbf{p}(t+1)'[\mathbf{A}(t)\mathbf{x}^*(t) + \mathbf{B}^1(t)\mathbf{u}_1 + \mathbf{B}^2(t)\mathbf{u}_2].$$

The saddle-point controls satisfy

$$\frac{\partial H}{\partial \mathbf{u}_j}\bigg|_{\substack{\mathbf{u}_1^*(t) \\ \mathbf{u}_2^*(t)}} = 0, \quad j = 1, 2$$

and the costate vector $\mathbf{p}(t)$ satisfies

$$\mathbf{p}(t) = \left[\frac{\partial H}{\partial \mathbf{x}}\bigg|_{\substack{\mathbf{u}_1^*(t) \\ \mathbf{u}_2^*(t)}}\right]' = \mathbf{A}(t)'\mathbf{p}(t+1) + \mathbf{Q}(t)\mathbf{x}^*(t), \qquad (6.213)$$

$$t = 0, \dots, T - 1, \qquad (6.214)$$

$$\mathbf{p}(T) = \mathbf{Q}(T)\mathbf{x}^*(T) = \mathbf{M}(T)\mathbf{x}^*(T). \qquad (6.215)$$

Start with $t = T - 1$. The saddle-point controls satisfy

$$\frac{\partial H}{\partial \mathbf{u}_j}\bigg|_{\substack{\mathbf{u}_1^*(T-1) \\ \mathbf{u}_2^*(T-1)}} = 0, \quad j = 1, 2.$$

Hence

$$\mathbf{u}_1^*(T-1) = \mathbf{B}_1(T-1)'\mathbf{p}(T) = \mathbf{B}_1(T-1)'\mathbf{M}(T)\mathbf{x}^*(T), \qquad (6.216)$$

$$\mathbf{u}_2^*(T-1) = -\mathbf{B}_2(T-1)'\mathbf{p}(T) = -\mathbf{B}_2(T-1)'\mathbf{M}(T)\mathbf{x}^*(T), \qquad (6.217)$$

Substituting (6.216), (6.217) into the state equation (6.200) yields the relation

$$\mathbf{x}^*(T) = \mathbf{A}(T-1)\mathbf{x}^*(T-1)+(\mathbf{B}_1(T-1)\mathbf{B}_1(T-1)'-\mathbf{B}_2(T-1)\mathbf{B}_2(T-1)')\mathbf{M}(T)\mathbf{x}^*(T) \qquad (6.218)$$

and so

$$\mathbf{A}(T-1)\mathbf{x}^*(T-1) = \mathbf{\Lambda}(T-1)\mathbf{x}^*(T) \qquad (6.219)$$

where

$$\mathbf{\Lambda}(T-1) = \mathbf{I} - (\mathbf{B}_1(T-1)\mathbf{B}_1(T-1)' - \mathbf{B}_2(T-1)\mathbf{B}_2(T-1)')\mathbf{M}(T). \qquad (6.220)$$

Since there exists a unique saddle-point solution, there necessarily exists a unique relation between $\mathbf{x}^*(T-1)$ and $\mathbf{x}^*(T)$. Hence, $\mathbf{\Lambda}(T-1)$ must be invertible and we will have

$$\mathbf{x}^*(T) = \mathbf{\Lambda}(T-1)^{-1}\mathbf{A}(T-1)\mathbf{x}^*(T-1), \qquad (6.221)$$

$$\mathbf{u}_j^*(T-1) = (-1)^{j-1}\mathbf{B}_j(T-1)'\mathbf{M}(T)\mathbf{\Lambda}(T-1)^{-1}\mathbf{A}(T-1)\mathbf{x}^*(T-1) \quad j = 1, 2. \qquad (6.222)$$

So the result is proved for $t = T - 1$.

Now consider $t = T - 2$. We obtain first, from the condition $\frac{\partial H}{\partial \mathbf{u}_j}\big|_{\substack{\mathbf{u}_1^*(T-2) \\ \mathbf{u}_2^*(T-2)}} = 0, j = 1, 2$

that

$$\mathbf{u}_j^*(T-2) = (-1)^{j-1}\mathbf{B}_j(T-2)'\mathbf{p}(T-1); \quad j = 1, 2. \qquad (6.223)$$

From (6.213) and using (6.208) we can write

$$\begin{aligned}
\mathbf{p}(T-1) &= \mathbf{A}(T-1)'\mathbf{p}(T) + \mathbf{Q}(T-1)\mathbf{x}^*(T-1) \\
&= \mathbf{A}(T-1)'\mathbf{Q}(T)\mathbf{x}^*(T) + \mathbf{Q}(T-1)\mathbf{x}^*(T-1) \\
&= [\mathbf{A}(T-1)'\mathbf{Q}(T)\mathbf{\Lambda}(T-1)^{-1}\mathbf{A}(T-1) + \mathbf{Q}(T-1)]\mathbf{x}^*(T-1)] \\
&= \mathbf{M}(T-1)\mathbf{x}^*(T-1).
\end{aligned}$$

Substituting

$$\mathbf{u}_j^*(T-2) = (-1)^{j-1}\mathbf{B}_j(T-2)'\mathbf{M}(T-1)\mathbf{x}^*(T-1); \quad j = 1,2, \qquad (6.224)$$

into the state equation (6.200) yields the relation

$$\mathbf{A}(T-2)\mathbf{x}^*(T-1) = \mathbf{\Lambda}(T-2)\mathbf{x}^*(T-1). \qquad (6.225)$$

Again, since there exists a unique saddle-point solution, the matrix $\mathbf{\Lambda}(T-2)$ must be invertible, and the following holds

$$\mathbf{x}^*(T-1) = \mathbf{\Lambda}(T-2)^{-1}\mathbf{A}(T-2)\mathbf{x}^*(T-2) \qquad (6.226)$$

$$\mathbf{u}_j^*(T-2) = (-1)^{j-1}\mathbf{B}_j(T-2)'\mathbf{M}(T-1)\mathbf{\Lambda}(T-2)^{-1}\mathbf{A}(T-2)\mathbf{x}^*(T-2) \quad j = 1,2. \qquad (6.227)$$

So the result is proved for $t = T - 2$. We proceed recurrently until $t = 0$. □

6.8.3 *Feedback saddle-point solution*

A feedback equilibrium solution to the game defined by (6.200), (6.202), will be a pair of feedback strategies $(\sigma_1^*, \sigma_2^*) = (\sigma_1^*(\tau, \mathbf{x}), \sigma_2^*(\tau, \mathbf{x}))$ satisfying the following equilibrium condition for any admissible strategies σ_1 and σ_2

$$J(\tau, x^\tau; \sigma_1^*, \sigma_2) \le J(\tau, x^\tau; \sigma_1^*, \sigma_2^*) \le J(\tau, x^\tau; \sigma_1, \sigma_2^*); \qquad (6.228)$$

where the normal form game payoff is defined by

$$J(\tau, \mathbf{x}^\tau; \sigma_1, \sigma_2) \equiv \sum_{t=0}^{T-1} \frac{1}{2}(\mathbf{x}(t)'\mathbf{Q}(t)\mathbf{x}(t) - \sigma_1(t, \mathbf{x}(t))'\sigma_1(t, \mathbf{x}(t))$$

$$+ \sigma_2(t, \mathbf{x}(t))'\sigma_2(t, \mathbf{x}(t))) + \frac{1}{2}\mathbf{x}(T)'\mathbf{Q}(T)\mathbf{x}(T),$$

with the state variables $\mathbf{x}(t)$, $t = \tau, \tau+1, \ldots, T$ determined from σ_1 and σ_2 by the state equations

$$\mathbf{x}(t+1) = \mathbf{A}(t)\mathbf{x}(t) + \sigma_1(t, \mathbf{x}(t)) + \sigma_2(t, \mathbf{x}(t)),$$

$$\mathbf{x}(\tau) = \mathbf{x}^\tau, \tau \in \{0, 1, \ldots, T-1\}.$$

Since we deal with a two-person zero-sum games, the equilibrium condition determines a feedback saddle point. The equilibrium solution can be obtained, using the approach detailed before for general linear-quadratic games.

Theorem 6.5. *The unique feedback saddle-point strategies are*

$$\sigma_1(t, \mathbf{x}) = \mathbf{B}_1(t)'\mathbf{M}(t+1)\mathbf{\Lambda}(t)^{-1}\mathbf{A}(t)\mathbf{x}, \tag{6.229}$$

$$\sigma_2(t, \mathbf{x}) = -\mathbf{B}_2(t)'\mathbf{M}(t+1)\mathbf{\Lambda}(t)^{-1}\mathbf{A}(t)\mathbf{x}, \tag{6.230}$$

$$t = 0, 2, \ldots, T-1,$$

provided the following matrices are negative definite

$$\mathbf{I} - \mathbf{B}_1(t)'\mathbf{M}(t+1)\mathbf{B}_1(t) < 0, \tag{6.231}$$

$$\mathbf{I} - \mathbf{B}_2(t)'\mathbf{M}(t+1)\mathbf{B}_2(t) < 0, \tag{6.232}$$

$$t = 0, 2, \ldots, T-1 \tag{6.233}$$

where the matrix $\mathbf{M}(t)$ *is a solution to the following Riccati-type equation*[46]

$$\mathbf{M}(t) = \mathbf{Q}(t) + \mathbf{A}(t)'\mathbf{M}(t+1)\mathbf{\Lambda}(t)^{-1}\mathbf{A}(t), \quad \mathbf{M}(T) = \mathbf{Q}(T), \tag{6.234}$$

$t = 0, \ldots, T-1,$ *with*

$$\mathbf{\Lambda}(t) = \mathbf{I} - (\mathbf{B}_1(t)\mathbf{B}_1(t)' - \mathbf{B}_2(t)\mathbf{B}_2(t)')\mathbf{M}(t+1). \tag{6.235}$$

The saddle-point value (6.228) is then given by

$$L(\mathbf{x}^0; \sigma_1^*, \sigma_2^*) = \mathbf{x}^{0'}\mathbf{M}(0)\mathbf{x}^0. \tag{6.236}$$

Proof.   Consider the dynamic-programming equation

$$V^*(t, \mathbf{x}) = \max_{u_1} \min_{u_2} \left\{ \frac{1}{2}(\mathbf{x}'\mathbf{Q}(t)\mathbf{x} - \mathbf{u}_1'\mathbf{u}_1 + \mathbf{u}_2'\mathbf{u}_2) + \right.$$
$$\left. V^*(t+1, \mathbf{A}(t)\mathbf{x} + \mathbf{B}_1(t)\mathbf{u}_1 + \mathbf{B}_2(t)\mathbf{u}_2) \right\}. \tag{6.237}$$

We will show that this equation is satisfied by the value function $V^*(t, \mathbf{x}) = \frac{1}{2}\mathbf{x}'\mathbf{M}(t)\mathbf{x}$, with saddle-point controls for the local game given by the linear feedback rules (6.229) and (6.230).

We proceed by recurrence. At stage T we have $V^*(T, \mathbf{x}) = \frac{1}{2}\mathbf{x}'\mathbf{M}(T)\mathbf{x}$, since by definition $\mathbf{M}(T) = \mathbf{Q}(T)$. Now assume that at stage $t+1$ one has $V^*(t+1, \mathbf{x}) = \frac{1}{2}\mathbf{x}'\mathbf{M}(t+1)\mathbf{x}$. The local zero-sum game in $\mathbf{u}_1$ and $\mathbf{u}_2$ that enters into the definition of the dynamic-programming equation (6.237) admits a saddle-point solution if and only if

$$\mathbf{I} - \mathbf{B}_1(t)'\mathbf{M}(t+1)\mathbf{B}_1(t) < 0, \tag{6.238}$$

$$\mathbf{I} - \mathbf{B}_2(t)'\mathbf{M}(t+1)\mathbf{B}_2(t) < 0. \tag{6.239}$$

Let us call $\mathbf{x}^+ \equiv \mathbf{A}(t)\mathbf{x} + \mathbf{B}_1(t)\mathbf{u}_1^+ + \mathbf{B}_2(t)\mathbf{u}_2^+$, where $\mathbf{u}_1^+$ and $\mathbf{u}_2^+$ are the saddle-point solution to the local game. It is clear that

$$\mathbf{u}_j^+ = (-1)^{j-1}\mathbf{B}_j(t)\mathbf{x}^+. \tag{6.240}$$

By the same argument as in the proof of Theorem 6.4, we can then show that

$$\mathbf{x}^+ = \mathbf{\Lambda}(t)^{-1}\mathbf{A}(t)\mathbf{x}, \tag{6.241}$$

---

[46]Compare equation (6.11) and item (2) on page 200.

where $\Lambda(t)$ is defined by (6.235). The local game saddle point solution is thus defined as the linear feedback rules ((6.229)–((6.230).

If $\mathbf{u}_1$ and $\mathbf{u}_2$ are given by (6.229) and (6.230), then

$$[\mathbf{A}(t)\mathbf{x} + \mathbf{B}_1(t)\mathbf{u}_1 + \mathbf{B}_2(t)\mathbf{u}_2]'\mathbf{M}(t+1)[\mathbf{A}(t)\mathbf{x} + \mathbf{B}_1(t)\mathbf{u}_1 + \mathbf{B}_2(t)\mathbf{u}_2]$$
$$= \mathbf{x}'\mathbf{A}'\Lambda(t)^{-1}\mathbf{M}(t+1)\Lambda(t)^{-1}\mathbf{A}\mathbf{x}.$$

Similarly the expression $-\mathbf{u}_1'\mathbf{u}_1 + \mathbf{u}_2'\mathbf{u}_2$, when $\mathbf{u}_1$ and $\mathbf{u}_2$ are given by (6.229) and (6.230) gives

$$\mathbf{x}'\mathbf{A}(t)'\Lambda(t)^{-1}\mathbf{M}(t+1)\left(-\mathbf{B}_1(t)'\mathbf{B}_1(t) + \mathbf{B}_2(t)'\mathbf{B}_2(t)\right)\mathbf{M}(t+1)\Lambda(t)^{-1}\mathbf{A}(t)\mathbf{x}$$
$$= \mathbf{x}'\mathbf{A}(t)'\Lambda(t)^{-1}\mathbf{M}(t+1)\left[\mathbf{I} - \mathbf{I} + \left(-\mathbf{B}_1(t)'\mathbf{B}_1(t) + \mathbf{B}_2(t)'\mathbf{B}_2(t)\right)\mathbf{M}(t+1)\right]$$
$$\Lambda(t)^{-1}\mathbf{A}(t)\mathbf{x}$$
$$= \mathbf{x}'\mathbf{A}(t)'\Lambda(t)^{-1}\mathbf{M}(t+1)\left[-\mathbf{I} + \Lambda(t)\right]\Lambda(t)^{-1}\mathbf{A}(t)\mathbf{x}$$
$$= -\mathbf{x}'\mathbf{A}(t)'\Lambda(t)^{-1}\mathbf{M}(t+1)\Lambda(t)^{-1}\mathbf{A}(t)\mathbf{x} + \mathbf{x}'\mathbf{A}(t)'\Lambda(t)^{-1}\mathbf{M}(t+1)\mathbf{A}(t)\mathbf{x}.$$

Then, by (6.237) we can write

$$V^*(t,\mathbf{x}) = \frac{1}{2}\left(\mathbf{x}'\mathbf{Q}(t)\mathbf{x} - \mathbf{u}_1'\mathbf{u}_1 + \mathbf{u}_2'\mathbf{u}_2\right) + V^*(t+1, \mathbf{A}(t)\mathbf{x} + \mathbf{B}_1(t)\mathbf{u}_1 + \mathbf{B}_2(t)\mathbf{u}_2)$$
$$= \frac{1}{2}\mathbf{x}'\left(\mathbf{Q}(t) + \mathbf{A}(t)'\Lambda(t)^{-1}\mathbf{M}(t+1)\mathbf{A}(t)\right)\mathbf{x}$$
$$= \frac{1}{2}\mathbf{x}'\mathbf{M}(t)\mathbf{x},$$

according to (6.234). Therefore we have obtained a solution of the dynamic-programming equations and (6.229)–(6.230) define a fedback equilibrium strategy pair, i.e., a feedback saddle point for the zero-sum game. □

Remark 6.20. We notice here that the open-loop and feedback saddle-point solutions are very similar. Both solutions will generate the same state, trajectory, and along this trajectory they are defined by the same linear feedback laws. More precisely, referring to the maximum principle for equilibrium solutions, Lemma 6.4, we notice that for zero-sum games, the saddle-point condition imposes that the partial derivative $\frac{\partial}{\partial \mathbf{u}_{-j}} H_j(\mathbf{p}_j(t+1), \mathbf{x}^*(t), \underline{\sigma}^*(t, \mathbf{x}^*(t)), t)$ be equal to 0. Therefore, the maximum principles are the same, for open-loop and for feedback saddle-point solutions, when the strategies are smooth.

6.9   Equilibria in a class of memory strategies for infinite-horizon games

6.9.1   *Other equilibria*

The feedback equilibrium obtained through dynamic programming is by no means the only subgame-perfect equilibrium of a multistage game, like, e.g., the fishery-management game, when played on an infinite horizon. It was shown in Sections 5.2.3 and 5.4.1, dedicated to repeated games, that other equilibria can be defined in terms of trigger strategies. Similar results have been shown to hold also in the class of multistage games played in the state space, which we study in this chapter.

We will use an example developed by Dutta and Radner in [66] and [67], to show how a trigger-strategy mechanism can be employed to compute a subgame-perfect equilibrium that can lead to Pareto-efficient outcomes. The game concerns a climate-negotiation process. Further in this section, we will consider the class of stationary multistage-games-without-discounting and characterize a memory-based equilibrium, using the overtaking-optimality criterion.

Later, in this chapter's $\mathbb{GE}$ illustration (beginning from page 231), we will show another multistage game for which a subgame-perfect memory-based strategy equilibrium is a solution. The particular game, for which we compute a *collusive equilibrium* (see Definition 5.2), models a macroeconomic conflict between workers and capitalists.

6.9.2 Example: Self-enforcing climate treaties

Dutta and Radner proposed in [66] a model of self-enforcing climate treaties, where the players use trigger strategies with threats. We reproduce below the main features of this model and show that there may exist equilibria that are both subgame perfect and Pareto optimal.[47]

6.9.2.1 *The basic symbols and definitions*

We begin by listing the model variables and parameters.

δ discount factor per period, common for all players
$a_j(t)$ emissions during period t by country j
$e_j(t)$ energy input into production and consumption associated
$f_j(t)$ emission factor[48] of country j in period t that can be related to the technology choice of country j
$v_j(t)$ reduction-of-emission factor of country j in period t which is a control variable of country j
m_j minimal possible value of the emission factor for country j
$A(t)$ global (total) emissions during period t
g_0 pre-industrial level of steady-state GHG stock
$g(t)$ the excess GHG stock due to human activity
ς proportion of the excess GHG stock, which survives from t to $t+1$ (so, $1 - \varsigma$ is naturally eliminated)
c_j unit damage cost in country j. The cost of the damage due to climate change is assumed linear in the global stock of excess GHG, i.e., equal to $c_j g(t)$, and is subtracted from country's GDP in that period
k_j unit cost of reduction of the emission factor Total cost in period t is $k_j(f_j(t) - f_j(t+1))$, when $f_j(t+1) \leq f_j(t)$

---

[47] In this section, we try to keep the same notations as in the original paper [66] and so we may differ from the rest of the book in the naming of players and state and control variables.
[48] Emissions are treated as a production factor in this model.

$Y_j(e_j(t))$ country j's GDP as a function of the energy input $e_j(t)$; a strictly concave, twice differentiable function with a maximum for some value of $e_j(t)$

$u_j(t)$ reward (utility derived from consumption and climate-change impact) of country i in period t

$s(t)$ state of the system in period t $s(t) = (\underline{f}(t), g(t))$, where $\underline{f}(t) = (f_j(t) : j = 1, \ldots, m)$.

The GHG emissions are directly related to the use of the energy input for each country j:

$$a_j(t) = f_j(t)e_j(t), \; j = 1, \ldots, m. \tag{6.242}$$

The total emissions are accumulated, as shown in equations (6.243)–(6.244),

$$A(t) = \sum_{j=1}^{m} a_j(t), \tag{6.243}$$

$$g(t+1) = A(t) + \varsigma g(t). \tag{6.244}$$

Each country can decrease the emission factor $f_j(t)$ by using the control variable $v_j(t)$. This defines the new emission factor, for next period $t + 1$:

$$f_j(t+1) = f_j(t) - v_j(t), \tag{6.245}$$

$$v_j(t) \geq 0. \tag{6.246}$$

There is a lower bound on the possible emission factors that can be achieved for each country

$$f_j(t) \geq m_j. \tag{6.247}$$

Thus, in each period, a country can only maintain or reduce its emission factor and there is a lower bound on the (eventual) level of its emission factor. The reward of country j in period t is

$$u_j(t) = Y_j(e_j(t)) - c_j g(t) - k_j[f_j(t) - f_j(t+1)], \; j = 1, \ldots, m. \tag{6.248}$$

Let δ denote the discount factor; then, the payoff for country j is defined as the discounted sum of rewards

$$u_j = \sum_{t=0}^{\infty} \delta^t u_j(t), \; j = 1, \ldots, m. \tag{6.249}$$

Finally, u is a global welfare function defined as a weighted sum of individual countries' payoffs u_j

$$u = \sum_{j=1}^{m} r_j u_j. \tag{6.250}$$

6.9.2.2 *Business-as-usual (BAU) equilibrium*

We assume that, in the absence an international treaty, the countries play a noncooperative game and implement a feedback (Markov) Nash equilibrium solution to the resulting dynamic game. This solution will be characterized by the dynamic-programming equations.

Basically, the game has two state variables: the GHG stock $g(t)$ and the vector of emission coefficients (technology) $\underline{f}(t)$. Let $\delta^t V_j^*(g(t), \underline{f}(t))$ denote the equilibrium value function for Player j when the game starts in period t at state $(g(t), \underline{f}(t))$. Therefore $V_j^*(g(t), \underline{f}(t))$ is the value function expressed in "current value" terms.

The dynamic-programming equations involve the local games defined by the local rewards at $(g(t), \underline{f}(t))$ defined below:

$$\mathfrak{h}_j(g(t), \underline{f}(t); \mathbf{e}(t), \mathbf{v}(t)) = Y_j(e_j(t)) - c_j g(t) - k_j v_j(t) + \delta V_j^*(g(t+1), \underline{f}(t+1)),$$

$$j = 1, \ldots m, \tag{6.251}$$

where $V_j^*(\cdot, \cdot)$ is country j's current valued equilibrium value function. In the dynamic-programming equations, this local game admits an equilibrium solution that defines the equilibrium strategy at $(g(t), \underline{f}(t))$.

The state equations are

$$g(t+1) = \sum_j f_j(t) e_j(t) + \varsigma g(t)$$

$$f_j(t+1) = f_j(t) - v_j(t).$$

Assume, for $t = 1, \ldots, \infty$, an affine stationary form for the value function

$$V_j^*(g(t+1), \underline{f}(t+1)) = \alpha_j g(t+1) + \sum_{i=1}^m \beta_{ij} f_i(t+1) + \gamma_j), \tag{6.252}$$

where α_j and β_{ij} and undetermined coefficients.

We will use the method of undetermined coefficients to solve (6.252). After substitutions, the right-hand side of (6.252) becomes

$$Y_j(e_j(t)) - c_j g(t) - k_j v_j(t) + \delta \left[\alpha_j \left(\sum_{i=1}^m f_i(t) e_i(t) + \varsigma g(t) \right) \right.$$

$$\left. + \sum_{i=1}^m \beta_{ij}(f_i(t) - v_i(t)) + \gamma_j \right].$$

Its derivative with respect to $e_j(t)$ is

$$Y_j'(e_j(t)) + \delta \alpha_j f_j(t)$$

and the derivative with respect to $v_j(t)$ is

$$-k_j - \delta \beta_{jj}.$$

From this last expression, we find that, in equilibrium,

$$v_j^*(t) = \begin{cases} 0 & \text{if } -k_j - \delta \beta_{jj} < 0 \\ f_j(t) - m_j & \text{if } -k_j - \delta \beta_{jj} \geq 0. \end{cases} \tag{6.253}$$

This implies that either $v_j^*(t) = 0$ and $f_j^*(t) \equiv f_j(0)$, $\forall t$, or $v_j^*(t) = f_j(t) - m_j$ and $f_j^*(t) \equiv m_j$, $\forall t \geq 1$. Thus $f_j^*(t)$ is constant and so is the control $e_j^*(t)$, which is now defined by

$$Y_j'(e_j^*) = -\delta \alpha_j f_j. \tag{6.254}$$

We will call $E_j^*[f_j]$ the value e_j^* that satisfies the above equation. Now we can identify the parameters defining the affine value function by writing the dynamic-programming (DP) equation

$$V_j^*(g, \underline{f}) = \alpha_j g + \sum_{i=1}^{m} \beta_{ij} f_i + \gamma_j$$

$$= Y_j(E_j^*[f_j]) - c_j g - k_j v_j^* + \delta \left[\alpha_j \left(\sum_{i=1}^{m} f_i E_i^*[f_j] + \varsigma g \right) \right.$$

$$\left. + \sum_{i=1}^{m} \beta_{ij} f_i + \gamma_j \right] \quad j = 1, \ldots, m.$$

and identify the factors on both sides, which yields, for $j = 1, \ldots, m$

$$0 = \alpha_j + c_j - \delta \alpha_j \varsigma,$$
$$0 = \beta_{ij} - \delta \alpha_j E_j^*[f_j],$$
$$0 = \gamma_j - Y_j(E_j^*[f_j]) + k_j v_j^* - \delta \gamma_j.$$

From which we get

$$\alpha_j = -\frac{c_j}{1 - \delta \varsigma}$$

$$\beta_{ij} = -\frac{\delta c_j}{1 - \delta \varsigma} E_j^*[f_j]$$

$$\gamma_j = \frac{1}{1 - \delta}(Y_j(E_j^*[f_j]) - k_j v_j^*).$$

From equation (6.254) we get

$$Y_j'(e_j^*) = \delta \frac{c_j}{1 - \delta \varsigma} f_j. \tag{6.255}$$

In addition, the following conditions must hold: either $v_j^* = 0$ and $f_j^* \equiv f_j^0$, which is the initial value of the emission factor or $v_j^* = f_j - m_j$ and $f_j^* \equiv m_j$. Therefore the equilibrium solution for a country is either to forever maintain its emission factor as it is at the beginning of the game, or to bring it down to its minimal value m_j and keep it at this level forever. Then, there is an energy input in the economy, which is constant and determined by equation (6.255).

Remark 6.21. We notice that because of the particular form of the model, the value function is linear in the state variables. As a consequence, the resulting feedback-Nash equilibrium solution is constant and so it is also an open-loop equilibrium solution. As it is a feedback-Nash equilibrium, this solution is subgame perfect in the sense of Selten.

6.9.2.3 *Global Pareto-optimal strategy profiles*

Let us compute a *global Pareto optimal* (efficient) (GPO) solution. This solution is obtained when a combined weighted payoff (with positive weights) is maximized.

The DP equation for the GPO solution is

$$\hat{V}(g(t), \underline{f}(t)) = \max_{(\mathbf{e}, \mathbf{v})} \left[\sum_{j=1}^{m} r_j (Y_j(e_j(t)) - c_j g(t) - k_j v_j(t)) + \delta \hat{V}(g(t+1), \underline{f}(t+1)) \right],$$

(6.256)

where $\hat{V}(\cdot, \cdot)$ is the current-valued Pareto value function.

The state equations are as before:

$$g(t+1) = \sum_j f_j(t) e_j(t) + \varsigma g(t),$$

$$f_j(t+1) = f_j(t) - v_j(t).$$

Assume, for $t = 1, \ldots, \infty$, an affine stationary form for the value function

$$\hat{V}(g(t+1), \underline{f}(t+1)) = \hat{\alpha} g(t+1) + \sum_{j=1}^{m} \hat{\beta}_j f_j(t+1) + \hat{\gamma}_j).$$

(6.257)

We use again the method of undetermined coefficients to solve the DP equation. After substitutions, the right-hand side of (6.257) becomes

$$\sum_{j=1}^{m} r_j (Y_j(e_j(t)) - c_j g(t) - k_j v_j(t)) + \delta \left[\hat{\alpha} (\sum_j f_j(t) e_j(t) + \varsigma g(t)) \right.$$

$$\left. + \sum_{j=1}^{m} \hat{\beta}_j (f_j(t) - v_j(t)) + \hat{\gamma}_j \right].$$

Its derivative with respect to $e_j(t)$ is

$$r_j Y_j'(e_j(t)) + \delta \hat{\alpha} f_j(t)$$

and the derivative with respect to $v_j(t)$ is

$$-r_j k_j - \delta \hat{\beta}_j.$$

From this last expression we find that, at equilibrium,

$$\hat{v}_j(t) = \begin{cases} 0 & \text{if } -r_j k_j - \delta \hat{\beta}_j < 0 \\ f_j(t) - m_j & \text{if } -r_j k_j - \delta \hat{\beta}_j \geq 0. \end{cases}$$

(6.258)

Again, this implies that, either $\hat{v}_j(t) = 0$ and $\hat{f}_j(t) \equiv f_j(0)$, $\forall t$, or $\hat{v}_j(t) = f_j(t) - m_j$ and $\hat{f}_j(t) \equiv m_j$, $\forall t \geq 1$. Thus $\hat{f}_j(t)$ is constant for $t \geq 1$ and so is the control $\hat{e}_j(t)$, which is now defined by

$$r_j Y_j'(\hat{e}_j) = -\delta \hat{\alpha} \hat{f}_j.$$

(6.259)

We will call $\hat{E}_j[f_j]$ the value $\hat{e}_j$ that satisfies the above equation (6.259). We can identify the parameters defining the affine value function by writing the maximized DP equation and equalizing the coefficients of the state variables f_j, g and the constants. This yields

$$\hat{V}(g, \underline{f}) = \hat{\alpha}g + \sum_{j=1...m} \hat{\beta}_j f_j + \hat{\gamma}$$

$$= \sum_{j=1}^{m} r_j \left(Y_j(\hat{E}_j[f_j]) - c_j g - k_j \hat{v}_j\right) + \delta \left[\hat{\alpha}\left(\sum_j f_j \hat{E}_j[f_j] + \varsigma g\right)\right.$$

$$\left. + \sum_{j=1}^{m} \hat{\beta}_j f_j + \hat{\gamma}\right].$$

We compare the similar terms on both sides and obtain

$$0 = \hat{\alpha} + \sum_{j=1}^{m} r_j c_j - \delta\hat{\alpha}\varsigma, \quad \text{or } \hat{\alpha} = \frac{\sum_{j=1}^{m} r_j c_j}{1 - \delta\varsigma},$$

$$0 = -(1-\delta)\hat{\beta}_j + \delta\hat{\alpha}\hat{E}_j[f_j],$$

$$0 = \hat{\gamma}_j - \sum_{j=1}^{m} r_j(Y_j(\hat{E}_j(f_j)) + k_j \hat{v}_j) - \delta\hat{\gamma}.$$

And, we easily obtain, from (6.259), that $\hat{e}_j$ is defined by the equation

$$Y_j'[\hat{e}_j] = \delta \frac{\sum_{i=1}^{m} r_i c_i}{r_j(1 - \delta\varsigma)} f_j \quad j = 1, \ldots, m. \tag{6.260}$$

Comparing (6.255) to (6.260), when the weights are equal, $r_j \equiv \frac{1}{m}$, we see that the energy input to the economy will be lower in the GPO than in the Nash equilibrium, for the same emission factor. Indeed, (6.260) becomes

$$Y_j'[\hat{e}_j] = \delta \frac{\sum_{i=1}^{m} c_i}{1 - \delta\varsigma} f_j \quad j = 1, \ldots, m. \tag{6.261}$$

whereas (6.255) gives

$$Y_j'(e_j^*) = \delta \frac{c_j}{1 - \delta\varsigma} f_j \quad j = 1, \ldots, m.$$

Since $Y_j(e_j)$ is positive, strictly increasing and concave, we see that necessarily $\hat{e}_j < e_j^*$ for the same value of f_j.

6.9.2.4 *Self-enforcing GPO solutions*

For simplicity, assume that all countries are identical. Assume that a GPO is negotiated with all weights equal ($r_j \equiv 1/m$), from an initial state $(g^0, \mathbf{f}^0)$. Assume that this GPO solution implies $\hat{f}_j \equiv m_j$ for all periods $t \geq 1$ and that it dominates the Nash-equilibrium solution in the sense that

$$\hat{V}_j(g, \underline{f}) > V_j^*(g, \underline{f}), \forall(g, \underline{f}), j = 1, \ldots m, \tag{6.262}$$

where $\hat{V}_j(g, \underline{f})$ is the current-value Pareto value-function for Player j, which is also equal to $\hat{V}(g, \underline{f})$, since all players are identical. Notice that v_j is necessarily equal to 0, since the f_j is at its lower bound m_j. Now assume that the following inequality holds for all g:

$$\hat{V}_j(g, \underline{m}) \geq \max_{e_j} \left\{ e_j - c_j\, g + \delta \left[\alpha_j^* \left(e_j + \sum_{i \neq j} f_i \hat{E}_i(m_i) + \varsigma g \right) \right. \right.$$
$$\left. \left. + \sum_{i=1}^m \beta_{ij}^* m_j + \gamma_j^* \right] \right\} \quad j = 1, \ldots, m. \tag{6.263}$$

Then the GPO solution *combined* with a threat to use the feedback-Nash ("simple") equilibrium as soon as a deviation from the GPO solution is detected, becomes a subgame-perfect equilibrium at $(g^0, \mathbf{f}^0)$. In that case, the climate treaty defined as: *use the GPO solution or face the feedback-Nash "simple" equilibrium* can be said to be *self-enforcing*.

Lemma 6.9. *Assume*[49] *that* $\lim_{\delta \to 1}(1-\delta)\hat{V}_j(g, \underline{m}) = \hat{\varphi}$ *and* $\lim_{\delta \to 1}(1-\delta)V_j^*(g, \underline{m}) = \hat{\varphi}^*$ *exist and are independent of g; assume also that there exists a value $\varepsilon > 0$ and a $\delta' \in [0, 1)$ such that*

$$(1 - \delta)V_j^*(g, \underline{m}) - \varepsilon \geq (1 - \delta) \max_{e_j} \left\{ e_j - c_j\, g + \delta \left[\alpha_j^*(e_j + \sum_{i \neq j} f_i \hat{E}_i(m_i) + \varsigma g) \right. \right.$$
$$\left. \left. + \sum_{i=1}^m \beta_{ij}^* m_j + \gamma_j^* \right] \right\} \quad j = 1, \ldots, m, \tag{6.264}$$

when $\delta \geq \delta'$. Then, there exists δ sufficiently close to 1 for which the inequality (6.263) holds true.

Proof. It suffices to multiply both sides of (6.263) by $1 - \delta$ and let delta tend 1. Then (6.263) is equivalent to the following inequality, satisfied for all admissible e_j:

$$(1 - \delta)V_j^*(g, \underline{m}) - \delta(1 - \delta)[\alpha_j^*(e_j + \sum_{i \neq j} f_j \hat{E}_i(m_i) + \varsigma g) + \sum_{i=1}^m \beta_{ij}^* m_j + \gamma_j^*]$$
$$\geq (1 - \delta)(e_j - c_j\, g) \quad j = 1, \ldots, m. \tag{6.265}$$

We can see that (6.264) implies (6.265) because when we let δ tend to 1, we get for both inequalities

$$\hat{\phi} - \phi^* - 3\varepsilon \geq (1 - \delta)(e_j - c_j\, g).$$

Since the RHS tends to zero when δ tends to 1, this inequality is satisfied for δ sufficiently close to 1. $\square$

Remark 6.22. The conditions under which $\lim_{\delta \to 1}(1 - \delta)\hat{V}_j(g, \underline{m}) = \hat{\varphi}$ and $\lim_{\delta \to 1}(1 - \delta)V_j^*(g, \underline{m}) = \hat{\varphi}^*$ exist and are independent of g, are Tauberian[50] conditions. They are satisfied in many infinite-horizon discounted-payoff models.

---

[49]Recall that $\hat{V}_j(g, \underline{m})$ and $V_j^*(g, \underline{m})$ are implicitly functions of the discount factor δ.
[50]Compare equation (5.8) on page 137 and the adjacent footnote.

In [67], a larger class of equilibria is characterized, with outcomes dominating the feedback-Nash equilibrium solution, which are "supported" by the threat to revert to the Nash subgame-perfect equilibrium.

6.9.3 *Memory-strategy equilibria in a non-discounted infinite-horizon multistage game*

6.9.3.1 *A motivation*

Remark 6.23. The consideration of the non-discounted sum of rewards is sometimes recommended as appropriate for dealing with global environmental models where the impacts will be felt in a distant future by the yet-unborn generations.

We saw in the previous subsection that a discount factor sufficiently close to 1 was necessary to construct a Pareto-optimal solution combined with a threat, being itself a feedback-Nash equilibrium, to form a subgame-perfect equilibrium for the original multistage game, at a given initial position. In this subsection, we consider a class of stationary multistage games where the payoffs are calculated as the non-discounted sum of the transition rewards. For this class, we develop below a rather complete theory of equilibria based on the use of threats and memory strategies. In these developments, we follow reference [117].

6.9.3.2 *Overtaking-optimality criteria*

Consider a system described by a state equation

$$\mathbf{x}(t+1) = \mathbf{f}(\mathbf{x}(t), \mathbf{u}_1(t), \dots, \mathbf{u}_m(t)), \quad t \in \mathbf{N} \quad \mathbf{x}(0) = \mathbf{x}^o, \tag{6.266}$$

where $\mathbf{x} \in X \subset \mathbf{R}^n, \mathbf{u}_j \in U_j \subset \mathbf{R}^{m_j}, \mathbf{f} : X \times \underline{U} \to X$, with $\underline{U} = U_1 \times \cdots \times U_m$.

A control sequence $\tilde{\underline{u}} : t \in \mathbf{N} \mapsto \underline{u}(t) \in \underline{U}$ generates a unique trajectory $\tilde{\mathbf{x}} : t \in \mathbf{N} \mapsto \mathbf{x}(t) \in X$, with $\mathbf{x}(0) = \mathbf{x}^o$. The payoffs accrued along a trajectory $\mathbf{x}(\cdot)$ generated by the control sequence $\tilde{\mathbf{u}}(\cdot)$, are defined by

$$y_j(t) = \sum_{\tau=0}^{t} L_j(\mathbf{x}(\tau), \mathbf{u}_1(\tau), \dots, \boldsymbol{u}_m(\tau)), \quad t \in \mathbf{N} \quad j \in M = \{1, \dots, m\}, \tag{6.267}$$

where $y_j(t)$ is the payoff accrued to Player j at stage t and $L_j : X \times U_1 \times \cdots \times U_m \to \mathbf{R}$, $i \in M$ are given functionals.

Since the system evolves over an infinite number of stages, the cumulative reward sequences $y_j(\cdot) : t \in \mathbf{N} \mapsto \mathbf{R}$ can be unbounded. As already indicated in previous encounters with games played over an infinite time horizon, we will use an overtaking criterion to compare payoffs that can be unbounded.

Definition 6.4. For Player j, the sequence $y(\cdot)$ of accrued payoffs is **overtaking** the sequence $y'(\cdot)$, denoted $y(\cdot) \succ y'(\cdot)$, if

$$\liminf_{t \to \infty} [y(t) - y'(t)] > 0. \tag{6.268}$$

We already defined the concepts of overtaking and weakly overtaking optimality in Section 6.2. The same concepts will be used here but with slightly changed notations, which will simplify the exposition.

The sequence $y(\cdot)$ is **weakly overtaking** the sequence $y'(\cdot)$, denoted $y(\cdot) \succeq y'(\cdot)$, if

$$\limsup_{t \to \infty}[y(t) - y'(t)] \geq 0. \tag{6.269}$$

These two comparison criteria are related in the following way:

(a) $y(\cdot) \succ y'(\cdot)$ implies $y(\cdot) \succeq y'(\cdot)$;
(b) $y(\cdot)$ is not overtaking $y'(\cdot)$ iff $y'(\cdot)$ is weakly overtaking $y(\cdot)$; similarly $y(\cdot)$ is not overtaking weakly $y'(\cdot)$ iff $y'(\cdot)$ is overtaking $y(\cdot)$.

6.9.3.3 *Strategies and equilibria*

A strategy for Player j is a sequence of policies $\tilde{\gamma}_j = (\gamma_j^t : t \in \mathbf{N})$, where each γ_j^t maps the history of play, up to time t, into the control set U_j. We call history of play the knowledge of initial state $\mathbf{x}^0$ and of the sequence of controls $\underline{\tilde{\mathbf{u}}}^t = (\mathbf{u}(0), \dots, \mathbf{u}(t-1))$. So, $\gamma_j^t : (\mathbf{x}^o, \underline{\tilde{\mathbf{u}}}^t) \in X \times \underline{U} \mapsto \mathbf{u}_j(t) = \gamma_j^t(\mathbf{x}^o, \underline{\tilde{\mathbf{u}}}^t) \in U_j$.

A strategy profile $\tilde{\gamma} = (\tilde{\gamma}_j)_{j \in M}$ generates a unique trajectory from the initial state $\mathbf{x}^o$, and an accrued payoff sequence $y_j(\cdot; \mathbf{x}^o, \tilde{\gamma})$.

Definition 6.5. A strategy profile $\tilde{\gamma}^* = (\gamma_j^*)_{j \in M}$ is a **weakly overtaking equilibrium** at $\mathbf{x}^o$ if, for any other strategy profile $\gamma^{*-j} = [\gamma_j, \gamma_{-j}^*]$ the following holds:

$$y_j(\cdot; \mathbf{x}^o, \gamma^*) \succeq y_j(\cdot; \mathbf{x}^o, \gamma^{*-j}). \tag{6.270}$$

6.9.3.4 *Feedback threats and trigger strategies*

A stationary-feedback law for Player j is defined by a mapping $\sigma_j : X \to U_j$. Note that a feedback law defines a strategy in the class introduced above, through the correspondence

$$\gamma_j^t(\mathbf{x}^o, \mathcal{U}_{[0,t-1]}) = \sigma_j(\mathbf{x}(t)).$$

Definition 6.6. A trigger strategy for Player j is defined by a nominal control sequence $\tilde{\mathbf{u}}^*(\cdot)$ and a feedback law σ_j, which are combined in the following sequence of policies:

$$\gamma_j^t(\mathbf{x}^o, \mathcal{U}_{[0,t-1]}) = \begin{cases} \mathbf{u}_j^*(t) & \text{if } \mathbf{u}^*(s) = \mathbf{u}(s), \quad s = 0, \dots, t-1 \\ \sigma_j(\mathbf{x}(t)) & \text{otherwise,} \end{cases}$$

where $\mathbf{x}(t)$ is the current state of the system controlled by the players.

So, as long as the control used by all players in the history of play agrees with the nominal control sequence, Player j decides to play $\mathbf{u}_j^*(t)$ at stage t. If there has been a deviation from the nominal control sequence in the past, Player j makes a decision about his control according to the feedback law σ_j. The feedback law is a *threat* that can be used by Player j to convince the other players to keep playing according to the nominal-control sequence.

Definition 6.7. The nominal control sequence $\tilde{\mathbf{u}}^*(\cdot)$ is said to be **supported** by the threat m-tuple $\underline{\sigma}$ if, for any stage $\tau \in \mathbf{N}$ and any admissible control $\mathbf{u}_j^\tau$, the following holds:

$$y_j^{\tau*}(\cdot) \succeq \hat{y}_j^\tau(\cdot) \quad j = 1, 2, \tag{6.271}$$

where

$$y_j^{\tau*}(t - \tau) = \sum_{s=\tau}^{t-1} L_j(\mathbf{x}^*(s), \mathbf{u}^*(s)), \quad t = \tau, \tau + 1, \ldots, \tag{6.272}$$

$$\hat{y}_j^\tau(\ell) = L_j(\mathbf{x}^*(\tau), [\mathbf{u}_j^\tau, \mathbf{u}_{-j}^*(\tau)]) + \sum_{s=1}^{\ell-1} L_j(\hat{\mathbf{x}}(s), \underline{\sigma}(\hat{\mathbf{x}}(s))), \quad \ell = 0, 1, \ldots. \tag{6.273}$$

At each stage τ, Player j compares the payoffs to come if the nominal control was used forever (equation (6.272)), to what he could get by playing $\mathbf{u}_j^\tau$ in stage τ and then receiving the payoff associated with the feedback law $\underline{\sigma}$, used by all players, (equation (6.273)). The nominal control sequence is supported by $\underline{\sigma}$ if (6.272) is weakly overtaking (6.273).

Lemma 6.10. *Given an initial state* $\mathbf{x}^o$, *a nominal control sequence* $u^*(\cdot)$ *that is supported by the threat m-tuple* $\underline{\sigma}$ *defines a trigger strategy m-tuple that is a weakly overtaking equilibrium at* $\mathbf{x}^o$.

Proof. Let $\underline{\gamma}^*$ be the trigger strategy m-tuple constructed by the given of $\mathbf{u}^*(\cdot)$ and the feedback law m-tuple $\underline{\sigma}$. equation (6.272) gives the payoff sequences obtained from any intermediate state along the trajectory $\mathbf{x}^*(\cdot)$ generated by $\mathbf{u}^*(\cdot)$ from $\mathbf{x}^o$. Along this trajectory, the payoff accrued to Player j is $y_j^*(\cdot)$.

Let γ_j be any strategy for Player j; the strategy m-tuple $[\gamma_j, \gamma_{-j}^*]$ will generate a control sequence that may first deviate from $\mathbf{u}^*(\cdot)$ at stage τ, where Player j plays $\mathbf{u}_j^\tau \neq \mathbf{u}_j^*(\tau)$. From that stage onward, the best payoff to come for Player j is given by equation (6.273), so the payoff sequence generated from $t = 0$ is $\tilde{y}^*(t) = \sum_{s=0}^{t-1} L_j(\mathbf{x}^*(s), \underline{\mathbf{u}}^*(s))$ if $t \leq \tau - 1$ and $\tilde{y}(t) = \sum_{s=0}^{t-1} L_j(\mathbf{x}^*(s), \underline{\mathbf{u}}^*(s)) + \hat{y}_j^\tau(\ell)$ if $t = \tau + \ell$, $\ell = 0, 1, 2, \ldots$. Now it is clear that (6.271) implies that $\tilde{y}^*(t) \succeq \tilde{y}(t)$ and hence, the conditions for a weakly overtaking equilibrium are satisfied. $\square$

Remark 6.24. This lemma shows that we can expect a very large set of equilibria at $\mathbf{x}^o$ in the class of memory strategies. However most of these equilibria are not subgame perfect. If the threat m-tuple is defined by a feedback weakly overtaking equilibrium $\underline{\sigma}^*$, valid for all $\mathbf{x} \in X$, then the trigger strategy constructed with a nominal-control sequence supported by $\underline{\sigma}^*$, is also a subgame perfect[51] equilibrium at $\mathbf{x}^o$.

---

[51]The proof of this result is left to the reader as an easy exercise.

6.9.3.5 *Most effective threats in a two-player game*

Consider the particular case of a two-player game.

Definition 6.8. A feedback law σ_j^{*j} is called most effective threat of Player j, if there exists another feedback law σ_{-j}^{*j} for the other Player $(-j)$, such that the pair $\underline{\sigma}^{*j} = (\sigma_j^{*j}, \sigma_{-j}^{*j})$ is a **feedback overtaking saddle point** solution of the zero-sum game with Player $(-j)$'s payoff, i.e., which satisfies, for all $\mathbf{x} \in X$,

$$y_{-j}(\cdot; \mathbf{x}, [\sigma_j, \sigma_{-j}^{*j}]) \succeq y_{-j}(\cdot; \mathbf{x}, \underline{\sigma}^{*j}), \tag{6.274}$$

$$y_{-j}(\cdot; \mathbf{x}, \sigma^{*j}) \succeq y_{-j}(\cdot; \mathbf{x}, [\sigma_j^{*j}, \sigma_{-j}]), \tag{6.275}$$

where we have used obvious notations to represent the accrued payoffs of Player $(-j)$ accrued payoffs and where σ_j or σ_{-j} represent any admissible feedback law for the two players.

In using such a threat, each player announces that he will try to destroy as much of his opponent's payoff as possible if there is a defection leading to a deviation from the nominal-control sequence.

Lemma 6.11. *In a two-player game, a pair of strategies $\boldsymbol{\gamma}^*$ constitutes a weakly overtaking equilibrium at a given initial state $\mathbf{x}^o$, only if they generate a control sequence $\mathbf{u}^*(\cdot)$ that is supported by the pair of most effective threats $(\sigma_1^{*2}, \sigma_2^{*1})$.*

Proof. Suppose the control sequence $\underline{\tilde{\mathbf{u}}}^*$ is not supported by the threat pair $(\sigma_1^{*2}, \sigma_2^{*1})$, then there must be, for a player, say Player 1, a stage τ and a control $\mathbf{u}_1^\tau$, such that condition (6.271), i.e., with obvious notations

$$\hat{y}_j^\tau(\cdot) \succ y_j^{\tau *}(\cdot) \quad j = 1, 2. \tag{6.276}$$

Now we construct a strategy $\tilde{\gamma}_1$ for Player 1 in the following way:

$$\gamma_1^s \equiv \gamma_1^{s*}, \quad s \leq \tau - 1, \tag{6.277}$$

$$\gamma_1^\tau(\mathbf{x}^o; \mathcal{U}_{[0, \tau - 1]}) = \mathbf{u}_1^\tau, \tag{6.278}$$

$$\gamma_1^t(\mathbf{x}^o; \mathcal{U}_{[0, t - 1]}) = \sigma_1^{*2}(\mathbf{x}(t)), \quad t \geq \tau + 1, \tag{6.279}$$

where $\mathbf{x}(t)$ is the state reached at stage t. From the saddle-point property of σ_1^{*2}, which is a weak saddle-point solution of the zero-sum game with Player 2's payoffs, we must have

$$y_1(\cdot; \mathbf{x}^o, (\gamma_1, \gamma_2^*)) \succ y_1(\cdot; \gamma^*), \tag{6.280}$$

which contradicts the supposed weak equilibrium property of γ^*. $\qquad\square$

6.9.3.6 *A sufficient condition for a feedback weak equilibrium*

We can extend the sufficient dynamic-programming conditions to the case of a feedback weakly overtaking equilibrium, by using a method called "reduction to finite rewards."

Theorem 6.6. *Let $\underline{\sigma}^*$ be a feedback strategy m-tuple such that:*

(a) *whatever be the initial state* $\mathbf{x}^o$, *the trajectory generated by* $\underline{\sigma}^*$ *is such that*

$$\lim_{t \to \infty} \mathbf{x}^*(t) = \bar{\mathbf{x}}^*,$$

uniformly on bounded subset of X, *where* $\bar{\mathbf{x}}^*$ *is an attractor called the "turnpike" of the game;*

(b) *there exist* m *constants* φ_j^* *and* m *potential functions* $v_j^*(\mathbf{x})$, *bounded on* X, *for* $j = 1, \ldots, m$, *such that*

$$\varphi_j^* + v_j^*(\mathbf{x}) = \max_{\mathbf{u}_j} \left\{ L_j(\mathbf{x}, [\mathbf{u}_j, \sigma_{-j}^*]) + v_j^*(f(\mathbf{x}, [\mathbf{u}_j, \sigma_{-j}^*])) \right\} \qquad (6.281)$$

$$= L_j(\mathbf{x}, \underline{\sigma}^*(\mathbf{x})) + v_j^*(f(\mathbf{x}, [\underline{\sigma}^*(\mathbf{x})])) \qquad (6.282)$$

(c) *the following holds along the trajectory generated from* $\boldsymbol{x}^o$ *by* $\underline{\sigma}^*$

$$\sum_{t=0}^{\infty} (L_j(\mathbf{x}^*(t), \underline{\sigma}^*(\mathbf{x}^*(t))) - \varphi_j^*) = v_j^*(\mathbf{x}^o). \qquad (6.283)$$

Then $\underline{\sigma}^*$ *is a feedback weakly overtaking equilibrium on* X.

Proof. We first note that the property (6.283) is implied by the conditions (a) and (b), since, by (6.283) we get

$$v_j^*(\mathbf{x}^o) = L_j(\mathbf{x}^o, \underline{\sigma}^*(\mathbf{x}^o)) - \varphi_j^* + v_j^*(f(\mathbf{x}^*(1)))$$

$$= \sum_{t=0}^{1} (L_j(\mathbf{x}^*(t), \underline{\sigma}^*(\mathbf{x}^*(t))) - \varphi_j^*) + v_j^*(f(\mathbf{x}^*(2)))$$

$$\cdots$$

$$= \sum_{t=0}^{T-1} (L_j(\mathbf{x}^*(t), \underline{\sigma}^*(\mathbf{x}^*(t))) - \varphi_j^*) + v_j^*(f(\mathbf{x}^*(T))). \qquad (6.284)$$

When $T \to \infty$, $\mathbf{x}^*(T)$ tends to $\bar{\mathbf{x}}^*$, and $v_j^*(\bar{\mathbf{x}}^*)$ can be set arbitrarily to 0. Then (6.283) holds true.

Now consider any control sequence $\tilde{u}_j(\cdot)$ for Player j and the trajectory $\tilde{x}(\cdot)$ generated from $\mathbf{x}^o$ by this control and the strategy σ_{-j}^* played by the other players. Using (6.283) repeatedly T times we get

$$v_j^*(\mathbf{x}^o) \geq L_j(\mathbf{x}^o, [\mathbf{u}_j(0), \sigma_{-j}^*(\mathbf{x}^o)]) - \varphi_j^* + v_j^*(f(\tilde{\mathbf{x}}(1)))$$

$$\geq \sum_{t=0}^{1} (L_j(\tilde{\mathbf{x}}(t), [\mathbf{u}_j(t), \sigma_{-j}^*(\mathbf{x}^*(t))]) - \varphi_j^*) + v_j^*(f(\tilde{\mathbf{x}}(2)))$$

$$\cdots$$

$$\geq \sum_{t=0}^{T-1} (L_j(\tilde{\mathbf{x}}(t), [\mathbf{u}_j(t), \sigma_{-j}^*(\mathbf{x}^*(t))]) - \varphi_j^*) + v_j^*(f(\tilde{\mathbf{x}}(T))). \qquad (6.285)$$

Now it can be proved that the condition (a) of the uniform convergence of equilibrium trajectories on bounded subsets of X implies that $\sum_{t=0}^{T-1} (L_j(\tilde{\mathbf{x}}(t), [\mathbf{u}_j(t), \sigma_{-j}^*(\mathbf{x}^*(t))]) - \varphi_j^*)$ tends to $-\infty$ when $T \to \infty$, if $\tilde{\mathbf{x}}(T)$ does not converge to the turnpike $\bar{\mathbf{x}}$. (See Proposition 5.2, p. 87 of [43].)

Therefore, (6.284) and (6.285) imply that

$$\limsup_{T \to \infty} \sum_{t=0}^{T-1} \left(L_j(\mathbf{x}^*(t), \sigma^*(\mathbf{x}^*(t))) - L_j(\tilde{\mathbf{x}}(t), [\mathbf{u}_j(t), \sigma^*_{-j}(\mathbf{x}^*(t))]) \right) \geq 0. \qquad (6.286)$$

Indeed, if $\tilde{\mathbf{x}}(T) \to \bar{\mathbf{x}}$, then $v_j^*(f(\tilde{\mathbf{x}}(T))) \to v_j^*(\bar{\mathbf{x}})$ and $v_j^*(f(\mathbf{x}^*(T))) \to v_j^*(\bar{\mathbf{x}})$ when $T \to \infty$, thus (6.284) and (6.285) imply (6.286). If $\tilde{\mathbf{x}}(T)$ does not converge to the turnpike $\bar{\mathbf{x}}$, then $\sum_{t=0}^{T-1}(L_j(\tilde{\mathbf{x}}(t), [\mathbf{u}_j(t), \sigma^*_{-j}(\mathbf{x}^*(t))]) - \varphi_j^*)$ tends to $-\infty$ and this also imply (6.286). Hence the weakly overtaking equilibrium property is satisfied. $\qquad \square$

Example 6.4. *Consider the Dutta-Radner model developed in section 6.9.2, when the discount factor δ tends to 1. We want to find a feedback law that is a weakly overtaking equilibrium for the limit game, when $\delta = 1$. We apply Theorem 6.6 with the following choices of turnpike value and potential functions:*

$$\varphi_j^* = Y_j(e_j^*) - c_j \bar{g}, \qquad (6.287)$$

$$v_j^*(g, \underline{f}) = -\frac{c_j(g - \bar{g})}{1 - \varsigma}, \qquad (6.288)$$

$$\bar{g} = \sum_{i=1}^{m} e_j^*, \qquad (6.289)$$

$$Y_j'(e_j^*) = \frac{c_f f_j^o}{1 - \varsigma}. \qquad (6.290)$$

The feedback law of Player j is a constant control, $vj^* \equiv 0$ and e_j^* defined by (6.287). A constant emission rate will drive the GHG concentration to the limit value $\bar{g}$ defined by (6.289). So Condition (a) of Theorem 6.6 is satisfied. Verify now that condition (b) also holds true. Writing equation (6.281) for our choice of values, we get

$$\varphi^* - \frac{c_j(g - \bar{g})}{1 - \varsigma} = Y_j(e_j^*) - c_j\, g - \frac{c_j(\varsigma\, g + (1 - \varsigma)\bar{g} - \bar{g})}{1 - \varsigma}.$$

In the above expression, we used the fact that the new state is given by $g^+ = \varsigma\, g + (1 - \varsigma)\bar{g}$. Rearranging terms in the RHS we obtain

$$\varphi^* - \frac{c_j(g - \bar{g})}{1 - \varsigma} = Y_j(e_j^*) - c_j\, \bar{g} - \frac{c_j(g - \bar{g})}{1 - \varsigma},$$

so the first part of Condition (b) is verified, i.e., equality holds in the DP equation when the feedback law is applied. The second part, i.e., the maximization w.r.t. control for Player j of the RHS is verified as in the case of discounted payoffs.

6.10 What have we learned in this chapter?

- Some typical economic conflict-situations were presented as dynamic games.
- A general model for a dynamic game was formulated in discrete time. In particular, difference equations were used to capture the system's dynamics.
- In contrast to optimal-control results, open-loop and feedback information structure lead to different solutions in dynamic games.

- Open-loop Nash equilibrium solutions can be obtained by implementing mathematical programming, or variational inequalities techniques.
- Feedback-Nash equilibria are subgame perfect, by construction.
- Dynamic programming was shown to be a constructive technique that can be used to compute a feedback-Nash equilibrium.
- Verification theorems establish that the dynamic-programming equations are sufficient conditions for a feedback-Nash equilibrium.
- The dynamic-programming equations for linear-quadratic games can be solved analytically. The feedback-Nash equilibrium takes the form of linear feedback rules, with gains defined through the solution of coupled Riccati equations.
- The two player zero-sum linear-quadratic game has been shown to admit similar saddle-point solutions, in open-loop or in feedback information structure.
- For multistage games played over an infinite time horizon, we can define equilibria in a class of memory based strategies. These strategies include retaliation threats that can themselves be defined as feedback-Nash equilibria. The resulting equilibrium can be both efficient (Pareto optimal) and subgame perfect.
- Later in this chapter's $\mathbb{GE}$ illustration, the growth-rate divergence puzzle will be (partially) explained as a result of a dynamic game played by two different economic subjects.

6.11 Exercises

Exercise 6.1. Open-loop NE for LQ games. Consider the open-loop Nash-equilibrium (ONE) for the linear-quadratic game defined by the state equation

$$\mathbf{x}(t+1) = \mathbf{A}(t)\mathbf{x}(t) + \sum_{j=1}^{m}\mathbf{B}_j(t)\mathbf{u}_j(t) + \mathbf{e}(t) \tag{6.291}$$

and **cost** functionals, for $j = 1, \ldots, m$,

$$J_j \triangleq \frac{1}{2}\sum_{t=0}^{T}(z^j(t) - \mathbf{C}^j(t)\mathbf{x}(t))'Q^j(t)(z^j(t) - \mathbf{C}^j(t)\mathbf{x}(t))$$

$$+ \frac{1}{2}\sum_{t=0}^{T-1}\sum_{i=1}^{m}\mathbf{u}_i(t)'R^{ji}(t)\mathbf{u}_i(t).$$

Using the equilibrium-maximum principle, show that the following conditions hold true for an ONE:

$$\mathbf{u}_j^*(t) = (\mathbf{R}^{jj}(t))^{-1}\mathbf{B}_i(t)' \left[\mathbf{P}^j(t+1)(\mathbf{W}^j(t)\mathbf{x}(t) + \mathbf{w}^j(t)) + \mathbf{p}^j(t+1)'\right],$$

where the symmetric matrices $\mathbf{P}^j(t+1)$ and row vector $p^j(t+1)$ are determined by the equations

$$\mathbf{P}^j(t) = -\mathbf{C}^j(t)'\mathbf{Q}^j(t)\mathbf{C}^j(t) + \mathbf{A}(t)'[(\mathbf{P}^j(t+1))^{-1}$$
$$-\mathbf{B}^j(t)((\mathbf{R}^{jj}(t))^{-1}\mathbf{B}_j(t)']^{-1}\mathbf{A}(t)t = 0, 1, \dots, T-1,$$
$$\mathbf{P}^j(T) = -\mathbf{C}^j(T)'\mathbf{Q}^j(T)\mathbf{C}^j(T),$$
$$\mathbf{p}^j(t) = \mathbf{z}^j(t)'\mathbf{Q}^j(t)\mathbf{C}^j(t) + [\mathbf{w}^j(t)\mathbf{P}^j(t+1) + \mathbf{p}^j(t+1)]\,\mathbf{A}(t),$$
$$t = 0, 1, \dots, T-1,$$
$$\mathbf{p}^j(T) = \mathbf{z}^j(T)'\mathbf{Q}^j(T)\mathbf{C}^j(T),$$

where we used the notation

$$\mathbf{W}^j(t) = [\mathbf{I} - \mathbf{B}_j(t)(\mathbf{R}^{jj}(t))^{-1}\mathbf{B}_j(t)'\mathbf{P}^j(t+1)]^{-1}\mathbf{A}(t), \quad t = 0, 1, \dots, T-1,$$

$$\mathbf{w}^j(t) = [\mathbf{I} - \mathbf{B}_j(t)(\mathbf{R}^{jj}(t))^{-1}\mathbf{B}_j(t)'\mathbf{P}^j(t+1)]^{-1}\left[\sum_{i \neq j}\mathbf{B}_i(t)\mathbf{u}_i^*(t)\right.$$

$$\left. +\mathbf{B}_j(t)(\mathbf{R}^{jj}(t))^{-1}\mathbf{B}_j(t)'\mathbf{p}^j(t+1) + \mathbf{e}(t)\right], \quad t = 0, 1, \dots, T-1.$$

Hint: see reference [62].

Exercise 6.2. *Feedback-NE for LQ games.* For the same system as in Exercise 6.1, consider the feedback-Nash equilibrium (FNE).

Using the dynamic-programming equations, show that the following conditions hold true for an FNE:

$$\sigma_j^*(t, \mathbf{x}) = (\mathbf{R}^{jj}(t))^{-1}\mathbf{B}_i(t)'\left[\mathbf{P}^j(t+1)(\mathbf{W}^j(t)\mathbf{x} + \mathbf{w}^j(t)) + \mathbf{p}^j(t+1)'\right],$$

where the symmetric matrices $\mathbf{P}^j(t+1)$ and row vector $p^j(t+1)$, $j = 1, \dots, m$ are determined by the equations

$$\mathbf{P}^j(t) = -\mathbf{C}^j(t)'\mathbf{Q}^j(t)\mathbf{C}^j(t) + \mathbf{W}(t)'[\mathbf{P}^j(t+1)$$
$$-\sum_{i=1}^m \mathbf{B}_i(t)(\mathbf{R}^{ii}(t))^{-1}\mathbf{R}^{ji}(t)(\mathbf{R}^{ii}(t))^{-1}\mathbf{B}_i(t)'\mathbf{P}^i(t+1)]\mathbf{W}(t),$$
$$t = 0, 1, \dots, T-1,$$
$$\mathbf{P}^j(T) = -\mathbf{C}^j(T)'\mathbf{Q}^j(T)\mathbf{C}^j(T),$$
$$\mathbf{p}^j(t) = \mathbf{z}^j(t)'\mathbf{Q}^j(t)\mathbf{C}^j(t) + [\mathbf{w}^j(t)\mathbf{P}^j(t+1) + \mathbf{p}^j(t+1)]\,\mathbf{W}(t)$$
$$-\sum_{i=1}^m [\mathbf{w}(t)\mathbf{P}^i(t+1) + \mathbf{p}^i(t)]\mathbf{B}_i(t)(\mathbf{R}^{ii}(t))^{-1}\mathbf{R}^{ji}(t)(\mathbf{R}^{ii}(t))^{-1}$$
$$\mathbf{B}_i(t)'\mathbf{P}^i(t+1)\mathbf{W}(t)t = 0, 1, \dots, T-1,$$
$$\mathbf{p}^j(T) = \mathbf{z}^j(T)'\mathbf{Q}^j(T)\mathbf{C}^j(T),$$

where we used the notation

$$\mathbf{W}^j(t) = \left[\mathbf{I} - \sum_{i=1}^m \mathbf{B}_i(t)(\mathbf{R}^{ii}(t))^{-1}\mathbf{B}_i(t)'\mathbf{P}^i(t+1)\right]^{-1}\mathbf{A}(t), \quad t = 0, 1, \dots, T-1,$$

$$\mathbf{w}^j(t) = \left[\mathbf{I} - \sum_{i=1}^m \mathbf{B}_i(t)(\mathbf{R}^{ii}(t))^{-1}\mathbf{B}_i(t)'\mathbf{P}^i(t+1)\right]^{-1}$$

$$\left[\mathbf{e}(t) + \sum_{i=1}^m \mathbf{B}_i(t)(\mathbf{R}^{ii}(t))^{-1}\mathbf{B}_i(t)'\mathbf{p}^i(t+1)\right], \quad t = 0, 1, \dots, T-1.$$

Hint: see reference [62].

Exercise 6.3. Consider the fishery-management model with the system dynamics governed by the state equation

$$x(t+1) = ax(t) - u_1(t) - u_2(t) + 1 \tag{6.292}$$

with the following payoffs to the players:

$$J_1 \triangleq \sum_{t=\tau}^{\infty} \varrho^t(x(t) + \ln u_1(t)), \tag{6.293}$$

$$J_2 \triangleq \sum_{t=\tau}^{\infty} \varrho^t(kx(t) + \ln u_2(t)), \tag{6.294}$$

where the constants are

$$0 < \varrho < 1,$$

$$0 < a < 1,$$

$$0 < k \leq 1.$$

(Notice that the constant k can model Player II's "carelessness" about the fish stock.)

(1) Compute the Pareto-efficient fishing (feedback) policy for each player when they maximize the sum of their individual payoffs.
(2) How does the "carelessness" of Player II influence each player's strategy (compare your result for $k < 1$ with the case of $k = 1$)?
(3) Assume $a = .9$. Compute the limit discount factor ϱ_c below which the stock will be eventually depleted for:

    a. $k = .5$;
    b. $k = 1$.

(4) Compute the Nash equilibrium (feedback) policy for each player.
(5) How does the "carelessness" of Player II influence now each player's strategy (compare your result for $k < 1$ with the case of $k = 1$)?

(6) Assume $a = .9$. Compute the limit discount factor ϱ_n below which the stock will be eventually depleted for:

   a. $k = .5$;
   b. $k = 1$.

(7) Overall, what can you say about the impact of the environmental "carelessness" of one of the players on the results you obtained in 3 and 6?

Exercise 6.4. Consider the fishery-management model with system dynamics governed by the state equation

$$x(t+1) = \mu x(t) - u(t) - v(t) + p, \tag{6.295}$$

with the following asymmetric payoff functions to players 1 and 2:

$$J_1 \triangleq \sum_{t=\tau}^{\infty} \varrho^t (x(t) + \ln(u(t))), \tag{6.296}$$

$$J_2 \triangleq \sum_{t=\tau}^{\infty} \varrho^t (x(t) + \ln(v(t) - q)). \tag{6.297}$$

The constants satisfy

$$0 < \varrho < 1, \quad 0 < \mu < 1, \quad p \geq 0, \quad q \geq 0$$

and the state x_τ is observable. As usual, $x(t)$ denotes stock, $u(t)$ and $v(t)$ is catch, p recruitment, ϱ discount factor, and $1 - \mu$ is mortality. The utility that the two players associate with the same catch $u(t)$ is different due to $q > 0$. This might reflect some differences in the countries' internal cost structures (e.g., transportation cost).

(1) Prove that the non-cooperative feedback-Nash equilibrium strategy for the second player is

$$\hat{u}_2(t) = \frac{1}{\varrho} - \mu + q.$$

   Hint: Try $W_1(x) = W_2(x) = W(x) = Ax + B$ as the value functions.
(2) What is the steady-state stock value of this fishery? Under which conditions is it positive? Hint: You need to compute the other player's equilibrium strategy.
(3) How does the discount factor influence the equilibrium strategies?

Exercise 6.5. Consider the infinite-horizon game without discounting, covered in Example 6.4 for feedback weakly overtaking equilibrium.

(1) Write the sufficient dynamic-programming conditions for a Pareto-optimal solution.
(2) Show that there exist Pareto-optimal solutions that can be supported by the feedback weakly overtaking equilibrium threat.

Exercise 6.6. Consider the state equation

$$x(t+1) = ax(t) - u_1(t) - u_2(t) + p, \quad x(0) = x^0, \tag{6.298}$$

and payoffs

$$J_j \triangleq \sum_{t=0}^{\infty} \beta^t [x(t) + \log(u_j(t))], \quad j = 1, 2, \tag{6.299}$$

where $\beta, 0 < \beta < 1$ is the common discount factor and $\tilde{u}_j = (u_j(t))_{t=0,1,...}$ here $x(t)$ is equal to the logarithm of the quantity of fish in a transboundary fishery at time period t; the fishing effort of country j is $u_j(t)$.

Show that the unique feedback-Nash equilibrium is defined by

$$\sigma_1(x) = \sigma_2(x) \equiv \frac{1 - a\beta}{\beta}, \tag{6.300}$$

with the corresponding equilibrium payoffs

$$V_1^*(x) = V_2^*(x) = V^N(x) = Ax + B, \tag{6.301}$$

where

$$A = \frac{1}{1 - a\beta}, \quad B = \frac{1}{1 - \beta}\left[\log\left(\frac{1 - a\beta}{\beta}\right) + \frac{\beta p}{1 - a\beta} - 2\right].$$

Exercise 6.7. The Great Fish War. Consider the Great Fish War model of Levhari and Mirman ([159]), summarized by the state equation

$$x(t+1) = (x(t) - u_1(t) - u_2(t))^\alpha. \tag{6.302}$$

and the payoffs

$$J_j \triangleq \sum_{t=0}^{\infty} \beta_j^t \log(u_j(t)), \quad j = 1, 2, \tag{6.303}$$

where $0 < \beta_j < 1$ is the discount factor of Player j.

(1) Assume that this game is played over one period only. The open-loop nprmal form payoffs are then $J_j(0, x^0, \underline{u}) = \log(u_j(0))$, and the controls are subject to the coupled constraint $u_1(0) + u_2(0) \leq x^0$. Show that any sharing of the fish stock, $u_j(0) \geq 0$, $j = 1, 2$ with $u_1(0) + u_2(0) = x^0$ is an equilibrium.

(2) Assume that this game is played over two periods, $t = 0, 1$ with feedback strategies. In period 1 the stock of fish is shared equally between the two players. Show that the local game in period $t = 0$ is defined by the payoffs

$$W_j(0, x^0; \underline{u}) = \log(u_j(0)) + \beta_j \log \frac{1}{2}(x^0 - u_1(0) - u_2(0)), \quad j = 1, 2. \tag{6.304}$$

Show that the equilibrium is defined by the linear-feedback laws

$$u_1^*(0) = \frac{\alpha\beta_2}{\alpha^2\beta_1\beta_2 + \alpha\beta_1 + \alpha\beta_2} x^0$$

$$u_2^*(0) = \frac{\alpha\beta_1}{\alpha^2\beta_1\beta_2 + \alpha\beta_1 + \alpha\beta_2} x^0.$$

Show that the equilibrium value function for Player j takes the form

$$V_j^*(0, x^0) = (1 + \alpha\beta_j)\log x^0 + A_j^1, \tag{6.305}$$

where A_j^1 is a constant, independent of x^0.

(3) Find the feedback-Nash equilibrium for the infinite-horizon model. For that, assume that the value function takes the form

$$V_j^*(0, x^0) = b_j^* \log x^0 + A_j^*, \quad j = 1, 2, \tag{6.306}$$

and use the method of undetermined coefficients. Show that the equilibrium feedback laws are

$$\sigma_1^*(x) = \frac{\alpha\beta_2(1 - \alpha\beta_1)}{1 - (1 - \alpha\beta_1)(1 - \alpha\beta_2)} x,$$

$$\sigma_2^*(x) = \frac{\alpha\beta_1(1 - \alpha\beta_2)}{1 - (1 - \alpha\beta_1)(1 - \alpha\beta_2)} x.$$

6.12 GE illustration: Growth-rate divergence between countries with the same fundamentals

In this section, we show how trade unions and capitalists in different countries could have *engineered* the wage negotiations to achieve different growth rates. The results described below were first reported in [141].

6.12.1 *A macroeconomic problem*

It has been one of the main interests of modern growth theory to examine why countries experience sharp divergences in their long-term growth rates. These divergences were first noted by Kaldor [134]. Since then, growth theorists have paid them a great deal of attention. In particular, Romer [212] heavily criticized the so-called S-S (Solow-Swan) theory of economic growth, which does not give a satisfactory explanation about divergences in long-term growth rates.

However, there has been almost no attempt to explain growth rates through the dynamic strategic interactions between capitalists and workers. The aim of this GE illustration is to show that the growth rate, and its divergences between different countries, can be a result of a dynamic game played between the two social classes.

Industrial relations and the bargaining power of workers versus capitalists differ from country to country. So does the growth rate. Bargaining power can be thought of as the weight of each class's utility in the social planner's objective. More objectively, the worker class's bargaining power can certainly be linked to the percentage of the labor force that belongs to trade unions and, by extension, to how frequently a Labor-like (socio-democratic) government prevails.

The argument about unionization level having an impact on a country's growth rate appears to be valid. Figure 6.6 shows a cross-sectional graph of the average 1980–1988 growth rate plotted against the corresponding percentage of union membership across 20 Asian and Latin American countries.[52] About 15% of the growth rate variability can be explained by union membership.

---

[52] Source [55].

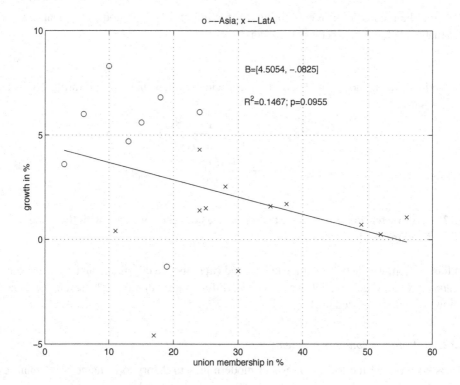

Fig. 6.6   Growth rates and unionisation levels

The above observations motivated our basic view that differences in countries' industrial relations are responsible, at least partly, for those in growth performance.

To accomplish our aim, we incorporate two new features into our game-theoretic growth model. First, we employ a non-cooperative (self-enforcing) equilibrium concept that we called *collusive equilibrium* in Section (see Definition 5.2). We recall that a collusive-equilibrium strategy is constituted by a pair of trigger policies, where actions are "triggered" by the changes in the (extended) state. Second, we assume that capitalists differ from workers in their utility functions: workers derive utility from consumption while capitalists derive utility from both consumption and possession of capital itself.

Among economists (or sociologists), it is perhaps Karl Marx [169] who first asserted that capitalists act to resolve *a Faustian conflict between the passion for accumulation and the desire for enjoyment*. Observation of the real business world leads us to agree with Marx's view and was a motivation for us to incorporate "the passion for accumulation" into our model and pursue its implications.

We build a simple model of endogenous growth that incorporates these two new features. This enables us to show that

(i) in such a model, there exist an uncountable large number of collusive equilibrium paths that have different long-term growth rates, and

(ii) the growth rate disparities can be a result of different levels of "bargaining power" of the workers' bargaining power compared to that of the capitalists.

These two facts will be sufficient for us to show that economies with the same economic fundamentals, but with distinctive industrial relations, can grow at different rates.

6.12.2 *An endogenous growth model*

A dynamic game problem. The economy consists of two players (social classes): *capitalists* and *workers*.

The state variable is capital $x(t) \geq 0$ that satisfies the state equation

$$x(t+1) = ax(t) - w(t) - c(t) \tag{6.307}$$

where a is capital productivity, $c(t) \geq 0$ represents capitalists' consumption and $w(t) \geq 0$ is the workers' wages.

The utility functions of the capitalists and workers as follows:

$$J_c \triangleq \sum_{t=\tau}^{\infty} \beta^t \left(B \cdot (x(t))^{\nu} + D \cdot (c(t))^{\nu} \right) \qquad B, D > 0, \tag{6.308}$$

$$J_w \triangleq \sum_{t=\tau}^{\infty} \beta^t F \cdot (w(t))^{\nu} \qquad F > 0 \tag{6.309}$$

where $\nu \in (0,1)$, and β is the discount factor[53] (also from interval (0,1)) that is identical for each player.

By saying that the capitalists and workers are "playing a dynamic game," we mean that the players choose feasible strategies $(\tilde{c}, \tilde{w})$ such that (6.308) and (6.309) are maximized for a historically given initial capital x^τ at period τ. We will write J_c and J_w to denote the *total utility measures,* which are the discounted sums of utility in period t for the capitalists and the workers, respectively.

Solution concepts. Suppose that players can observe the state variable k. This means that we can expect the game defined above to be solved in a feedback, or Markovian, pair of stationary strategies for capitalists' consumption and workers' wages, as follows:

$$\sigma_c(x), \quad \sigma_w(x). \tag{6.310}$$

Depending on how the players organize, we can expect the game to admit

- a feedback Nash equilibrium (non-cooperative), and
- a Pareto-efficient solution (cooperative);

or, assuming that the players can observe one another's actions

---

[53] We use β here for a discount factor because we use β to denote a strategy.

- a collusive equilibrium (non-cooperative) in a pair of stationary trigger strategies

$$\delta_c(x,y), \quad \delta_w(x,y),$$

where y is an auxiliary state variable to be defined below. This solution concept can produce a solution that is Pareto-efficient and a subgame-perfect equilibrium. In reality, this requires some negotiations between the players before the game starts. In "real life," the negotiations needed for the definition of a collusive equilibrium could be wage negotiations.

6.12.3 *Game solutions*

A feedback-equilibrium solution. Following Theorem 6.2, the pair of strategies $(\sigma_c(x), \sigma_w(x))$ provides a feedback (Markovian) Nash equilibrium solution if and only if there exist value functions $V_c(x)$ and $V_w(x)$ that satisfy the Bellman equations:

$$V_c(x) = \max_c \left\{ Bx^\nu + Dc^\nu + \beta V_c(ak - c - \sigma_w(x)) \right\}, \tag{6.311}$$

$$V_w(x) = \max_w \left\{ Fw^\nu + \beta V_w(ak - \sigma_c(x) - w) \right\}. \tag{6.312}$$

We need to solve the above equations to determine the feedback-Nash equilibrium.

Let us assume the value functions for the players as follows:

$$V_c(x) = \Gamma_c x^\nu \quad V_w(x) = \Gamma_w x^\nu. \tag{6.313}$$

We will show that these functions satisfy equations (6.311) and (6.312).

Substituting the value functions in (6.311)–(6.312) and using the first-order condition for the maxima of the right-hand sides (that are concave functions of the players' decisions), we obtain the following Markovian-equilibrium strategies:

$$\sigma_c(x(t)) = \left(\frac{\beta}{D}\Gamma_c\right)^{\frac{1}{\nu-1}} x(t) \equiv \bar{c}x(t) \quad \text{and} \quad \sigma_w(x(t)) = \left(\frac{\beta}{F}\Gamma_w\right)^{\frac{1}{\nu-1}} x(t) \equiv \bar{w}x(t). \tag{6.314}$$

If we establish the existence of nonnegative $\bar{c}, \bar{w}, \Gamma_c$ and Γ_w, then we will know that the feedback-Nash (Markovian) equilibrium (6.314) exists.

Allowing for (6.314), the first-order conditions and the envelop conditions[54] are obtained

$$\nu D\bar{c}^{\nu-1}x^{\nu-1} = \beta\nu\Gamma_c(a - \bar{c} - \bar{w})^{\nu-1}x^{\nu-1}, \tag{6.315}$$

$$\nu F\bar{w}^{\nu-1}x^{\nu-1} = \beta\nu\Gamma_w(a - \bar{c} - \bar{w})^{\nu-1}x^{\nu-1}, \tag{6.316}$$

and

$$\nu\Gamma_c x^{\nu-1} = \nu Bx^{\nu-1} + \beta\nu(a - \bar{w})\Gamma_c(a - \bar{c} - \bar{w})^{\nu-1}x^{\nu-1}, \tag{6.317}$$

$$\nu\Gamma_w x^{\nu-1} = \beta\nu(a - \bar{c})\Gamma_w(a - \bar{c} - \bar{w})^{\nu-1}x^{\nu-1}, \tag{6.318}$$

---

[54]The *envelope condition* is obtained by differentiating both sides of the Bellman equation in state variable, after the right-hand side has been maximized in strategies.

respectively. Assuming that the undetermined constants are nonzero, we derive from from (6.315)–(6.318) the system of three equations with the three constants as unknown:

$$D\bar{c}^{\nu-1} = \beta\Gamma_c(a - \bar{c} - \bar{w})^{\nu-1}, \tag{6.319}$$

$$F\bar{w}^{\nu-1} = \beta\Gamma_w(a - \bar{c} - \bar{w})^{\nu-1}, \tag{6.320}$$

$$\Gamma_c = B + \beta(a - \bar{w})\Gamma_c(a - \bar{c} - \bar{w})^{\nu-1}, \tag{6.321}$$

$$1 = \beta(a - \bar{c})(a - \bar{c} - \bar{w})^{\nu-1}. \tag{6.322}$$

Repeating substitutions yields the following equation, which the capitalists' gain coefficient $\bar{c}$ has to satisfy:

$$Q(\bar{c}) \equiv a - 2\bar{c} - \left(\frac{B}{D}\right)\bar{c}^{1-\nu} - \beta^{\frac{1}{1-\nu}}(a - \bar{c})^{\frac{1}{1-\nu}} = 0. \tag{6.323}$$

Notice that $Q(0)$ has to be positive for (6.323) to have a solution in $[0, a]$. Requesting $Q(0) = a - (\beta a)^{\frac{1}{1-\nu}} > 0$ yields

$$\beta a^{\nu} < 1. \tag{6.324}$$

For a unique feedback-Nash equilibrium to exist in the capitalist game, we require a unique solution to (6.323). For large $\bar{c}$ (e.g., $\bar{c} > \frac{a}{2}$), $Q(\bar{c}) < 0$ and the graph of $Q(\bar{c})$ crosses 0 at least once. The intersection will be unique if the derivative

$$\frac{dQ(\bar{c})}{d\bar{c}} = -2 - (1 - \nu)\left(\frac{B}{D}\right)\bar{c}^{-\nu} + \frac{1}{1-\nu}\beta^{\frac{1}{1-\nu}}(a - \bar{c})^{\frac{1}{1-\nu}-1} < 0 \tag{6.325}$$

for all $\bar{c} \in (0, a)$. The above inequality is difficult to solve. However, notice that the second term in the above expression is always negative and can only help to achieve the desired monotonic decrease of $Q(\bar{c})$; in fact, (6.325) will always be negative for sufficiently large B/D. We will drop this term from (6.325) and prove a stronger (and sufficient) condition for a unique solution to (6.323). Using (6.324) we get

$$\frac{dQ(\bar{c})}{d\bar{c}} < -2 + \frac{1}{1-\nu}\beta^{\frac{1}{1-\nu}}(a - \bar{c})^{\frac{1}{1-\nu}-1} < -2 + \frac{1}{1-\nu}(\beta a^{\nu})^{\frac{1}{1-\nu}} < -2 + \frac{1}{1-\nu} < 0. \tag{6.326}$$

The last inequality is true for

$$\nu < \frac{1}{2}. \tag{6.327}$$

Hence there uniquely exists a positive $\bar{c} \in (0, a)$ such that $Q(\bar{c}) = 0$, if conditions (6.324) and (6.327) are satisfied.

Once $\bar{c}$ has been obtained, $\bar{w}$ is uniquely determined by (6.322). In turn, Γ_c and Γ_w are uniquely determined by (6.319) and (6.320), respectively. Note that all the above constants thus determined are positive under (6.324).

Let us consider the condition for a positive growth rate $\mu = \dfrac{x(t+1)}{x(t)}$. We get from (6.307) and (6.314),

$$x(t + 1) = (a - \bar{c} - \bar{w})x(t). \tag{6.328}$$

Consequently, $x(t+1) - x(t) > 0$ if and only if $a - \bar{c} - \bar{w} > 1$, that is, equivalent (from (6.322)) to

$$\beta(a - \bar{c}) > 1 \tag{6.329}$$

or

$$\bar{c} < a - \frac{1}{\beta}. \tag{6.330}$$

Expression (6.330) holds if $Q(\cdot)$ evaluated at $a - \dfrac{1}{\beta}$ is negative. This is so if B/D is sufficiently large (i.e., as stated earlier, when the capitalists *truly* value the capital in their utility function). Indeed, it is evident from (6.323) that

$$Q\left(a - \frac{1}{\beta}\right) < 0 \tag{6.331}$$

if B/D is large. In fact, we just proved a lemma as follows.

Lemma 6.12. *A pair of linear feedback-Nash equilibrium strategies (6.314) and a pair of value functions (6.313), which satisfy equations (6.311) and (6.312), uniquely exist if*

$$a^{\nu} < \frac{1}{\beta} < a \qquad \text{and} \qquad \nu < \frac{1}{2}.$$

Moreover, if B/D is sufficiently large the growth rate $\mu = \dfrac{x(t+1)}{x(t)}, \quad t = 0, 1, 2, \dots$ corresponding to the equilibrium strategy pair is positive and time-invariant.

We can interpret this lemma as establishing a relationship between capital productivity, the discount factor and the utility function shape (i.e., the degree of agent risk aversion) for an equilibrium to exist. For example, a feedback-Nash equilibrium in our capitalistic game would be *unlikely* to exist in economies with very high capital productivity ($a >> 1$), a high discount factor ($\beta \approx 1$; obviously, $\beta < 1$) and almost risk-neutral agents (≈ 1). However, the conditions given in the above lemma are sufficient and may be weakened for large B to D ratios. Furthermore, the economic interpretation of a large B/D is that the capitalists *truly* value the capital in their utility function. So, the "greedier" the capitalists, the stronger the growth.

Equation (6.323) and the remaining conditions determining $\bar{c}, \bar{w}, \Gamma_c, \Gamma_w$ and the growth rate can be solved numerically only. We need these constants for the definition of a collusive equilibrium in Section 6.12.3 and have computed them here for a few discount-factor values. We have assumed the rest of our economy parameters to be as follows:

$$
\begin{array}{ll}
a = 1.07 & \text{7\% capital productivity} \\
\nu = 0.2 & \text{utility function exponent} \\
B = 1 & \text{capitalists' weight for "passion for accumulation"} \\
D = 1 & \text{capitalists' weight for "desire for enjoyment"} \\
F = 1 & \text{workers' weight of utility from consumption} \\
.98 \leq \beta \leq .984 & \text{discount-factor range corresponding to "true"} \\
& \text{discount rate} \in [1.6, 2]\% \\
x_0 = 1 & \text{capital stock}
\end{array}
$$

Evidently, each pair of strategies (6.314) depends on the discount rate. Furthermore, for each discount rate, the strategy pair (6.314) generates a pair of different utility measures to the players (see Figure 6.7).

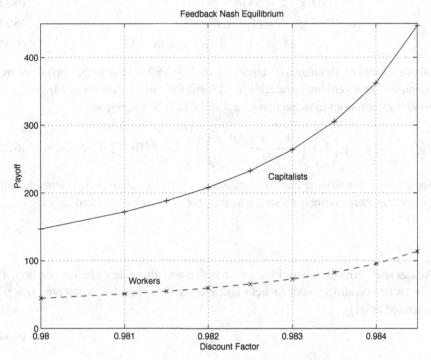

Fig. 6.7 Feedback-Nash equilibrium utility measures

Pareto-efficient solutions. Suppose that the players have agreed to act jointly to maximize

$$\alpha J_c^\alpha + (1 - \alpha) J_w^\alpha$$

where the parameter $\alpha \in [0, 1]$ is a "proxy" for what we call "bargaining power." Symbols J_c^α, J_w^α are defined as in (6.308) and (6.309) with the superindex α added[55] in anticipation that the objective values thus obtained will depend on bargaining power.

Let us derive a Pareto-optimal pair of strategies, which can be computed as a joint maximizer of the following Bellman equation:

$$U(x) = \max_{c,w} \left\{ \alpha(Bx^\nu + Dc^\nu) + (1 - \alpha)Fw^\nu + \beta W_1(ax - c - w) \right\}. \qquad (6.332)$$

We specify the Pareto-efficient strategies for c and w and the value function $U(x)$ as the following functions of state $x(t)$ as follows:

$$c(x(t)) = \underline{c}x(t) \quad w(x(t)) = \underline{w}x(t), \qquad (6.333)$$

$$U^\alpha(x) = \Gamma x^\nu. \qquad (6.334)$$

---

[55] In fact, the coefficients $\underline{c}$, $\underline{w}$ and Γ will all depend on α. However, to simplify notation, we will keep these symbols non indexed.

From the duopoly theory, we expect $\underline{c} \leq \overline{c}$ and $\underline{w} \leq \overline{w}$.

From the first-order and envelope conditions we obtain

$$\alpha D \underline{c}^{\nu-1} = \beta \Gamma (a - \underline{c} - \underline{w})^{\nu-1}, \tag{6.335}$$

$$\alpha D \underline{c}^{\nu-1} = (1 - \alpha) F \underline{w}^{\nu-1}, \tag{6.336}$$

$$\Gamma = \alpha B + \beta a (a - \underline{c} - \underline{w})^{\nu-1} \Gamma. \tag{6.337}$$

The above system of simultaneous equations (6.335)–(6.337) has to be solved for us to determine the Pareto-efficient strategies (6.333) and the value function (6.334).

Similar substitutions as in Section 6.12.3 yield the following equation:

$$R(\underline{c}) \equiv D \left(a - \underline{c} \left(1 + \left(\frac{(1-\alpha)F}{\alpha D} \right)^{\frac{1}{1-\nu}} \right) \right)^{1-\nu} - \beta (B \underline{c}^{1-\nu} + aD) = 0. \tag{6.338}$$

The analogous reasoning that we conducted to solve equation (6.323) leads us to the conclusion that there uniquely exists a positive $\underline{c}$ for that $R(\underline{c}) = 0$, if and only if

$$\beta a^{\nu} < 1 \tag{6.339}$$

holds.

We can also derive the conditions for a positive-growth μ rate under cooperation. The positive-growth condition, under Pareto efficiency and the players' bargaining power as characterized by α, is

$$a - \underline{c} - \underline{w} > 1, \tag{6.340}$$

which is equivalent (because of (6.338)) to

$$\beta \frac{B}{D} \underline{c} > 1 - a. \tag{6.341}$$

In fact, we have just proved a lemma as follows.

Lemma 6.13. *A pair of linear Pareto-optimal strategies uniquely exists if and only if (6.339) is satisfied. Moreover, the growth rate μ corresponding to the strategy pair is positive and time-invariant, even if $B = 0$.*

Notice that condition (6.339) is also needed for the existence of a feedback-Nash equilibrium. However, Pareto efficient strategies can generate positive growth even capitalists are not "firm" in their passion for accumulation.

We have numerically solved equation (6.338) and the remaining conditions determining $\underline{c}, \underline{w}, \Gamma$, and the growth rate.

Each pair of strategies depends on α (bargaining power) and β (discount factor) and results in a different growth rate as shown in Figure 6.8. These growth rates generate different pairs of utility measures for the players $(J_w^{\alpha}, J_c^{\alpha}) = (\text{Workers, Capitalists})$; see Figure 6.9 (where the "+" represents the growth rate obtained for the feedback-Nash equilibrium).

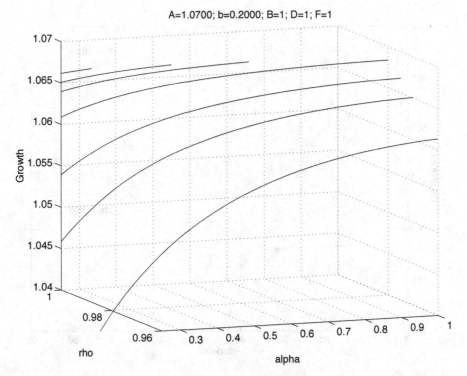

Fig. 6.8 Pareto efficient growth rates

Collusive equilibria. Now, we define a pair of strategies (compare Definition 5.2, page 144)

$$\delta_c^\alpha(x, y), \quad \delta_w^\alpha(x, y)$$

that aim at enforcing the payoffs associated with the pair of Pareto-efficient strategies $\beta_c^\alpha(x)$, $\beta_w^\alpha(x)$. Subsequently, we show that $(\delta_c^\alpha, \delta_w^\alpha)$ can be a subgame-perfect equilibrium strategy pair.

Assume that the game is played under the *feedback with observable actions information structure* and define an auxiliary state variable y, called the mode of the game:

$$\begin{cases} y(0) \;\; = 1 \\ y(t+1) = \begin{cases} 1 & \text{if } y(t) = 1 \text{ and } c(t) = \underline{c}x(t) \text{ and } w(t) = \underline{w}x(t) \\ 0 & \text{otherwise.} \end{cases} \end{cases} \tag{6.342}$$

Notice that once the *mode of the game* changes from 1 (cooperative) to 0 (non-cooperative), the players remain in this mode indefinitely; so, there is no return to negotiations in this case.

Define a new pair of strategies $(\delta_c^\alpha, \delta_w^\alpha)$ in the following way:

$$\begin{cases} \delta_c^\alpha(x(t), y(t)) = \begin{cases} \underline{c}x(t) & \text{if } y(t) = 1 \\ \overline{c}x(t) & \text{otherwise} \end{cases} \\ \delta_w^\alpha(x(t), y(t)) = \begin{cases} \underline{w}x(t) & \text{if } y(t) = 1 \\ \overline{w}x(t) & \text{otherwise.} \end{cases} \end{cases} \tag{6.343}$$

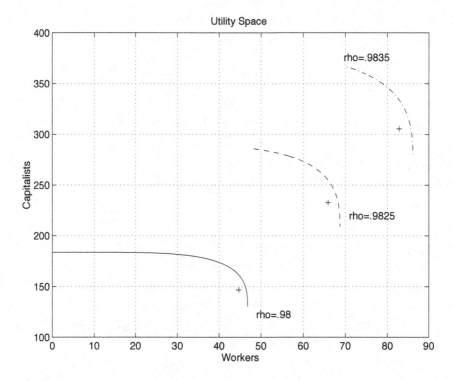

Fig. 6.9   Pareto efficient utility frontiers

This definition implies that each player uses a Pareto-efficient strategy so long as the other player does the same.

To support the claim of Section 6.12.3 that the payoffs obtained in Figure 6.9, and the strategies that generate them, are plausible outcomes, we need to show that strategies (6.343) constitute a subgame-perfect equilibrium. To prove this argument, we will show that $(\delta_c^\alpha, \delta_w^\alpha)$ is an equilibrium at every point (x, y).

It is easy to see that the pair is an equilibrium if $y(\tau) = 0$ for $\tau > 0$. Indeed, if one of the players has cheated before τ, the players are using feedback-Nash equilibrium strategies $\bar{c}k, \bar{w}k$, which are subgame-perfect equilibrium strategies.

Moreover, $(\delta_c^\alpha, \delta_w^\alpha)$ will be an equilibrium for $(x, 1)$ if the maximum gain from "cheating", i.e., breaching the agreement concerning consumption streams $\bar{c}k, \bar{w}k$, does not exceed the agreed-upon Pareto-efficient utility levels J_c^α, J_w^α for either player. We will solve the following system of inequalities to examine the ranges of the discount factor and bargaining power for which the above claim is true.

$$\max_{c^+} \left\{ Bx^\nu + D \cdot (c^+)^\nu + \beta V_c\big(ax - c^+ - \beta_w^\alpha(x)\big) \right\} < W_c^\alpha(x), \qquad (6.344)$$

$$\max_{w^+} \left\{ F \cdot (w^+)^\nu + \beta V_w\big(ax - \beta_c^\alpha(x) - w^+\big) \right\} < W_w^\alpha(x). \qquad (6.345)$$

Here, c^+ and w^+ are the best cheating policies, $V_c(\kappa), V_w(\kappa)$ is the pair of utilities resulting from the application of the punishment strategies $(\bar{c}\kappa, \bar{w}\kappa)$ from any state $\kappa > 0$, and

$W_c^\alpha(\kappa), W_w^\alpha(\kappa)$ is the pair of utilities resulting from using the Pareto-efficient strategies $(\underline{c}\,\kappa, \underline{w}\,\kappa)$.

The maxima of the left-hand sides of the above inequalities are achieved at

$$c^+ = \frac{(a - \underline{w}^\alpha)\,x}{1 + \left(\frac{D}{\beta \Gamma_c}\right)^{\frac{1}{\nu-1}}}, \tag{6.346}$$

$$w^+ = \frac{(a - \underline{c}^\alpha)\,x}{1 + \left(\frac{F}{\beta \Gamma_w}\right)^{\frac{1}{\nu-1}}}, \tag{6.347}$$

respectively.[56] Then, by substituting c^+ and w^+ in the expressions under the max operator in (6.344), (6.345) and by simplifying the result, we obtain

$$\left(B + \frac{(a - \underline{w}^\alpha)^\nu D}{\left(1 + \left(\frac{D}{\beta \Gamma_c}\right)^{\frac{1}{\nu-1}}\right)^{\nu-1}} \right) x < W_c^\alpha(x), \tag{6.348}$$

$$\frac{(a - \underline{c}^\alpha)^\nu D}{\qquad} x < W^\alpha(x) \tag{6.349}$$

We examir

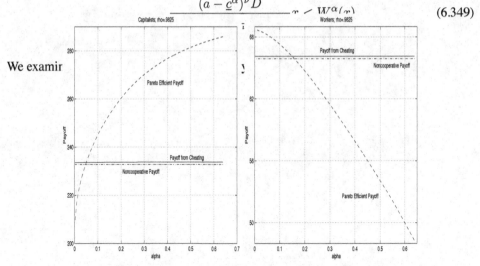

Fig. 6.10 Where is cheating non-profitable for $\beta = .9825$?

We observe existence of a nonempty interval of α, approximately from 0.05 (left panel) to 0.15, for which cheating is profitable for *neither* player. Consequently, there exists a "bargaining power" range and a discount factor for that (6.343) is a subgame-perfect equilibrium.

Concluding remarks. We have obtained a *continuum of growth rates that correspond to the same set of fundamentals*. Moreover, we can characterize this continuum. For example,

---

[56]Where all $\underline{c}^\alpha, \underline{w}^\alpha, \Gamma_c, \Gamma_w$ depend on β and α; see Sections 6.12.3 and 6.12.3.

Figure 6.8 suggests that, for a given value of β, countries with higher α experience higher growth rates.

In other words, we have *engineered* a plausible dynamic game that trade unions and "capitalists" can play and, as a result, obtained a continuum of growth paths. Each path is Pareto efficient and arises as a consequence of a feedback (Markov) equilibrium. We believe that our dynamic game model captures some of the reality of the economic growth process in which the capitalists are characterized by their passion for accumulation ($D > 0$).

Chapter 7

Differential Games

7.1 Introduction

In this chapter, we will deal with games where the time runs continuously and the evolution of the system (i.e., of the state variables) is described by differential equations (hence the name *differential games*). As with the multistage games considered in Chapter 5.7.5, the techniques used here are extensions of optimal-control and dynamic-programming techniques. More specifically, we will use a maximum principle and Hamilton-Jacobi-Bellman equations to identify noncooperative equilibria for these games.

Differential games were initiated by Rufus Isaacs at the Rand Corporation, in a series of memoranda in the 1950s and early 1960s. His book [128] is considered the field's starting point. Initially, the focal points of differential-games theory were military applications and antagonistic zero-sum games. Since then, the theory of differential games has developed significantly. These days, applications of nonzero-sum differential games are found in many areas, most notably in management science, engineering, economics and operations research.

This chapter is an introduction to the theory of differential games and only their basic elements will be explained. We will also state the main results obtained in this area and illustrate them through examples and a $\mathbb{GE}$. A number of specialized books on differential games are available. The early books are Blaquière et al. [30], A. Friedman [85], Leitmann [152] and Krassovski and Subbotin [138]. More recent books include Başar and Olsder [18], Mehlmann [175], Petrosjan [198], Petrosjan and Zenkevitch [199] and Dockner et al. [59]. This last book offers a detailed treatment of differential games and their applications in economics and management science. Engwerda [68] is specialized in linear-quadratic differential, as well as multistage games. Jørgensen and Zaccour [131] deal with applications of differential games in marketing and Yeung and Petrosjan [247] concentrate on cooperative differential games. For a recent survey of differential games, see Buckdahn et al. [37].

243

7.2   Elements of a differential game

The description of a deterministic differential game (DG) played on a time interval $[t^0, T]$ involves the following elements:[1]

(1) A set of players $M = \{1, \ldots, m\}$;
(2) For each player $j \in M$, a vector of controls $\mathbf{u}_j(t) \in U_j \subseteq \mathbb{R}^{m_j}$, where U_j is the set of admissible control values for Player j;
(3) A vector of state variables $\mathbf{x}(t) \in X \subseteq \mathbb{R}^n$, where X is the set of admissible states. The evolution of the state variables is governed by a system of differential equations, called the state equations:

$$\dot{\mathbf{x}}(t) = \frac{d\mathbf{x}}{dt}(t) = f(\mathbf{x}(t), \underline{\mathbf{u}}(t), t), \quad \mathbf{x}(t^0) = \mathbf{x}^0, \tag{7.1}$$

where $\underline{\mathbf{u}}(t) \triangleq (\mathbf{u}_1(t), \ldots, \mathbf{u}_m(t))$;
(4) A payoff for Player $j, j \in M$,

$$J_j \triangleq \int_{t^0}^{T} g_j(\mathbf{x}(t), \underline{\mathbf{u}}(t), t)\, dt + S_j(\mathbf{x}(T)) \tag{7.2}$$

where function g_j is Player j's instantaneous payoff and function S_j is his terminal payoff (or reward, or bequest function);
(5) An information structure, which defines the information that is available to Player j, $j \in M$ when he selects his control vector $\mathbf{u}_j(t)$ at any time t;
(6) For each player $j \in M$, a strategy set Γ_j, where a strategy $\gamma_j \in \Gamma_j$ is a decision rule that defines the control $\mathbf{u}_j(t) \in U_j$ as a function of the information available at time t.

To avoid having to deal with certain technical difficulties that are far beyond the scope of this introductory chapter on differential games, we make the following assumptions:

Assumption 7.1. All feasible state trajectories remain in the interior of the set of admissible states X.

Assumption 7.2. Functions f and g are continuously differentiable in $\mathbf{x}, \underline{\mathbf{u}}$ and t. The S_j functions are continuously differentiable in $\mathbf{x}$.

We briefly comment on the elements involved in the description of a deterministic differential game.

Control set.   The choice of an action by Player j must satisfy the constraint $\mathbf{u}_j(t) \in U_j$. Player j's set U_j of admissible controls is referred to as the control set. This set may be time-invariant and independent of the state of the system or may depend on the *position of the game*, i.e., $(t, \mathbf{x}(t))$. In the latter case, the choice of a control must respect the constraint $\mathbf{u}_j(t) \in U_j(t, \mathbf{x}(t)))$, where the point-to-set correspondence $(t, \mathbf{x}(t)) \rightarrow U_j(t, \mathbf{x}(t)))$ is assumed to be upper-semicontinuous.[2]

---

[1]Chapter 5.7.5 introduced the control and multistage-game counterparts of most of the definitions formulated in this section.
[2]Upper semi-continuous correspondences were introduced in Section 3.4.2, Definition 3.7; see page 58.

More generally, we could also consider coupled constraints on the controls of the players. However, for the sake of clarity, in this chapter, we deal with the simplest case where the control sets are constant and decoupled.

Information structure.   In Chapter 5.7.5, we introduced different information structures that the players may adopt when making their decisions. As we saw, the equilibrium outcomes crucially depend on the selected information structure. In this chapter, we focus on the two information structures that are most frequently used in DG applications, namely, *open-loop* and *feedback* (or *Markovian*) information structures. Recall that open loop means that the players base their decision only on time and an initial condition; whereas, they use the *position of the game* $(t, \mathbf{x}(t))$ as information basis in a Markovian context. A feature that is common to these two information structures is that the players do not need to remember the whole history of the game when making a decision: only running time and the *initial* position are relevant for the open-loop information structure while, for the feedback structure, only information on the current position is relevant. In Section 7.12, we introduce more complex information structures, and more specifically, those that allow for the use of memory strategies.

Strategies.   A Markovian (or feedback) strategy selects the control action according to a feedback rule $\mathbf{u}_j(t) = \sigma_j\,(t, \mathbf{x}(t))$, that is, Player j observes the system's position $(t, \mathbf{x}(t))$ and chooses his action as prescribed by decision rule σ_j. For example, this rule can be a linear or affine function of $\mathbf{x}$ with coefficients depending on t.

However, there is no reason to assume that a feedback strategy is *always* a smooth function of x and t. The so-called "bang-bang" controls, for instance, are examples of feedback laws that have discontinuities. This triggers a delicate theoretical problem, which has been the object of important developments in the early theory of differential games. Introducing a discontinuous function of $\mathbf{x}$ and t into the right-hand side of the differential equation (7.1) implies that we may lose the uniqueness property of the trajectory generated by such strategies, according to the state equations. A good part of the theory of *pursuit-evasion games*, where using bang-bang control is often appropriate, therefore focuses on analyzing *singular surfaces* in the state space, which are manifolds where several trajectories merge to a single one or where a single trajectory splits into several different ones. This difficulty remains when we deal with general nonzero-sum differential games. Fortunately, for some categories of differential games that prove to be useful for engineering, economic and managerial applications, the equilibrium strategies can be obtained in a class of smooth functions of the state vector, which eliminates this problem. In particular, these categories include the *linear-quadratic differential games*.

An open-loop strategy selects the control action according to a decision rule μ_j, which is a function of the initial state $\mathbf{x}^0$:

$$\mathbf{u}_j(t) = \mu_j(\mathbf{x}^0, t).$$

In this case, because the initial state is fixed, there is no need to distinguish between $\mathbf{u}_j(t)$ and $\mu_j(\mathbf{x}^0, t)$.

Using an open-loop strategy means that the player commits, at the initial time, to a fixed time path for his control, that is, his choice of control at each instant of time is predetermined.

When using a Markovian strategy, a player commits to the use of a well-defined servomechanism[3] to manage the system, that is, his reaction to a new position of the system is predetermined.

State equations. The equations in (7.1), which are also called the *system dynamics, evolution equations* or the *equations of motion*, show that, in general, the state vector's rate of change depends on time, the value of the state variables, and the choice of controls by all m players.

When players use open-loop strategies, which are piecewise continuous in time, the fundamental theorem of differential equations guarantees that a unique trajectory will be generated from the given initial state.

When they use feedback strategies, we must make a simplifying assumption to avoid dealing with difficult technicalities.

Assumption 7.3. For every admissible strategy vector $\underline{\sigma} = (\sigma_j : j \in M)$, the differential equations (7.1) admit a unique solution, that is, a unique state trajectory, which is an absolutely continuous[4] function of time.[5]

Time horizon. In this book, we assume that the time horizon T can be finite or infinite but is prespecified for the game we solve. However, this is not the general case. Differential games where T is naturally an endogenous variable include, e.g., pursuit-evasion games and patent-race games. The termination date in a pursuit-evasion game is the minimal amount of time needed for the pursuer to capture the evader or for the evader to reach a safe region. In a patent-race game, T corresponds to the date on which one of the competing firms discovers the sought-after new product and wins the patent. We will also consider the case where T tends to infinity (infinite-horizon games).

7.3 Nash equilibrium

7.3.1 *The definition*

Given an information structure, we define the set of the players' admissible strategies and express the payoffs as functions of these strategies rather than actions. This means that for any initial position $(t^0, \mathbf{x}^0)$, the differential game will have a normal-form representation (see Section 2.4), for which the Nash equilibrium concept applies.

---

[3]See Figure 6.1 on page 159.

[4]A function is absolutely continuous if it is differentiable almost everywhere (i.e., except on a set of measure 0).

[5]Roughly speaking, this assumption is met when the following conditions are satisfied: (i) $f(\mathbf{x}(t), \underline{u}(t))$ is continuous in t for each x and $\mathbf{u}_j, j \in M$; (ii) $f(\mathbf{x}(t), \mathbf{u}(t), t)$ is uniformly Lipschitz in x, $\mathbf{u}_1, \ldots, \mathbf{u}_m$; and (iii) $\sigma_j(t, \mathbf{x})$ is continuous in t for each $\mathbf{x}$ and uniformly Lipschitz in $\mathbf{x}$. (See Başar and Olsder ([18], p. 212) for additional details.)

Although we deal in this section with finite-horizon differential games, we slightly modify the form of the payoff functionals 7.2 and assume (to preserve compatibility with the main-stream economic literature) that Player $j, j \in M$, maximizes a stream of discounted gains, that is,

$$J_j \triangleq \int_{t^0}^{T} e^{-\rho_j t} g_j(\mathbf{x}(t), \underline{\mathbf{u}}(t), t) \, dt + e^{-\rho_j T} S_j(\mathbf{x}(T)), \tag{7.3}$$

where ρ_j is the discount rate satisfying $\rho_j \geq 0$. If, for any reason, discounting is not desirable, then it suffices to set $\rho_j = 0, j \in M$, in the following two theorems and the results remain valid.

Open-loop Nash equilibrium. The payoff functions 7.3 , with the state equations 7.1, and initial data $(t^0, \mathbf{x}^0)$ define the normal form of an open-loop differential game:

$$\underline{\mathbf{u}}(\cdot) = (\mathbf{u}_1(\cdot), \ldots, \mathbf{u}_j(\cdot), \ldots, \mathbf{u}_m(\cdot)) \mapsto J_j(t^0, \mathbf{x}^0; \underline{\mathbf{u}}(\cdot)), \quad j \in M. \tag{7.4}$$

Definition 7.1. The control m-tuple $\underline{\mathbf{u}}^*(\cdot) = (\mathbf{u}_1^*(\cdot), \ldots, \mathbf{u}_m^*(\cdot))$ is an **open-loop Nash equilibrium** (OLNE) at $(t^0, \mathbf{x}^0)$ if the following holds:

$$J_j(t^0, \mathbf{x}^0; \underline{\mathbf{u}}^*(\cdot)) \geq J_j(t^0, \mathbf{x}^0; [\mathbf{u}_j(\cdot), \underline{\mathbf{u}}_{-j}^*(\cdot)]), \quad \forall \mathbf{u}_j(\cdot), j \in M,$$

where $\mathbf{u}_j(\cdot)$ is any admissible control of Player j and $[\mathbf{u}_j(\cdot), \underline{\mathbf{u}}_{-j}^*(\cdot)]$ is the m-vector of controls obtained by replacing the j-th component in $\underline{\mathbf{u}}^*(\cdot)$ by $\mathbf{u}_j(\cdot)$.

In other words, for each Player j, $\mathbf{u}_j^*(\cdot)$ solves the optimal control problem

$$\max_{\mathbf{u}_j(\cdot)} \left\{ \int_{t^0}^{T} e^{-\rho_j t} g_j\left(\mathbf{x}(t), [\mathbf{u}_j(t), \underline{\mathbf{u}}_{-j}^*(t)], t\right) dt + e^{-\rho_j T} S_j(\mathbf{x}(T)) \right\},$$

subject to the state equations

$$\dot{\mathbf{x}}(t) = \frac{d\mathbf{x}}{dt}(t) = f\left(\mathbf{x}(t), [\mathbf{u}_j(t), \underline{\mathbf{u}}_{-j}^*(t)], t\right), \quad \mathbf{x}(t^0) = \mathbf{x}^0. \tag{7.5}$$

Markovian (feedback)-Nash equilibrium. Now consider the case of a Markovian information structure, where the players use feedback strategies $\underline{\sigma}(t, \mathbf{x}) = (\sigma_j(t, \mathbf{x}) : j \in M)$. The normal form of the game, at $(t^0, \mathbf{x}^0)$ is now defined by

$$J_j(t^0, \mathbf{x}^0; \underline{\sigma}) = \int_{t^0}^{T} e^{-\rho_j t} g_j\left(\underline{\sigma}(t, \mathbf{x}), t\right) dt + e^{-\rho_j T} S_j(\mathbf{x}(T)) \tag{7.6}$$

subject to the state equations

$$\dot{\mathbf{x}}(t) = \frac{d\mathbf{x}}{dt}(t) = f\left(\mathbf{x}(t), \underline{\sigma}(t, \mathbf{x}(t)), t\right), \quad \mathbf{x}(t^0) = \mathbf{x}^0. \tag{7.7}$$

To save on notation, define the $(m-1)$−vector

$$\underline{\sigma}_{-j}(t, \mathbf{x}(t)) \triangleq (\sigma_1(t, \mathbf{x}(t)), \ldots, \sigma_{j-1}(t, \mathbf{x}(t)), \sigma_{j+1}(t, \mathbf{x}(t)), \ldots, \sigma_m(t, \mathbf{x}(t))).$$

Definition 7.2. The feedback m-tuple $\underline{\sigma}^*(\cdot) = (\sigma_1^*(\cdot), \ldots, \sigma_m^*(\cdot))$ is a **feedback or Markovian-Nash equilibrium** (MNE) on $[0, T] \times X$ if for each $(t^0, \mathbf{x}^0)$ in $[0, T] \times X$, the following holds:

$$J_j(t^0, \mathbf{x}^0; \underline{\sigma}^*(\cdot)) \geq J_j(t^0, \mathbf{x}^0; [\sigma_j(\cdot), \underline{\sigma}_{-j}^*(\cdot)];), \quad \forall \sigma_j(\cdot), j \in M,$$

where $\sigma_j(\cdot)$ is any admissible feedback law for Player j and $[\sigma_j(\cdot), \underline{\sigma}_{-j}^*(\cdot)]$ is the m-vector of controls obtained by replacing the j-th component in $\sigma^*(\cdot)$ by $\sigma_j(\cdot)$.

In other words, $\mathbf{u}_j^*(t) \equiv \sigma_j^*(t, \mathbf{x}^*(t))$, where $\mathbf{x}^*(\cdot)$ is the equilibrium trajectory generated by $\underline{\sigma}^*$ from $(t^0, \mathbf{x}^0)$, solves the optimal control problem

$$\max_{\mathbf{u}_j(\cdot)} \left\{ \int_{t^0}^T e^{-\rho_j t} g_j \left(\mathbf{x}(t), \left[\mathbf{u}_j(t), \boldsymbol{\sigma}_{-j}^*(t, \mathbf{x}(t)) \right], t \right) dt + e^{-\rho_j T} S_j(\mathbf{x}(T)) \right\}, \qquad (7.8)$$

subject to

$$\dot{\mathbf{x}}(t) = f(\mathbf{x}(t), \left[\mathbf{u}_j(t), \boldsymbol{\sigma}_{-j}^*(t, \mathbf{x}(t)) \right], t), \quad \mathbf{x}(t^0) = \mathbf{x}^0. \qquad (7.9)$$

We can also say that σ_j^* synthesizes the optimal control $\mathbf{u}_j^*(\cdot)$ for the best response problem (7.8) - (7.9).

7.3.2 *Identifying an MNE using a sufficient maximum principle*

Suppose that a differential game is played using Markovian strategies. To solve the optimal control problem of Player j, formulated by (7.8) and (7.9), the solution "toolbox" contains two standard approaches: one is based on the maximum principle (uses Hamiltonians and costate variables); the other relies on a dynamic programming argument (uses value functions and the Hamilton-Jacobi-Bellman equations). We begin by a theorem based on the maximum principle of optimal control theory.

Theorem 7.1. *Assume that the salvage value functions S_j are continuously differentiable and concave.*

Suppose that an m-tuple $\underline{\sigma}^* = (\sigma_1^*, \ldots, \sigma_m^*)$ of feedback law functions $\sigma_j : X \times [t^0, T] \mapsto R^{m_j}, j \in \{1, \ldots, m\}$, is such that

(i) *$\underline{\sigma}^*(t, \mathbf{x})$ is continuously differentiable in $\mathbf{x}$ almost everywhere, and piecewise continuous in t;*

(ii) *$\underline{\sigma}^*(t, \mathbf{x})$ generates at $(t^0, \mathbf{x}^0)$ a unique trajectory $\mathbf{x}^*(\cdot) : [t^0, T] \to X$, solution of*

$$\dot{\mathbf{x}}(t) = f(\mathbf{x}(t), \underline{\sigma}^*(t, \mathbf{x}), t), \quad \mathbf{x}(t^0) = \mathbf{x}^0,$$

   which is absolutely continuous and remains in the interior of X;

(iii) *there exist m **costate vector** functions $\lambda_j(\cdot) = [t^0, T] \to \mathbf{R}^n$, which are absolutely continuous and such that, for all $j \in \{1, \ldots, m\}$, if we define the Hamiltonians[6]*

$$H_j(\lambda_j(t), \mathbf{x}^*(t), [\mathbf{u}_j, \underline{\mathbf{u}}_{-j}], t) = g_j \left(\mathbf{x}^*(t), [\mathbf{u}_j, \underline{\mathbf{u}}_{-j}], t \right) + \lambda_j'(t) f(\mathbf{x}^*(t), [\mathbf{u}_j, \underline{\mathbf{u}}_{-j}], t),$$

   and the maximized Hamiltonians

$$H_j^*(\lambda_j(t), \mathbf{x}^*(t), t) = \max_{\mathbf{u}_j \in U_j} \left\{ g_j \left(\mathbf{x}^*(t) \right), [\mathbf{u}_j, \boldsymbol{\sigma}_{-j}^*(t, \mathbf{x}^*(t))], t \right)$$
$$+ \lambda_j'(t) f(\mathbf{x}^*(t), [\mathbf{u}_j, \boldsymbol{\sigma}_{-j}^*(t, \mathbf{x}^*(t))], t) \right\}, \qquad (7.10)$$

   then the maximum is reached at $\sigma_j^*(t, \mathbf{x}^*(t))$, i.e.,

$$H_j^*(\lambda_j(t), \mathbf{x}^*(t), t) = g_j \left(\mathbf{x}^*(t) \right), \underline{\sigma}^*(t, \mathbf{x}(t)), t)$$
$$+ \lambda_j'(t) f(\mathbf{x}^*(t), \underline{\sigma}^*(t, \mathbf{x}(t)), t); \qquad (7.11)$$

---

[6]Product $\lambda_j'(t) f(\cdots)$ is the scalar product of two n dimensional vectors $\lambda_j(t)$ and $f(\cdots)$.

(iv) *that the functions* $\mathbf{x} \mapsto H_j^*(\lambda_j(t), \mathbf{x}, t)$ *where* H_j^* *is defined as in (7.10), but at position* $(t, \mathbf{x})$, *are continuously differentiable and concave for all* $t \in [t^0, T]$ *and* $j \in M$;

(v) *that the following adjoint differential equations are satisfied for almost all* $t \in [t^0, T]$:

$$\dot{\lambda}_j(t) = \rho_j \lambda_j(t) - \frac{\partial H_j^*(\lambda_j(t), \mathbf{x}^*(t), t)}{\partial \mathbf{x}'}, \tag{7.12}$$

and the so-called **transversality conditions**

$$\lambda_j(T) = \frac{\partial S_j}{\partial \mathbf{x}'(T)} (\mathbf{x}(T), T), \tag{7.13}$$

are also satisfied.

Then $(\sigma_1^*, \dots, \sigma_m^*)$ *is an MNE at* $(t^0, \mathbf{x}^0)$.

Proof. To reply optimally to the strategies $\sigma_{-j}^*(\cdot)$ at $(t^0, \mathbf{x}^0)$, Player j solves a standard optimal control problem formulated by (7.8) and (7.9).

Let $\mathbf{u}_j^*(\cdot)$ be the control generated by $\sigma_j^*(\cdot)$, from $(t^0, \mathbf{x}^0)$, i.e., $\mathbf{u}_j^*(t) = \sigma_j(t, \mathbf{x}^*(t))$. Denote by $\mathbf{u}_j(t)$ any feasible control of Player j and by $\mathbf{x}(\cdot)$ the state trajectory generated by $\mathbf{u}_j(\cdot)$ and $\sigma_{-j}^*(\cdot)$, from $(t^0, \mathbf{x}^0)$.

We want to show that $\mathbf{u}_j^*(\cdot)$ is the optimal reply amongst all admissible $\mathbf{u}_j(\cdot)$, i.e., that for every $j \in M$,

$$J_j(t^0, \mathbf{x}^0; [\mathbf{u}_j^*(\cdot), \sigma_{-j}^*(\cdot)]) - J_j(t^0, \mathbf{x}^0; [\mathbf{u}_j(\cdot), \sigma_{-j}^*(\cdot)]) \geq 0. \tag{7.14}$$

For any admissible control $\mathbf{u}_j \in U_j$, define the Hamiltonians[7]

$$H_j(\lambda_j(t), \mathbf{x}^*(t), [\mathbf{u}_j, \sigma_{-j}^*(t, \mathbf{x}^*(t))], t)$$
$$= g_j(\mathbf{x}^*(t), [\mathbf{u}_j, \sigma_{-j}^*(t, \mathbf{x}^*(t))], t) + \lambda_j'(t) f(\mathbf{x}^*(t), [\mathbf{u}_j, \sigma_{-j}^*(t, \mathbf{x}^*(t))], t)$$

and

$$H_j(\lambda_j(t), \mathbf{x}(t), [\mathbf{u}_j, \sigma_{-j}^*(t, \mathbf{x}^*(t))], t)$$
$$= g_j(\mathbf{x}(t), [\mathbf{u}_j, \sigma_{-j}^*(t, \mathbf{x}^*(t))], t) + \lambda_j'(t) f(\mathbf{x}^*(t), [\mathbf{u}_j, \sigma_{-j}^*(t, \mathbf{x}^*(t))], t).$$

By definition, the LHS of (7.14) is equal to

$$\int_{t^0}^T e^{-\rho_j t} \left(g_j(\mathbf{x}(t), [\mathbf{u}_j^*(t), \sigma_{-j}^*(t, \mathbf{x}^*(t))], t) \right.$$
$$\left. - g_j(\mathbf{x}(t), [\mathbf{u}_j(t), \sigma_{-j}^*(t, \mathbf{x}^*(t))], t) \right) dt + e^{-\rho_j T}(S_j(\mathbf{x}^*(T)) - S_j(\mathbf{x}(T))).$$

Using the Hamiltonians defined above, the payoff difference can be written as

$$J_j(t^0, \mathbf{x}^0; \mathbf{u}^*(\cdot)) - J_j(t^0, \mathbf{x}^0; \mathbf{u}(\cdot)) =$$

$$\int_{t^0}^T e^{-\rho_j t} \left(H_j(\lambda_j(t), \mathbf{x}^*(t), [\mathbf{u}_j^*(t), \sigma_{-j}^*(t, \mathbf{x}^*(t))], t) - \lambda_j(t)\dot{\mathbf{x}}^*(t) \right) dt$$

$$- \int_{t^0}^T e^{-\rho_j t} \left(H_j(\lambda_j(t), \mathbf{x}(t), [\mathbf{u}_j(t), \sigma_{-j}^*(t, \mathbf{x}^*(t))], t) - \lambda_j(t)\dot{\mathbf{x}}(t) \right) dt$$

$$+ e^{-\rho_j T}(S_j(\mathbf{x}^*(T)) - S_j(\mathbf{x}(T))). \tag{7.15}$$

---

[7]Actually this function of $\mathbf{u}_j$ should be called the "pre-Hamiltonian."

By the assumption of feasibility of $\mathbf{u}_j(\cdot)$ and the definition of the maximized Hamiltonian, we have
$$H_j^*(\lambda_j(t), \mathbf{x}(t), t) \geq H_j(\lambda_j(t), \mathbf{x}(t), [\mathbf{u}_j(t), \boldsymbol{\sigma}_{-j}^*(t, \mathbf{x}^*(t))], t).$$
Using the above inequality and the maximizing condition in (7.11) in (7.15), we get
$$J_j(t^0, \mathbf{x}^0; [\mathbf{u}_j^*(\cdot), \boldsymbol{\sigma}_{-j}^*(\cdot)]) - J_j(t^0, \mathbf{x}^0; [\mathbf{u}_j(\cdot), \boldsymbol{\sigma}_{-j}^*(\cdot)])$$
$$\geq \int_0^T e^{-\rho_j t} \left\{ H_j^*(\mathbf{x}^*(t), \lambda_j(t), t) - H_j^*(\mathbf{x}(t), \lambda_j(t), t) \right.$$
$$\left. -\lambda_j(t)'(\dot{\mathbf{x}}^*(t) - \dot{\mathbf{x}}(t)) \right\} dt + e^{-\rho_j T}(S_j(\mathbf{x}^*(T)) - S_j(\mathbf{x}(T))).$$
By the concavity and differentiability of $H_j^*(\lambda_j(t), \mathbf{x}(t), t)$ with respect to $\mathbf{x}$, we have
$$H_j^*(\lambda_j(t), \mathbf{x}^*(t), t) - H_j^*(\lambda_j(t), \mathbf{x}(t), t) \geq \frac{\partial H_j^*(\lambda_j(t), \mathbf{x}^*(t), t)}{\partial \mathbf{x}}(x^*(t) - \mathbf{x}(t)).$$
Using (7.12), the above inequality becomes
$$H_j^*(\lambda_j(t), \mathbf{x}^*(t), t) - H_j^*(\lambda_j(t), \mathbf{x}(t), t) \geq \left(\rho_j \lambda_j(t) - \dot{\lambda}_j(t)\right)'(x^*(t) - \mathbf{x}(t)).$$
By substitution, we obtain
$$J_j(t^0, \mathbf{x}^0; [\mathbf{u}_j^*(\cdot), \boldsymbol{\sigma}_{-j}^*(\cdot)]) - J_j(t^0, \mathbf{x}^0; [\mathbf{u}_j(\cdot), \boldsymbol{\sigma}_{-j}^*(\cdot)])$$
$$\geq \int_{t^0}^T e^{-\rho_j t}\left(\left(\rho_j \lambda_j(t) - \dot{\lambda}_j(t)\right)(x^*(t) - \mathbf{x}(t)) - \lambda_j'(t)(x^*(t) - \mathbf{x}(t))\right) dt$$
$$+e^{-\rho_j T}(S_j(\mathbf{x}^*(T)) - S_j(\mathbf{x}(T))),$$
$$= \int_{t^0}^T \frac{d}{dt} e^{-\rho_j t} \lambda_j(t)(\mathbf{x}(t) - \mathbf{x}^*(t)) dt + e^{-\rho_j T}(S_j(\mathbf{x}^*(T), T) - S_j(\mathbf{x}(T))),$$
$$= e^{-\rho_j T}(\lambda_j(T)(\mathbf{x}(T) - \mathbf{x}^*(T)) + S_j(\mathbf{x}^*(T)) - S_j(\mathbf{x}(T)))$$
$$-\lambda_j'(t^0)(\mathbf{x}(t^0) - \mathbf{x}^*(t^0)),$$
$$= e^{-\rho_j T}\left(\frac{\partial S_j(\mathbf{x}(T))}{\partial \mathbf{x}(T)}(\mathbf{x}(T) - \mathbf{x}^*(T)) + S_j(\mathbf{x}^*(T), T) - S_j(\mathbf{x}(T))\right) \geq 0.$$
The non-negativity of the above term follows from the assumptions of differentiability and concavity[8] of $S_j(\mathbf{x})$. $\qquad\square$

The main difficulty in applying the above theorem lies in solving the adjoint (or costate) equations (7.12). Although these equations look like ordinary differential equations, they are not, because the right-hand side of (7.12) depends on the state through the unknown strategies $\boldsymbol{\sigma}_{-j}^*(t, \mathbf{x}(t)))$ of the rival players. More precisely, when written out in long form, the adjoint equations (7.12), read as
$$\dot{\lambda}_j(t) = \rho_j \lambda_j(t) - \frac{\partial H_j^*(\lambda_j(t), \mathbf{x}^*(t), t)}{\partial \mathbf{x}'}$$
$$= \rho_j \lambda_j(t) - \frac{\partial}{\partial \mathbf{x}'} H_j(\lambda_j(t), \mathbf{x}^*(t), [\mathbf{u}_j^*(t), \boldsymbol{\sigma}_{-j}^*(t, \mathbf{x}^*(t))], t)$$
$$- \frac{\partial}{\partial \underline{\mathbf{u}}_{-j}} H_j(\lambda_j(t), \mathbf{x}^*(t), [\mathbf{u}_j^*(t), \boldsymbol{\sigma}_{-j}^*(t, \mathbf{x}^*(t))], t) \frac{\partial}{\partial \mathbf{x}'} \boldsymbol{\sigma}_{-j}^*(t, \mathbf{x}^*(t)). \quad (7.16)$$

---

[8]Recall that, if a function $F(\mathbf{x})$ is concave, then it satisfies $F(\mathbf{x} + \mathbf{h}) - F(\mathbf{x}) \geq \frac{\partial F}{\partial \mathbf{x}} \mathbf{h}$.

The complicating factor in these equations is the term involving the partial derivatives of the unknown strategies of the $-j$ players, i.e., $\frac{\partial}{\partial \mathbf{x}'} \sigma^*_{-j}(t, \mathbf{x}^*(t))$.

7.3.3 *Identifying an OLNE using the maximum principle*

We exploit the fact that an open-loop strategy is a special case of a feedback law, which does not vary with $\mathbf{x}$. We can then apply Theorem 7.1, as is. However in the open-loop case, the adjoint or costate equations are much more tractable. Indeed, the derivatives $\frac{\partial \sigma_{-j}(t, \mathbf{x}(t)))}{\partial \mathbf{x}'}$ vanish and the costate equations[9] read as follows:

$$\dot{\lambda}_j(t) = \rho_j \lambda_j(t) - \frac{\partial}{\partial \mathbf{x}'} g_j\left(\mathbf{x}(t), \underline{\mathbf{u}}^*(t), t\right) - \lambda'_j(t) \frac{\partial}{\partial \mathbf{x}'} f\left(\mathbf{x}(t), \underline{\mathbf{u}}^*(t), t\right),$$

$$\lambda_j(T) = \frac{\partial S_j}{\partial \mathbf{x}'(T)}\left(\mathbf{x}(T)\right).$$

These adjoint equations define, together with the state equation,

$$\dot{\mathbf{x}}(t) = f(\mathbf{x}(t), \underline{\mathbf{u}}^*(t), t), \quad \mathbf{x}(t^0) = \mathbf{x}^0,$$

a two-point boundary value problem (TPBVP), which consists of $2n$ ordinary differential equations with n initial and n terminal conditions.[10]

An adjoint or costate variable has an interesting *shadow-price* interpretation; it measures the impact of a marginal variation in the state-variable value on payoff. The terminal conditions $\lambda_j(T) = \frac{\partial S_j}{\partial \mathbf{x}(T)}\left(\mathbf{x}(T)\right)$ reflect this idea. Indeed, if we change marginally the value of $\mathbf{x}(T)$, its consequent marginal impact on the total payoff is captured by $\frac{\partial S_j}{\partial \mathbf{x}(T)}\left(\mathbf{x}(T)\right)$. Then, the costate or adjoint variational equations indicate how these marginal values are evolving along the state trajectory.

Notice that, if there is no salvage value, then the transversality conditions become

$$\lambda_j(T) = 0.$$

7.3.4 *Identifying an MNE using Hamilton-Jacobi-Bellman equations*

The following theorem relies on dynamic programming.

Theorem 7.2. *Suppose that an m-tuple* $\underline{\sigma}^* = (\sigma^*_1, \ldots, \sigma^*_m)$ *of feedback laws is given, such that*

(i) *for any admissible initial point* $(t^0, \mathbf{x}^0)$ *there is a unique, absolutely continuous solution* $t \in [t^0, T] \mapsto \mathbf{x}^*(t) \in X \subset \mathbf{R}^n$ *of the differential equation*

$$\dot{\mathbf{x}}^*(t) = f(\mathbf{x}^*(t), \sigma^*_1(t, \mathbf{x}^*(t)), \ldots, \sigma^*_m(t, \mathbf{x}^*(t)), t), \quad \mathbf{x}^*(t^0) = \mathbf{x}^0;$$

---

[9]Often called adjoint variational equations.

[10]The computation of OLNE through a TPBVP solution has been exploited by, e.g., Hämäläinen et al. [100, 99] in a series of papers dealing with fishery games, a topic dealt with at different places in this book.

(ii) *there exist continuously differentiable functionals* $V_j^* : [t^0, T] \times X \to \mathbf{R}$, *such that the following Hamilton-Jacobi-Bellman partial differential equations are satisfied for all* $(t, \mathbf{x}) \in [t^0, T] \times X$

$$\rho_j V_j^*(t, \mathbf{x}) - \frac{\partial V_j^*(t, \mathbf{x})}{\partial t} = \max_{\mathbf{u}_j \in U_j} \left\{ g_j \left(\mathbf{x}, [\mathbf{u}_j, \sigma_{-j}^*(t, \mathbf{x})], t \right) \right.$$

$$\left. + \frac{\partial V_j^*(t, \mathbf{x})}{\partial \mathbf{x}} f(\mathbf{x}, [\mathbf{u}_j, \sigma_{-j}^*(t, \mathbf{x})], t) \right\} \quad (7.17)$$

$$= g_j \left(\mathbf{x}, \underline{\sigma}^*(t, \mathbf{x}), t \right)$$

$$+ \frac{\partial V_j^*(t, \mathbf{x})}{\partial \mathbf{x}} f(\mathbf{x}, \underline{\sigma}^*(t, \mathbf{x}), t); \quad (7.18)$$

(iii) *the boundary conditions*

$$V_j^*(T, \mathbf{x}) = S_j(\mathbf{x}) \quad (7.19)$$

are satisfied for all $\mathbf{x} \in X$ *and* $j \in M$.

Then, $\sigma_j^*(t, \mathbf{x})$, *is a maximizer of the right-hand side of the Hamilton-Jacobi-Bellman equation for Player* j *and the* $m-$*tuple* $\underline{\sigma}^* = (\sigma_1^*, \ldots, \sigma_m^*)$ *is an MNE at every initial point* $(t^0, \mathbf{x}^0) \in [t^0, T] \times X$.

Proof. Given the feedback laws adopted by the other players, $\sigma_{-j}(t, \mathbf{x}(t))$, Player j, looking for his best reply, has to solve a standard optimal control problem. He must find $\mathbf{u}_j^*(t)$ such that

$$\int_{t^0}^T e^{-\rho_j t} g_j(\mathbf{x}(t), [\mathbf{u}_j(t), \sigma_{-j}^*(t, \mathbf{x}(t))], t) dt + e^{-\rho_j T} S_j(\mathbf{x}(T)),$$

is maximized subject to (7.9). Note that $\sigma_{-j}^*(t, \mathbf{x}^*(t))$ is given in the optimization problem of Player j.

Let $\mathbf{u}_j^*(\cdot)$ be an optimal control path generated by $\sigma_j^*(t, \mathbf{x}^*(t))$, i.e.,

$$\mathbf{u}_j^*(t) = \sigma_j(t, \mathbf{x}^*(t)),$$

and denote by $\mathbf{u}_j(\cdot)$ any feasible control for Player j and by $\mathbf{x}(\cdot)$ the corresponding state trajectory generated by $[\mathbf{u}_j(\cdot), \sigma_{-j}^*(\cdot)]$.

We want to show that, amongst all admissible controls $\mathbf{u}_j(\cdot)$, $\mathbf{u}_j^*(\cdot)$ is an optimal reply to $\sigma_{-j}^*(\cdot)$, at $(t^0, \mathbf{x}^0)$, i.e., for every $j \in M$

$$J_j([\mathbf{u}_j^*(\cdot), \sigma_{-j}^*(\cdot)]; t^0, \mathbf{x}^0) - J_j([\mathbf{u}_j(\cdot), \sigma_{-j}^*(\cdot)]; t^0, \mathbf{x}^0) \geq 0. \quad (7.20)$$

By the feasibility of $\mathbf{u}_j(t)$ and (7.17), we have

$$g_j(t, \mathbf{x}(t)), [\mathbf{u}_j(t), \sigma_{-j}^*(t, \mathbf{x}(t))], t) \leq$$

$$\rho_j V_j(t, \mathbf{x}(t)) - \frac{\partial V_j(t, \mathbf{x}(t))}{\partial t} - \frac{\partial V_j(t, \mathbf{x}(t))}{\partial \mathbf{x}} f(\mathbf{x}(t), [\mathbf{u}_j(t), \sigma_{-j}^*(t, \mathbf{x}(t))], t).$$

Multiplying by $e^{-\rho_j t}$ and using (7.9), this inequality can be written equivalently as

$$e^{-\rho_j t} g_j \left(\mathbf{x}(t), [\mathbf{u}_j(t), \sigma_{-j}^*(t, \mathbf{x}(t))], t \right) \leq -\frac{d}{dt} \left(e^{-\rho_j t} V_j(t, \mathbf{x}(t)) \right).$$

Repeating the same two steps considering $\mathbf{u}_j^*(\cdot)$ (which is also feasible) and the corresponding trajectory $\mathbf{x}^*(\cdot)$, we obtain, this time, the following equality:

$$e^{-\rho_j t} g_j \left(\mathbf{x}(t), [\mathbf{u}_j^*(t), \boldsymbol{\sigma}_{-j}^*(t, \mathbf{x}(t))], t\right) = -\frac{d}{dt}\left(e^{-\rho_j t} V_j(t, x^*(t))\right).$$

We can thus write

$$J_j(t^0, \mathbf{x}^0; [\mathbf{u}_j^*(\cdot), \boldsymbol{\sigma}_{-j}^*(\cdot)]) - J_j(t^0, \mathbf{x}^0; [\mathbf{u}_j(\cdot), \boldsymbol{\sigma}_{-j}^*(\cdot)])$$

$$= \int_{t^0}^T e^{-\rho_j t}\left(g_j\left(x^*(t), [\mathbf{u}_j^*(t), \boldsymbol{\sigma}_{-j}^*(t, \mathbf{x}(t))], t\right)\right.$$

$$\left. -g_j\left(\mathbf{x}(t), [\mathbf{u}_j(t), \boldsymbol{\sigma}_{-j}^*(t, \mathbf{x}(t))], t\right)\right) dt + e^{-\rho_j T}\left(S_j(\mathbf{x}^*(T)) - S_j(\mathbf{x}(T))\right)$$

$$\geq \int_{t^0}^T \frac{d}{dt} e^{-\rho_j t}\left(V_j(t, \mathbf{x}(t)) - V_j(t, \mathbf{x}^*(t))\right) dt + e^{-\rho_j T}\left(S_j(\mathbf{x}^*(T)) - S_j(\mathbf{x}(T))\right)$$

$$= V_j(t^0, \mathbf{x}^*(t^0)) - V_j(t^0, \mathbf{x}(t^0)) = 0$$

since $\mathbf{x}^*(t^0) = \mathbf{x}(t^0) = \mathbf{x}^0$. This establishes, for each $j \in M$, the inequality

$$J_j(t^0, \mathbf{x}^0; [\mathbf{u}_j^*(\cdot), \boldsymbol{\sigma}_{-j}^*(\cdot)]) - J_j(t^0, \mathbf{x}^0; [\mathbf{u}_j(\cdot), \boldsymbol{\sigma}_{-j}^*(\cdot)]) \geq 0.$$

This establishes the equilibrium property at $(t^0, \mathbf{x}^0)$. The same reasoning can be used from any other initial point and this completes the proof. □

The functions $V_1(t, \mathbf{x}(t)), \ldots, V_m(t, \mathbf{x}(t))$, are the players' *equilibrium value functions*, which give the payoff-to-go at position $(t, \mathbf{x}(t))$ for each player when the equilibrium strategy is played. The main difficulty in applying the above theorem lies in solving the Hamilton-Jacobi-Bellman (HJB) equations given by (7.17). These are partial (and possibly highly nonlinear) differential equations involving the functions $V_j(t, \mathbf{x}(t)), j \in M$. A closed-form solution of these equations is rarely obtainable; therefore, we frequently can only define these value functions implicitly. In the absence of a general theory of partial differential equations, we have to rely on the available collection of solution methods and solutions to specific equations (see, e.g., [175, 68] and [132]). Additionally, to obtain a solution for V_j, we need to know the strategies of all other players "$-j$" and these strategies are usually not available.

Note that when the state space is one-dimensional, the Hamilton-Jacobi-Bellman in (7.17) becomes an ordinary differential equation, which is easier to solve than a partial differential equation. Further, in infinite-horizon differential games with stationary strategies, the HJB equations are algebraic equations, when expressed in current time, and are, therefore, easier to solve (see coming examples). This explains why, in many applications of differential games, one-dimensional state space is popular and players employ (in infinite-horizon games) stationary strategies $\sigma_j(\mathbf{x})$.

Remark 7.1. Theorems 7.1 and 7.2 provide **sufficient conditions**, that is, given that the assumptions of a theorem can be verified, we have identified an MNE.

Remark 7.2. Under the additional regularity condition that the value functions $V_j(t, \mathbf{x})$ are twice differentiable in $\mathbf{x}$, it can be established that costate variables $\lambda_j(t)$ are equal to the partial derivatives of the value function along the equilibrium trajectory, i.e.,

$$\frac{\partial V_j(t, \mathbf{x})}{\partial \mathbf{x}}\Big|_{\mathbf{x}=\mathbf{x}^*(t)} = \lambda_j(t)'.$$

7.4 The infinite-horizon case

7.4.1 *Several performance criteria*

Theorems 7.1 and 7.2 are formulated under the assumption that the time horizon is finite. If the planning horizon is infinite, then the transversality or boundary conditions, that is, $\lambda_j(T) = \frac{\partial S_j(\mathbf{x}(T))}{\partial \mathbf{x}'(T)}$ in Theorem 7.1 and $V_j(T, \mathbf{x}(T)) = S_j(\mathbf{x}(T))$ in Theorem 7.2, have to be modified. Unfortunately, the required modifications are not straightforward. An important issue in an infinite-horizon dynamic optimization problem is that the integral payoff

$$J_j \triangleq \int_{t^0}^{\infty} g_j(\mathbf{x}(t), \underline{\mathbf{u}}(t), t) \, dt, \tag{7.21}$$

may not converge for all feasible control and trajectory paths $(\underline{\mathbf{u}}(\cdot), \mathbf{x}(\cdot))$. To see intuitively why this causes a problem, let us go back to the definition of optimality used in Theorems 7.1 and 7.2, i.e., a control path $\mathbf{u}^*(\cdot)$ is optimal if $J_j(t^0, \mathbf{x}^0; \mathbf{u}^*(\cdot)) \geq J_j(t^0, \mathbf{x}^0; \mathbf{u}(\cdot))$ for all feasible $\mathbf{u}(\cdot)$. According to the definitions, to select the optimal path(s), we need to compare every pair of feasible control and trajectory paths in terms of the payoff (or performance) criterion. Now, if for some feasible pairs, the integral does not converge, then it becomes unclear how we can still carry out this comparison. Two things have been done in the literature to avoid this problem. One is to adopt a different payoff functional, and the other is to modify the definition of what is meant by an optimal path.

As seen in Chapter 5.7.5, several alternative objective functionals can be considered when dealing with infinite-horizon problems. The most popular ones are the discounted payoff

$$J_j \triangleq \int_0^{\infty} e^{-\rho_j t} g_j(\mathbf{x}(t), \mathbf{u}(t)) \, dt, \tag{7.22}$$

and the limit of average payoff, when the reward does not explicitly depend on t

$$J_j \triangleq \lim_{T \to \infty} \inf \frac{1}{T} \int_0^T g_j(\mathbf{x}(t), \mathbf{u}(t)) \, dt. \tag{7.23}$$

Remark 7.3. Following a long tradition in infinite-horizon dynamic optimization, we focus on autonomous problems, that is, we consider functions f (system's dynamics) and g_j (instantaneous utility or payoff) that do not depend explicitly on time, and will confine our interest to stationary feedback strategies: $\mathbf{u}_j(t) = \sigma_j(\mathbf{x}(t))$. The reason is that irrespective of for how long the game has been played, the players face, at any instant of time, the same game for the remaining (also infinite-horizon) part of the time horizon, which is **infinity**.

Proposition 7.1. *When Player* $j, j \in M$, *optimizes the discounted payoff functional in* *(7.22), then the (sufficient) transversality conditions in (7.19) and (7.13) become, respectively,*

$$\lim_{T \to \infty} e^{-\rho_j t} V_j(t, \mathbf{x}(t)) = 0, \tag{7.24}$$

$$\lim_{T \to \infty} e^{-\rho_j t} \lambda_j(t) = 0. \tag{7.25}$$

Proof. It suffices to set the salvage value equal to zero and to take the limit in the final steps in the proofs of Theorems 7.2 and 7.1 to get the results. $\qquad \square$

The other way out has been to (re)define what is meant by an optimal path in an infinite-horizon setting. A number of definitions of optimality have been proposed in the literature (see, e.g., Carlson and Haurie [43], Carlson et al. [43] and Léonard and Long [157]), each leading to its own sufficient transversality conditions. A rigorous treatment of these conditions in infinite-horizon dynamic optimization and differential games is beyond the scope of this book. Here we will only give some definitions and results that are helpful in extending and applying Theorems 7.1 and 7.2 to the infinite-horizon differential games.[11]

Consider an optimal control problem with the following set of performance criterions defined for each terminal time θ as a functional of state and control trajectories over the time interval $[0, \theta]$:

$$J^{\theta}(\mathbf{x}(\cdot), \mathbf{u}(\cdot)) = \int_0^{\theta} g(\mathbf{x}(t), \underline{\mathbf{u}}(t), t) \, dt,$$

defined for every $\theta \geq 0$.

Definition 7.3. A trajectory $\mathbf{x}^*(\cdot)$ emanating from $\mathbf{x}^0$ is said to be:

(1) **Strongly optimal** at $\mathbf{x}^0$ if it is generated by $\mathbf{u}^*(\cdot)$ such that

$$\lim_{\theta \to \infty} J^{\theta}(\mathbf{x}^*(\cdot), \mathbf{u}^*(\cdot)) < \infty, \tag{7.26}$$

and for any other trajectory $\mathbf{x}(\cdot)$ emanating from $\mathbf{x}^0$ and generated by $\mathbf{u}(\cdot)$, it holds that

$$\lim_{\theta \to \infty} \left(J^{\theta}(\mathbf{x}^*(\cdot), \mathbf{u}^*(\cdot)) - J^{\theta}(\mathbf{x}(\cdot), \mathbf{u}(\cdot)) \right) \geq 0.$$

(2) **Overtaking optimal** at $\mathbf{x}^0$ if (7.26) does not necessarily hold and

$$\lim_{\theta \to \infty} \inf \left(J^{\theta}(\mathbf{x}^*(\cdot), \mathbf{u}^*(\cdot)) - J^{\theta}(\mathbf{x}(\cdot), \mathbf{u}(\cdot)) \right) \geq 0.$$

(3) **Weakly overtaking optimal** at $\mathbf{x}^0$ if (7.26) does not necessarily hold and

$$\lim_{\theta \to \infty} \sup \left(J^{\theta}(\mathbf{x}^*(\cdot), \mathbf{u}^*(\cdot)) - J^{\theta}(\mathbf{x}(\cdot), \mathbf{u}(\cdot)) \right) \geq 0.$$

---

[11] The interested reader may consult, i.e., Carlson and Haurie [42] and Carlson et al. [43] for complete coverage of infinite-horizon optimal control theory, and Carlson and Haurie [40] for a treatment of of infinite-horizon open-loop differential games.

(4) **Finitely optimal** at $\mathbf{x}^0$ if for every $\theta > 0$ and every admissible pair $(\mathbf{x}(\cdot), \mathbf{u}(\cdot))$
defined on $[0, \theta]$ and satisfying $\mathbf{x}(\theta) = \mathbf{x}^*(\theta)$ we have

$$J^\theta(\mathbf{x}^*(\cdot), \mathbf{u}^*(\cdot)) \geq J^\theta(\mathbf{x}(\cdot), \mathbf{u}(\cdot)).$$

The simplest definition is the one of strong optimality. It is meaningful only when
the payoff functional remains bounded for all trajectories. For instance, if the func-
tion $g_j(\mathbf{x}(t), \underline{\mathbf{u}}(t), t) = e^{-\rho_j t} \tilde{g}_j(\mathbf{x}(t), \underline{\mathbf{u}}(t))$, where $\tilde{g}_j(\mathbf{x}(t), \underline{\mathbf{u}}(t))$ is uniformly bounded,
then, obviously, when ρ_j is positive, the integral is convergent for all feasible $\mathbf{u}(\cdot)$.

A trajectory is overtaking optimal if the performance index "catches up to" the performance index of any other trajectory emanating from the initial state $\mathbf{x}^0$. Figure 7.1
illustrates this situation.

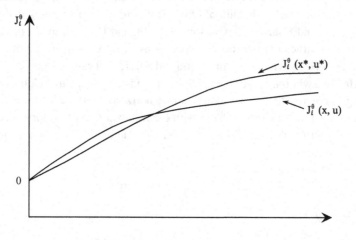

Fig. 7.1   Overtaking optimality

A trajectory is weakly overtaking optimal if, for no other feasible trajectory emanating
from the initial state $\mathbf{x}^0$, the performance index "catches up to" the performance index of
$(\mathbf{x}^*(\cdot), \underline{\mathbf{u}}^*(\cdot))$.

Figure 7.2 illustrates the definition of weakly overtaking. As we can see, here
the performance trajectory $J^\theta(\mathbf{x}(\cdot), \mathbf{u}\cdot))$ can never catch up to the performance index
$J^\theta(\mathbf{x}^*(\cdot), \mathbf{u}^*(\cdot))$.

Remark 7.4. We note the following:

(1) The above optimality criteria are listed in decreasing order of strength, that is,

strongly optimal $\Rightarrow$ overtaking optimality $\Rightarrow$ weakly overtaking optimality
$\Rightarrow$ finite optimality.

(2) If the integral payoff converges for all feasible paths, then all of the above optimality
definitions reduce to the usual one, namely, $u^*(\cdot)$ is optimal if $J(\mathbf{x}^*(\cdot), \mathbf{u}^*(\cdot)) \geq$
$J(\mathbf{x}(\cdot), \mathbf{u}(\cdot))$ for all feasible $(\mathbf{x}(\cdot), \mathbf{u}(\cdot))$.

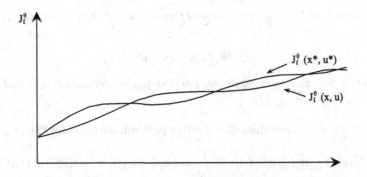

Fig. 7.2   Weak overtaking optimality

(3) A maximum principle (necessary condition) holds true for finite optimality.

The following two propositions provide the sufficient transversality conditions for the infinite-horizon case.

7.4.2   Sufficient infinite-horizon transversality conditions for the maximum principle

We extend here the sufficient maximum principle for a feedback Nash-equilibrium over an infinite time horizon.

Proposition 7.2. *Let the payoff function of Player j be given by*

$$J_j(t^0, \mathbf{x}^0; [\mathbf{u}_j(\cdot), \boldsymbol{\sigma}^*_{-j}(\cdot)]) = \int_0^\infty g_j\left(\mathbf{x}(t), [\mathbf{u}_j(t), \boldsymbol{\sigma}^*_{-j}(t, \mathbf{x}(t))], t\right) dt.$$

The results in Theorem 7.1 remain valid, provided that the transversality conditions in (7.13) are modified as follows:

(1) If optimality is understood in the sense of strongly optimal, then the boundary conditions are

$$\lim_{\theta \to \infty} \lambda'_j(\theta)\, (\mathbf{x}(\theta) - \mathbf{x}^*(\theta)) \geq 0,$$

for all $\mathbf{x} \in X$ and $j \in M$.

(2) If optimality is understood in the sense of overtaking, then the boundary conditions are

$$\lim_{\theta \to \infty} \inf \lambda'_j(\theta)\, (\mathbf{x}(\theta) - \mathbf{x}^*(\theta)) \geq 0,$$

for all $\mathbf{x} \in X$ and $j \in M$.

(3) If optimality is understood in the sense of weakly overtaking, then the boundary conditions are

$$\lim_{\theta \to \infty} \sup \lambda'_j(\theta)\, (\mathbf{x}(\theta) - \mathbf{x}^*(\theta)) \geq 0,$$

for all $\mathbf{x} \in X$ and $j \in M$.

(4) *If optimality is understood in the sense of finitely optimal, then the boundary conditions are*

$$\lambda'_j(\theta)\left(\mathbf{x}(\theta) - \mathbf{x}^*(\theta)\right) \geq 0,$$

for every $\theta > 0$ and every admissible pair $(\mathbf{x}(\cdot), \mathbf{u}(\cdot))$ defined on $[0, \theta]$ and satisfying $\mathbf{x}(\theta) = \mathbf{x}^*(\theta)$, for $j \in M$.

Proof.   For any $\theta > 0$, introduce the following performance index for Player j:

$$J_j^\theta(t^0, \mathbf{x}^0; [\mathbf{u}_j(\cdot), \boldsymbol{\sigma}^*_{-j}(\cdot)]) = \int_0^\theta g_j\left(\mathbf{x}(t), [\mathbf{u}_j(t), \boldsymbol{\sigma}^*_{-j}(t, \mathbf{x}(t))], t\right) dt.$$

Following the first steps in the proof of Theorem 7.1, we get

$$J_j^\theta(\mathbf{x}^0; [\mathbf{u}_j^*(\cdot), \boldsymbol{\sigma}^*_{-j}(\cdot)]) - J_j^\theta(\mathbf{x}^0; [\mathbf{u}_j(\cdot), \boldsymbol{\sigma}^*_{-j}(\cdot)])$$

$$\geq \int_0^\theta \left(\left(\rho_j \lambda_j(t) - \dot{\lambda}_j(t)\right)'(\mathbf{x}^*(t) - \mathbf{x}(t)) - \lambda_j(t)'(\mathbf{x}^*(t) - \mathbf{x}(t)) \right) dt,$$

$$= \int_0^\theta \frac{d}{dt} \lambda'_j(t)(\mathbf{x}(t) - x^*(t)) dt,$$

$$= \lambda'_j(\theta)(\mathbf{x}(\theta) - \mathbf{x}^*(\theta)) - \lambda(0)'(\mathbf{x}(0) - \mathbf{x}^*(0)),$$

that is,

$$J_j^\theta(t^0, \mathbf{x}^0; [\mathbf{u}_j^*(\cdot), \boldsymbol{\sigma}^*_{-j}(\cdot)]) - J_j^\theta(t^0, \mathbf{x}^0; [\mathbf{u}_j(\cdot), \boldsymbol{\sigma}^*_{-j}(\cdot)]) \geq \lambda'_j(\theta)(\mathbf{x}(\theta) - \mathbf{x}^*(\theta)).$$

Applying, when $\theta \to \infty$, the definitions of optimality leads to the different sufficient boundary conditions.   □

7.4.3   *Sufficient infinite-horizon boundary conditions for the HJB equations*

We provide in the next proposition the boundary conditions for the HJB approach. These conditions extend to a dynamic-game setting those available for optimal control problems. Note that it is not easy in practice to check the asymptotic inequalities on value functions that are all growing to infinity. We give these conditions for completeness.

Proposition 7.3. *Let the payoff function of Player j be given by*

$$J_j(t^0, \mathbf{x}^0; [\mathbf{u}_j(\cdot), \boldsymbol{\sigma}^*_{-j}(\cdot)]) = \int_0^\infty g_j\left(\mathbf{x}(t), [\mathbf{u}_j(t), \boldsymbol{\sigma}_{-j}(t, \mathbf{x}(t))], t\right) dt.$$

The results in Theorem 7.2 remain valid, provided that the boundary conditions (7.19) are modified as follows:

(1) *If optimality is understood in the sense of strongly optimal, then the boundary conditions are*

$$\lim_{\theta \to \infty} \left(V_j(\theta, \mathbf{x}) - V_j(\theta, \mathbf{x}^*)\right) \geq 0,$$

for all $x \in X$ and $j \in \{1, \ldots, m\}$.

(2) If optimality is understood in the sense of overtaking, then the boundary conditions are

$$\lim_{\theta \to \infty} \inf \left(V_j(\theta, \mathbf{x}) - V_j(\theta, \mathbf{x}^*) \right) \geq 0,$$

for all $x \in X$ and $j \in \{1, \ldots, m\}$.

(3) If optimality is understood in the sense of weakly overtaking, then the boundary conditions are

$$\lim_{\theta \to \infty} \sup \left(V_j(\theta, \mathbf{x}) - V_j(\theta, \mathbf{x}^*) \right) \geq 0,$$

for all $x \in X$ and $j \in \{1, \ldots, m\}$.

(4) If optimality is understood in the sense of finitely optimal, then the boundary conditions are

$$V_j(\theta, \mathbf{x}(\theta)) - V_j(\theta, \mathbf{x}^*(\theta)) \geq 0,$$

for every $\theta > 0$ and every admissible pair $(\mathbf{x}(\cdot), \mathbf{u}(\cdot))$ defined on $[0, \theta]$ and satisfying $\mathbf{x}(\theta) = \mathbf{x}^*(\theta)$, for $j \in M$.

Proof.   For any $\theta > 0$, introduce the following performance index for Player j:

$$J_j(t^0, \mathbf{x}^0; [\mathbf{u}_j(\cdot), \sigma^*_{-j}(\cdot)]) = \int_0^\theta g_j(\mathbf{x}(t), [\mathbf{u}_j(t), \sigma^*_{-j}(t, \mathbf{x}(t))], t) dt.$$

Following the first steps in the proof of Theorem 7.2, we get

$$J_j(t^0, \mathbf{x}^0; [\mathbf{u}_j^*(\cdot), \sigma^*_{-j}(\cdot)]) - J_j(t^0, \mathbf{x}^0; [\mathbf{u}_j(\cdot), \sigma^*_{-j}(\cdot)])$$

$$= \int_0^\theta \left(g_j\left(\mathbf{x}^*(t), [\mathbf{u}_j^*(t), \sigma^*_{-j}(t, \mathbf{x}(t))], t \right) - g_j\left(\mathbf{x}(t), [\mathbf{u}_j(t), \sigma^*_{-j}(t, \mathbf{x}(t))], t \right) \right) dt$$

$$\geq \int_0^\theta \frac{d}{dt} (V_j(t, \mathbf{x}) - V_j(t, \mathbf{x}^*)) dt$$

$$= V_j(\theta, \mathbf{x}(\theta)) - V_j(\theta, \mathbf{x}^*(\theta)) + V_j(0, \mathbf{x}^*(0)) - V_j(0, \mathbf{x}(0)),$$

that is,

$$J_j(t^0, \mathbf{x}^0; [\mathbf{u}_j^*(\cdot), \sigma^*_{-j}(\cdot)]) - J_j(t^0, \mathbf{x}^0; [\mathbf{u}_j(\cdot), \sigma^*_{-j}(\cdot)]) \geq V_j(\theta, \mathbf{x}(\theta)) - V_j(\theta, \mathbf{x}^*(\theta)).$$

Applying the definitions of optimality leads to the different sufficient boundary conditions. $\qquad\square$

Remark 7.5. If the time horizon is infinite, the game is autonomous and if the objective functional of Player j is as in (7.22), then the Hamilton-Jacobi-Bellman equations become:

$$\rho_j V_j(\mathbf{x}) = \max_{\mathbf{u}_j \in U_j} \left\{ g_j\left(\mathbf{x}, [\mathbf{u}_j, \sigma^*_{-j}(t, \mathbf{x})] \right) + \right.$$

$$\left. \frac{\partial V_j(\mathbf{x})}{\partial \mathbf{x}} f(\mathbf{x}, [\mathbf{u}_j, \sigma^*_{-j}(t, \mathbf{x})] \right\} \quad j \in M. \quad (7.27)$$

These equations are simpler to solve than their finite-horizon counterparts in (7.17), because the value functions are now time-independent (hence the absence of their derivatives with respect to time in (7.27)).

7.5   Time consistency and subgame perfectness

In this section, we explain the difference between two important concepts in differential games, namely, *time consistency* and *subgame perfectness*. A similar topic was discussed in Section 6.4.1.2 in the context of multistage games.

To define these concepts some additional notations are needed.

Let $(t^i, \mathbf{x}^i) \in [0, T] \times X$ be any initial position.

We call *subgame* $\Phi(t^i, \mathbf{x}^i)$, the differential game defined on the time interval $[t^i, T]$ by the following data:

- For $s \in [t^i, T]$, the system dynamics are described by

$$\dot{\mathbf{x}}(s) = f(\mathbf{x}(s), \underline{\mathbf{u}}(s), s), \ \mathbf{x}(t^i) = \mathbf{x}^i;$$

- The payoff functional of Player j is given by

$$\int_{t^i}^{T} e^{-\rho_j(s-t)} g_j\left(\mathbf{x}(s), \underline{\mathbf{u}}(s), s\right) ds + e^{-\rho_j(T-t)} S_j(\mathbf{x}(T)), \quad j \in M.$$

Thus, the game introduced in Section 7.2 can be now be considered as the subgame $\Phi(0, \mathbf{x}^0)$.

By construction, an MNE, or feedback-Nash equilibrium $\underline{\sigma}^*$ is an equilibrium for all subgames $\Phi(t^i, \mathbf{x}^i)$ at all $(t^i, \mathbf{x}^i) \in [0, T] \times X$. We say that an MNE is a *subgame perfect equilibrium*.

Now consider an OLNE for the game $\Phi(0, \mathbf{x}^0)$, and the equilibrium trajectory $\mathbf{x}^*(\cdot)$ that is generated by these equilibrium open-loop strategies. Then for every position $(t, \mathbf{x}^*(t))$ along the equilibrium trajectory, the restriction of the equilibrium open-loop controls on the time interval $[t, T]$ is still an OLNE for the game $\Phi(t, \mathbf{x}^*(t))$. We say that an OLNE is a *time-consistent equilibrium*. The proof of this property has already been given for multi-stage games and will be the same for differential games (see Lemma 6.1 in Section 6.4.1.2.)

Remark 7.6. The following statements are obvious:

(1) Every subgame-perfect equilibrium is time consistent;
(2) An MNE is subgame perfect and hence also time consistent;
(3) An OLNE is time consistent but, in general, not subgame perfect.

These properties are linked to the *issue of credibility* of the announced equilibrium strategies. We have implicitly assumed that, at the initial date, all players simultaneously announce the strategies that they will use to play the game. A natural question is whether this announcement is *credible*, that is, whether each player will indeed implement his announced strategy. Suppose that Player j, for some reason, has an interest in playing a different strategy in subgame $\Phi(t, \mathbf{x}^*(t))$ starting at any intermediate position along the equilibrium trajectory. Then, at time zero, the other players should (and would) not believe Player j's announced strategy σ_j. Actually, they will be better off taking into account the expected deviation. This would generally lead to equilibrium strategies, say, $\psi_j, j \neq i$, other than those of game $\Phi(0, \mathbf{x}^0)$. In terms of credibility, time consistency can be seen as a minimal requirement and subgame perfectness a maximal one. Even if the players found themselves, for whatever reason, at position $(t, \mathbf{x})$, which is off the equilibrium-state trajectory, subgame perfectness provides the (strong) insurance that the equilibrium of the game $\Phi(t, \mathbf{x})$ is still the one announced at initial position $(0, \mathbf{x}^0)$ for game $\Phi(0, \mathbf{x}^0)$.

7.6 Differential games with special structures

Identifying a Nash equilibrium for a differential game, using either the Hamilton-Jacobi-Bellman equations or the maximum principle involves huge difficulties related to solving the HJB equations (7.17) or the system of adjoint equations in (7.12). In both approaches, it turns out that the required work is considerably simplified when the game belongs to the class of so-called *tractable differential games*.[12] This class includes linear-quadratic, linear-state and exponential differential games. To simplify the exposition, we assume in this section that each player has one control variable and that the state of the system is one-dimensional. [13] Let the payoff of Player $j \in M$, be given by

$$J_j \triangleq \int_{t^0}^{T} g_j(x(t), \underline{u}(t), t) \, dt + S_j(x(T)),$$

and the state dynamics by

$$\dot{x}(t) = \frac{dx}{dt}(t) = f(x(t), \underline{u}(t), t), \quad x(t^0) = x^0,$$

with $\underline{u}(t) \triangleq (u_1(t), \dots, u_m(t))$.

Linear-quadratic differential games (LQDG). As the linear quadratic multistage games studied in Chapter 5.7.5, this class of games has the following two characteristics:

(1) The instantaneous payoff function $g_j(\cdot)$ and the salvage-value function $S_j(\cdot)$ are quadratic in the state and control variables;
(2) The function $f(\cdot)$ is linear in the state and control variables.

For such games, it holds that Player j's MNE equilibrium strategy is a linear feedback law, i.e.,

$$u_j(t) = \sigma_j(t, x(t))) = a_j(t)x(t) + b_j(t),$$

and the value function is quadratic in x,

$$V_j(t, x(t)) = \frac{1}{2} A_j(t)x^2(t) + B_j(t)x(t) + C_j(t),$$

with coefficients $a_j(t), b_j(t), A_j(t), B_j(t)$ and $C_j(t)$ depending on time. The solution of HJB equations involves a set of coupled *Riccati differential equations*, as shown in Example 7.7.2. Additionally, if the LQDG is homogenous, i.e., there are no linear terms in the

---

[12]See references [58, 59], which provide a detailed treatment of this class of differential games.
[13]So we do not use boldface letters to represent scalars.

objective functional, then Player j's strategy is reduced to $\sigma_j(t, x(t))) = a_j(t)x(t)$ and the value function to $V_j(t, x(t)) = \frac{1}{2}A_j(t)x^2(t)$.

If the game is autonomous and played over an infinite horizon, then the coefficients

$$a_j(t), b_j(t), A_j(t), B_j(t) \text{ and } C_j(t)$$

are constant (see Example 7.7.1).

Linear-state differential games. When Player j's instantaneous payoff and salvage-value function and the state dynamics are affine (or linear)[14] functions of x, the game is called a *linear-state differential game*. For this class of games, it holds that Player j's equilibrium strategy is constant (that is, a degenerate feedback or independent of the state) and the value function is linear, $V_j(t, x(t)) = D_j(t)x(t) + E_j(t)$, where $D_j(t)$ and $E_j(t)$ are time functions that can be determined. If moreover the time horizon is infinite and the game is autonomous, then the coefficients $D_j(t)$ and $E_j(t)$ are independent of time, that is, constant. Additionally, because the feedback strategies are degenerate, open-loop and feedback-Nash equilibria coincide.

Exponential differential games. These games have two particular features: (i) the state variable does not enter the right-hand side of the system dynamics; and (ii) the state variable enters the objective in an exponential way. In such a game, the problem of Player j is given by

$$J_j \triangleq \int_{t^0}^{T} g_j(\underline{u}(t), t) e^{-\eta_j x(t)} \, dt + S_j x(T),$$

$$\dot{x}(t) = f(\underline{u}(t), t), \quad x(t^0) = x^0,$$

where η_j is a parameter. An exponential game can be transformed into a linear-state one by introducing a new state variable

$$y(t) = e^{-\eta_j x(t)}.$$

Differentiating with respect to time yields

$$\dot{y}(t) = -\eta_j \dot{x}(t) e^{-\eta_j x(t)} = -y(t)\dot{x}(t) = -y(t) f(\underline{u}(t), t).$$

Therefore, we get the new expression for the payoff functionals

$$J_j \triangleq \int_{t^0}^{T} g_j(\underline{u}(t), t) e^{-\eta_j x(t)} \, dt + S_j x(T) = \int_{t^0}^{T} g_j(\underline{u}(t), t) y(t) \, dt + S_j x(T), \quad j \in M,$$

which is linear in the state variable y. Therefore, all results that hold for linear-state games also apply to exponential ones.

If one of the above tractable differential-game formulations is appropriate for modeling the strategic situation at hand, then the problem of determining equilibrium strategies and

---

[14]The function $f(x)$ is affine in x if it takes the form

$$y = f(x) = ax + b.$$

We can also say that $f(x)$ is linear in x.

payoffs is much simpler. It amounts to identifying the coefficients appearing in the strategies and the value functions as shown above. Indeed, when using the Hamilton-Jacobi-Bellman approach, we know from the outset the functional form of the value functions, and no guessing is involved. Note, however, that this task can still be demanding if the LQDG has many state variables. The tractability of these games explains their huge popularity in applications (see [132] for a review of these applications in economics and management science). Although an LQDG may look too special at first glance, it still has some attractive features, namely, it allows for non-constant returns to scale, as well as for interactions between players' controls and between the control and state variables. A drawback of LQDG is the impossibility of accounting for interactions between control and state variables in the state dynamics (which must be linear in state and control).

State-redundant differential games. In linear-state differential games, open-loop and feedback-Nash equilibria coincide. This is a by-product of the result that the strategies are constant (independent of the state and of time). Therefore, the credibility of OLNE is not an issue in this context. The class of *state-redundant differential games* shares the same property of OLNE perfectness, as linear-state games, without however being necessarily tractable. The following condition provides a test to verify if a game is state redundant.

Condition 7.1. *The game is said* **state-redundant** *if the Hamiltonian-maximization conditions (7.10) are independent of the state variables and their initial values,* **after** *substituting the solution of the adjoint equation (7.12) for* $\lambda_j(t)$ *in (7.10).*

To illustrate how we can check for this condition, we consider the following example taken from Calzolari and Lambertini ([39]). Denote by u_j the control variable and by x_j the state variable of Player $j, j = 1, 2,$. The state dynamics are decoupled and given by

$$\dot{x}_j(t) = f(x_j(t)) - u_j(t) - \delta x_j(t), \quad x_j(0) = x_j^0, \quad j = 1, 2, \qquad (7.28)$$

where $f(x_j)$ is a strictly increasing concave function and δ a positive parameter. The payoff functionals are given by

$$J_j \triangleq \int_0^\infty e^{-\rho t} \left(a - u_j(t) - s u_{3-j}(t)\right) u_j(t) dt, \quad j = 1, 2,$$

where a and s are positive constants.[15] Omitting the time argument, we introduce the current-value Hamiltonian

$$H_j\left(\lambda_{jj}, \lambda_{j3-j}, x_j, x_{3-j}, u_j, u_{3-j}\right) = \left(a - u_j - s u_{3-j}\right) u_j + \lambda_{jj} \left[f(x_j) - u_j - \delta x_j\right]$$
$$+ \lambda_{j3-j} \left[f(x_{3-j}) - u_{3-j} - \delta x_{3-j}\right],$$

where λ_{jj} and λ_{j3-j} are adjoint variables appended by Player j to the state equations in (7.28). Note that the state equations are decoupled because the dynamics of variable x_j

---

[15]The reader will have remarked that $3 - j$ is equal to 1 if $j = 2$ and is equal to 2 if $j = 1$. In other parts of this book we use the convention $-j$ to designate the opponent player.

are independent of x_i and the state variables do not appear in the payoff functions. In particular, this implies that

$$\frac{\partial u_j}{\partial x_{3-j}} = 0, \text{ and } \lambda_{j3-j} \equiv 0, \text{ for } j = 1, 2.$$

The current-value Hamiltonian of Player j can be rewritten as

$$H_j \left(\lambda_{jj}, x_j, u_j, u_{3-j} \right) = \left(a - u_j - su_{3-j} \right) u_j + \lambda_{jj} \left[f \left(x_j \right) - u_j - \delta x_j \right].$$

Partial differentiation leads to

$$\frac{\partial H_j}{\partial u_j} = a - 2u_j - su_{3-j} - \lambda_{jj}, \quad \frac{\partial H_j}{\partial x_j} = \lambda_{jj} \left(f' \left(x_j \right) - \delta \right),$$

$$\frac{\partial^2 H_j}{\partial u_j \partial x_j} = \frac{\partial^2 H_j}{\partial u_j \partial x_{3-j}} = 0, \quad j = 1, 2,$$

$$\frac{\partial^2 H_j}{\partial x_j^2} = \lambda_{jj} f'' \left(x_j \right), \quad \frac{\partial^2 H_j}{\partial x_j \partial x_{3-j}} = 0, \quad j = 1, 2.$$

The derivatives in the third line show that the Hamiltonian is not linear in the state variables. The derivatives in the second line prove that the Hamiltonian maximization conditions are independent of the state variables and, therefore, the players' strategies can eventually be expressed as functions of the adjoint variables only.

The costate equation reads

$$\dot{\lambda}_{jj} = \rho - \frac{\partial H_j}{\partial x_j} = \lambda_{jj} \left(\rho + \delta - f' \left(x_j \right) \right), \quad \lim_{t \to \infty} e^{-\rho t} \lambda_{jj} = 0.$$

The above differential equation has as its solution $\lambda_{jj} \left(t \right) = 0$ for all $t \in [0, \infty)$. Substituting for $\lambda_{jj} \left(t \right)$ in the Hamiltonian-maximization condition $\frac{\partial H_j}{\partial u_j} = 0$ yields

$$a - 2u_j - su_{3-j} = 0, \quad j = 1, 2,$$

leading to the conclusion that the equilibrium strategies are independent of the state variables and of their initial values. Hence, the game is state-redundant, and the open-loop equilibrium is subgame perfect.

To wrap up, whereas a linear-state differential game is state redundant, the reverse is not necessarily true. Indeed, as the above example shows, it is possible to have non-linear dynamics and still have state redundancy.

The study of the class of state-redundant differential games was initiated by Leitmann and Schmitendorf [153], whereas Clemhout and Wan's contribution [46] is considered the origin of the differential-games literature seeking an OLNE, which is subgame perfect. (See [131] for a review of state-redundant differential games.)

7.7 Examples of the identification of Nash equilibria

We illustrate the use of Theorems 7.1 and 7.2 for both: infinite- and finite-horizon differential games.

7.7.1 An infinite-horizon example

Denote by $u_j(t)$ the control variable of Player $j = 1, 2$, and by $x(t)$ the state variable, for $t \in [0, \infty)$. Let the differential game data be given by

$$J_j \triangleq \int_0^\infty e^{-\rho t} \left(u_j(t) \left(\kappa - \frac{1}{2} u_j(t) \right) - \frac{1}{2} \varphi x^2(t) \right) dt$$

s.t.

$$\dot{x}(t) = u_1(t) + u_2(t) - \alpha x(t), \quad x(0) = x^0,$$

where φ and κ are positive parameters and $0 < \alpha < 1$. This game has the following features: (i) the objective functional of Player j is quadratic in the control and state variables and only depends on Player j's own control variable; (ii) there is no interaction either between the two players' control variables or between the control and the state variables; (iii) the game is autonomous, i.e., the integrand and the right-hand side of the state dynamics do not explicitly depend on time; and (iv) the game is fully symmetric in the state and control variables.

We omit the time argument in the rest of the example when no ambiguity may arise.

Open-loop Nash equilibrium. Introduce the Hamiltonian of Player j:

$$H_j(\lambda_j, x, u_1, u_2) = u_j \left(\kappa - \frac{1}{2} u_j \right) - \frac{1}{2} \varphi x^2 + \lambda_j (u_1 + u_2 - \alpha x), \quad j = 1, 2,$$

where λ_j is the costate variable associated with the state variable x. Assuming an interior solution, the necessary and sufficient (verify that the maximized Hamiltonian is indeed concave in x) equilibrium conditions read

$$\frac{\partial H_j}{\partial u_j} = 0 \Leftrightarrow u_j = \kappa + \lambda_j,$$

$$\dot{\lambda}_j = \rho \lambda_j - \frac{\partial H_j}{\partial x} = (\rho + \alpha)\lambda_j + \varphi x, \quad \lim_{t \to \infty} e^{-\rho t} \lambda_j(t) = 0,$$

$$\dot{x} = u_1 + u_2 - \alpha x, \quad x(0) = x^0.$$

It is easy to see that $\lambda_1(t) = \lambda_2(t), \forall t \in [0, \infty)$, and therefore, $u_1(t) = u_2(t), \forall t \in [0, \infty)$. This is not surprising given the symmetry of the game. Replacing the expression

$$u_1 = u_2 = \kappa + \lambda, \tag{7.29}$$

in the differential equations for variables x and λ, we get a two-equation differential system. In matrix form, this system reads:

$$\begin{pmatrix} \dot{x} \\ \dot{\lambda} \end{pmatrix} = \begin{pmatrix} -\alpha & 2 \\ \varphi & \rho + \alpha \end{pmatrix} \begin{pmatrix} x \\ \lambda \end{pmatrix} + \begin{pmatrix} 2\gamma \\ 0 \end{pmatrix}.$$

We look for the solution of this system converging to the steady state which is given by

$$(x_{ss}, \lambda_{ss}) = \left(\frac{2\kappa(\alpha + \rho)}{\alpha^2 + \alpha \rho + 2\varphi}, -\frac{2\kappa\varphi}{\alpha^2 + \alpha \rho + 2\varphi} \right).$$

The solution can be written as:

$$x(t) = (x^0 - x_{ss})e^{\omega_1 t} + x_{ss},$$

$$\lambda(t) = -(x^0 - x_{ss})\frac{2\varphi}{2\alpha + \rho + \sqrt{(2\alpha + \rho)^2 + 8\varphi}}e^{\omega_1 t} + \lambda_{ss},$$

where ω_1 is the negative eigenvalue of the matrix associated with the differential-equations system, and is given by

$$\omega_1 = \frac{1}{2}(\rho - \sqrt{(2\alpha + \rho)^2 + 8\varphi}).$$

It suffices to replace $\lambda(t)$ by its value in (7.29) to obtain the equilibrium strategies.

Feedback-Nash equilibrium. We use Theorem 7.2 and Remark 7.27 to determine feedback-Nash equilibrium strategies. The HJB equation for Player j is

$$\rho V_j(x) = \max_{u_j \geq 0}\left[u_j\left(\kappa - \frac{1}{2}u_j\right) - \frac{1}{2}\varphi x^2 + V_j'(x)(u_1 + u_2 - \alpha x)\right]. \qquad (7.30)$$

Differentiating the RHS and equating to zero yields

$$u_j(x) = \kappa + V_j'(x). \qquad (7.31)$$

Given the symmetric feature of this game, we focus on symmetric equilibrium strategies. Taking into account the linear-quadratic specification of the differential game, we make the informed guess that the value function is quadratic. Additionally, because the game is symmetric, and we focus on symmetric solutions, the value function is the same for both players and is given by

$$V_j(x) = \frac{A}{2}x^2 + Bx + C, \quad i = 1, 2.$$

Using (7.31) then leads to the linear feedback form

$$u_j(x) = \kappa + Ax + B.$$

Replacing this expression into the RHS of (7.30) we obtain

$$\frac{1}{2}\left(3A^2 - 2A\alpha - \varphi\right)x^2 + (3AB - B\alpha + 2A\kappa)x + \frac{1}{2}\left(3B^2 + 4B\kappa + \kappa^2\right).$$

The LHS of (7.30) reads

$$\rho\left(\frac{A}{2}x^2 + Bx + C\right).$$

Equating the coefficients of x^2, x and the constant term, we obtain[16] three equations with three unknowns, A, B and C.

---

[16]This is another instance of application of the method of undetermined coefficients explained in Chapter 5.7.5; see page 184.

Solving these equations we get the following coefficients for the noncooperative equilibrium value functions:

$$A = \frac{\rho + 2\alpha \pm \sqrt{(\rho + 2\alpha)^2 + 12\varphi}}{6},$$

$$B = \frac{-2A\kappa}{3A - (\rho + \alpha)},$$

$$C = \frac{\kappa^2 + 4B\kappa + 3B^2}{2\rho}.$$

Remark 7.7. The coefficient A is the root of a second-degree polynomial having one positive and one negative root. The selection of the negative root

$$A = \frac{\rho + 2\alpha - \sqrt{(\rho + 2\alpha)^2 + 12\varphi}}{6},$$

guarantees the global stability of the state trajectory. The noncooperative equilibrium state trajectory is given by

$$x(t) = \left[x^0 + \frac{2(\kappa + B)}{2A - \alpha}\right] e^{(2A - \alpha)t} - \frac{2(\kappa + B)}{2A - \alpha}.$$

The state dynamics of the game have a globally asymptotically stable steady state if $2A - \alpha < 0$. It can be shown that, in order to guarantee this inequality and therefore, to guarantee the globally asymptotically stable steady state, the only possibility is to choose $A < 0$.

Remark 7.8. We could have used Theorem 7.1 to determine a feedback-Nash equilibrium. To illustrate this, we reconsider the Hamiltonian of Player j:

$$H_j(\lambda_j, x, u_1, u_2) = u_j \left(\kappa - \frac{1}{2}u_j\right) - \frac{1}{2}\varphi x^2 + \lambda_j(u_1 + u_2 - \alpha x), \quad j = 1, 2,$$

where λ_j is the costate variable associated with the state variable x. Assuming an interior solution, the necessary equilibrium conditions

$$\frac{\partial H_j}{\partial u_j} = 0 \Leftrightarrow u_j = \kappa + \lambda_j,$$

$$\dot{x} = u_1 + u_2 - \alpha x, \quad x(0) = x^0,$$

are as before. However, the costate equations now read (see Eq. (7.12))

$$\dot{\lambda}_j = \rho \lambda_j - \frac{\partial H_j}{\partial x} - \frac{\partial H_j}{\partial u_i} \frac{\partial u_i^*}{\partial x} = (\rho + \alpha)\lambda_j + \varphi x - \lambda_j \frac{\partial u_i^*}{\partial x},$$

with the transversality conditions

$$\lim_{t \to \infty} e^{-\rho t} \lambda_j(t) = 0, \quad j = 1, 2.$$

The term $\frac{\partial H_j}{\partial u_i} \frac{\partial u_i^*}{\partial x}$, which does not appear in the open-loop case, shows the interaction between the control of the other player and the state variable. In general, we do not know the shape $u_i^*(x)$ and hence the value of the derivative $\frac{\partial u_i^*}{\partial x}$. However, as we did when we applied the HJB approach to determine a feedback equilibrium, we make the informed guess that Player i's strategy is linear in the state and hence its derivative is a constant Z_i. Invoking symmetry, we assume that $Z_1 = Z_2 = Z$. Therefore, the costate equations become

$$\dot{\lambda}_j = \rho\lambda_j - \frac{\partial H_j}{\partial x} - \frac{\partial H_j}{\partial u_i}\frac{\partial u_i^*}{\partial x} = (\rho + \alpha - Z)\lambda_j + \varphi x, \quad j = 1, 2.$$

Solving the above differential equations gives λ_j as function of x, say $k(x)$. Substituting in the maximizing condition $\frac{\partial H_j}{\partial u_j} = 0$, yields

$$u_j = \kappa + k(x),$$

which is the desired markovian equilibrium strategy.

7.7.2 A finite-horizon example

Let the differential game data be given by

$$J_j \triangleq \int_{t^0}^{T} \left(\frac{\kappa}{2}u_j^2(t) - \frac{\varphi}{2}x^2(t) \right) dt, \quad j \in M,$$

s.t.

$$\dot{x}(t) = u_1(t) + u_2(t) - \alpha x(t), \quad x(0) = x^0 > 0,$$

where κ and φ are positive parameters and $0 < \alpha < 1$. Note that the objective does not involve a linear term and that the model is fully symmetric. We still omit the time argument when no ambiguity may arise.

Open-loop Nash equilibrium. We use Theorem 7.1 to determine open-loop Nash-equilibrium strategies. Introduce the Hamiltonian of Player $j = 1, 2$:

$$H_j(\lambda_1, x, u_1, u_2) = \frac{\kappa}{2}u_j^2 - \frac{\varphi}{2}x^2 + \lambda_j(u_1 + u_2 - \alpha x),$$

where $\lambda_j, i = 1, 2$, are the costate variables. Assuming an interior solution, maximizing the Hamiltonians with respect to the controls u_1 and u_2, respectively, yields

$$u_j = -\frac{\lambda_j}{\kappa}, \quad j = 1, 2. \tag{7.32}$$

The costate equations and their transversality conditions are

$$\dot{\lambda}_j = \varphi x + \alpha\lambda_j, \quad \lambda_j(T) = 0, \ j = 1, 2.$$

The above differential equations are independent of the control variables. Also, it is easy to verify that $\lambda_1(t) = \lambda_2(t) \equiv \lambda(t), \forall t \in [0, T]$, and hence, we need to solve only one adjoint equation:

$$\dot{\lambda} = \varphi x + \alpha\lambda, \quad \lambda(T) = 0.$$

Therefore, $u_1(t) = u_2(t) \equiv u(t), \forall t \in [0, T]$. Substituting the equilibrium strategies u into the state equation yields

$$\dot{x} = -\frac{2\lambda}{\kappa} - \alpha x, \quad x(0) = x^0.$$

The equilibrium controls u_j^* are independent of the state x, and the Hamiltonians are concave in x. Then the maximized Hamiltonians are concave in x and, by Theorem 7.1, strategies $u_j(t)$ provide an OLNE.

The state and costate equations (as well as their boundary conditions) define a two-point boundary-value problem (TPBVP). In matrix form, this system reads

$$\begin{pmatrix} \dot{x} \\ \dot{\lambda} \end{pmatrix} = \begin{pmatrix} -\alpha & -2/\kappa \\ \varphi & \alpha \end{pmatrix} \begin{pmatrix} x \\ \lambda \end{pmatrix},$$

with $x(0) = x^0$ and $\lambda(T) = 0$. This problem can be solved explicitly by standard methods of linear differential equations. It is easy to verify (for instance, using any of the available software) that the solution is given by

$$x(t) = x^0 \frac{(\alpha + Z) e^{(T-t)Z} - (\alpha - Z)}{(\alpha + Z) e^{TZ} - (\alpha - Z) e^{-tZ}},$$

$$\lambda(t) = \frac{x^0 (\alpha + Z)(\alpha - Z) \left(e^{(t-T)Z} - 1 \right)}{\beta \left[(\alpha + Z) e^{TZ} - (\alpha - Z) e^{-TZ} \right]},$$

where $Z = \sqrt{\alpha^2 - \beta\varphi}$. It suffices to substitute for $\lambda_j(t)$ in (7.32) to obtain the equilibrium strategies $u_j(t)$.

Feedback-Nash equilibrium. We use Theorem 7.2 to identify an equilibrium with non-degenerate Markovian strategies $\sigma_1(t, x(t))$ and $\sigma_2(t, x(t))$. Introduce value functions $V_j(t, x(t))$ and the HJB equations:

$$-\frac{\partial V_j(t,x)}{\partial t} = \max_{u_j} \left\{ \frac{\kappa}{2} u_j^2 - \frac{\varphi}{2} x^2 + \frac{\partial V_j(t,x)}{\partial x} (u_1 + u_2(t) - \alpha x) \right\},$$

$$V_j(T, x(T)) = 0.$$

Maximization of the RHS yields

$$\sigma_j(t, x) = -\frac{1}{\kappa} \frac{\partial V_j(t,x)}{\partial x}, \quad j = 1, 2. \tag{7.33}$$

Notice that costates $\lambda_j(t)$ of the open-loop case have now been replaced by the derivatives $\frac{\partial V_j(t,x)}{\partial x}$. This was expected since both quantities represent a shadow price of the stock x.

Inserting the equilibrium strategies $\sigma_j(t, x)$ into the RHS of the HJB equations yields two partial differential equations for the value functions $V_j(t, x), j = 1, 2$:

$$-\frac{\partial V_j(t,x)}{\partial t} = \frac{1}{2\kappa} \left(\frac{\partial V_j(t,x)}{\partial x} \right)^2 - \frac{\varphi}{2} x^2 +$$

$$\frac{\partial V_j(t,x)}{\partial x} \left(-\frac{1}{\kappa} \frac{\partial V_1(t,x)}{\partial x} - \frac{1}{\kappa} \frac{\partial V_2(t,x)}{\partial x} - \alpha x \right),$$

$$V_j(T, x(T)) = 0.$$

As the game is symmetric, we focus on symmetric strategies $\sigma_1(t, x) = \sigma_2(t, x) = \sigma(t, x)$ and symmetric value functions. The above equations now read

$$-\frac{\partial V(t,x)}{\partial t} = \left\{ -\frac{3}{4\kappa} \left(\frac{\partial V(t,x)}{\partial x} \right)^2 - \frac{\varphi}{2} x^2 - \alpha x \frac{\partial V(t,x)}{\partial x} \right\},$$

$$V(x, T) = 0.$$

The differential game at hand is homogenous linear-quadratic, and it can be shown that the quadratic value function

$$V(t, x) = \frac{1}{2}vx^2,$$

solves the HJB equations, provided that the time function v satisfies the following Riccati differential equation:

$$\dot{v}(t) = \frac{3}{2\kappa}v^2(t) + \varphi + 2\alpha v(t), \quad v(T) = 0.$$

It suffices to substitute the solution of the above differential equation in (7.33) to get the strategy of Player $j, j = 1, 2$. As expected, in a finite-horizon setting, the decision rule changes over time via the time-dependency of coefficient v.

7.8 On the non-uniqueness of feedback equilibria

In this section, we discuss the issue of the non-uniqueness of feedback equilibria in infinite-horizon games. In the proof of Theorem 6.2 in Section 6.4.3 (see page 190), we said that when $T \rightarrow \infty$, in general, the transversality conditions may be satisfied by many functions. We can see in Proposition 7.1 that the same situation can happen in continuous time. Potentially, this creates non-uniqueness for the equilibria. We will see in this section that even the well-studied two-person linear-quadratic differential game can have multiple equilibria when $T \rightarrow \infty$.

7.8.1 *A linear-quadratic infinite-horizon differential game in* **R**

Consider the two-player infinite-horizon differential game formulated in Section 7.7.1 (also, see the discrete-time formulations in Section 6.7, 197). The game, solved in that section in *linear-feedback strategies*, can have an environmental-economics interpretation, as shown below. This interpretation is not crucial for the understanding of the arising non-unique non-linear feedback strategies; however, it helped [61] to rank these strategies.

We follow [61] and attach the following meaning to the model variables and parameters: state $x(t)$ is pollution stock; controls $u_1(t), u_2(t)$ are emissions; $\rho > 0$ is the discount factor; $\alpha \in (0, 1)$ is the self-cleaning coefficient; $\varphi > 0$ is the coefficient responsible for the quadratic damage due to pollution; $\kappa > 0$ is the marginal revenue from emissions (suppose emissions are proportional to output); $t \in [0, \infty)$ is the time horizon. Assuming that there are two symmetric players $j = 1, 2$, the game is formulated as follows:

$$\max_{u_j \geq 0} \left\{ J_j = \int_0^\infty e^{-\rho t} \left(u_j(t) \left(\kappa - \frac{1}{2}u_j(t) \right) - \frac{1}{2}\varphi x^2(t) \right) dt \right\}, \quad (7.34)$$

s.t.:

$$\dot{x}(t) = u_1(t) + u_2(t) - \alpha x(t), \quad x(0) = x^0, \quad (7.35)$$

The Hamilton-Jacobi-Bellman (HJB) equation for Player j, obtained as (7.30) (see page 266), is

$$\rho V_j(x) = \max_{u_j \geq 0} \left[u_j \left(\kappa - \frac{1}{2} u_j \right) - \frac{1}{2} \varphi x^2 + V_j'(x) \left(u_1 + u_2 - \alpha x \right) \right]. \qquad (7.36)$$

The equilibrium strategies must satisfy the first-order conditions

$$u_j(x) = \kappa + V_j'(x). \qquad (7.37)$$

7.8.2 *Linear feedback-equilibrium strategies*

Given the informed *guess* that the value function is quadratic

$$V_j(x) = \frac{A}{2} x^2 + Bx + C, \quad j = 1, 2, \qquad (7.38)$$

the feedback-equilibrium strategy for each player is *linear:*

$$u_j^*(x) = Ax + \kappa + B. \qquad (7.39)$$

Using the method of undetermined coefficients, the constants A, B, C are determined in Section 7.7.1 (see page 266). In particular, the coefficient A satisfies a quadratic equation with two roots and is chosen as the negative root. Coefficient B is also negative. (Constant C does not intervene in the equilibrium strategy (7.39) or the expression for the strategy and we do not comment on this constant here.) After substitutions, we obtain the linear feedback strategy for each player as

$$u_j^*(x) = \frac{\rho + 2\alpha - \sqrt{(\rho + 2\alpha)^2 + 12\varphi}}{6} x + \frac{\kappa \left(A - \rho - \alpha \right)}{3 A - \rho - \alpha}. \qquad (7.40)$$

The equilibrium state trajectory, which corresponds to (7.40), is

$$x^*(t) = \left[x^0 + \frac{2 \kappa (\alpha - A + \rho)}{(\alpha - 3 A + \rho)(2A - \alpha)} \right] e^{(2A-\alpha)t} - \frac{2 \kappa (\alpha - A + \rho)}{(\alpha - 3 A + \rho)(2 A - \alpha)}. \qquad (7.41)$$

It follows from Remark 7.7, and after simple algebra, that the globally asymptotically stable steady state is

$$x_\infty^* = \frac{2 \kappa (\alpha - 3 A + \rho)}{(\alpha - 3 A + \rho)(\alpha - 2A)} \geq 0. \qquad (7.42)$$

We notice that expression (7.42) matches the result reported in [61]. In their interpretation, this is the steady-state pollution stock, which, as expected, is positively correlated with marginal revenue κ and negatively with the self-cleaning coefficient α.

7.8.3 *Nonlinear feedback-equilibrium strategies*

7.8.3.1 *An envelope condition*

Consider the HJB equations (7.36), for $j = 1, 2$, and rewrite the first-order conditions (7.37) to express the marginal value function's dependence on equilibrium strategy:

$$V_j'(x) = u_j(x) - \kappa. \qquad (7.43)$$

Notice that both $V_j(\cdot)$ and $u_j(\cdot)$ are unknown functions and that, in general, we do not have to adjust strategies to value functions, as in (7.37) (or (7.31), see page 266) might suggest. In particular, the equilibrium first-order condition (7.43) does not preclude the possibility that value function $V_j(x)$ can have a form other than linear-quadratic.

Let us use the first-order condition (7.43) and substitute $V_j'(x)$ in (7.36) to see which equations $V_j(x)$ have to satisfy, irrespectively of their functional form. Additionally, we will assume that we are looking for a symmetric equilibrium (so $u_1 + u_2 = 2u_j$). We obtain

$$\rho V_j(x) = \kappa u_j - \frac{u_j^2}{2} - \frac{\varphi}{2}x^2 + (u_j - \kappa)(2u_j - \alpha x), \tag{7.44}$$

where we dropped argument x from the players' controls $u_j(x)$ to lighten the notation. The problem now consists of computing strategies $u_j(x)$, not necessarily linear, using (7.44).

Let us solve (7.44) for u, to see which general form strategies $\tilde{u}_j$ can have. This is a quadratic equation in u and we get two roots:

$$\tilde{u}_{j,1}(x) = \frac{\kappa}{3} + \frac{\alpha x}{3} - \frac{\sqrt{\alpha^2 x^2 - 4\alpha\kappa x + \kappa^2 + 3\varphi x^2 + 6V(x)\rho}}{3}$$

$$\tilde{u}_{j,2}(x) = \frac{\kappa}{3} + \frac{\alpha x}{3} + \frac{\sqrt{\alpha^2 x^2 - 4\alpha\kappa x + \kappa^2 + 3\varphi x^2 + 6V(x)\rho}}{3}.$$

We can see that, indeed, a nonlinear function of state is to be added to the affine function of state. In particular, for the non-discounted problem (i.e., when $\rho = 0$), the nonlinear feedback strategy has the following form:

$$\tilde{u}_j(x) = \frac{\alpha x + \kappa}{3} + h(x), \tag{7.45}$$

where $h(x)$ is a nonlinear function of state x. We will assume that $\tilde{u}_j(x)$ will be in this class of functions for any $\rho > 0$, and thus, replace u in (7.44) by $\tilde{u}_j(x)$; this yields

$$\rho V(x) = \frac{3h^2(x)}{2} - \frac{\kappa^2}{6} + \frac{2\alpha\kappa}{3}x - \left(\frac{\alpha^2}{6} + \frac{\varphi}{2}\right)x^2. \tag{7.46}$$

Now, we will analyze the *envelope condition* that we obtain after differentiating the above HJB equation in *state*, and into which the *equilibrium strategies* have been substituted. We get

$$\rho\left(\frac{\alpha}{3}x - \frac{2\kappa}{3} + h(x)\right) = 3h(x)h'(x) + \frac{2\alpha\kappa}{3} - \left(\frac{\alpha^2}{3} + \varphi\right)x, \tag{7.47}$$

where the left-hand side comes from (7.43).

Expression (7.47) is a non-homogenous differential equation that defines $h(x)$ and, if the solution is not unique, a plethora of nonlinear equilibrium strategies (7.45) will be obtained.

7.8.3.2 Solutions

A solution procedure for (7.47) is provided in [61]. Below, the main idea of that procedure is summarized.

Integration by substitution will be used at some stage. Keeping this in mind, it is convenient to have the derivative on the left-hand side and to group the problem parameters into two constants as follows:

$$3h(x)\,h'(x) = \rho\,h(x) + \underbrace{\left(\frac{\alpha^2}{3} + \frac{\rho\alpha}{3} + \varphi\right)}_{\equiv D} x - \underbrace{\frac{2\alpha}{3}(\kappa + \rho)}_{\equiv C}. \qquad (7.48)$$

This enables us to present the equation in a simpler form:

$$h'(x) = \frac{\rho\,h(x) + Dx - C}{3h(x)}. \qquad (7.49)$$

The state (independent) variable x will now be shifted by $\dfrac{C}{D}$ to become

$$y = x - \frac{C}{D}, \qquad (7.50)$$

which simplifies (7.49) to

$$\frac{dh}{dy} = \frac{Dy + \rho h}{3h}. \qquad (7.51)$$

Now, we will substitute $h = z\,y$ so $dh = y\,dz + z\,dy$ and

$$\frac{dh}{dy} = y\frac{dz}{dy} + z.$$

This makes (7.51) evolve to

$$y\frac{dz}{dy} + z = \frac{D + \rho z}{3z}, \qquad (7.52)$$

which can be rewritten with separated variables as

$$\frac{1}{y}\,dy = \frac{3z}{D + \rho z - 3z^2}\,dz. \qquad (7.53)$$

Both sides of this differential equation are analytically integrable and can relatively easily[17] obtain y as a function of z

$$y = \frac{\sqrt{3}\,K\,\exp\left(\dfrac{\arctan\left(\dfrac{(\rho - 6z)\sqrt{\rho^2}}{\rho\sqrt{-\rho^2 - 12D}}\right)\sqrt{\rho^2}}{\sqrt{-\rho^2 - 12D}}\right)}{\sqrt{3z^2 - \rho z - D}}, \qquad (7.54)$$

where K is an integration constant.

Inverting the above function and "reverting" to the original model parameters (to obtain $h(x)$) is *not* an elementary operation. However, it is clear that the right-hand side of (7.54)

---

[17]Using e.g., Matlab Symbolic Mathematics Toolbox.

is not constant and that the inverse exists, at least for some intervals of y (easily convertible to x). In particular, for $\rho = 0$, a family of hyperbolas

$$3h^2 - Dy^2 = \sqrt{3}\,K$$

is easily obtained. Given (7.45), which was a solution to (7.44) when $\rho = 0$, we can expect that the relationships between h and y (or x) will also be hyperbolic.

Indeed, [61] integrated (7.52) directly and obtained an implicit solution

$$K = (h - Z_a\,y)^{\xi_a}\,(h - Z_b\,y)^{\xi_b}, \tag{7.55}$$

where K is an(other) integration constant and Z_a, Z_b, ξ_a, ξ_b are constants related to the problem parameters. Shifting the state variable back by $\dfrac{C}{D}$ gives us the following implicit definition of the nonlinear terms $h(x)$ in the equilibrium strategies (7.45):

$$K = \left(u - \left(\frac{\alpha}{3} + Z_a\right)x + Z_a\frac{C}{D} - \frac{\kappa}{3}\right)^{\xi_a}$$

$$\left(u - \left(\frac{\alpha}{3} + Z_b\right)x + Z_b\frac{C}{D} - \frac{\kappa}{3}\right)^{\xi_b}, \tag{7.56}$$

which represents a family of hyperbolic relations between u and x.

7.8.3.3 *Discussion*

A question that naturally arises is how good are the nonlinear strategies defined by (7.56), relative to linear feedback (7.40)?

Unfortunately, integrating the performance criterion (7.34) for nonlinear strategy $\tilde{u}_j(x)$ and comparing it with the criterion value obtained for linear strategy $u_j^*(x)$ is *analytically* impossible. An approximate *numerical* comparison of these values is feasible but extremely complicated[18] and, to our knowledge, no author (including [61]; or [215] and [234], the authors of two other classics in this area) has attempted it.

However, some comparisons between the strategies' *outcomes* have been attempted. In [61], steady state x_∞^* represents a pollution level. For the linear feedback strategies, the steady state is computed in (7.42). It is interesting to note that, after linearizing (sic) the state equation obtained for the nonlinear strategies, [61] show that a nonlinear feedback-equilibrium strategies will lead the steady-state pollution level to values that can be inferior to x_∞^*, for very low discount rates. So, such a non-linear strategy appears better than its linear counterpart. Nevertheless, this result cannot imply the superiority of nonlinear feedback strategies over the linear strategies because minimizing the steady-state pollution level is *not* a goal included in the performance criteria. Presumably, if it were included, all strategies would be different that those obtained for (7.34), the criterion considered in this section.

---

[18]In particular, not only would integral (7.34) need to be numerically calculated but so would its integrant (in the first round).

7.9 Stackelberg equilibria

In the previous sections, the assumption was that the players select their actions simultaneously. As we already did in section 3.6, we consider here a two-player game where one player, the leader, makes his decision before the other player, the follower.[19] Such a sequence of moves was first introduced by Stackelberg in the context of a duopoly-output game, see [241].

Denote by L the leader and by F the follower. Suppose that $\mathbf{u}_L(t)$ and $\mathbf{u}_F(t)$, respectively, are the control vectors of the two players. The control constraints $\mathbf{u}_L(t) \in U_L$ and $\mathbf{u}_F(t) \in U_F$ must be satisfied for all t. The state dynamics and the payoff functionals are as before. As for the Nash equilibrium, we will define an open-loop Stackelberg equilibrium (OLSE) and a Markovian (or feedback)-Stackelberg equilibrium (MSE).

7.9.1 *Open-loop Stackelberg equilibria*

When both players use open-loop strategies, their control paths are determined by $\mathbf{u}_L(t) = \mu_L(t, \mathbf{x}^0)$ and $\mathbf{u}_F(t) = \mu_F(t, \mathbf{x}^0)$, respectively. Here μ_j denotes the open-loop strategy of Player j.

The game proceeds as follows. At time $t = 0$, the leader announces his control path $\mathbf{u}_L(\cdot)$ for $t \in [0, T]$. Suppose, for the moment, that the follower believes in this announcement. The best he can do is then to select his own control path $\mathbf{u}_F(\cdot)$ to maximize the objective functional

$$J_F \triangleq \int_0^T e^{-\rho_F t} g_F\left(\mathbf{x}(t), \mathbf{u}_L(t), \mathbf{u}_F(t), t\right) dt + e^{-\rho_F T} S_F(\mathbf{x}(T)),$$

subject to the state dynamics

$$\dot{\mathbf{x}}(t) = f\left(\mathbf{x}(t), \mathbf{u}_L(t), \mathbf{u}_F(t), t\right) \quad x(0) = x^0,$$

and

$$\mathbf{u}_F(t) \in U_F.$$

This is a standard optimal control problem. To solve it, introduce the follower's Hamiltonian

$$H_F(\lambda_F(t), \mathbf{x}(t), \mathbf{u}_F(t), \mathbf{u}_L(t), t)$$
$$= g_F(\mathbf{x}(t), \mathbf{u}_F(t), \mathbf{u}_L(t), t) + \lambda_F' f(\mathbf{x}(t), \mathbf{u}_F(t), \mathbf{u}_L(t), t),$$

where the adjoint variable $\lambda_F = \lambda_F(t)$ is an $n-$vector. Suppose that the Hamiltonian H_F is concave in $\mathbf{u}_F$. Then the maximization of H_F with respect to $\mathbf{u}_F$, for $t \in [0, T]$, uniquely determines $\mathbf{u}_F(t)$ as a function of $t, \mathbf{x}, u_L,$ and λ_F is denoted

$$\mathbf{u}_F(t) = R(t, \mathbf{x}(t), \mathbf{u}_L(t), \lambda_F(t)).$$

It defines the follower's *best reply* to the leader's announced time path $\mathbf{u}_L(\cdot)$.

---

[19]The setup can be easily extended to a case of several followers. A standard assumption is then that the followers play a (Nash) simultaneous-move game vis-a-vis each other, and a sequential game vis-a-vis the leader.

The follower's costate equations and their boundary conditions are given by

$$\dot{\lambda}_F(t) = \rho_F - \frac{\partial H_F}{\partial \mathbf{x}'},$$

$$\lambda_F(T) = \frac{\partial S_j}{\partial \mathbf{x}'(T)}(\mathbf{x}(T)).$$

Substituting function R into the state and costate equations yields a two-point boundary value problem. The solution of this problem, $(\mathbf{x}(t), \lambda_F(t))$, can be inserted into function R. This represents the follower's optimal behavior, given the leader's announced time path $\mathbf{u}_L(\cdot)$.

The leader can replicate the follower's arguments. This means that the leader can calculate the follower's best reply R to any $\mathbf{u}_L(\cdot)$ that the leader may announce. The leader's problem is then to select a control path $\mathbf{u}_L(\cdot)$ that maximizes the payoff

$$J_L = \int_0^T e^{-\rho_L t} g_L\left(\mathbf{x}(t), \mathbf{u}_L(t), R\left(t, \mathbf{x}(t), \mathbf{u}_L(t), \lambda_F(t)\right)\right), t) dt + e^{-\rho_L T} S_L(\mathbf{x}(T))$$

subject to

$$\dot{\mathbf{x}}(t) = f(\mathbf{x}(t), \mathbf{u}_L(t), R(t, \mathbf{x}(t), \mathbf{u}_L(t), \lambda_F(t)), t), \quad x(0) = x^0,$$

$$\dot{\lambda}_F(t) = \rho_F - \frac{\partial}{\partial \mathbf{x}'} H_F\left(\mathbf{x}(t), \mathbf{u}_L(t), R\left(t, \mathbf{x}(t), \mathbf{u}_L(t), \lambda_F(t)\right)\right), t),$$

$$\lambda_F(T) = \frac{\partial S_F}{\partial \mathbf{x}'(T)}(\mathbf{x}(T)),$$

and

$$\mathbf{u}_L(t) \in U_L.$$

Note that the leader's dynamics include two state equations, one governing the evolution of the original state variables $\mathbf{x}$ and a second one accounting for the evolution of λ_F, the adjoint variables of the follower, which are now treated as state variables. Again, we have an optimal control problem that can be solved using the maximum principle. To do so, we introduce the leader's Hamiltonian

$$H_L\left(\mathbf{x}(t), \mathbf{u}_L(t), R\left(t, \mathbf{x}(t), \mathbf{u}_L(t), \lambda_L(t), \lambda_F(t), \theta(t)\right)\right)$$
$$= g_L\left(\mathbf{x}(t), \mathbf{u}_L(t), R\left(t, \mathbf{x}(t), \mathbf{u}_L(t), \lambda_F(t)\right)\right), t)$$
$$+ \lambda_L(t) f(x(t), \mathbf{u}_L(t), R(t, \mathbf{x}(t), \mathbf{u}_L(t), \lambda_F(t)), t)$$
$$+ \theta'(t) \left(\rho_F - \frac{\partial H_F\left(\mathbf{x}(t), \mathbf{u}_L(t), R\left(t, \mathbf{x}(t), \mathbf{u}_L(t), \lambda_F(t)\right)\right), t)}{\partial \mathbf{x}'} \right),$$

where $\lambda_L = \lambda_L(t)$ is the n vector of costate variables appended to the state equation $\dot{\mathbf{x}}(t)$, with the boundary conditions

$$\lambda_L(T) = \frac{\partial S_L}{\partial \mathbf{x}'(T)}(\mathbf{x}(T)),$$

and $\theta = \theta(t)$ is the vector of n costate variables appended to the state equation $\dot{\lambda}_F(t)$, satisfying the initial conditions

$$\theta(0) = 0.$$

The initial condition $\theta(0) = 0$ is a consequence of the fact that $\lambda_F(0)$ is "free," i.e., unrestricted. This result is a counterpart of the one we used before. Recall that if there is no salvage value in the objective, and the state value $\mathbf{x}(T)$ is unconstrained, then the costate $\lambda(t)$ must satisfy the condition $\lambda(T) = 0$. The determination of an OLSE will be illustrated with an example as well as in the GE.

7.9.2 *Markovian Stackelberg Equilibria*

When the players use Markovian strategies, their control paths are generated by $\mathbf{u}_L(t) = \sigma_L(t, \mathbf{x}(t))$ and $\mathbf{u}_F(t) = \sigma_F(t, \mathbf{x}(t))$, respectively. The leader announces at time zero the strategy $\sigma(t, \mathbf{x}(t))$ and commits to using this strategy. Then the follower knows, that at any position $(t, \mathbf{x}(t))$ of the game, the leader will determine his control action as $\mathbf{u}_L(t) = \sigma_L(t, \mathbf{x}(t))$. The follower then reacts rationally to the leader's announcement. Knowing this, the leader selects a strategy that maximizes her payoff functional.

First we look at the follower's optimal control problem. Using the dynamic programming approach, we get the HJB equation for the follower's problem

$$\rho_F V_F(t, \mathbf{x}) - \frac{\partial V_F(t, \mathbf{x})}{\partial t} = \max_{\mathbf{u}_F \in U_F} \{ g_F(t, \mathbf{x}, \mathbf{u}_F(t), \sigma_L(t, \mathbf{x}))$$
$$+ \frac{\partial V_F(t, \mathbf{x})}{\partial \mathbf{x}} f(\mathbf{x}(t), \mathbf{u}_F(t), \sigma_L(t, \mathbf{x})) \},$$

which has the terminal condition $V_F(T, \mathbf{x}) = S_F(\mathbf{x}(T))$.¿From the HJB equation, we can determine, at least in principle, the follower's best reply in the form of (another) function $\tilde{R}$

$$\sigma_F(\cdot) = \tilde{R}(\cdot; \sigma_L(\cdot, \cdot)). \tag{7.57}$$

The leader makes this calculation too, and selects a strategy $\sigma_L(t, \mathbf{x})$) that maximizes his objective functional, subject to (7.57) and the system dynamics. This problem, however, is a nonstandard optimal-control problem. It is not obvious how to obtain a solution, given that σ_L can be *any* decision rule. One option is to restrict the leader's strategy choice to the set of *linear, stationary strategies*:

$$\sigma_L(t, \mathbf{x}) \equiv \sigma_L(\mathbf{x}) = a + b\mathbf{x}, \tag{7.58}$$

in which a and $b = (b_1, \ldots, b_n)$ are constants. The leader's problem then is to choose the constants a and b optimally. The follower's best reply becomes $\mathbf{u}_F = \tilde{R}(t, \mathbf{x}; a, b)$.

The above approach constitutes a way out of a difficult problem and has been used in applications. However, it is conceptually debatable because it means that the leader is essentially using an open-loop strategy. The reason is that the leader at time zero determines, announces, and commits to using the constants a and b.

Notice that, so far, the leader has had a *global* first-mover advantage over the follower, in the sense that the leader selects and announces his strategy at time zero given by (7.58), or an open-loop strategy. The equilibrium discussed above is called a *global Stackelberg solution* (see [18]).

An alternative option is to give the leader only a *stagewise* first-mover advantage; in continuous time, this translates into an instantaneous advantage at each time t. Then one can define what Başar and Olsder [18] called a *feedback-Stackelberg equilibrium (FSE)*. The characterization of such an equilibrium for $j \in \{L, F\}$ involves the HJB equations

$$\rho_j V_j(t, \mathbf{x}) - \frac{\partial V_j(t, \mathbf{x})}{\partial t} = g_j(\mathbf{x}, [\sigma_F(t, \mathbf{x}), \sigma_L(t, , \mathbf{x})])$$

$$+ \frac{\partial V_j(t, \mathbf{x})}{\partial \mathbf{x}} f(\mathbf{x}, [\sigma_F(t, \mathbf{x}), \sigma_L(t, , \mathbf{x})], t) \qquad (7.59)$$

in which $[\sigma_F(t, \mathbf{x}), \sigma_L(t, , \mathbf{x})]$ is a feedback (or Markovian) strategy pair. The best reply of the follower is now a function $\hat{R}$, which is defined by

$$\hat{R}(t, \mathbf{x}; \sigma_L(t, \mathbf{x})) =$$

$$\arg \max_{\mathbf{u}_F \in U_F} \left\{ g_F(t, \mathbf{x}, [\mathbf{u}_F, \sigma_L(t, \mathbf{x})], t) + \frac{\partial V_F(t, \mathbf{x})}{\partial \mathbf{x}} f(\mathbf{x}, [\mathbf{u}_F, \sigma_L(t, \mathbf{x})], t) \right\}.$$

The equilibrium strategy pair is given by

$$\hat{\sigma}_L(t, \mathbf{x}) = \arg \max_{\mathbf{u}_L \in U_L} \left\{ g_L(\mathbf{x}), ([\hat{R}(t, \mathbf{x}; \mathbf{u}_L), \mathbf{u}_L], t) + \right.$$

$$\left. \frac{\partial V_L(t, \mathbf{x})}{\partial \mathbf{x}} f(x, [\hat{R}(t, \mathbf{x}; \mathbf{u}_L), \mathbf{u}_L], t)) \right\},$$

$$\hat{\sigma}_F(t, \mathbf{x}) = \hat{R}(t, \mathbf{x}, \hat{\sigma}_L(t, \mathbf{x})).$$

As expected, the main obstacle in the characterization of an FSE is in obtaining a solution to the partial differential equations in (7.59).

7.10   An example of the identification of Stackelberg equilibria

Consider the example in Section 7.7.1 but assume that Player 1 is the leader (from now on referred to as Player L) and Player 2 is the follower (Player F). Denote by u_j the control variable of Player j, $j = L, F$, and by x the state variable. Player j's optimization problem is given by

$$\max_{u_j \geq 0} \left\{ J_j = \int_0^\infty e^{-\rho t} \left(u_j(t) \left(\kappa - \frac{1}{2} u_j(t) \right) - \frac{1}{2} \varphi x^2(t) \right) dt \right\},$$

s.t.

$$\dot{x}(t) = u_L(t) + u_F(t) - \alpha x(t), \quad x(0) = x^0,$$

where φ and κ are positive parameters and $0 < \alpha < 1$. We omit the time argument in the rest of the example when no ambiguity may arise.

Open-loop Stackelberg strategies.   To obtain the best reply of the follower to the leader's announcement of the path $u_L(t)$, we introduce the Hamiltonian of Player F:

$$H_F(\lambda_F, x, u_L, u_F) = u_F \left(\kappa - \frac{1}{2} u_F \right) - \frac{1}{2} \varphi x^2 + \lambda_F(u_L + u_F - \alpha x),$$

where λ_F is the follower's costate variable associated with the state variable x. Assuming an interior solution, the first-order equilibrium conditions are

$$\frac{\partial H_F}{\partial u_F} = 0 \Leftrightarrow \mathbf{u}_F = \kappa + \lambda_F, \tag{7.60}$$

$$\dot{\lambda}_F = \rho\lambda_F - \frac{\partial H_F}{\partial x} = (\rho + \alpha)\lambda_F + \varphi x, \quad \lim_{t\to\infty} e^{-\rho t}\lambda_F(t) = 0, \tag{7.61}$$

$$\dot{x} = u_L + u_F - \alpha x, \quad x(0) = x^0. \tag{7.62}$$

The algorithm is to first solve the two differential equations,[20] and next, to substitute the results in optimality condition (7.60) to get follower's best reply, i.e., $u_F(t) = R(x(t), u_L(t), \lambda_F(t))$. Another option is to postpone the resolution of these differential equations and to substitute (7.60) for u_F in (7.61)–(7.62); this yields

$$\dot{\lambda}_F = (\rho + \alpha)\lambda_F + \varphi x, \quad \lim_{t\to\infty} e^{-\rho t}\lambda_F(t) = 0,$$

$$\dot{x} = u_L + \kappa + \lambda_F - \alpha x, \quad x(0) = x^0.$$

These two equations are then added, as state equations, to the leader's problem, i.e.,

$$\max_{u_L \geq 0}\left\{ J_L = \int_0^\infty e^{-\rho t}\left(u_L\left(\kappa - \frac{1}{2}u_L\right) - \frac{1}{2}\varphi x^2\right) dt\right\}$$

$$\dot{\lambda}_F = (\rho + \alpha)\lambda_F + \varphi x, \quad \lim_{t\to\infty} e^{-\rho t}\lambda_F(t) = 0,$$

$$\dot{x} = u_L + \kappa + \lambda_F - \alpha x, \quad x(0) = x^0.$$

This is an optimal-control problem with two state variables (λ_F and x) and one control variable (u_L). Introduce the leader's Hamiltonian:

$$H_L(\lambda_F, x, \lambda_L, u_L, \theta) = u_L\left(\kappa - \frac{1}{2}u_L\right) - \frac{1}{2}\varphi x^2 + \theta\left((\rho + \alpha)\lambda_F + \varphi x\right) +$$
$$\lambda_L\left(u_L + \kappa + \lambda_F - \alpha x\right),$$

where θ and λ_L are adjoint variables associated with the two state equations in the leader's optimization problem. Assuming an interior solution, the first-order optimality conditions are

$$\frac{\partial H_L}{\partial u_L} = 0 \Leftrightarrow u_L = \kappa + \lambda_L,$$

$$\dot{\theta} = \rho\theta - \frac{\partial H_L}{\partial \lambda_F} = -\theta\alpha - \lambda_L, \quad \theta(0) = 0,$$

$$\dot{\lambda}_L = \rho\lambda_L - \frac{\partial H_L}{\partial x} = (\rho + \alpha)\lambda_L + \varphi(x - \theta), \quad \lim_{t\to\infty} e^{-\rho t}\lambda_L(t) = 0,$$

$$\dot{x} = \frac{\partial H_L}{\partial \lambda_L} = u_L + \kappa + \lambda_F - \alpha x, \quad x(0) = x^0,$$

$$\dot{\lambda}_F = \frac{\partial H_L}{\partial \theta} = (\rho + \alpha)\lambda_F + \varphi x, \quad \lim_{t\to\infty} e^{-\rho t}\lambda_F(t) = 0.$$

---

[20] Here u_F is independent of x. This is not the case in general.

Substituting u_L in $\dot{x}$, we receive the following system of four differential equations, written in matrix form as follows:

$$\begin{pmatrix} \dot{\theta} \\ \dot{\lambda}_L \\ \dot{x} \\ \dot{\lambda}_F \end{pmatrix} = \begin{pmatrix} -\alpha & -1 & 0 & 0 \\ -\varphi & \rho + \alpha & \varphi & 0 \\ 0 & 1 & -\alpha & 1 \\ 0 & 0 & \varphi & \rho + \alpha \end{pmatrix} \begin{pmatrix} \theta \\ \lambda_L \\ x \\ \lambda_F \end{pmatrix} + \begin{pmatrix} 0 \\ 0 \\ 2\kappa \\ 0 \end{pmatrix}.$$

Solving the above system yields $(\theta, \lambda_L, x, \lambda_F)$. The last step is to insert the result in the equilibrium conditions

$$u_F = \kappa + \lambda_F, \quad u_L = \kappa + \lambda_L,$$

to get the Stackelberg equilibrium controls u_L and u_F.

Feedback-Stackelberg strategies.   To determine the follower's best reply, we introduce the HJB equation of Player F:

$$\rho V_F(x) = \max_{u_F \geq 0} \left[u_F \left(\kappa - \frac{1}{2} u_F \right) - \frac{1}{2} \varphi x^2 + V_F'(x) \left(u_L + u_F - \alpha x \right) \right], \qquad (7.63)$$

where $V_F(x)$ is the follower's value function. Maximization of the RHS yields:

$$u_F = \kappa + V_F'(x). \qquad (7.64)$$

Note that the above reaction function of the follower does not directly depend on the leader's control u_L, but only indirectly, through the state variable.

Accounting for the follower's response, the leader's HJB equation is

$$\rho V_L(x) = \max_{u_L \geq 0} \left[u_L \left(\kappa - \frac{1}{2} u_L \right) - \frac{1}{2} \varphi x^2 + V_L'(x) \left(u_L + \kappa + V_F'(x) - \alpha x \right) \right], \qquad (7.65)$$

where $V_L(x)$ denotes the leader's value function. Maximizing the RHS gives

$$u_L = \kappa + V_L'(x).$$

Substituting in (7.65) leads to

$$\rho V_L(x) = \left(\kappa + V_L'(x) \right) \left(\kappa - \frac{1}{2} \left(\kappa + V_L'(x) \right) \right) \qquad (7.66)$$

$$- \frac{1}{2} \varphi x^2 + V_L'(x) \left(V_L'(x) + V_F'(x) + 2\kappa - \alpha x \right).$$

As the game at hand is of the linear-quadratic variety, we can assume that the value functions are quadratic. Let

$$V_L(x) = \frac{A_L}{2} x^2 + B_L x + C_L, \qquad (7.67)$$

$$V_F(x) = \frac{A_F}{2} x^2 + B_F x + C_F, \qquad (7.68)$$

be the leader's and follower's value function, respectively. Substituting the value functions and their derivatives in (7.66) yields

$$\rho \left(\frac{A_L}{2} x^2 + B_L x + C_L \right) = \frac{1}{2} \left(A_L^2 - \varphi + 2 \left(A_F - \alpha \right) A_L \right) x^2 + \left(A_L \left(B_L + B_F + 2\kappa \right) \right.$$

$$\left. + \left(A_F - \alpha \right) B_L \right) x + \frac{1}{2} \left(\kappa^2 + B_L^2 \right) + \left(B_F + 2\kappa \right) B_L.$$

Using (7.63), (7.64) and (7.67)–(7.68), we get the following algebraic equation for the follower:

$$\rho \left(\frac{A_F}{2} x^2 + B_F x + C_F \right) = \frac{1}{2} \left(A_F^2 - \varphi + 2 \left(A_L - \alpha \right) A_F \right) x^2 + \left(A_F \left(B_F + B_L + 2\kappa \right) \right.$$

$$+ \left(A_L - \alpha \right) B_F \right) x + \frac{1}{2} \left(\kappa^2 + B_F^2 \right) + \left(B_L + 2\kappa \right) B_F.$$

By identification,[21] the coefficients of the leader's and the follower's value functions satisfy the following six-equation, non-linear algebraic system:

$$0 = A_L^2 + \left(2A_F - 2\alpha - \rho \right) A_L - \varphi,$$

$$0 = A_L \left(B_L + B_F + 2\kappa \right) + \left(A_F - \alpha - \rho \right) B_L,$$

$$0 = \frac{1}{2} \left(\kappa^2 + B_L^2 \right) + \left(B_F + 2\kappa \right) B_L - \rho C_L,$$

$$0 = A_F^2 - \varphi + \left(2A_L - 2\alpha - \rho \right) A_F,$$

$$0 = A_F \left(B_F + B_L + 2\kappa \right) + \left(A_L - \alpha - \rho \right) B_F,$$

$$0 = \frac{1}{2} \left(\kappa^2 + B_F^2 \right) + \left(B_L + 2\kappa \right) B_F - \rho C_F.$$

The above system admits multiple solutions. Conditions for, e.g., an interior equilibrium, convergence to an asymptotically globally stable steady state, can be invoked to disregard some of these solutions. Denote by $\left(A_L^S, B_L^S, C_L^S, A_F^S, B_F^S, C_F^S \right)$ a solution to the above system, where the superscript S stands for Stackelberg. A pair of feedback-Stackelberg equilibrium strategies is then given by

$$u_F = \kappa + V_F' \left(x \right) = A_F^S x + B_F^S,$$

$$u_L = \kappa + V_L' \left(x \right) = A_L^S x + B_L^S.$$

7.11   Time consistency of Stackelberg equilibria

When, at an initial instant of time, the leader announces the strategy he will use throughout the game, his goal is to influence the follower's strategy choice in a way that will be beneficial to himself. Time consistency deals with the following question: given the option to reoptimize at a later date, will the leader stick to his original plan, i.e., the announced time path for his control variables? If it is in his best interest to deviate, then the leader will do so and the equilibrium is said to be *time inconsistent*. An inherently related question is then why would the follower, who is a rational player, believe the announcement made by the leader at time zero? The answer is clearly that he would not.

In most Stackelberg differential games, it turns out that the OLSE is time inconsistent, that is, the leader's announced control path $u_L(\cdot)$ is not credible. Markovian or feedback-Stackelberg equilibria are subgame perfect and hence time consistent.

The OLSE in Section 7.10 is time inconsistent. To see this, suppose that the leader has the option of revising his plan at time $\tau > 0$ and to choose a new decision rule $u_L(\cdot)$ for

---

[21]This is another instance of the method of undetermined coefficients (see page 184).

the remaining time span $[\tau, \infty)$.. Then he will select a rule that satisfies $\theta(\tau) = 0$ (because this choice will fulfill the initial condition on the costate θ). It can be shown, by using the four state and costate equations $\left(\dot{x}, \dot{\lambda}_F, \dot{\lambda}_L, \dot{\theta}\right)$, that for some instant of time, $\tau > 0$, it will hold that $\theta(\tau) \neq 0$. Therefore, the leader will want to announce a new strategy at time τ and this makes the original strategy time inconsistent, i.e., the new strategy does not coincide with the restriction of the original strategy to the interval $[\tau, \infty)$.

We can formulate the following two remarks clarifying the usefulness of Stackelberg equilibria in the social sciences.

Remark 7.9. The important feature of the time consistency of Stackelberg feedback (Markovian) equilibria relies on the hypothesis that, at time t, players do **not** remember the history of the game.

Remark 7.10. Albeit time-inconsistent, the open-loop Stackelberg equilibrium can still be a useful solution concept for some short-term-horizon problems where it makes sense to assume that the leader will not be tempted to reoptimize at an intermediate instant of time. The GE at the end of this chapter examines a case where such an assumption is justified. Another case would be a game between a government (leader) and the business sector (follower). Here, the government can forgo the possibility of re-optimization for political or moral reasons. For example, even if a taxation policy $u_L(t)$, announced during the pre-election period, proves non-optimal for the government at some time in the inter-election period, the government may choose to continue implementing $u_L(t)$. This would be because the government would lose credibility among its electorate if it revoked its pre-election promise.

7.12 Memory strategies and collusive equilibria

7.12.1 *Engineering a subgame-perfect equilibrium*

We introduced the concept of memory strategies in dynamic games in Chapters 5 and 5.7.5; in particular, see Sections 5.4 and 6.9 and the GE illustration in Chapter 5.7.5. In this section, we extend this concept to differential games.

By memory strategies we mean that the players can, at any instant of time, recall any specific past information. The motivation for using memory strategies in differential games is the same impetus behind using history-dependent strategies in repeated games, namely, reaching a desirable outcome that is not obtainable noncooperatively.[22] Loosely speaking, this requires that the players *agree* (implicitly, or without taking on any binding agreement) on a desired trajectory to follow throughout the game (typically a cooperative solution),

---

[22]If cooperation is allowed and binding agreements are possible, then players can adopt the Nash bargaining solution (NBS) and select a Pareto-efficient point. Haurie, in references [103, 104] showed that this is hardly acceptable in dynamic games. The reason is that the NBS depends on initial data and it may well happen that, at some intermediate instant of time, one player is better off (in terms of payoff-to-go) switching to a noncooperative strategy. The use of memory strategies can be seen as an alternative approach to implementing a cooperative agreement without requiring the strong assumption of binding commitments.

and are willing to implement a punishment strategy if a deviation is observed. This is what was obtained in the game solved in the $\mathbb{GE}$ illustration in Chapter 5.7.5. As long as the workers and capitalists keep their promises (regarding wages and consumption), the economy follows a Pareto-efficient growth path. If one party realizes, or remembers, that, in a previous period, the other party deviated from an agreed-upon strategy, it implements some pre-calculated punishment. Out of the fear of punishment, the players adhere to the Pareto-efficient path, which would be unobtainable in a strictly non-cooperative game.

A punishment is conceptually and practically attractive only if it is *effective*, i.e., it deprives a player of the benefits of a defection, and *credible*, i.e., it is in the best interest of the player(s) who did not defect to implement this punishment. We first define the concept of non-Markovian strategies and the resulting Nash equilibrium, and next illustrate these concepts through a simple example.

Consider a two-player infinite-horizon differential game. The state equation is

$$\dot{\mathbf{x}}(t) = f\left(\mathbf{x}(t), \mathbf{u}_1(t), \mathbf{u}_2(t), t\right), \quad \mathbf{x}(0) = \mathbf{x}^0.$$

To a pair of controls $(\mathbf{u}_1(t), \mathbf{u}_2(t))$, there corresponds a unique trajectory $\mathbf{x}(\cdot)$ emanating from $\mathbf{x}^0$. Player j's payoff is given by

$$J_j \triangleq \int_0^\infty e^{-\rho_j t} g_j(\mathbf{x}(t), \mathbf{u}_1(t), \mathbf{u}_2(t), t)\, dt,$$

where $g_j(\mathbf{x}(t), \mathbf{u}_1(t), \mathbf{u}_2(t), t)$ is assumed to be bounded and continuously differentiable. (This assumption allows us to use the strong-optimality concept and avoid introducing additional technicalities.) As before, the control set of Player j is U_j and the state set is identical to $\mathbb{R}^n$.

Up to now, a strategy has been defined as a mapping from player's information space to his control set. Unfortunately, this direct approach poses formidable mathematical difficulties in our context; therefore, we will define a strategy as an infinite sequence of approximate constructions, called δ-strategies. For Player j, consider the sequence of times $t_i = i\delta, i = 0, 1, \ldots$, where δ is a fixed positive number. For any time interval $[t_i, t_{i+1})$ call $\mathcal{U}_j^i$ the set of measurable control functions $\mathbf{u}_{j,i} : [t_i, t_{i+1}) \to U_j$, and let $\mathcal{U}^i = \mathcal{U}_1^i \times \mathcal{U}_2^i$. A δ-strategy for Player j is a sequence $\Delta_j^\delta = (\Delta_{j,i})_{i=0,1,\ldots}$ of mappings

$$\Delta_{j,0} \in \mathcal{U}_j^0,$$
$$\Delta_{j,i} = \mathcal{U}^0 \times \mathcal{U}^1 \times \ldots \times \mathcal{U}^{i-1} \to \mathcal{U}_i^j \text{ for } j = 1, 2, \ldots.$$

A *strategy for Player* j is an infinite sequence of δ-strategies where $\delta \to 0$

$$\Delta_j = \left\{ \Delta_j^{\delta_n} : \delta_n \to 0, n = 1, 2, \ldots \right\}.$$

Note that this definition implies that the information set of Player j at time t is

$$\{(\mathbf{u}_1(s), \mathbf{u}_2(s)), 0 \le s < t\},$$

that is, the whole control history up to time t. So when players choose δ-*strategies*, they are using, at successive sample times t_i, the accumulated information to generate a pair of measurable controls $\left(\mathbf{u}_1^\delta(\cdot), \mathbf{u}_2^{\delta'}(\cdot)\right)$ which, in turn, generate a unique trajectory $x^{\bar\delta}(\cdot)$ and thus, a unique outcome $w^{\bar\delta} = \left(w_1^{\bar\delta}, w_2^{\bar\delta}\right) \in \mathbb{R}^2$, where $\bar\delta = (\delta, \delta')$ and

$$w_j^{\bar\delta} = \int_0^\infty e^{-\rho_j t} g_j(\mathbf{x}^{\bar\delta}(t), \mathbf{u}_1^\delta(t), \mathbf{u}_2^{\delta'}(t), t)\, dt.$$

An *outcome* of the strategy pair $\bar\Delta$ is a pair $\bar w \in \mathbb{R}^2$, which is a *limit* of the sequence $\left\{w^{\bar\delta_n}\right\}$ of the outcomes of δ-strategy pairs $\bar\Delta^{\bar\delta_n} = \left(\Delta_1^{\delta_n}, \Delta_2^{\delta'_n}\right)$ when n tends to infinity. With a strategy pair, the initial state and time are thus associated to a set $v\left(t^0, \mathbf{x}^0; \bar\Delta\right)$ of possible outcomes. (Note that we have used the obvious extension to a non-zero initial time t^0.) The game is well defined if, for any strategy pair $\bar\Delta$ and any initial conditions $\left(t^0, x^0\right)$, the set of outcomes $v\left(t^0, \mathbf{x}^0; \bar\Delta\right)$ is nonempty.

Definition 7.4. A strategy pair $\bar\Delta^*$ is a Nash equilibrium at $\left(t^0, \mathbf{x}^0\right)$ iff

(1) the outcome set $v\left(t^0, \mathbf{x}^0; \bar\Delta\right)$ reduces to a singleton $w^* = (w_1^*, w_2^*)$;
(2) for all strategy pairs $\bar\Delta^{(1)} \triangleq (\Delta_1, \Delta_2^*)$ and $\bar\Delta^{(2)} \triangleq (\Delta_1^*, \Delta_2)$, the following holds for $j = 1, 2$:

$$(w_1, w_2) \in v\left(t^0, \mathbf{x}^0; \bar\Delta^{(j)}\right) \Rightarrow w_j \leq w_j^*.$$

The equilibrium condition for the strategy pair is valid only at $\left(t^0, \mathbf{x}^0\right)$. This implies, in general, that the Nash equilibrium that was just defined is not subgame perfect.

Definition 7.5. A strategy pair $\bar\Delta^*$ is a subgame-perfect Nash equilibrium at $\left(t^0, \mathbf{x}^0\right)$ iff

(1) given a control pair $\bar u(\cdot) : [t^0, t) \to U_1 \times U_2$ and the state $\mathbf{x}(t)$ reached at time t, we define the prolongation of $\bar\Delta^*$ at $(t, \mathbf{x}(t))$ as $\left\{\Delta'^{*\delta_n} : \delta_n \to 0, n = 1, 2, \dots\right\}$ defined by

$$\Delta'^{\delta_n}\left(\bar{\mathbf{u}}_{[t,t+\delta_n]}, \dots, \bar{\mathbf{u}}_{[t+i\delta_n, t+(i+1)\delta_n]}\right)$$
$$= \Delta^{\delta_n}\left(\bar{\mathbf{u}}_{[0,\delta_n]}, \bar{\mathbf{u}}_{[\delta_n, 2\delta_n]} \dots, \bar{\mathbf{u}}_{[t+i\delta_n, t+(i+1)\delta_n]}\right), \quad j = 1, 2;$$

(2) the prolongation of $\bar\Delta^*$ at $(t, x(t))$ is again an equilibrium at $(t, \mathbf{x}(t))$.

Before providing an illustrative example, let us formulate the following remark.

Remark 7.11. Notice the following:

(1) The information set was defined as the whole control history. An alternative definition is $\{x(s), 0 \leq s < t\}$, that is, each player bases his decision on the whole state trajectory. Clearly, this definition requires less memory capacity and hence may be an attractive option when the differential game involves more than two players. (See Tolwinski et al. [230] for details.)

(2) The consideration of memory strategies in differential games can be traced back to Varaiya and Lin [236], A. Friedman [85] and Krassovski and Subbotin [138]. Their setting was (mainly) zero-sum differential games, and they used memory strategies as a convenient tool for proving the existence of a solution. The exposition above follows Tolwinski et al. [230] and Haurie and Pohjola [115], where the setting is non-zero-sum differential games and the focus is on the construction of cooperative equilibria.

7.12.2 *An example*

Consider a two-player differential game where the state equation is given by

$$\dot{x}(t) = (1 - u_1(t)) u_2(t), \quad x(0) = x^0 > 0, \tag{7.69}$$

where $0 < u_j(t) < 1$. The players strive to maximize the following objective functionals:

$$J_1 \triangleq \alpha \int_0^\infty e^{-\rho t} \left(\ln u_1(t) + x(t) \right) dt,$$

$$J_2 \triangleq (1 - \alpha) \int_0^\infty e^{-\rho t} \left(\ln (1 - u_1(t)) (1 - u_2(t)) + x(t) \right) dt,$$

where $0 < \alpha < 1$ and $0 < \rho \le 1/4$.

Suppose that the two players wish to implement a cooperative solution noncooperatively by using non-Markovian strategies and threats.

Step 1: Determine cooperative outcomes. Assume that these outcomes are given by the joint maximization of the players' payoffs. To solve this optimal-control problem, we introduce the current-value Hamiltonian (we skip the time argument):

$$H(\lambda, x, u_1, u_2) = \alpha \ln u_1 + (1 - \alpha) \ln (1 - u_1)(1 - u_2) + x + \lambda (1 - u_1) u_2,$$

where λ is the adjoint variable associated with the state equation (7.69). Necessary and sufficient optimality conditions are

$$\dot{x} = (1 - u_1) u_2, \quad x(0) = x^0 > 0,$$

$$\dot{\lambda} = \rho \lambda - 1, \quad \lim_{t \to \infty} e^{-\rho t} \lambda = 0,$$

$$\frac{\partial H}{\partial u_1} = \frac{\alpha}{u_1} - \frac{(1 - \alpha)}{(1 - u_1)} - \lambda u_2 = 0,$$

$$\frac{\partial H}{\partial u_2} = -\frac{(1 - \alpha)}{(1 - u_2)} - \lambda (1 - u_1) = 0.$$

It is easy to verify that the unique optimal solution is given by

$$(u_1^*, u_2^*) = \left(\alpha \rho, \frac{1 - \rho}{1 - \alpha \rho} \right), \quad x^* = x^0 + (1 - \rho) t,$$

$$J_1 \left(\mathbf{x}^0; u_1^*(\cdot), u_2^*(\cdot) \right) = \frac{\alpha}{\rho} \left(\ln \alpha \rho + x^0 + \frac{1 - \rho}{\rho} \right),$$

$$J_2 \left(\mathbf{x}^0; u_1^*(\cdot), u_2^*(\cdot) \right) = \frac{1 - \alpha}{\rho} \left(\ln (1 - \alpha) \rho + x^0 + \frac{1 - \rho}{\rho} \right).$$

Note that both optimal controls satisfy the constraint $0 < u_j(t) < 1$.

Step 2: Compute Nash-equilibrium outcomes. As the game is of the linear-state variety, open-loop and feedback-Nash equilibria coincide.[23] To determine this equilibrium, let the players' Hamiltonians be given by

$$H_1 (\lambda_1, x, u_1, u_2) = \alpha (\ln u_1 + x) + \lambda_1 (1 - u_1) u_2,$$
$$H_2 (\lambda_2, x, u_1, u_2) = (1 - \alpha) (\ln (1 - u_1) (1 - \mathbf{u}_2) + x) + \lambda_2 (1 - u_1) u_2,$$

where λ_j is the costate variable associated by Player j to the state equation (7.69). The necessary conditions for a Nash equilibrium are

$$\dot{x} = (1 - u_1) u_2, \quad x(0) = x^0 > 0,$$
$$\dot{\lambda}_1 = \rho \lambda_1 - \alpha, \quad \lim_{t \to \infty} e^{-\rho t} \lambda_1 = 0,$$
$$\dot{\lambda}_2 = \rho \lambda_2 - (1 - \alpha), \quad \lim_{t \to \infty} e^{-\rho t} \lambda_2 = 0,$$
$$\frac{\partial H_1}{\partial u_1} = \frac{\alpha}{u_1} - \lambda_1 u_2 = 0,$$
$$\frac{\partial H_2}{\partial u_2} = -\frac{(1 - \alpha)}{(1 - u_2)} - \lambda_2 (1 - u_1) = 0.$$

It is easy to check that the unique Nash equilibrium is given by

$$(\bar{u}_1, \bar{u}_2) = \left(\frac{1-k}{2}, \frac{1+k}{2} \right), \quad \bar{x} = x^0 + \left(\frac{1+k}{4} \right) t,$$

with payoffs

$$J_1 = \frac{\alpha}{\rho} \left(\ln \left(\frac{1-k}{2} \right) + x^0 + \frac{1+k}{2\rho} - 1 \right),$$
$$J_2 = \frac{1-\alpha}{\rho} \left(\ln \rho + x^0 + \frac{1+k}{2\rho} - 1 \right),$$

where $k = \sqrt{1 - 4\rho}$. Observe that the equilibrium controls satisfy the $0 < u_j(t) < 1$, and, as expected in view of the game structure, they are constant over time.

Step 3: Construct a collusive equilibrium.[24] Using the above notation, we have so far obtained

$$(w_1^*, w_2^*) = \left(J_1 \left(t^0, x^0; u_1^*(\cdot), u_2^*(\cdot) \right), J_2 \left(t^0, x^0; u_1^*(\cdot), u_2^*(\cdot) \right) \right),$$
$$(\bar{w}_1, \bar{w}_2) = \left(J_1 \left(t^0, x^0; \bar{u}_1 (\cdot), \bar{u}_2(\cdot) \right), J_2 \left(t^0, x^0; \bar{u}_1 (\cdot), \bar{u}_2(\cdot) \right) \right).$$

Compute the differences

$$w_1^* - \bar{w}_1 = \frac{\alpha}{\rho} \left(\ln \left(\frac{1-k}{2\alpha\rho} \right) + \frac{1 + \rho^2 - 3\rho + k}{2\rho} \right),$$
$$w_2^* - \bar{w}_2 = \frac{1-\alpha}{\rho} \left(\ln (1 - \alpha) + \frac{1 - 3\rho + k}{2\rho} \right).$$

---

[23] We use the maximum principle, which is an easier procedure than the one based on the solutions to the Hamilton-Jacobi-Bellman equations.

[24] Recall Definition 5.2 of a "collusive equilibrium", on page 144.

Note that these differences are independent of the initial state x^0 and that their signs depend on the parameter values. For instance, if the following restriction on the parameter values is satisfied,

$$0 < \alpha < \min \left(\frac{1-k}{2\rho \exp\left(\frac{3\rho-1-\rho^2-k}{2\rho}\right)}, 1 - \exp\left(\frac{3\rho-1+k}{2\rho}\right) \right),$$

then $w_1^* > \bar{w}_1$ and $w_2^* > \bar{w}_2$. Suppose that this is true. What remains to be shown is that by combining the cooperative (Pareto-optimal) controls with the feedback Nash strategy pair,

$$(\sigma_1(x), \sigma_2(x)) = (\bar{u}_1, \bar{u}_2) = \left(\frac{1-k}{2}, \frac{1+k}{2} \right),$$

we can construct a subgame-perfect equilibrium strategy in the sense of Definition 7.5.

Consider a strategy pair

$$\bar{\Delta}_j = \left\{ \bar{\Delta}_j^{*\delta_n} : \delta_n \to 0, n = 1, 2, \dots \right\},$$

where, for $j = 1, 2, \bar{\Delta}_j^{*\delta}$ is defined as follows:

$$\Delta_j^{*\delta} = (\Delta_{j,i})_{i=0,1,2,\dots},$$

with

$$\Delta_{j,0}^* = u_{j,0}^*(\cdot),$$

$$\Delta_{j,i}^* = \begin{cases} u_{j,i}^*(\cdot), & \text{if } \bar{u}(s) = \bar{u}^*(s) \text{ for almost } s \leq i\delta, \\ \varphi_j(x(j\delta)) = \bar{u}_j, & \text{otherwise,} \end{cases}$$

for $i = 1, 2, \dots$, where $u_{j,i}^*(\cdot)$ denotes the restriction of $u_{j}^*(\cdot)$ to the subinterval $[i\delta, (i+1)\delta]$, $i = 0, 1, 2, \dots$, and $x(i\delta)$ denotes the state observed at time $t = i\delta$.

The strategy just defined is known as a *trigger strategy*. A statement of the trigger strategy, as it would be made by a player, is this: "At time t, I implement my part of the optimal solution if the other player has never cheated up to now. If he cheats at t, then I will retaliate by playing the feedback-Nash strategy from t onward." It is left up to the reader, as an exercise, to show that this trigger strategy is a subgame-perfect equilibrium.

We conclude this section with two observations. First, as in repeated games, it is also possible in a differential-game setting to define a retaliation period of finite length, following a deviation. Actually, the duration of this period can be designed to discourage any player from actually defecting. Second, in the above development and example, we assumed that a deviation is instantaneously detected. This is not necessarily the case, and therefore, we can consider a detection lag. For an example of a trigger strategy with a finite retaliation period and detection lag, see Hämäläinen et al. [100].

7.13 Remarks on zero-sum and pursuit-evasion games

As was the case of the classical theory of games, the first breakthrough results in differential games concerned zero-sum games and, more specifically, *pursuit-evasion* games. In such games, one player (the *pursuer*) attempts to catch the other player (the *evader*). In military applications, the players could be jet fighters where their position and velocity are state variables and thrust (acceleration) and direction are control variables. The founder of pursuit-evasion games, and by the same token the initiator of the theory of differential games, was Rufus Isaacs.[25]

As was mentioned in the introduction to this chapter, Isaacs published in the 1950s and 1960s a series of memoranda on pursuit-evasion differential games while he was working at the Rand Corporation; he also continued this line of research after he left the corporation. His contribution is commemorated in the *Isaacs equation*, which is an extension of the Hamilton-Jacobi equation known from calculus of variations. The Isaacs equation provides a sufficient condition for a solution to the pursuit-evasion problem.

Leonard D. Berkovitz and Wendell H. Fleming, who worked at Rand in the 1950s (i.e., during the same period as Isaacs) and Lev Pontryagin, from the former USSR, are also considered great contributors to the foundations of pursuit-evasion games.[26] Over the last few decades, many other researchers have contributed to the mathematical developments of the theory of pursuit-evasion games and to the numerical methods for their solution.[27] Their contributions can be easily found on the Internet and will not be included here, mainly because confirming their relevance for the dynamic games discussed in this book, surpasses its scope.

In a pursuit-evasion game, the players do not have (in general) a quantitative performance index (see Section 6.1.4 on page 162) to be optimized, so these games are called *qualitative games* or also *games of kind*. In broad terms, in a game of kind the evader is looking for a set of positions that will enable him to escape (i.e., reach a safe region) and the pursuer is searching for a set of positions that will allow him to capture the evader in finite time. Quite naturally in this context, the players compare the different strategies, which they may use, in terms of their qualitative merits, i.e., fail or succeed to escape for the evader, and capture or fail to capture for the pursuer. For this reason, these games are also referred to as *qualitative differential games*. Games where strategies are compared in terms of their corresponding quantitative rewards are named, unsurprisingly, *quantitative differential games*. Actually, this difference was pointed out by Blaquière, Gérard and Leitmann in the title of their book *Quantitative and Qualitative Games* [30], which is an early and very influential book on differential games.

---

[25]This is acknowledged by the International Society of Dynamic Games whose highest award for a contribution to the field is named the Isaacs Award. Since 2004, two scientists receive this award at the symposium of the Society held every even year.

[26]During the period when Isaacs was developing zero-sum differential games, important advances took place in dynamic optimization. In particular, the Soviet school led by Pontryagin was developing optimal control methods while Richard Bellman from Rand was setting foundations for dynamic programming. We mention these developments because zero-sum differential games, optimal control and dynamic programming are related areas of applied mathematics. Who inspired whom during that burgeoning period is a matter of debate. The early history of differential games is told by Michael Breitner in a well documented paper [32] published in the *Journal of Optimization Theory and Applications*, an outlet that played an important historical role in disseminating new results in differential games.

[27]Among the scientists of the first and second generations, we can mention Austin Blaquière, Pierre Bernhard, John Breakwell, Nicolai Krasovskii, George Leitmann, Arik Melikyan, Valerii S. Patsko, Leon Petrosyan, Josef Shinar and Andrei Subbotin.

As an interesting observation, we note a proximity between qualitative differential games and viability theory, initiated by Jean-Pierre Aubin (see [8] and [9]), where the objective is to determine the set of initial states (called the viability kernel), such that controlled trajectories emanating from these points remain in a desirable set or reach a target (if defined), whatever be the disturbances affecting the dynamic system .

Quantitative two-player zero-sum differential games have proven to be a useful formalism to deal with many problems in economics, management science and engineering. One important class of such games that needs be mentioned here is the class of games against nature. As the name suggests, these games involve, as protagonists, a rational decision maker (controller or player) having a quantitative objective functional and a feasible strategy set and nature, whose behavior can be described in several ways. In one formulation, the player aims at optimizing his performance index against the worst possible choice of strategy by nature. This is robust-control approach that is, clearly, relevant for numerous applications in finance, production systems and networks, to name only a few. We skipped dealing in this chapter with such zero-sum differential games, amenable to an analysis by robust-control and H-infinity methods, because this topic is expertly covered in the book by Başar and Bernhard [17].

7.14 What have we learned in this chapter?

In this chapter, we have studied dynamic conflict models that run in continuous time. In particular, we have learned

- to distinguish between different information structures and how to define a differential game in normal form;
- how to identify open-loop Nash equilibria using the maximum principle;
- how to identify Markovian (feedback)-Nash equilibria also using the maximum principle and how these equilibria differ, among other things, in terms of the adjoint variables;
- how to identify Markovian (feedback)-Nash equilibria using the HJB equations;
- how to identify open-loop and feedback-Stackelberg equilibria;
- about the sufficient transversality conditions for the infinite-horizon case, and that these conditions vary with the definition of optimality;
- that equilibria in linear-quadratic and linear-state games are relatively easy to compute and that this class of games is widely used in practice;
- that even in infinite-horizon linear-quadratic differential games, the feedback-Nash equilibrium may not be unique;
- that open-loop and feedback-Nash equilibria coincide in differential games of the state-redundant variety;
- to distinguish between two important concepts in dynamic settings, namely, time consistency and subgame perfectness;
- that an open-loop Nash equilibrium is time consistent, whereas an open-loop Stackel-

berg equilibrium is, in general, time inconsistent;
- how to define non-Markovian memory strategies in differential games;
- how the players can use these memory strategies and threats to reach a cooperative outcome;
- how a deterministic differential game can be used to provide useful qualitative insights into a realistic situation as in the GE below.

7.15 Exercises

Exercise 7.1. Denote by u_j the control variable of Player j, $j = 1, 2$ and by x the state variable. Let Player j's optimization problem be given by

$$\max_{u_j \geq 0} \left\{ J_j = \int_0^\infty e^{-\rho t} \left(u_j(t) \left(\beta_j - \frac{1}{2} u_j(t) \right) - \frac{1}{2} \varphi_j x^2(t) \right) dt \right\},$$

s.t.

$$\dot{x}(t) = u_1(t) + \mathbf{u}_2(t) - \alpha x(t),$$
$$x(t^0) = \mathbf{x}^0,$$

where φ_j and β_j are positive parameters and $0 < \alpha < 1$.

Determine the open-loop and feedback-Nash equilibria of this game. (This is an asymmetric version of the example worked out in the chapter).

Exercise 7.2. Leitmann-Schmitendorf advertising model [153]. Denote by u_j the advertising rate (control variable) of Player j, $j = 1, 2$ and by x_j the sales rate of Player j. The sales rate evolves according to

$$\dot{x}_j(t) = u_j(t) - \frac{c_j}{2} u_j^2(t) - k_j u_j(t) x_j(t) - \delta_j x_j(t) \tag{7.70}$$

where c_j, δ_j, and k_j are positive parameters. Player j's objective is to maximize his total profit over the planning horizon $[0, T]$, given by

$$J_j(u_j) = \int_0^T (p_j x_j(t) - u_j(t)) \, dt \tag{7.71}$$

where $p_j > 0$ is a constant price per unit sold.

(1) Determine an open-loop equilibrium. Is this equilibrium unique?
(2) Show that this game is state redundant.

Exercise 7.3. Consider a two-player finite-horizon differential game where u_j denotes the control of Player j, and x the state variable. The objective functional of Player j is given by

$$J_j = \int_0^T \left(u_j(t) - \frac{u_j^2(t)}{2} - x(t) \right) dt.$$

The state evolves according to the following differential equation:

$$\dot{x}(t) = u_1(t) + u_2(t) - \delta x(t), \quad x(0) = x^0,$$

where δ is a parameter satisfying $0 < \delta < 1$.

(1) Suppose that Player 1 is the leader and Player 2 is the follower. Determine an open-loop Stackelberg equilibrium.
(2) Is this equilibrium time consistent?

Exercise 7.4. Consider a supply chain formed by one manufacturer, player M, and one retailer, player R. The manufacturer controls his margin $m_M(t)$ and rate of advertising $a_M(t)$ at time $t \in [0, \infty)$. The retailer controls his margin $m_R(t)$.. The retail price is the sum of the margins, i.e.,

$$p(t) = m_M(t) + m_R(t).$$

Denote by $G(t)$ the brand goodwill at t. The evolution of this state variable is governed by the following differential equation:

$$\dot{G}(t) = a_M(t) - \delta G(t), \quad G(0) = G_0 > 0,$$

where δ is the decay rate $(0 < \delta < 1)$.

The rate of consumer demand $q(t)$ depends on the retail price $p(t)$ and on the brand goodwill $G(t)$. The following functional form is assumed:

$$q(t) = G(t) - \beta p(t)),$$

where β is a positive parameter.

The manufacturer's advertising cost is given by $\frac{1}{2}wa_M^2(t)$ where w is a positive parameter. The players are assumed to maximize their profits given by

$$\pi_M = \int_0^\infty e^{-\rho t}\left[m_M(t)(G(t) - \beta p(t)) - \frac{1}{2}wa_M^2(t)\right]dt,$$

$$\pi_R = \int_0^\infty e^{-\rho t}m_R(t)(G(t) - \beta p(t))\,dt,$$

where ρ is the discount rate $(0 < \rho < 1)$.

(1) Determine MNE strategies (i.e., advertising and margins strategies) and the steady-state value of the brand goodwill.
(2) Determine OLNE strategies (i.e., advertising and margins strategies) and the steady-state value of the brand goodwill.
(3) Compare the MNE and OLNE steady-state values. Interpret the result.
(4) Suppose that the manufacturer is the leader and the retailer is the follower. Determine OLSE strategies. Is this equilibrium time consistent?
(5) Suppose now that the manufacturer and the retailer vertically integrate and maximize the sum of their profits. Determine the joint-optimization solution and compare it to the equilibria identified above. Will the consumer benefit from this merger by paying a lower price?

Exercise 7.5. Considering the example in Section 7.12.2,

(1) Redo the analysis assuming that the Pareto solution is obtained by maximizing a weighted sum of the players' objectives, i.e., $\beta J_1 + (1 - \beta)J_2$, with $0 < \beta < 1$.

(2) Show that the proposed trigger strategy is part of a subgame-perfect Nash equilibrium.

Exercise 7.6. Assume that the market share $x_i(t)$ of firm i evolves according to the dynamics

$$\dot{x}_i(t) = -\left[\alpha_i - \beta_i p_i(t)\right] x_i(t) + \left[\alpha_j - \beta_j p_j(t)\right] x_j(t), \quad i, j =\in \{1, 2\}, i \neq j,$$

where $p_i(t)$ is the price charged by firm i and α_i, β_i are positive parameters. Firm i wishes to maximize the payoff

$$J_i = \int_0^\infty e^{-\rho t} \left[p_i(t) - c_i\right] \dot{x}_i(t) dt.$$

(1) Interpret the state dynamics.
(2) Reformulate the model to have a single state variable and characterize its OLNE pricing strategies.

Exercise 7.7. The model is a two-player version of the game in Feichtinger and Dockner [76]. Consider a duopoly where the sales rate of firm i is given by

$$\dot{x}_i(t) = \left[\alpha_i - \beta_i p_i(t) + \zeta(p_j(t) - p_i(t))\right] \left[a_i + b_i x_i(t)\right], \quad i, j \in \{1, 2\}, i \neq j$$

where $p_i(t)$ is the price charged by firm i and $\alpha, \beta, \zeta, a_i$, and b_i are positive parameters. Suppose that the unit production cost of firm i is constant, equal to c_i, and assume that the firm maximizes its discounted stream of profits, given by

$$J_i = \int_0^T e^{-\rho t} \left[p_i(t) - c_i\right] \dot{x}_i(t) dt.$$

(1) Compute open-loop Nash equilibrium pricing strategies and characterize their evolution over time.
(2) Assume that firm 1 is the leader and firm 2 the follower in a Stackelberg game. Compute open-loop Stackelberg equilibrium pricing strategies.
(3) Is the open-loop Stackelberg equilibrium time consistent?

Exercise 7.8. This exercise is based on Buratto and Zaccour [38]. Consider a licensing arrangement in which the owner of a famous fashion brand (the licensor, e.g., *Roberto Cavalli*) contracts to another firm (the licensee, e.g., *Dressing s.p.a.*) the rights to produce and market its second-line brand (e.g., *CLASS*) against monetary compensation. Let L refer to the *L*icensor, and l to the *l*icensee. The duration of a licensing contract in the fashion industry is typically long, and may actually last for decades. However, our focus here is on the strategic interactions between the two entities in terms of advertising campaigns, which have to be planned for each selling season. Let this season be given by the time interval $[0, T], T > 0$.

Denote by S the cumulative sales revenues of the licensed brand by $t \in [0, T]$. Let $\dot{S}(t) = \frac{S(t)}{dt}$ be the rate of sales at $t \in [0, T]$. Denote by $a_i(t)$ the advertising expenditures of player $i \in \{L, l\}$, at $t \in [0, T]$. The advertising costs are given by :

$$K_i(a_i(t)) = \frac{k_i}{2} a_i^2(t), \quad k_i > 0, i \in \{L, l\}.$$

For granting the rights to produce and market its brand to the licensee, the licensor obtains a financial counterpart, i.e., a *royalty*. Denote by $r \in (0, 1)$ the share of the licensor in the sales revenues. The licensor collects, during a selling season, the total royalty given by

$$R_L = \int_0^T r \dot{S}_j(t) dt = r S_j(T),$$

and the licensee keeps

$$R_l = \int_0^T (1 - r) \dot{S}_j(t) dt = (1 - r) S_j(T).$$

We suppose that r is given exogenously, i.e., that it corresponds to a certain industry standard.

Denote by $G_L(t)$ the licensor's front-line brand goodwill, and by $G_l(t)$ its licensed second-line brand goodwill. We suppose that player L (resp. l) controls the advertising expenditures for brand L (resp. l). The evolution of the front-line brand goodwill is governed by the following linear differential equation,

$$\dot{G}_L(t) = \eta_L a_L(t) - \delta G_L(t), \quad G_L(0) = \bar{G}_L, \tag{7.72}$$

where $\delta > 0$ is the decay rate, $\bar{G}_L > 0$ is the initial goodwill's value, and $\eta_L > 0$ represents the efficiency of the licensor's advertising. The second-line brand goodwill also evolves according to the linear differential equation

$$\dot{G}_l(t) = \beta_l a_l(t) + \beta_L G_L(t) - \delta G_l(t), \quad G_l(0) = \bar{G}_l, \tag{7.73}$$

where $\delta > 0$ is again the decay rate; $\bar{G}_l > 0$ is the initial goodwill's value; $\beta_l > 0$ is the efficiency of the second-line brand advertising; and $\beta_L > 0$ represents the spillover from the front-line goodwill to the reputation of the second-line.

The rate of sales revenue depends linearly on the goodwill stocks, i.e.,

$$\dot{S}_B(t) = \alpha_L G_L(t) + \alpha_l G_l(t), \quad S_B(0) = 0, \tag{7.74}$$

where $\alpha_i > 0, i \in \{L, l\}$, is a scaling parameter transforming the corresponding goodwill stock into sales. The assumption that α_L is strictly positive reflects the idea that the reputation of the front-line brand matters, in terms of sales of the second-line brand.

The licensee's objective is to maximize his profit over the selling season, and hence, faces the following optimization problem:

$$\max_{a_l(t) \geq 0} \pi_l = \sigma_l G_l(T) + (1 - r) S_B(T) - \int_0^T \frac{k_l}{2} a_l^2(t) dt, \tag{7.75}$$

subject to (7.72), (7.73) and (7.74)

where $\sigma_l > 0$. The term $\sigma_l G_l(T)$ approximates, in a simple way, the future payoffs that the licensee can obtain starting from a goodwill stock $G_l(T)$.

Similarly, the licensor aims at maximizing its profit from the sales of the second-line brand, i.e., its share of sales revenues minus advertising expenditures, plus the value of the first-line brand at horizon date T. Again, we assume that this last item can be approximated by a linear function of $G_L(T)$. Formally, its optimization problem reads as follows:

$$\max_{a_L(t) \geq 0} \pi_L = \sigma_L G_L(T) + r S_B(T) - \int_0^T \frac{k_L}{2} a_L^2(t) dt, \qquad (7.76)$$

$$\text{subject to } (7.72), (7.73) \text{ and } (7.74).$$

where σ_L is a strictly positive parameter.

By (7.72)–(7.74) and (7.75)–(7.76), we have defined a two-player finite-horizon differential game, with two control variables, $a_i(t) \geq 0, i \in \{L, l\}$, and three state variables, $G_L(t), G_l(t)$ and $S_B(t)$.

(1) Determine an open-loop Nash equilibrium.
(2) Determine a feedback Nash equilibrium.
(3) Determine an open-loop Stackelberg equilibrium with the licensor as leader.
(4) Determine a feedback Stackelberg equilibrium with the licensor as leader.
(5) Compare the equilibrium strategies and outcomes obtained above.
(6) Why open-loop Stackelberg equilibrium is time consistent?

Exercise 7.9. This exercise is based on Amrouche et al. [6]. Consider a supply chain formed of a manufacturer (player M) of a national brand and a retailer (player R). The retailer offers the national manufacturer's brand (NB) as well as its own store brand (SB) or private label in its outlet. The SB is supplied to the retailer at a constant wholesale price w_s by another manufacturer who does not play any strategic role in this case. Denote by $p_n(t)$ the price-to-consumer of the national brand at time $t \in [0, \infty)$ and by $p_s(t)$ the retail price of the SB. Both prices are controlled by the retailer. Denote by $w_n(t)$ the manufacturer NB's wholesale price to the retailer.

Each player advertises its brand in order to increase its brand equity (or market potential). Let $A_i(t)$ represent the advertising investment at time t for brand $i, i \in \{n, s\}$, where n stands for national brand and s for store brand. The advertising cost $C_i(A_i)$ is assumed convex increasing and given by:

$$C_i(A_i) = c_{i1} A_i + \frac{c_{i2}}{2} A_i^2, \quad i \in \{n, s\},$$

where c_{i1} and c_{i2} are positive constants.

Denote by $G_i(t)$ brand i's market potential whose evolution is governed by the following differential equation:

$$\dot{G}_i(t) = A_i(t) - \delta G_i(t) - k_i A_j(t), \ G_i(0) = G_{i0} \geq 0, \ i, j \in \{n, s\}, i \neq j,$$

where $\delta > 0$ is the decay rate. The above specification modifies the classical Nerlove-Arrow dynamics by adding the term $-k_i A_j(t)$, where $0 \leq k_i \leq 1, i \in \{n, s\}$. This means that there are two sources of depreciation of brand i's equity; the standard one, given by $\delta G_i(t)$ which captures the idea that consumers forget to some extent the brand, and the one due to the advertising by the other brand. Thus, the parameter k_i represents the vulnerability of brand i to the advertising made by brand j.

Each brand demand depends on its equity and on both prices as follows:

$$D_i(t) = \alpha_i G_i(t) - p_i(t) + \psi_i p_j(t), \quad i, j \in \{n, s\}, i \neq j,$$

where $\psi_i, i \in \{n, s\}$, is a positive parameter capturing the cross-price effect on demand.

Assuming that both players are profit maximizers, their optimization problems read as follows:

$$\max_{w_n, A_n} \pi_M = \int_0^\infty e^{-\rho t} \{(w_n(t) - d_n)D_n(t) - C_n(A_n(t))\} \, dt,$$

$$\max_{A_s, p_n, p_s} \pi_R = \int_0^\infty e^{-\rho t} \{(p_n(t) - w_n(t))D_n(t) + (p_s(t) - w_s)D_s(t) - C_s(A_s(t))\} \, dt,$$

where ρ is the common discount rate and d_n is the constant unit production cost of the NB.

Determine feedback and open-loop Nash equilibrium strategies and outcomes.

Exercise 7.10. This exercise is based on Breton et al. [34]. Consider two firms producing one variety of differentiated but substitutable goods. Firms independently undertake cost-reduction research and development (R&D) and sell their products on the market. Let q_j be firm j's output. The market inverse demands are linear and given by

$$P_j(t) = A - \omega q_j(t) - \eta q_i(t), \quad i, j = 1, 2, i \neq j, \tag{7.77}$$

where η and ω are parameters satisfying $0 \leq \eta \leq \omega$. Their ratio $\frac{\eta}{\omega}$ represents the degree of product differentiation; the closer this term is to 1, the more substitutable the products are.

Let K_j be firm j's accumulated capital stock of R&D. The total production cost of firm i is supposed to depend on the quantity produced and, to take into account spillovers in knowledge, on both players' R&D stocks. The following quadratic form is adopted:

$$C_j(q_j(t), K_j(t), K_i(t)) = (c_j + \sigma q_j(t) - \psi(K_j(t) + \frac{\eta}{\omega}\beta K_i(t)))q_j(t),$$

$$i, j = 1, 2, i \neq j \tag{7.78}$$

where $c_j < A, 0 \leq \psi \leq 1, \sigma \geq 0$, and $0 \leq \beta \leq 1$ The degree of spillover of the process R&D is thus given by $\frac{\eta}{\omega}\beta$, where the spillover parameter β is assumed to be exogenous. Equation (7.78) is assumed to be strictly positive. Note that $C_j(\cdot)$ is convex increasing in q_j and linear decreasing in capital stocks.

Denote by I_j the investment of firm j in R&D. The resulting investment cost is assumed to be quadratic convex and increasing (in I_j):

$$F_j(I_j(t)) = \theta_j I_j(t) + \frac{\phi}{2}I_j^2(t), \tag{7.79}$$

where $\theta_j, \phi > 0$.

Firm j's capital stock evolves over time according to the following differential equation:

$$\dot{K}_j(t) = I_j(t) - \delta K_j(t), \quad K_j(0) = K_{j0}, \tag{7.80}$$

where $\delta, 0 \leq \delta < 1$ is the depreciation rate.

Assuming that each player maximizes his total discounted stream of profits over an infinite horizon, then player j's objective functional is given by

$$\pi_j = \int_0^\infty e^{-\rho t}[P_j(t)q_j(t) - C_j\left(q_j(t), K_j(t), K_i(t)\right) - F_j(I_j(t))]dt \qquad (7.81)$$

where $\rho, 0 < \rho < 1$ is the discount rate.

In addition to deciding on the level of its R&D investment, firm j chooses either output (in the Cournot game) or price (in the Bertrand game) levels to maximize π_j subject to the state dynamics given by (7.80). In differential game terminology, $K_j, i = 1, 2$, are the state variables, and I_i and q_j (or P_j) are the control variables.

(1) Suppose that competition is à la Cournot. Compute open-loop and Markovian Nash-equilibrium strategies (quantities and investments in R&D).
(2) Suppose that competition is à la Bertrand. Compute open-loop and Markovian Nash-equilibrium strategies (prices and investments in R&D).

Note: This exercise involves two state variables, and the characterization of equilibria requires heavy analytical work. One option is to give some values to the parameters and use software to derive the results.

Exercise 7.11. In the GE to follow, we made some simplifying assumptions which can be relaxed. In this exercise we ask you to assess how relaxing them impacts the results obtained in the GE.

(1) How would the equilibrium results be affected if the two players adopted a positive discount rate?
(2) How would the equilibrium results be affected if the price of the alternative technology was not constant but given by $p(t)$ with $p'(t) < 0$?
(3) How would the equilibrium results be affected if the government faced an inequality budget constraint $(y(T) \le B)$?
(4) How would the equilibrium results be affected if the unit-production cost was given by a strictly convex function?

7.16   GE illustration: Subsidies and guaranteed buys of a new technology

This GE is based on Jørgensen and Zaccour [129]. We consider a two-player differential game involving a firm selling a new technology with desirable features (e.g., more efficient and/or less polluting than an available alternative technology) and a government wishing to accelerate the consumer adoption of this technology. The main objective is to show how a differential-game formalism can be used to derive qualitative economic insights in a public-policy context.

Background: After the two oil crises of the 1970s, many developed countries' governments made a great deal of effort to make their economies less dependent on imported oil.

The promotion energy conservation and the development of oil alternatives ranked high on the agendas of concerned governmental agencies. Support programs were designed to help companies develop and market new less-oil-consuming technologies, and a series of measures incited consumers to adopt these technologies. More recently, governments have implemented similar programs, however with an environmental focus, e.g., subsidies for the production of electricity from renewable sources and subsidies to households to replace their old polluting cars by new less-polluting ones.

As pointed out by Kalish and Lilien [135], a government has a collection of options at its disposal to support the market penetration of new technologies. Examples of such options include (i) support of fundamental R&D to lower long-term production costs (of, for instance, renewable energies); (ii) demonstration programs and advertising to remove perceptual barriers related to the use of the new technology; (iii) price subsidies (or tax rebates) to consumers and firms to lower the cost of adoption; and (iv) guaranteed buys by the government. Taking photovoltaics (solar batteries) as an example, the US government established a program aimed at accelerating private market penetration. The program had three components, namely, the reduction of photovoltaic (PV)-system costs, the increase of consumer acceptance of PV by building working demonstration sites, and the creation of early awareness among potential customers by disseminating information. In Quebec, the government offered a $2 000 tax reduction to households converting to a dual-energy heating system during a given period of time.

In this GE, we consider a government that has a budget to support the diffusion of a new technology developed by a local firm. This budget can be used to subsidize households that buy this technology and to purchase, through a guaranteed-buys contract, some units to equip the government's own institutions. The government's aim in using *guaranteed buys* is threefold: (i) to lower the marketing and financial risks of the innovating company by ensuring a minimum market size; (ii) to accelerate the decline in the unit production cost through learning-by-doing, which might give the producer an incentive to charge lower consumer prices; and (iii) to achieve certain environmental objectives (e.g., reducing pollution emissions) and possibly other objectives (e.g., supporting a local company and stimulating employment).

Kalish and Lilien [135] were the first to approach the problem of government subsidies using an intertemporal marketing model. They studied governmental optimal-control problem but emphasized the need to develop a dynamic game model to handle the interactions between the government's subsidizing policies and the firms' pricing strategies. Zaccour [249] and Dockner et al. [60] studied the subsidy problem as a differential game, employing a model of new product diffusion. Both these papers deal with the firm's pricing decisions and the government's subsidy policy. Zaccour [249] identified Nash equilibria with open-loop strategies, whereas Dockner et al. [60] studied a case in which government acts as the Stackelberg leader, and found feedback as well as open-loop equilibria.

7.16.1   *The model*

We refer to the government as Player g and to the firm as Player f, and denote by B the total budget available to the government to support the market penetration of the new technology. Let T be the planning horizon corresponding to the duration of the government program. Denote by $p(t)$ the price charged by the firm per unit sold to households at time t, and by $s(t)$ the subsidy a household receives from the government. Therefore, it costs to consumer $p(t) - s(t)$ to buy one unit of the firm's product. Denote by $x_h(t)$ the cumulative sales to households, and by $x_g(t)$ the cumulative sales to the public sector by time $t \in [0, T]$.

Production cost.   It is well accepted that the unit-production cost of new durable products decreases with experience. This phenomenon is known as learning-by-doing. In the absence of inventories, experience in production is fairly well captured by cumulative sales. Denote by $C(x(t))$ the unit-production cost, where $x(t) = x_h(t) + x_g(t)$. with $C'(x(t)) \leq 0$. We assume that $C(t, x(t))$ can be approximated by

$$C(t, x(t)) = \max \left\{ 0, c^0 - cx(t) \right\},$$

where c^0 is the cost before any experience has been achieved and c is the learning-speed parameter. (Our assumption of an affine production-cost function is a simplifying one, to keep the model tractable. In general, $C(t, x(t))$ is most likely strictly convex.)

Demand.   Denote by p_a the constant price of the available alternative technology. We suppose that the household demand rate at time t is given by

$$\dot{x}_h(t) = \alpha - \beta \left(p(t) - s(t) - p_a \right), \quad x_h(0) = 0, \tag{7.82}$$

in which α and β are positive constants. Our hypothesis is that the household-demand rate depends on the difference between the (net) consumer price of the new technology and the price of the existing technology. The price level of the new technology is not important in of itself: a potential adopter considers the price differential between the two products and purchases the cheaper one.

The demand by the public sector is managed by a contract stipulating that the government pays to the firm a price of

$$p_g(t) = (1 + \mu) \left(c^0 - cx(t) \right),$$

where μ is a constant mark-up on the firm's unit-production cost. The demand rate $q(t) = \dot{x}_g(t)$ is restricted by $q_{\min} \leq q(t) \leq q_{\max}$. The lower bound, $q_{\min} > 0$, is the purchase rate that the government guarantees the firm. The upper bound corresponds to the maximum level at which government institutions can switch to the new technology.

Players' objectives.   We suppose that the firm maximizes its discounted stream of profits over the planning horizon. This is rather a classical assumption. It is not as easy, however, to choose an objective function for the government that is both conceptually acceptable and easy to implement. Loosely speaking, we will evaluate the effects of the government's subsidy program using two yardsticks: the new technology's consumer-price level, and the number of households that have adopted the new technology by the end of the government-subsidy program. Such objectives have the merit of being clear and quantifiable, which is in line with what the voters and policy makers want to see when it comes to investing public money.

Optimization problems. The players' optimization problems are:

$$\max J^f = \int_0^T \left(\left(p\left(t \right) - c^0 + cx\left(t \right) \right) \dot{x}_h\left(t \right) + \mu \left(c^0 - cx\left(t \right) \right) q\left(t \right) \right) dt,$$

$$\text{subject to (7.82) and } p\left(t \right) \geq 0,$$

$$\max J^g = x_h\left(T \right),$$

$$\text{subject to (7.82),}$$

$$\dot{x}_g\left(t \right) = q\left(t \right), \quad x_g\left(0 \right) = 0,$$

$$B = \int_0^T \left(p_g\left(t \right) q\left(t \right) + s\left(t \right) \dot{x}_h\left(t \right) \right) dt, \tag{7.83}$$

$$s\left(t \right) \geq 0, \quad q_{min} \leq q\left(t \right) \leq q_{max}.$$

where $B > 0$ the budget.

The government's budget constraint (7.83) can be written differently and equivalently by introducing a new state variable, $y\left(t \right)$, evolving according to the following differential equation:

$$\dot{y}\left(t \right) = p_g\left(t \right) q\left(t \right) + s\left(t \right) \dot{x}_h\left(t \right), \quad y\left(0 \right) = 0, \quad y\left(T \right) = B.$$

Note that we have assumed away discounting and that the government must use its total budget. Relaxing both assumptions does not cause many difficulties. (See Exercise 7.11.)

Information structure and equilibrium. As in [60], the game is played in the Stackelberg leader-follower fashion. We suppose that the government takes the leader's role and announces its strategy before the firm makes its decisions. The players employ open-loop strategies, which means that, at the initial instant of time, each player decides upon a strategy that depends only on time. An open-loop Stackelberg equilibrium only makes sense if the leader can credibly pre-commit to his strategy. In our context of a short-horizon game, it seems plausible to assume government pre-commitment: in practice, a subsidy scheme is determined and announced from the outset and once the government's decision is irrevocable, the announcement becomes credible; see Remark 7.10. However, there may, of course, be rare instances in which the government changes, or even abandons, a program after its initiation. (This could happen if, for some unforeseen reason, the very grounds of the program turn out to be unjustified.)

To wrap up, we have defined a two-player, finite-horizon differential game with three control variables $\left(p\left(t \right), s\left(t \right), q\left(t \right) \right)$, and three state variables $x_h\left(t \right)$, $x_g\left(t \right)$ and $y\left(t \right)$.

7.16.2 *An open-loop Stackelberg equilibrium*

The first step is to determine the follower's price $p(t)$ as the best reply to any feasible subsidy rate $s(t)$ and any feasible purchase quantity $q(t)$. The Hamiltonian function of the firm is given by

$$H^f = \left(p(t) - c^0 + cx(t) + \lambda_h^f(t) \right) (\alpha - \beta(p(t) - s(t) - p_a)) + \mu \left(c^0 - cx(t) \right) q(t),$$

where $\lambda_h^f(t)$ is the adjoint variable associated with equation (7.82). The Hamiltonian is concave in p and the maximization of the Hamiltonian will provide a unique price trajectory $p(t)$ that is continuous in t. Assuming an interior solution (i.e., $p(t) > 0$ for all t), a sufficient optimality condition is satisfied in the optimal-control problem of the firm since for each $\left(t, \lambda_h^f \right)$, we maximize a concave function of p and x_h, subject to the linear constraint (7.82). Hence, the Hamiltonian is concave, and the following necessary conditions are also sufficient:

$$\frac{\partial H^f}{\partial p} = 0 \Leftrightarrow p^*(t) = \frac{\alpha}{2\beta} + \frac{1}{2} \left[s(t) + p_a + c^0 - cx(t) - \lambda_h^f(t) \right], \qquad (7.84)$$

$$\dot{\lambda}_h^f(t) = -c\dot{x}_h(t) + \mu cq(t), \quad \lambda_h^f(T) = 0, \qquad (7.85)$$

in which $\dot{x}_h(t)$ is given by (7.82).

Two remarks on the government's purchasing policy are in order. First, we have not considered the government's state equation $\dot{x}_g(t) = q(t)$ in the follower's problem. This is a private constraint to the leader. Second, although the best reply in (7.84) does not depend on the quantity purchased by the government,, $q(t)$ still has an influence on the firm's pricing strategy through $\lambda_h^f(t)$.

From now on, we omit time argument when no ambiguity may arise. Introduce the government's Hamiltonian:

$$\begin{aligned} H^g &= \lambda_h^g \dot{x}_h + \lambda_g^g \dot{x}_g + \phi \dot{y} + \nu \dot{\lambda}_h^f, \\ &= (\alpha - \beta(p^* - s - p_a))(\lambda_h^g + \phi s - cv) \\ &\quad + q \left(\lambda_g^g + \phi((1 + \mu)(c^0 - cx) - p_a) + \nu \mu c \right), \end{aligned}$$

where $\lambda_h^g, \lambda_g^g, \phi$ and ν are the adjoint variables corresponding to x_h, x_g, y and λ_h^f. Note that the adjoint variable λ_h^f in the follower's problem now enters the leader's problem as a state variable. We confine our interest to the case where the optimal subsidy rate is strictly positive for all t. Necessary optimality conditions for the government's problem include

$$\dot{x}_h(t) = \alpha - \beta(p^* - s - p_a), \quad x_h(0) = 0, \qquad (7.86)$$

$$\dot{x}_g(t) = q(t), \quad x_g(0) = 0, \qquad (7.87)$$

$$\dot{\lambda}_h^g = \dot{\lambda}_g^g = cq\phi(1 + \mu) - \frac{\beta c}{2}(\lambda_h^g + \phi s - cv), \qquad (7.88)$$

$$\lambda_h^g(T) = 1, \ \lambda_g^g(T) = 0, \qquad (7.89)$$

$$\dot{\lambda}_h^f = -c\dot{x}_h + \mu cq, \quad \lambda_h^f(T) = 0, \qquad (7.90)$$

$$\dot{\nu} = -\frac{\beta}{2}\left(\lambda_h^g + \phi s - c\nu\right), \quad \nu\left(0\right) = 0 \tag{7.91}$$

$$\dot{y}\left(t\right) = p_g\left(t\right)q\left(t\right) + s\left(t\right)\dot{x}_h\left(t\right), \quad y\left(0\right) = 0, \quad y\left(T\right) = B, \tag{7.92}$$

$$\dot{\phi} = 0, \tag{7.93}$$

$$\frac{\partial H^g}{\partial s} = \frac{\beta}{2}\left(\lambda_h^g + \phi s - c\nu\right) + \phi\left(\alpha - \beta\left(p^* - s - p_a\right)\right) = 0, \tag{7.94}$$

$$q\left(t\right)\begin{cases} = q_{min} & \text{if } D\left(t\right) < 0, \\ \in [q_{min}, q_{max}] & \text{if } D\left(t\right) = 0, \\ = q_{max} & \text{if } D\left(t\right) > 0, \end{cases} \tag{7.95}$$

$$D\left(t\right) = \lambda_g^g + \phi\left(\left(1+\mu\right)\left(c^0 - cx\right) - p_a\right) + \nu\mu c. \tag{7.96}$$

Without fully solving the above system, we can derive some qualitative and economic insights from the equilibrium conditions,

(1) The costate variable ϕ is constant over time. This means that the marginal contribution of any dollar spent by the government is the same. It makes economic sense to assume that ϕ is negative, i.e., the more money has been spent, the less remains available in the budget.

(2) Assume that the demand rate $\dot{x}_h$ is positive for all t. ¿From (7.88) and (7.94), we have

$$\dot{\lambda}_h^g = \dot{\lambda}_g^g = c\phi\left[q\left(1+\mu\right) + \dot{x}_h\right] < 0. \tag{7.97}$$

This and the transversality conditions, $\lambda_h^g\left(T\right) = 1$ and $\lambda_g^g\left(T\right) = 0$, lead to the conclusion that the shadow prices of x_h and x_g are positive and decreasing over time. The positivity of λ_h^g and λ_g^g is not surprising as x_h and x_g are "good" stocks. Indeed, they both contribute to the decline in the unit production cost and, in view of the government's objective, adoption of the new technology is desirable .

(3) Differentiate (7.94) with respect to time yields

$$\frac{d}{dt}\left(\frac{\partial H^g}{\partial s}\right) = -\beta\left(\dot{s} + c\left(1+\mu\right)q\right) = 0,$$

which implies

$$\dot{s} = -c\left(1+\mu\right)q < 0.$$

Therefore, the government's optimal subsidy is decreasing over time. As the initial unit-production cost is high, heavily subsidizing early adopters is very necessary to boost demand, which in turn, will help to decrease this cost.

(4) Differentiating

$$p^* = \frac{\alpha}{2\beta} + \frac{1}{2}\left[s + p_a + c^0 - cx - \lambda_h^f\right],$$

with respect to time and using (7.90) and (7.97), we get

$$\dot{p}^* = \frac{1}{2}\left[\dot{s} + c\left(1+\mu\right)q\right] = \dot{s}.$$

This shows that the consumer price decreases at the same rate as the subsidy rate. The implication is that the net consumer price, that is, $p\left(t\right) - s\left(t\right)$, remains constant over time. Consequently, the demand rate by household is also constant over time $\left(\ddot{x}_h\left(t\right) = -\beta\left(\dot{p}^* - \dot{s}\right) = 0\right)$.

(5) Differentiating the government's expenditure rate over time yields

$$\ddot{y} = -c\left(1 + \mu\right)q\left(2\dot{x}_h + q\right) < 0,$$

which shows that the government's overall expenditure rate decreases monotonically over time.

(6) Finally, it can be shown that the government's purchasing policy is to buy at a constant rate given by

$$q^* = \begin{cases} q_{\max} \ \forall t \in [0, T] \ \text{if} \ \lambda_g^g\left(0\right) > \left(p_g\left(0\right) - p_a\right), \\ q_{\min} \ \forall t \in [0, T] \ \text{if} \ \lambda_g^g\left(0\right) < \left(p_g\left(0\right) - p_a\right). \end{cases}$$

The interpretation of the purchasing policy is as follows. Recall that

$$p_g\left(t\right) = \left(1 + \mu\right)\left(c^0 - cx(t)\right)$$

is the markup price charged by the firm for its sales to the government. Due to the initial condition $x\left(0\right) = 0$ and the fact that $x(t)$ is monotonically increasing over time, this markup is monotonically decreasing over time. Therefore, $p_g\left(0\right)$ is the maximal price that the government may face. If the shadow price λ_g^g of x_g, evaluated at the initial instant of time, is greater than the difference between the initial markup and the price of the alternative technology, then the government will buy at maximum rate. If not, then the government buys at the guaranteed level.

In this GE, we characterized how a government should allocate its budget to support the market penetration of a new desirable technology, in a setting where the innovating firm reacts strategically to public policy. An open-loop Stackelberg equilibrium was derived and some qualitative properties of the subsidy, purchasing and pricing strategies were established.

PART 3
Stochastic Games

Chapter 8

Equilibria in Games Played over Event Trees

8.1 Introduction

In this chapter,[1] we deal with a class of stochastic games where the uncertainty is described by an event tree. A typical example of such games is an oligopoly where firms (e.g., electricity producers) invest in different production capacities (nuclear, thermal, hydro, etc.) and compete in one or more market segments (peak-load, base-load, local market, export market, etc.) over time. At each period, the price in each market depends on the total available supply in this market, and on the realization of some random event (e.g., weather conditions or the state of the economy). In the terminology of multistage games, discussed in Chapter VI, the quantities committed to each market segment and the investments in the different production technologies are the player's control variables and the installed production capacities are the state variables. The difference here is that the decisions (quantities and investments) must be taken under uncertainty. Note that it may happen, as in the example above, that some decisions are made once the realization of the random variable is known (e.g., electricity companies know today's weather when they decide how much to produce today, subject to their capacity constraints); whereas, other decisions cannot wait for this information to be revealed. For instance, due to the long delay before an investment becomes productive, the decision to build a power plant must be made long before demand is known.

In Ref. [124] a solution concept for this class of games was introduced under the name of *S*-adapted equilibrium, where the *S* stands for *sample* of realizations of the random process. This solution concept was clearly related to the concept of *stochastic variational inequality* as shown in Refs. [98] and [114].

In this chapter, we develop a general theory of games that are played on *uncontrolled event trees*, i.e., games where the transition from one node to another is nature's decision and cannot be influenced by the players' actions. In Section 8.2, we start from the basics and interpret the concept of *S*-adapted strategies using the classical context of games in extensive form. In Section 8.3, we study the relationships between *S*-adapted equilibria and the concept of stochastic variational inequalities. Next, we discuss the existence and

---

[1]This chapter draws heavily from Ref. [121].

uniqueness of S-adapted equilibria. In Section 8.4, we extend the concept of S-adapted strategies to the context of multistage games. For such games, we provide a maximum principle and observe that it is close to the maximum principle formulated for *open-loop equilibria*, for deterministic multistage games (see Section 6.4.1.3 on page 174). In Section 8.5, we address the issues of time consistency and subgame perfectness. The chapter ends with the $\mathbb{GE}$ illustration where we model the European gas market as an S-adapted game.[2]

8.2 S-adapted strategies and classical game theory

In this section, we present S-adapted strategies using the terminology of games in extensive form. To simplify the exposition, we restrict ourselves to the two-player case.

8.2.1 *A two-player game in extensive form*

We consider a two-player game with sequential simultaneous moves (see Section 2.2.2 on page 12). Nature intervenes randomly after every decision run, consisting of an action chosen by Player 1 and an action chosen by Player 2.

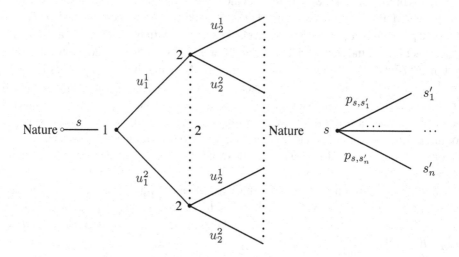

Fig. 8.1 A "run" in the extensive form game

---

[2]It is interesting to note that it was a research mandate from the European Commission's energy department, to analyze contractual issues in this market, that led to the theoretical developments and the first application of this equilibrium concept in [248, 123] and [122]. This market also served as a motivating example in [98]. More recent contributions to the theory and computation of S-adapted equilibria and their applications to energy markets include, e.g., [93, 112, 114, 201, 200, 210, 92]..

Players cannot observe their opponent's selected action. Nature selects an event randomly and independently of the players' moves. Nature's moves can be sequentially dependent, like, e.g., in a Markov chain, or can be a representation of a more general stochastic process. Nature's possible moves are known to each player. Players use VNM utilities (see Section 2.2.3 on page 13), so they strive to maximize their expected payoffs. A run is a sequence consisting of one decision move by each player, followed by Nature's move. The game is played over T runs. Payoffs depend on the terminal node reached after the last run.

A run is illustrated in Figure 8.1, where, as usual, the information structure is represented by the dotted lines that connect together all the elements belonging to the same information set. At the beginning of a run, Player 1, knowing that the game is in node S, chooses an action. Player 2 selects his action without knowing what choice was made by Player 1, as indicated by the dotted line. Nature then makes a random move, without taking into account the actions made by the players (this is also indicated by a dotted line crossing all the nodes at that decision level). The next run is played similarly, Player 1 knowing the new node s', which is the updated history of nature's moves, etc.

The tree graph in Figure 8.1 shows the situation at a node S that corresponds to a given history of the random process. Player 1 chooses an action, u_1^1 or u_1^2 knowing this history. Player 2 chooses an action without knowing what was the action chosen by Player 1, but knowing the history of the process. (See the dotted line connecting the two nodes labelled P_2 corresponding to the possible action choices of Player 1.) Now nature plays, randomly, using transition probabilities $p_{s,s'}$, to define a transition to a new process history node s'. (The nodes labelled s_1', s_i', s_n' are different in the different terminating nodes of this graph.) The fact that these transition probabilities do not depend on the players' actions is represented by the dotted line connecting all the nodes labelled E.

8.2.2 *Normal-form representation*

We index the runs over a set of time periods $t \in \{0, 1, \ldots, T-1\}$. Starting from an initial state $\xi(0)$, nature's moves are described by the stochastic process $\{\xi(t) : t = 0, 1, \ldots, T\}$. At each time t, players select actions from a finite set U_j^t where j is the player's index. Denote by $h^t = \{\xi(0), \ldots, \xi(t)\}$ the history of the $\xi(\cdot)$ process up to time t. Player j chooses action $u_j^t \in U_j^t$ using a strategy defined by the following mappings

$$\gamma_j^t : h^t \mapsto \gamma_j^t(h^t) \in \mathcal{P}[U_j^t], \; j = 1, 2, \; t = 0, 1, \ldots, T-1, \tag{8.1}$$

where $\mathcal{P}[U_j^t]$ is the simplex of probability measures on the finite set U_j^t.

The rewards are defined at terminal nodes $n(T)$ of the game tree. We represent them by two functions of the terminal nodes $G_j(n(T))$, $j = 1, 2\ldots$ With this in mind, when the players adopt strategies $\gamma_j = \{\gamma_{jt} : t = 0, 1, \ldots, T-1\}, j = 1, 2$, respectively, the rewards become random variables $\tilde{G}_j(T)$ with probability laws depending on the strategies pair $\gamma = (\gamma_1, \gamma_2)$. Because the players are VNM maximizers we can define the game payoffs as

$$V_j(\gamma_1, \gamma_2) = \mathbb{E}_\gamma \left[\tilde{G}_j(T) \right], \quad j = 1, 2. \tag{8.2}$$

Definition 8.1. An equilibrium is an admissible strategy pair (γ_1^*, γ_2^*) such that

$$V_1(\gamma_1, \gamma_2^*) \leq V_1(\gamma_1^*, \gamma_2^*),$$
$$V_2(\gamma_1^*, \gamma_2) \leq V_2(\gamma_1^*, \gamma_2^*),$$

for any admissible strategy γ_1 or γ_2.

We notice that, with our (implicit) assumption of the finiteness of state and action sets, and using Kuhn's theorem [146], we can formulate this game as a large-scale bimatrix game. However, if we are to model a realistic economic competition, the state and action spaces will be large and the resulting matrix-game formulation will be inconvenient. Furthermore, the equilibria will most probably be obtained as mixed-strategies equilibria whose interpretation is difficult. This is why, in the following section, we will reformulate the above model as a stochastic concave game with actions in vector spaces.

8.3 *S*-adapted strategies in dynamic games

In this section, we propose a framework capable of describing dynamic games in a random environment. We introduce a formalism that is akin to *stochastic programming*.[3] We will also use the term *period* for what we have defined as a *run*.

8.3.1 *Strategies as variables indexed on an event tree*

The set of periods is $\mathcal{T} = \{0, 1, \ldots, T\}$. Denote by $(\xi(t) : t \in \mathcal{T})$ the exogenous stochastic process represented by an *event tree*. This tree has a root node n_0 in period 0 and has a set of nodes $\mathcal{N}^t$ in period $t = 1, \ldots, T$. Each node $n^t \in \mathcal{N}^t$ represents a possible sample value of the history h^t of the $\xi(\cdot)$ process up to time t. The tree graph structure represents the nesting of information as one time period succeeds the other. We introduce the following notations:

- $a(n^t) \in \mathcal{N}^{t-1}$ is the unique predecessor of node $n^t \in \mathcal{N}^t$ on the event-tree graph for $t = 1, \ldots, T$;
- $\mathcal{S}(n^t) \in \mathcal{N}^{t+1}$ is the set of all possible direct successors of node $n^t \in \mathcal{N}^t$ for $t = 0, \ldots, T-1$;
- a complete path from the root node n_0 to a terminal node n^T is called a *scenario*. Each scenario has a probability and the probabilities of all scenarios sum up to 1;
- symbol $\pi(n^t)$ denotes the *probability of passing through this node*, which is associated with each node n^t; this is the sum of the probabilities of all scenarios that contain this node. In particular, $\pi(n_0) = 1$ and $\pi(n^T)$ is equal to the probability of the single scenario that terminates in (leaf) node n^T.

---

[3]For an introduction to the domain of optimization called stochastic programming, we refer the reader to the book of Birge & Louveaux, Ref. [28].

We consider the set $M = \{1, \ldots, m\}$ of players. For each player $j \in M$, we define a set of decision variables indexed over the set of nodes. We call $u_j^{n^t} \in \mathbf{R}^{m_j}$ the decision variables of player j at node n^t.

Remark 8.1. The information structure in an S-adapted game is subsumed by the indexing of the decision variables over the set of nodes on the event tree. Indeed, each node of the event tree is an exhaustive summary of the history of the $\xi(\cdot)$-process. Making the decision variables depend on the nodes in the event tree is therefore equivalent to saying that the decisions are adapted to the history of the $\xi(\cdot)$-process, which is exactly what is meant by S-adapted strategies.

8.3.2 *Rewards and constraints*

8.3.2.1 *Transition and terminal rewards*

Define, for each node $n^t, t = 1, \ldots . T$, a *transition reward* function for player $j \in M$

$$L_j^{n^t}(\underline{u}^{n^t}, \underline{u}^{a(n^t)}),$$

where $L_j^{n^t}(\cdot, \cdot)$ is twice continuously differentiable. Notice that the rewards depend on the decision made in the preceding period at antecedent node $a(n^t)$ and and on decisions made in the current period and at node n^t. Therefore, $L_j^{n^t}(\cdot, \cdot)$ is a transition reward that should be associated with time period $t - 1$ and, hence, discounted by factor β_j^{t-1} if a discount factor $\beta_j \in [0, 1]$ were used by Player j. However, we will not introduce discounting in our formalism, in order to keep it as simple as possible.

At each terminal node n^T a terminal reward $\Phi_j^{n^T}(\underline{u}^{n^T})$ is defined, and is also supposed to be twice continuously differentiable.

8.3.2.2 *Constraints*

We introduce two groups of constraints for each player

$$f_j^{n^t}(u_j^{n^t}) \geq 0, \tag{8.3}$$

$$g_j^{n^t}(u_j^{n^t}, u_j^{a(n^t)}) \geq 0, \tag{8.4}$$

where f_j and g_j are twice continuously differentiable mappings from Euclidean spaces to Euclidean spaces.

Remark 8.2. These constraints do not couple the different players' decisions.

Remark 8.3. It is possible to also introduce constraints that involve ancestors of the current node n^t, like, e.g., $h_j(u_j^{n^t}, u_j^{a(n^t)}, \ldots, u_j^{a^p(n^t)}) \geq 0$, where $a^\ell(n^t)$ represents the operator that associates the ℓth-period ancestor in the event tree with node n^t.[4] However, we will retain the formulation (8.3)–(8.4) for the sake of simplifying the notation.

---

[4]These types of constraints could be used to represent capacity accumulation in technology models. A technology with life τ has, at time t, a capacity that corresponds to the sum of all investments made over the past τ years $(t - \tau + 1, \ldots, t)$.

8.3.3 *Normal form game and S-adapted equilibrium*

Definition 8.2. An admissible S-adapted strategy for player j, is a vector $\gamma_j = \{u_j^{n^t} : n^t \in \mathcal{N}^t, t = 0, \ldots, T-1\}$ that satisfies constraints (8.3)–(8.4). We call Γ_j the set of all the admissible S-adapted strategies of Player j.

Associated with an admissible S-adapted strategy vector $\underline{\gamma} = \{\gamma_j\}_{j \in M}$ are the following payoffs:

$$V_j(\underline{\gamma}) = \sum_{t=1,\ldots,T-1} \sum_{n^t \in \mathcal{N}^t} \pi(n^t) L_j^{n^t}(\underline{u}^{n^t}, \underline{u}^{a(n^t)}) + \sum_{n^T \in \mathcal{N}_T} \pi^{n^T} \Phi_j^{n^T}(\underline{u}^{n^T}), \quad j \in M.$$

The strategy sets Γ_j and the payoff functions $V_j : \Gamma_1 \times \ldots \times \Gamma_j \times \ldots \times \Gamma_m \to \mathbf{R}$ define a game in normal form.

An S-adapted equilibrium is an equilibrium for this particular normal form game.

Definition 8.3. An **S-adapted equilibrium** is an admissible S-adapted strategy vector $\underline{\gamma}^*$ such that

$$V_j([\gamma_j, \underline{\gamma}^*_{-j}]) \leq V_j(\underline{\gamma}^*), \quad j = 1, \ldots, m,$$

where

$$[\gamma_j, \underline{\gamma}^*_{-j}] = (\gamma_1^*, \ldots, \gamma_j, \ldots, \gamma_m^*),$$

represents a unilateral deviation by Player j from the equilibrium.

8.3.4 *Concave-game properties and variational-inequality reformulation*

The game defined above is a concave game as defined by Rosen [213]. We can therefore adapt the existence and uniqueness theorems proved by Rosen.

8.3.4.1 *Existence and uniqueness conditions*

Lemma 8.1. *Assume the functions* $L_j^{n^t}(\underline{u}^{n^t}, \underline{u}^{a(n^t)})$ *and* $g_j^{n^t}(u_j^{n^t}, u_j^{a(n^t)})$ *are concave in* $(u_j^{n^t}, u_j^{a(n^t)})$; *assume the functions* $f_j^{n^t}(u_j^{n^t})$ *are concave in* $u_j^{n^t}$ *and assume the functions* $\Phi_j^{n^T}(\underline{u}^{n^T})$ *are concave in* $u_j^{n^T}$. *Assume that the set of admissible strategies is compact. Then there exists an equilibrium.*

Proof. Using the classical result of Rosen [213] (see Theorem 3.5) which is based on Kakutani's fixed-point theorem, we can easily prove the existence of equilibria. □

We can also adapt the results proved in [213] to provide conditions under which this equilibrium is unique. It is convenient to introduce the notations $\mathbf{u}_j^{n^t} = (u_j^{n^t}, u_j^{a(n^t)})$ and $\underline{\mathbf{u}}^{n^t} = (\underline{u}^{n^t}, \underline{u}^{a(n^t)})$, respectively. Now, define the *pseudo-gradients*

$$\mathcal{G}^{n^t}(\underline{\mathbf{u}}^{n^t}) = \left(\frac{\partial L_1^{n^t}(\underline{\mathbf{u}}^{n^t})}{\partial \, \mathbf{u}_1^{n^t}} \, \cdots \, \frac{\partial L_j^{n^t}(\underline{\mathbf{u}}^{n^t})}{\partial \, \mathbf{u}_j^{n^t}} \, \cdots \, \frac{\partial L_m^{n^t}(\underline{\mathbf{u}}^{n^t})}{\partial \, \mathbf{u}_m^{n^t}} \right), \quad t = 1, \ldots, T-1,$$

$$\hat{\mathcal{G}}^{n^T}(\underline{u}^{n^T}) = \left(\frac{\partial \Phi_1^{n^T}(\underline{u}^{n^T})}{\partial u_1^{n^T}} \, \cdots \, \frac{\partial \Phi_j^{n^T}(\underline{u}^{n^T})}{\partial u_j^{n^T}} \, \cdots \, \frac{\partial \Phi_m^{n^T}(\underline{u}^{n^T})}{\partial u_m^{n^T}} \right),$$

and the Jacobian matrices

$$\mathcal{J}^{n^t}(\underline{\mathbf{u}}^{n^t}) = \frac{\partial \mathcal{G}^{n^t}(\underline{\mathbf{u}}^{n^t})}{\partial \underline{\mathbf{u}}^{n^t}}, \quad t = 1, \dots, T,$$

$$\hat{\mathcal{J}}^{n^T}(\underline{u}^{n^T}) = \frac{\partial \hat{\mathcal{G}}^{n^T}(\underline{u}^{n^T})}{\partial \underline{u}^{n^T}}.$$

Lemma 8.2. *If, for all $\underline{\mathbf{u}}^{n^t}$, the matrices $\mathcal{Q}^{n^t}(\underline{\mathbf{u}}^{n^t}) = \frac{1}{2}[\mathcal{J}^{n^t}(\underline{\mathbf{u}}^{n^t}) + (\mathcal{J}^{n^t}(\underline{\mathbf{u}}^{n^t}))']$ and $\hat{\mathcal{Q}}^{n^T}(\underline{u}^{n^T}) = \frac{1}{2}[\mathcal{J}^{n^T}(\underline{u}^{n^T}) + (\mathcal{J}^{n^T}(\underline{u}^{n^T}))']$ are negative definite, then the equilibrium is unique.*

Proof. This is the setting of Theorem-5 in Ref. [213] (see Theorem 3.6 in Chapter 3). The negative definiteness of the matrices $\mathcal{Q}^{n^t}(\underline{\mathbf{u}}^{n^t})$ and $\hat{\mathcal{Q}}^{n^T}(\underline{u}^{n^T})$ implies the strict-diagonal concavity of the function $\sum_{j=1}^{m} V_j(\gamma)$, which then implies the uniqueness of the equilibrium. $\square$

Remark 8.4. Another result found in Rosen's paper [213] could be easily extended to this stochastic-equilibrium framework. It is related to the existence of normalized equilibria when there is a coupled constraint $h(\underline{\mathbf{u}}^{n^t}) \geq 0$ at each node, where h is a concave function satisfying the constraint-qualification conditions. We shall not go into this aspect any further here, but we refer interested readers to [41, 108, 120] for more developments on dynamic game models with coupled constraints.

Remark 8.5. As shown in Chapter III, the Nash-equilibria for concave games can be characterized as solutions of a variational inequality (*VI*). This type of characterization extends readily to the case of equilibria in the class of *S*-adapted strategies. See [98, 112] and [114] for details and the design of efficient numerical methods.

8.4 A stochastic-control formulation

In this section, we modify slightly the formalism that was previously used in order to explore the possibility of characterizing the *S*-adapted equilibria through maximum principles and establish a link with the theory of open-loop multistage games.

8.4.1 *A multistage-game formulation*

Consider an event tree with nodes $n^t \in \mathcal{N}^t, t = 0, 1, \dots, T$. Let $\pi(n^t)$ be the probability of passing through node $n^t \in \mathcal{N}^t, t = 0, 1, \dots, T$. A system of state equations (compare Definition 6.1 on page 158) is defined over the event tree. Let $X \subset \mathbf{R}^q$, with q a given positive integer, be a state set. For each node $n^t \in \mathcal{N}^t, t = 0, 1, \dots, T$, let $U_j^{n^t} \subset \mathbf{R}^{\mu_j^{n^t}}$, with $\mu_j^{n^t}$, a given positive integer, be the control set of player j. Denote by $\underline{U}^{n^t} = U_1^{n^t} \times \cdots \times U_j^{n^t} \times \cdots \times U_m^{n^t}$ the product control sets. A transition function $f^{n^t}(\cdot, \cdot) : X \times \underline{U}^{n^t} \mapsto$

X is associated with each node n^t. The state equations[5] are given as

$$x(n^t) = f^{a(n^t)}(x(a(n^t)), \underline{u}(a(n^t))), \tag{8.5}$$

$$\underline{u}(a(n^t)) \in \underline{U}^{a(n^t)}, \quad n^t \in \mathcal{N}^t, t = 1, \ldots, T. \tag{8.6}$$

According to the state equation (8.5), at each node n^t, a vector of controls $\underline{u}(n^t)$ will determine, in association with the current state $x(n^t)$, the state $x(\nu)$ for **all** descendent nodes $\nu \in \mathcal{D}(n^t)$. This is illustrated in Figure 8.2.

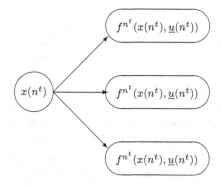

Fig. 8.2   The stochastic state equations

In this state-equation formalism, the decisions are the controls $u_j(n^t)$ that are independently chosen by the players. Once the controls have been chosen, the state variables $x(n^t)$ are determined. The state variables are shared by all the players and enter their reward functions as shown below in equation (8.7).

At each node n^t, $t = 0, \ldots, T-1$, the reward to Player j is a function of the state and of the controls of all players, given by $\phi_j^{n^t}(x(n^t), \underline{u}(n^t))$. At a terminal node n^T, the reward to Player j is given by the function $\Phi_j^{n^T}(x(n^T))$.

The state equations and the reward functions define the following multistage game, where we denote $\tilde{x} = \{x(n^t) : n^t \in \mathcal{N}^t, t = 0, \ldots, T\}$ and $\tilde{\underline{u}} = \{\underline{u}(n^t) : n^t \in \mathcal{N}^t, t = 0, \ldots, T-1\}$ respectively, and where $J_j(\tilde{x}, \tilde{\underline{u}})$ is the payoff to Player j:

$$J_j(\tilde{x}, \tilde{\underline{u}}) = \sum_{t=0}^{T-1} \sum_{n^t \in \mathcal{N}^t} \pi(n^t) \phi_j^{n^t}(x(n^t), \underline{u}(n^t)) +$$
$$\sum_{n^t \in \mathcal{N}_T} \pi(n^T) \Phi_j^{n^T}(x(n^T)), \quad j \in M, \tag{8.7}$$

---

[5] To distinguish *this* treatment of dynamic games from the preceding formulation as concave game s, we will use different notations to index the variables over the event tree. Specifically, we will now use $x(n^t)$ instead of $x(n^t)$.

s.t.

$$x(n^t) = f^{a(n^t)}(x(a(n^t)), \underline{u}(a(n^t))), \tag{8.8}$$

$$\underline{u}(a(n^t)) \in \underline{U}^{a(n^t)}, \quad n^t \in \mathcal{N}^t, t = 1, \ldots, T,$$

$$x(n_0) = x^0 \text{ given.} \tag{8.9}$$

Definition 8.4. An admissible S-adapted strategy for Player j is defined by a vector $\tilde{u}_j = \{u_j(n^t) : n^t \in \mathcal{N}^t, t = 0, \ldots, T - 1\}$, i.e., a plan of actions adapted to the history of the random process represented by the event tree .

We have already called $\underline{\tilde{u}} = (\tilde{u}_j : j \in M)$ the S-adapted strategy vector of the m players. We can thus define a game in normal form, with payoffs $V_j(\underline{\tilde{u}}, x^0) = J_j(\tilde{x}, \underline{\tilde{u}}), j \in M$, where $\tilde{x}$ is obtained from $\underline{\tilde{u}}$ as the unique solution of the state equations that emanates from the initial state x^0.

Definition 8.5. An S-adaped equilibrium is an admissible S-adaped strategy $\underline{\tilde{u}}^*$ such that for every player j the following holds

$$V_j(\underline{\tilde{u}}^*, x^0) \geq V_j([\tilde{u}_j, \tilde{\mathbf{u}}_{-j}^*], x^0),$$

where $\tilde{\mathbf{u}}_{-j}^*$ is defined as usual, as the equilibrium policy vector of all players $i \neq j$.

Remark 8.6. In the definition of the S-adapted equilibrium , we notice a close resemblance to the open-loop information structure . The important difference lies essentially in the definition of state equations over an exogenous event tree, and of the controls as vectors indexed over the set of nodes of the event tree.

8.4.2 *Lagrange multipliers and the maximum principle*

As for open-loop multistage games, we can formulate the necessary conditions for an S-adapted equilibrium, in the form of the maximum principle.

For each player j, we form the Lagrangian

$$\mathcal{L}_j(\tilde{\lambda}_j, \tilde{x}, \underline{\tilde{u}}) = \phi_j^{n_0}(x^{n_0}, \underline{u}^{n_0}) + \sum_{t=1}^{T-1} \sum_{n^t \in \mathcal{N}^t} \pi(n^t) \left\{ \phi_j^{n^t}(x(n^t), \underline{u}(n^t)) + \right.$$

$$\left. \lambda_j(n^t)'(f^{a(n^t)}(x(a(n^t)), \underline{u}(a(n^t))) - x(n^t)) \right\} +$$

$$\sum_{n^t \in \mathcal{N}_T} \pi(n^T) \left\{ \Phi_j^{n^T}(x(n^T)) + \right.$$

$$\left. \lambda_j(n^T)'(f^{a(n^T)}(x(a(n^T)), \underline{u}(a(n^T))) - x(n^T)) \right\}. \tag{8.10}$$

In this expression, we have introduced, for each player j, a *costate variable*, sometimes called an *adjoint state variable*, $\lambda_j(n)$, also indexed over the set of nodes and having the same dimension as $x(n)$. Then, for each player j and each node $n \in \mathcal{N}^t$, $t = 0, 1, \ldots, T - 1$, we can define the pre-Hamiltonian function

$$H_j^n\left(\underline{\lambda}_j(\mathcal{S}(n)), x(n), \underline{u}(n)\right) = \phi_j^n\left(x(n), \underline{u}(n)\right) + \sum_{\nu \in \mathcal{S}(n)} \frac{\pi^\nu}{\pi(n)} \lambda_j(\nu)' f^n(x(n), \underline{u}(n)),$$

(8.11)

where $\underline{\lambda}_j(\mathcal{S}(n))$ stands for all the vectors λ_j^ν with $\nu \in \mathcal{S}(n)$.

Remark 8.7. The main difference between the usual Hamiltonian of a multistage game with the open-loop information structure and the pre-Hamiltonian, is that the latter allows for an average "sensitivity vector" $\sum_{\nu \in \mathcal{S}(n)} \frac{\pi^\nu}{\pi(n)} \lambda_j(\nu))$. The interpretation of this average sensitivity becomes clear when we notice that the ratio $\frac{\pi^\nu}{\pi(n)}$ is the transition probability from node n to node ν.

Theorem 8.1. *Assume that $\tilde{\underline{u}}^*$ is an S-adapted equilibrium at x^0, generating the state trajectory $\tilde{x} = \{x^*(n^t) : n^t \in \mathcal{N}^t, t = 0, \ldots, T - 1\}$ over the event tree. Then there exists, for each player j a costate trajectory $\lambda_j(n^t)$ such that the following conditions hold for $j \in M$, $u_j(n^t) = u_j^*(n^t)$, $x(n^t) = x^*(n^t)$*

$$0 = \frac{\partial H_j^n}{\partial u_j(n)} \quad n \in \mathcal{N}^t, \ t = 0, 1, \ldots, T - 1, \tag{8.12}$$

$$\lambda_j(n) = \left[\frac{\partial H_j^n}{\partial x(n)}\right]' \quad n \in \mathcal{N}^t, \ t = 0, 1, \ldots, T - 1, \tag{8.13}$$

$$\lambda_j(n^T) = \left[\frac{\partial \Phi_j^{n^T}(x(n^T))}{\partial x(n^T)}\right]' \quad n^T \in \mathcal{N}_T. \tag{8.14}$$

Proof. In the expression of the Lagrangian (8.10), we group together the terms that contain $x(n)$, to get

$$\mathcal{L}_j(\tilde{\lambda}_j, \tilde{x}, \tilde{\underline{u}}) = \sum_{t=0}^{T-1} \sum_{n^t \in \mathcal{N}^t} \pi(n^t) \left\{ H_j^{n^t}(\underline{\lambda}_j^{\mathcal{S}(n^t, x(n^t), \underline{u}(n^t))}) - \lambda_j(n^t)' x(n^t) \right\}$$

$$+ \sum_{n^T \in \mathcal{N}_T} \pi(n^T) \left\{ \Phi_j^{n^T}(x(n^T)) - \lambda_j(n^T)' x(n^T) \right\}. \tag{8.15}$$

Equations (8.11) and (8.12)–(8.14) are then obtained by equalling to 0 the partial derivatives of the Lagrangian w.r.t. $x(n)$ and $u(n)$. $\square$

8.5 Properties of S-adapted equilibria

8.5.1 General properties

We know from Sections 5.4.2 and 6.4.1.2 (see pages 144 and 173, respectively) that dynamic games admit equilibria that may have two desirable properties:

subgame perfectness This property says that, if we have defined an equilibrium strategy γ^* and that the players have not played accordingly for some time and a position $(t, x(t))$ is reached, then resuming the play according to γ^* from initial condition $(t, \underline{x}(t))$ will still be an equilibrium. We repeat Section 5.4.2 in saying that this property (see [218]) holds true for feedback-Nash equilibria.

time consistency This property says that, along an *equilibrium trajectory*, if the game is stopped at any given time t and then the play resumes from the position $(t, \underline{x}^*(t))$ reached along the equilibrium trajectory, then the continuation of the same equilibrium will still be an admissible equilibrium for this new game. Recall that we proved in Section 6.4.1.2 that this property is verified for open-loop Nash equilibria in multistage games.

8.5.2 Subgame perfectness

S-adapted equilibria are not subgame perfect. The reason is that these strategies are very much akin to open-loop ones for deterministic games. If players do not play correctly at a node n^t, the control vector is $\underline{u}(n^t)$ instead of $\underline{u}^*(n^t)$. Then the trajectory is perturbed and the initial S-adapted strategy is no longer an equilibrium for the subgames that would start out of the nodes $\nu \in \mathcal{S}(n^t)$ with the state trajectory $\underline{x}' = f^{n^t}(\underline{x}^*(n^t), \underline{u}(n^t))$. In short, an error in the play will destroy the equilibrium property of the subsequent subgames.

8.5.3 Time consistency

However, we have the following result:

Lemma 8.3. S-*adapted equilibria are time consistent.*

Proof. Consider the multistage game as defined by equations (8.7)–(8.9), which is played on a finite event tree. Given an S-adapted strategy defined by the controls $\underline{u}(n^t)$ indexed over the nodes n^t and an initial state $x(n^0)$, we can define an *extended trajectory* that associates a state $x(n^t)$ with each node n^t. Now, let $\underline{\tilde{u}}^* = \{\underline{u}^*(n^t) : n^t \in \mathcal{N}^t, t = 0, 1 \ldots, T-1\}$ be an S-adapted equilibrium, with the associated extended trajectory $\tilde{x}^* = \{x^*(n^t) : n^t \in \mathcal{N}^t, t = 0, 1 \ldots, T\}$ emanating from $\underline{x}(n^0)$, and let us consider the subgame that starts at a node n^τ, with initial state $x^*(n^\tau)$. The S-adapted strategy $\underline{u}^*(\cdot)$ restricted to the nodes that are descendants of n^τ, is still an equilibrium for this subgame. We call $\mathcal{D}[n^\tau]$ the set of all nodes that are descendants of n^τ.

The proof of this result is, as usual, a direct application of the *tenet of transition* or *Bellman principle* (see Section 6.4.3 on page 178). If the restricted strategy $\{\underline{u}^*(n^t) : n^t \in \mathcal{D}[n^\tau]\}$ were not an equilibrium, a player j could improve his payoff by unilaterally changing his strategy on this portion of the event tree. The new strategy for the subgame could be combined with the vector $\{\underline{u}^*(n^t) : n^t \in \mathcal{N} \setminus \mathcal{D}[n^\tau]\}$ to define a new S-adapted strategy for the whole game, where Player j would unilaterally change his strategy and improve his payoff. However, this would contradict the equilibrium property assumed for $\underline{u}^*(\cdot)$.   □

8.6   What have we learned in this chapter?

In this chapter, we have studied the various facets of the information structure subsumed by the S-adapted strategy concept with the help of different paradigms used in the theory of dynamic games. In particular, we have learned

- That S-adapted strategies are related to a particular type of information structure for a game in extensive form which is halfway between open-loop and closed-loop, since the players will only remind the history of random disturbances, and not the history of actions or the whole state of dynamic system.
- The normal-form representation of a game played on an uncontrolled event tree, corresponds to the given of payoff functions that depend on control or action sequences indexed over the set of nodes of the event tree.
- Conditions for existence and uniqueness of an S-adapted Nash equilibrium can be obtained from the theory of concave games.
- The necessary conditions for an S-adapted equilibrium in a multistage context can be formulated as an extended coupled-maximum principle.
- That the S-adapted Nash equilibrium is time consistent, but not subgame perfect.
- Furthermore, we will see in this chapter's $\mathbb{GE}$ illustration that the S-adapted formalism is very useful for analyzing oligopolistic markets subjected to random disturbances that can be described by an event tree.

8.7   Exercises

Exercise 8.1. Consider an industry formed by n firms competing in a two-period Cournot-type game. At period 0, the demand is certain. At period 1, the demand can be either high (H), with probability p, or low (L), with probability $1 - p$.

Each firm is endowed with an initial given production capacity $K_j(0), j = 1, \ldots, m$. At period 1, the capacity is given by

$$K_j(1) = K_j(0) + I_j(0),$$

where $I_j(0) \geq 0$ is the investment made by firm j at the initial period. This means that it takes one period before the investment becomes productive. The investment cost is given by the convex function $c_j(I_j) = \frac{1}{2}\alpha_j I_j^2, j = 1, \ldots, m$.

Denote by $q_j(t)$ the quantity produced by firm j, and by $Q(t)$ the total quantity produced by the m firms at period $t = 0, 1$. The production-cost function is assumed to be quadratic and given by $g_j(q_j) = \frac{1}{2}\beta_j q_j^2, j = 1, \ldots, m$. Denote by $P(0)$ the price at period 0 and by $P(1, H)$ and $P(1, L)$ the price at period 1 when demand is high and low, respectively. The inverse-demand laws at the different nodes are given by

$$P(0) = \max\{(a_0 - b_0 Q(0)); 0\},$$
$$P(1, H) = \max\{(A_H - BQ(0)); 0\},$$
$$P(1, L) = \max\{(A_L - BQ(0)); 0\},$$

where a_0, b_0, A_H, A_L and B are strictly positive parameters, with $A_H > A_L$. Assume that each player maximizes the sum of his profits over the two periods.

(1) Show that there exists a unique S-adapted Nash equilibrium for this game.
(2) Is this equilibrium subgame perfect?

Exercise 8.2. Consider a two-player, three-stage game. The players are producers of a homogenous commodity sold on a competitive market. The market is represented by a stochastic inverse-demand law

$$p(t, s^t) = P\left(q_1(t, s^t) + q_2(t, s^t), s^t\right),$$

where s^t is the sample value at period t of a random perturbation in the market at period t; $q_1(t, s^t) + q_2(t, s^t)$ is the total quantity put on the market; and $p(t, s^t)$ is the clearing market price at period t for the realization s^t of the random perturbation. The function $P(\cdot, \cdot)$ is assumed to be affine, with a negative slope, w.r.t. its first argument, and random perturbations are described by an event tree.

Let S^t denote the set of possible realizations of the random perturbation at period t; let $a(s^t) \in S^{t-1}$ denote the unique predecessor of $s^t \in S^t, t = 1, 2$; and let $B(s^t) \subset S^{t+1}, t = 0, 1$, denote the set of successors of s^t on the event tree. Let $\theta(s^t/a(s^t)) \geq 0$ be the conditional probability associated with the arc $(a(s^t), s^t)$ in the event tree, with

$$\sum_{s^{t+1} \in B(s^t)} \theta(s^{t+1}/s^t) = 1.$$

The set S^0 reduces to the singleton s^0, the root of the event tree.

The players' actions correspond to the quantities they put on the market at each of the three stages, together with their investment decisions to increase their production capacities. Each Player $j = 1, 2$ is described by the following data:

(i) The production capacity $K_j(t, s^t), t = 0, 1, 2$, which evolves according the following difference equation:

$$K_j(t, s^t) = K_j(t - 1, a(s^t)) + I_j(t, a(s^t)),$$

with initial capacities given by $K_1(0, s^0) = 5$ and $K_2(0, s^0) = 4$.

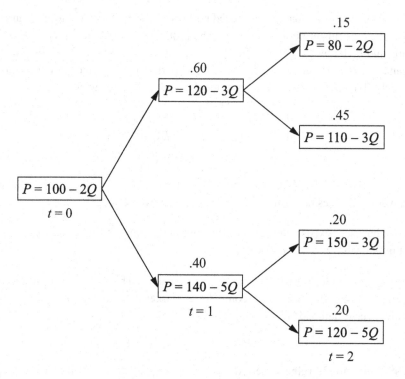

Fig. 8.3   Representation of the demand laws at each node of the event tree. Numbers above boxes denote probabilities

(ii) A production cost function $C_j\left(q_j\left(t, s^t\right)\right)$, where $q_2\left(t, s^t\right)$ denotes the quantity put on the market at period t and for sample value s^t. The function $C_j\left(\cdot\right)$ is assumed strictly convex, increasing, and twice continuously differentiable.

(iii) An investment-cost function $F_j\left(I_j\left(t, s^t\right)\right)$, where $I_j\left(t, s^t\right)$ denotes the physical capacity installed at period t and for sample value s^t. The function $F_j\left(\cdot\right)$ is assumed strictly convex, increasing, and twice continuously differentiable.

Assume that Player j strives to maximize a discounted stream of profits, with a discount factor β_j, over the 3-period time horizon.

(1) Write the dynamic optimization problem of Player $j, j = 1, 2$.
(2) Show that this game admits at least one S-adapted Nash equilibrium. Is this equilibrium unique?
(3) Let the demand laws be represented by the event tree in Figure 8.3.
(4) Finally, suppose that the cost functions are given by
$$C_1\left(q_1\right) = 3q_1^2, \quad C_2\left(q_2\right) = 3q_2^2,$$
$$F_1\left(I_1\right) = 8I_1^2, \quad F_2\left(I_2\right) = 7I_2^2,$$
and the discount factor $\beta_j = 0.9, j = 1, 2$. Calculate the S-adapted open-loop Nash equilibrium of this game.

8.8   GE illustration: The European gas market

In this section, we provide an application of the S-adapted equilibrium concept to the European gas market. As mentioned in the introduction, it was modeling this market that led to the formulation of this solution concept (see [248, 123] and [122]). The numerical analysis was carried out in the mid-1980s. Some twenty years later, Haurie & Zaccour in Ref. [121] compared the simulations obtained from the model with their realizations in 1985–2000. This gives us a rather unique opportunity to assess the predictive power of the S-adapted equilibrium formalism.

8.8.1   *The structure of the European gas market*

The European gas market is organized as a multilevel structure involving producers, distributors and consumers as schematized in Figure 8.4.

| Producers | Distributors | Consuming regions |

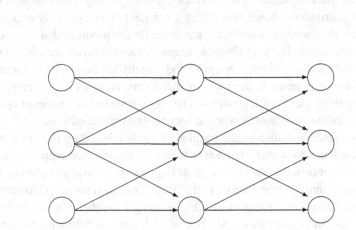

Fig. 8.4   A multilevel market structure

A modeling exercise was undertaken in 1985 to represent the evolution of this market over the four periods indicated in Table 8.1. The model does not consider the distributors

Table 8.1   The four time periods of the gas-market model

| 1985–1989 | 1990–1994 | 1995–1999 | 2000–2019 |

to be active players and concentrates on the producers and their production facilities,[6] as indicated in Table 8.2. The nine gas-consuming regions are Belgium-Luxembourg, France

Table 8.2   The players and their production units

Algeria	→	Arzew, Skikda, Algeria Pipe
Holland	→	Groningen
Norway	→	EKOFISK, Troll
Russia	→	Russia1, Russia2

North, France South, Italy North, Italy South, Holland, UK, Germany North and Germany South.

We assume that the market is oligopolistic and that the producers behave as Nash-Cournot equilibrium seekers.

8.8.2   Modeling the gas contracts

In the European markets, natural gas is exchanged through long-term contracts between producers and distributors. Since developing a gas field is an extremely expensive and lengthy venture, long-term contracts are a guarantee for the producer that a sufficient demand exists for this field. For a distributing company, a long-term contract is synonymous to a reliable supply at a certain price. So, price and quantity are the two basic parameters of a contract. However, in practice, contracts are much more complex arrangements, involving more than a single pair (price, quantity). They usually include a number of operational clauses, a price-indexation clause, lower and upper bounds on quantity, etc.

The price-indexation clause in gas contracts is mainly based on the price of oil (or a basket of petroleum products). The lower bound on quantity represents a take-or-pay clause.[7] Since, obviously, the future price of oil is a random event, a gas contract between a seller and a buyer may take the form of a set of pairs (price, quantity), each corresponding to one possible realization of the price of oil. This indexation clause is translated into the game-theoretic model through the introduction of the S-adapted information structure.

To achieve this, we represent uncertainty as an event tree, where the paths correspond to different oil-price scenarios. The gas demand law in each consuming region is affected by the random oil price scenario. A player's decision-vector specifies, for each node of the event tree, the quantities to be exchanged with the distributors under that node's price-of-oil characteristics. The tree structure can be interpreted in contractual terms between a producer and a distributor. The above-mentioned set of pairs (price, quantity) is made up of the results for all nodes in the event tree that correspond to the same period. The

---

[6]When the research was conducted, the two Russian gas fields were referred to as FUSSR1 and FUSSR2, where FUSSR stands for former USSR.

[7]This means that the buyer must pay for this minimum quantity, whether he actually takes it or not.

lowest quantity in this set can be interpreted as a take-or-pay quantity clause. The highest one provides the upper bound on the volume of gas that can be lifted (e.g., annually). Therefore, these results correspond to quantities of gas that are contingent on the price of oil.

8.8.3 *A model of competition under uncertainty*

We now provide a mathematical formulation of the model sketched out above. Consider a set of time periods $t \in \{0, \ldots, T\}$. Let C be the set of consuming regions. In each region $c \in C$, we consider a set of demand laws indexed over a set of scenarios Ω, which represent different evolutions of the gas market. A scenario $\omega \in \Omega$ is represented as a path in the event tree, as indicated in Figure 8.5.

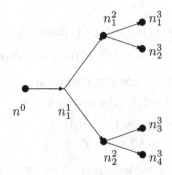

Fig. 8.5 An event tree

The set $\mathcal{N}^t$ contains the nodes n^t of the event tree at period t. These nodes represent all the different possible histories of the random events up to time t. We say that a scenario passes through node n^t if it has the history represented by node n^t up to and including time t. The cumulative probability of all the scenarios passing through the node $n^t \in \mathcal{N}^t$ is denoted $\pi(n^t)$.

The set of players (producers) is denoted $M = \{1, \ldots, m\}$. Each player j controls a set U_j of production units. Denote by q_ℓ the quantity produced at $l \in U_j$ and by $q_{\ell c}$ the quantity shipped from this unit to consuming region $c \in C$. Each unit $\ell \in U_j$ is described by

(1) the reserves it exploits $R_\ell(t)$ (with initial condition $R_\ell(0) = R_\ell^0$ given);
(2) the production capacity $K_\ell(t)$ (with initial condition $K_\ell(0) = K_\ell^0$ given);
(3) the production cost function $G_\ell(q_\ell(t))$, which is defined as a convex, twice-differentiable, monotone increasing function;
(4) the investment cost function $\Gamma_\ell(q_\ell(t))$, also defined as a convex, twice-differentiable, monotone increasing function.

At each period, the players have to choose how much to produce and how much to invest in capacity expansion. When they make their decisions, they know the node of the event tree that has been reached, i.e., they know the history of the gas market. The players' decisions are therefore indexed over the nodes of the event tree. These contingent decisions made by the players are part of a Nash equilibrium, i.e., each player's contingency plan is his best reply, in terms of expected profit, to the contingency plans adopted by the other players.

The stochastic optimization problem of player j can then be summarized as follows:

$$\max \Pi_j = \sum_{t=0}^{T} \beta_j^t \left\{ \sum_{c \in C} \sum_{\ell \in U_j} \sum_{n^t \in \mathcal{N}^t} \pi(n^t) \left[q_{\ell c}(n^t)(P_c(n^t)) \right. \right.$$
$$\left. \left. - G_\ell(q_\ell(n^t)) - \Gamma_\ell(I_\ell(n^t)) \right] \right\}_{j \in M} \tag{8.16}$$

s.t.

$$R_\ell(n^t) = R_\ell(a(n^t)) - q_\ell(a(n^t)) \text{ reserve depletion,} \tag{8.17}$$

$$K_\ell(n^t) = K_\ell(a(n^t)) + I_\ell(a(n^t)) \text{ capacity expansion,} \tag{8.18}$$

$$q_\ell(n^t) \leq K_\ell(n^t) \quad \text{capacity constraints,}$$

$$P_c(n^t) = f_c(Q_c(n^t)) \quad \text{demand laws,}$$

$$0 \leq I_\ell(a(n^t)), \ q_{\ell c}(n^t), \ R_\ell(n^t)$$

$$\forall t, n^t \in \mathcal{N}^t, \quad c \in C, \ \ell \in U_j, \quad j \in M.$$

In the profit function (8.16), the term $P_c(n^t)$ represents the market clearing price of gas in consuming region j, given the economic condition represented by the node n^t. This price depends on the total quantity supplied to the market and will be further discussed in the next subsection.

In the constraints (8.17) and (8.18), we use $a(n^t)$ to denote the unique predecessor of the node n^t in the event tree. These constraints represent the state equations. As indicated, the investment and reserve-commitment decisions made at the antecedent node $a(n^t)$ determine the state variables $R_\ell(n^t)$ and $K_\ell(n^t)$.

We will now specify the functional forms of the cost and demand.

Production-cost functions: We assume that the marginal production cost[8] increases when the production q gets closer to the capacity K

$$G'(q_\ell) = \alpha_\ell + \frac{\gamma_\ell}{K_\ell - q_\ell}.$$

Investment-cost functions: We assume constant marginal investment cost with differences among the producing plants

$$\Gamma'(I_\ell) = \begin{cases} \$15/\text{MMBTU Troll} \\ \$16.9/\text{MMBTU Russia2} \\ \equiv 0 \text{ Algeria, Holland.} \end{cases}$$

---

[8]The parameter values that we use in the cost and demand functions are taken from report [122].

Stochastic-demand law. The natural-gas (inverse) demand law of each consuming region, defines the price of gas as an affine function of the total quantity delivered to this market by the producers. Both the slope and intercept are random parameters that depend on the oil price. The event tree in Figure 8.6 shows the possible time evolutions of the oil price. It shows a starting value of $30/bbl[9] in 1985–1989 with a predicted decline to $15/bbl in 1990–1994, followed in 1995–1999 by two possible price increases to $20/bbl or $30/bbl, respectively. Further branching occurs after that period, leading to prices that range from $20/bbl to $60/bbl in 2000–2019. At each node, the probability of each branch is also indicated.

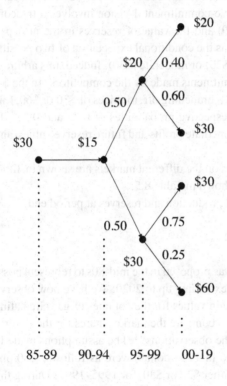

Fig. 8.6 Stochastic oil prices

8.8.4 *The equilibrium strategies*

The computation of the equilibrium strategies, adapted to the event tree describing the demand uncertainty, leads to the regional consumption flows shown in Table 8.3, and the total production decisions shown in rows (a) of Table 8.6 (rows (b) give total reserves).

---

[9]Barrel.

The equilibrium strategy becomes interesting for periods that have more than one price node; e.g., see the columns that correspond to the years 1995–1999. As expected, the investment decisions will differ depending on the prevailing oil price ($20 or $30). Consider, for example, the reserve commitments shown in Table 8.6. In 1995–1999, Algeria commits 22.94 $mtoe$[10] or 27.85 $mtoe$ of its reserves if the price of oil is $20 or $30, respectively.

Investment in production capacity the choice of a reserve commitment are intertemporal decisions, i.e., they determine the evolution of the reserves, which are another type of (nonrenewable) capital. In Figure 8.6, we see the different situations corresponding to the two nodes for 1995–1999. In the $20 node, the perspective is a branching to prices of either $20 or $30 for the future (final) 2000–2019 period, with probabilities of 0.40 and 0.60, respectively. The reserve-commitment decision involves a tradeoff between an immediate profit (in 1995–1999) and the value of reserves in the next period, 2000–2019. This future value is obtained as the conditional expectation of two possible reserve values: when the oil price is as low ($20) or moderate ($30). Indeed, this arbitrage is also made as the best response to the commitments made by the competitors. In the $30 node of period 1995-1999, the perspective is a branching off, to prices of $30 or $60, for the future (final) period of 2000–2019, with respective probabilities of 0.75 and 0.25. This is a different perspective, with different immediate profits and future reserve values, and hence, different commitments.

The resulting prices of gas on the different markets are shown in Table 8.4.

The total gas demand is shown in Table 8.5.

Table 8.6 provides annual production and reserves at period end.

8.8.5 *Twenty years later*

As stated before, this model was proposed in the mid-80s to represent possible market share evolutions on a time horizon extending up to 2020. We have now observed how oil prices have evolved, and we can obtain values for the volumes of gas circulating in Europe from energy statistics. We may thus compare the market shares for the different producers that were forecasted in 1985, to the observations.[11] The assumptions made in 1985 regarding the price of oil and expressed in 1983 dollars were as follows: $30/bbl for 1985–1989, $15/bbl for 1990–1994 and either $20 or $30, for 1995–1999. During this last period, the price of oil was actually closer to $20/bbl than $30/bbl, and therefore, we shall compare the realizations with the forecast made under this assumption. By comparing the predicted and observed market shares,[12] we obtain the Root Mean Square Errors (RMSE) reported in Table 8.7

As can be seen, the quantities predicted by the model for Holland, Germany and Belgium-Luxembourg, are quite close to the realized quantities. The model performance is also good for France and Italy. In the case of France, the model forecasted an important

---

[10]Million tonnes of oil equivalent.

[11]We could have based this comparison on quantities instead of market shares. The advantage of this formulation is that all root mean square errors are easily comparable with market shares as a unit of measurement.

[12]The market-share values are provided in the report [122].

Table 8.3 Flows in mtoe (NHV)

	Period	1985–1989	1990–1994	1995–1999			2000–2019		
Source	Destination	$30	$15	$20	$30	$20	$30	$30	$60
U.K.	U.K.	33.70	35.00	37.00	37.00	38.00	38.00	38.00	38.00
FRG	FRG (North)	10.00	13.70	12.74	11.40	12.40	11.62	11.71	8.39
FRG	FRG (South)	3.70	0.10	1.06	2.40	1.41	2.18	2.09	5.41
France	France (North)	1.86	2.61	3.00	3.00	3.00	3.00	3.00	1.58
France	France (South)	2.94	0.29	0.00	0.00	0.00	0.00	0.00	1.42
Italy	Italy (North)	2.03	0.00	0.04	0.69	0.45	1.06	1.10	1.39
Italy	Italy (South)	8.37	10.00	9.96	9.32	9.55	8.94	8.90	8.61
Alg. pipe	Italy (South)	5.06	4.25	7.55	9.00	9.30	10.82	10.84	9.66
Arzew	France (North)	3.24	2.43	4.48	5.47	5.53	7.00	7.03	6.28
Arzew	Belgium Lux.	3.55	2.79	3.98	4.55	4.61	5.30	5.32	4.52
Skikda	France (South)	5.28	3.97	6.93	8.83	8.18	10.55	10.55	10.25
Russia1	FRG (South)	13.67	16.01	16.90	15.88	17.88	17.26	17.30	12.82
Russia1	France (South)	7.92	12.56	15.14	14.14	17.45	16.75	16.74	13.22
Russia1	Italy (North)	9.72	14.54	17.15	16.38	18.90	18.60	18.62	14.70
Russia2	U.K.	0.00	0.00	22.38	22.42	22.44	22.51	22.51	22.09
Holland	Holland	22.93	20.96	24.11	23.97	24.35	23.97	23.90	22.08
Holland	FRG (North)	16.66	13.77	13.79	14.60	12.63	12.99	12.90	13.00
Holland	France (North)	3.31	4.04	3.93	3.71	3.50	3.14	3.10	3.52
Holland	Italy (North)	5.34	4.13	5.11	6.25	4.63	5.50	5.41	7.17
Holland	Belgium Lux.	3.86	4.67	4.22	3.88	4.25	3.77	3.75	3.29
Ekofisk	U.K.	9.07	8.70	4.52	4.70	0.00	0.00	0.00	0.00
Ekofisk	FRG (North)	7.06	3.82	8.34	9.50	0.00	0.00	0.00	0.00
Ekofisk	France (North)	1.49	1.23	11.88	1.65	0.00	0.00	0.00	0.00
Ekofisk	Belgium Lux.	2.18	2.58	3.06	2.83	0.27	0.27	0.00	0.00
Troll	U.K.	0.00	0.00	0.00	0.00	9.23	9.10	8.97	7.50
Troll	FRG (North)	0.00	0.00	0.00	0.00	12.34	13.11	13.03	13.99
Troll	France (North)	0.00	0.00	0.00	0.00	3.17	3.07	3.03	3.88
Troll	Belgium Lux.	0.00	0.00	0.00	0.00	3.82	3.47	3.72	3.47

Table 8.4 Price of gas $83/MMBtu (million of British thermal units) on different markets

Period	1985–1989	1990–1994	1995–1999			2000–2019		
Oil price	$30	$15	$20	$30	$20	$30	$30	$60
Holland	6.29	3.26	4.11	5.88	4.06	5.85	5.85	10.60
U.K.	4.72	2.61	2.66	3.50	2.66	3.51	3.515	4.715
Germany (North)	4.35	2.59	3.23	4.41	3.24	4.44	4.45	7.29
Germany (South)	4.35	2.59	3.23	4.42	3.24	4.44	4.45	7.29
France (North)	4.57	2.71	3.11	4.02	3.04	3.95	3.46	6.18
France (South)	4.57	2.66	3.06	4.00	2.97	3.82	3.82	6.18
Italy (North)	4.22	2.66	3.28	4.39	3.35	4.45	4.45	7.33
Italy (South)	4.22	2.71	3.28	4.39	3.25	4.45	4.45	7.33
Belgium-Luxemb.	5.11	3.09	3.71	4.95	3.07	4.93	4.94	8.04

Table 8.5   Total annual gas demand in mtoe (PCI)

Period	1985–1989	1990–1994	1995–1999			2000–2019		
Oil price	$30	$15	$20	$30	$20	$30	$30	$60
Gas demand	182.9	182.1	227.3	231.7	247.2	252.0	251.5	236.3

Table 8.6   Annual production and reserves at period end

Period		1985–1989	1990–1994	1995–1999			2000–2019		
Oil price $83/bbl		$30	$15	$20	$30	$20	$30	$30	$60
Algeria	(a)	17.14	13.44	22.94	27.85	27.62	33.66	33.74	30.71
	(b)	2745	2678	2564	2539	2011	1890	1864	1925
Holland	(a)	52.09	47.56	51.17	52.40	49.37	49.37	49.06	49.06
	(b)	1481	1243	987	981				
Norway	(a)	19.80	16.32	17.80	18.87	28.82	29.01	28.75	28.85
	(b)	176[1]	94	5					
		1733[2]	1733	1733	1733	1202	1179	1199	1197
Russia	(a)	31.31	43.10	71.58	68.83	76.63	75.12	75.16	62.85
	(b)	3280	3065	2707	2721	1174	1205	1218	1464

(1)   Ekofisk
(2)   Troll
(a)   production
(b)   reserves

Table 8.7   Root mean square errors

Holland	0,03113
UK	0,15103
Germany	0,04296
France	0,09296
Italy	0,09296
Belgium-Luxembourg	0,05461

role for Norway; and, in the case of Italy, the model underestimated Russia's market share. The model's relatively poor performance in the case of the UK is due to the prediction that Russian gas would play an important role in 1995–1999, a role that did not materialize at all.

These comparisons show us that the *S*-adapted information structure, when used to model stochastic resource markets, can provide useful insights. We could also re-run the model for the realized scenario and assess the value of information by comparing the deterministic solution with the uncertain one.

8.8.6 *In summary*

We have used the S-adapted equilibrium concept to model the European gas market. Our model displays the following features:

- It models real-life contingent contracts that have indexation clauses triggered by random evolutions in the price of oil.
- It uses the formalism of a Nash-Cournot game played over an event tree that represents the possible evolutions (scenarios) of the price of oil.
- The equilibrium is computed as a solution to an extended variational inequality, for which we have existence and uniqueness results, as shown in Section 8.3.
- The model has dynamic constraints representing the capacity-expansion and reserve-depletion processes; the decision to invest and produce (reserve commitment) are therefore similar to control variables that influence the evolution of a state variable. An economic interpretation of the decision taken at each node can be obtained by considering the tradeoffs for the agents between the immediate gain and the impact of the current decision on the conditional expected value of the state variables in the next periods.
- Finally a comparison of the scenario simulations and the actual data realizations shows an acceptable fit.

Chapter 9

Markov Games

9.1 Overview

In the first part of this chapter, we will present the formalism of stochastic games, initially proposed by L.S. Shapley in a seminal paper [219] where he defined *Markov games* in a two-player zero-sum context. This was the first paper to introduce the concept of a *controlled Markov chain* in the case of two antagonistic controllers. It so preceded the development of the theory of Markov decision processes (MDPs), as later initiated by Howard [127]. The very clever idea of Shapley was to combine, in a *Markov game*, the following two concepts of: (i) a matrix game and (ii) a controlled Markov chain. This provided a paradigm for competition in stochastic systems and gave an impetus to the development of stochastic games theory.

In the second part of this chapter, we will show how to extend the Shapley-Markov game formalism to the case of nonzero-sum stochastic games. This extension can also be seen as a generalization of the multistage game paradigm discussed in Chapter 5.7.5. Following the work of W. Whitt [245], we will use a dynamic-programming formalism to characterize infinite-horizon discounted payoff feedback (Markov)-Nash equilibrium solutions in the class of stationary Markov (feedback) strategies. We will give a proof of existence of Markov (feedback) Nash equilibrium, when the state and action sets are *finite*. Then, we will consider the possibility to further extend the theory to multistage stochastic games with *continuous* state and action spaces.[1] The mathematical apparatus necessary to deal with this type of stochastic games becomes challenging, so we content ourselves to provide the main existence theorems without proofs. We pay special attention to stochastic *supermodular* games because of their importance for economic applications.

We also provide two applications of the use of memory strategies in stochastic games. In Section 9.8, we show how the stochastic-game formalism can be exploited to characterize a subgame-perfect *collusive equilibrium* in a repeated-duopoly game with uncertainty and threats. And, in the $\mathbb{GE}$ section, we explain how to construct a collusive equilibrium in a discounted stochastic sequential game played over an infinite-time horizon.

---

[1] We will deal with continuous-time stochastic (diffusion) games in Chapter 11.

9.2 Shapley's Markov games

9.2.1 *Controlled transition probabilities*

Let us define the basic elements of a Markov game, as introduced by L. Shapley in [219].
We consider an infinite discrete-time horizon $t = 0, 1, 2, \ldots, \infty$, and let $S = \{1, 2, \ldots, n\}$
be the set of possible states of a discrete-time stochastic process $\{x(t) : t = 0, 1, \ldots\}$. Let
$U_j = \{1, 2, \ldots, m_j\}$ be the finite action (or control) sets of Player j, $j = 1, 2$, at each time
t. The process dynamics is described by the given of *transition probabilities*

$$p_{s,s'}(\underline{u}) = \mathrm{P}[x(t+1) = s' | x(t) = s, \underline{u}], \quad s, s' \in S, \quad \underline{u} = (u_1, u_2) \in U_1 \times U_2,$$

which satisfy, for all $\underline{u} \in U_1 \times U_2$,

$$p_{s,s'}(\underline{u}) \geq 0,$$

$$\sum_{s' \in S} p_{s,s'}(\underline{u}) = 1, \quad s \in S,$$

where $p_{s,s'}(\underline{u})$ is the conditional probability that the system state be s' at period $t + 1$,
given that it was in state s at time t and that the control $\underline{u}$ has been applied by the two
players together. As the transition probabilities depend on the players' actions we speak
of a *controlled Markov chain*, formed by the stochastic process $\{x(t) : t = 0, 1, \ldots\}$. A
transition-reward function

$$r(s, \underline{u}), \quad s \in S, \quad \underline{u} \in U_1 \times U_2,$$

defines the gain of Player 1 when the process is in state s and the two players take the action
pair $\underline{u}$. Since the game is *zero-sum*, Player 2's reward is given by $-r(s, \underline{u})$.

Notice that we have assumed stationary probabilities of transition and rewards, i.e., they
are defined by the same functions at each period of time.

9.2.2 *Numerical example*

In Figure 9.1, we show the data input in spreadsheet form for a Markov game with two
states $s \in \{1, 2\}$, two actions for Player 1 at each time $u_1 \in \{1, 2\}$, and three actions for
Player 2 $u_2 \in \{1, 2, 3\}$.

Later in Exercise 9.1 (on page 370), we ask the reader to replicate the spreadsheet on
his own Excel spreadsheet installation. The spreadsheet will be used to illustrate Shapley's
method and algorithm. In Table 9.1 we report the rewards corresponding to each state (1 or
2) and each choice of action by the players. Next period's rewards will be discounted. In
this example, we assume a discount factor of 0.75.

The values of the transition probabilities, which depend on the current state and the
actions chosen by the two players are given in Table 9.2. Notice that, for a given initial
state and when the actions of the two players are kept fixed, the transition probabilities
toward the two possible states always sum up to 1. This is a fundamental property of
Markov chains.

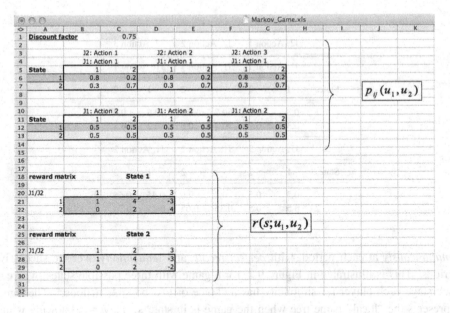

Fig. 9.1   Markov game data

Table 9.1   Reward data

Reward matrix		State 1	
J1 / J2	1	2	3
1	1	4	-3
2	0	2	4
Reward matrix		State 2	
J1 / J2	1	2	3
1	1	2	-4
2	0	1	-2

9.3   Information structure and strategies

The game structure introduced by Shapley can be related to the definition of a game in extensive form. This, in turn, will permit us to define the information structure, the strategy sets for each player and the payoffs related to strategy choices. That will provide a description of this game in normal form.

9.3.1   *The extensive form of the game at a given state s*

A Shapley-Markov game corresponds to a game in extensive form with an infinite number of moves. We assume an information structure where the players choose their controls

Table 9.2   Transition probabilities

	J2: Action 1		J2: Action 2		J2: Action 3	
	J1: Action 1		J1: Action 1		J1: Action 1	
States	1	2	1	2	1	2
1	0.8	0.2	0.7	0.3	0.6	0.4
2	0.3	0.7	0.4	0.6	0.2	0.8
	J1: Action 2		J1: Action 2		J1: Action 2	
States	1	2	1	2	1	2
1	0.5	0.5	0.4	0.6	0.45	0.55
2	0.5	0.5	0.35	0.65	0.25	0.75

simultaneously at each period, with *perfect recall*, and *sequentially* over the successive periods. This is illustrated in Figure 9.2, for a game with three possible states, where at each stage each player for simplicity has a choice of two possible actions.[2] The figure represents the "local" game tree when the game is in state s. Player 1, knowing which state has been reached, selects a control. Player 2, knowing which state has been reached (i.e., s), but ignoring which control action has been selected by Player 1, chooses his own control action. Then, nature plays (this is represented here by the symbol E, which is used to represent "event") and a random transition occurs to a new state, with probabilities depending on the current state s and the control actions chosen by the two players. This elementary building block will be repeated, over and over for all successive periods and for all discrete states that could be reached along histories of the game.

This extensive form representation is purely formal since the game tree will grow to infinity as the game progresses over time. The extensive form will not be convenient to explore the possible solutions to this game. In particular there is no possibility to define an associated matrix game. However, once we define the proper concept of strategy, we will be able to use probability calculus and some developments of the theory of Markov chains to obtain a normal-form description of the game. Then, a solution method will be proposed in the spirit of dynamic programming as in the previous chapter where we treated multistage games.

9.3.2   *Strategies, normal form and saddle-point solution*

A strategy is a device which transforms the information into action. Since the dynamics is a stochastic process, it is difficult to envision an open-loop information structure in which, at each time or stage, an action is selected independently of the current state of the process. On the other hand, since the players may recall the past observations, the general form of a strategy can be quite complex if it incorporates the whole history of the game in

---

[2]So, the graph does not directly correspond to our numerical example, in which the second player has three choices.

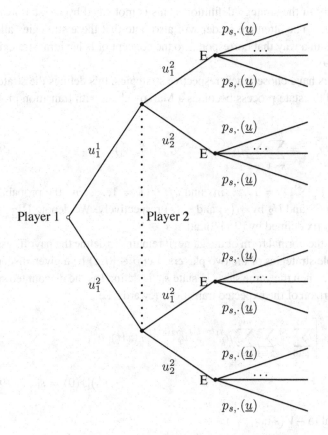

Fig. 9.2   A Markov game in extensive form

its definition. Conveniently, it is possible to prove that, when dealing with two-player zero-sum Markov games, we can limit our attention to the so-called Markov or feedback strategies.[3]

9.3.2.1   *Markov strategies*

These strategies are also called *feedback strategies*. The assumed information structure allows the use of strategies defined as mappings

$$\gamma_j : S \mapsto \mathcal{P}(U_j), \quad j = 1, 2,$$

where $\mathcal{P}(U_j)$ is the class of probability distributions over the action set U_j. Since the strategies are only based on the information conveyed by the current state of the x-process, they are called stationary *Markov strategies*. Notice that we use *stationary strategies* when time

---

[3]This result will not be proved in this book. We refer the reader to the excellent book of Filar & Vrieze [82] for more details on the theory of stochastic games.

t does not enter explicitly in the strategy definition. This is motivated by the stationarity of the Markov-game data. The attentive reader will also note that these strategies allow randomization of actions in a way that corresponds to the concept of behavioral strategies, (see Section 4.4, page 99).

When the two players have chosen their respective strategies, this defines the strategy vector $\underline{\gamma} = (\gamma_1, \gamma_2)$ and the state process becomes a Markov chain with transition probabilities

$$p_{s,s'}^{\underline{\gamma}} = \sum_{k=1}^{m_1} \sum_{\ell=1}^{m_2} \mu_k^{\gamma_1(s)} \nu_\ell^{\gamma_2(s)} p_{s,s'}(u_1^k, u_2^\ell), \tag{9.1}$$

where we have denoted $\mu_k^{\gamma_1(s)}, k = 1, \ldots, m_1$ and $\nu_\ell^{\gamma_2(s)}, \ell = 1, \ldots, m_2$ the probability distributions induced on U_1 and U_2 by $\gamma_1(s)$ and $\gamma_2(s)$ respectively. We denote $\mathbf{P}(\underline{\gamma})$ the transition probability matrix defined by (9.1) for all $s, s' \in S$.

In order to formulate the normal form of the game, it remains to define the payoffs associated with the admissible strategies of the two players. Let $\beta \in [0, 1)$ be a given discount factor. Player 1's payoff, when the game starts in state s_0, is defined as the discounted sum over the infinite-time horizon of the expected transitions rewards, i.e.,

$$V(s_0; \gamma_1, \gamma_2) = \mathbb{E}_{\gamma_1, \gamma_2} \left[\sum_{t=0}^{\infty} \beta^t \sum_{k=1}^{m_1} \sum_{\ell=1}^{m_2} \mu_k^{\gamma_1(x(t))} \nu_\ell^{\gamma_2(x(t))} r(x(t), u_1^k, \right.$$

$$\left. u_2^\ell) | x(0) = s_0 \right]. \tag{9.2}$$

Player 2's payoff is equal to $-V(s_0; \gamma_1, \gamma_2)$.

We denote

$$r(s, \underline{\gamma}) = \sum_{k=1}^{m_1} \sum_{\ell=1}^{m_2} \mu_k^{\gamma_1(s)} \nu_\ell^{\gamma_2(s)} r(s, u_1^k, u_2^\ell)$$

the expected transition reward when the current state is s and the strategy pair $\underline{\gamma}$ is used, and $\mathbf{r}(\underline{\gamma}) = (r(s, \underline{\gamma}) : s \in S)$ the vector of expected transition rewards. Now if we call $v(\underline{\gamma}) = (V(s; \gamma_1, \gamma_2) : s \in S)$ the vector of expected payoffs from any possible initial state, we claim that (9.2) can also be written as

$$v(\underline{\gamma}) = \sum_{t=0}^{\infty} \mathbf{P}(\underline{\gamma})^t \mathbf{r}(\underline{\gamma}) = (I - \mathbf{P}(\underline{\gamma}))^{-1} \mathbf{r}(\underline{\gamma}). \tag{9.3}$$

For that, we used the fact that in a Markov chain the transition probabilities in t steps are given by the power t of the one-step transition probability matrix.[4]

---

[4]We also use the matrix calculus identity

$$\sum_{t=0}^{\infty} \mathbf{P}(\underline{\gamma})^t = (I - \mathbf{P}(\underline{\gamma}))^{-1}.$$

Definition 9.1. A pair of strategies (γ_1^*, γ_2^*) is a **saddle-point** solution if, for all strategies γ_1 and γ_2 of players 1 and 2 and for all $s \in S$

$$V(s; \gamma_1, \gamma_2^*) \leq V(s; \gamma_1^*, \gamma_2^*) \leq V(s; \gamma_1^*, \gamma_2). \tag{9.4}$$

The number

$$v^*(s) = V(s; \gamma_1^*, \gamma_2^*)$$

is called the *value of the game* at state s.

Remark 9.1. Notice that the saddle-point property is required to hold at every possible initial state s. This is due to the fact that we are now working in a random process framework and the strategy pair will have to remain a saddle-point solution for all the states, which are possibly visited.

9.3.2.2 *Memory strategies:*

In a closed-loop information structure each player can base his decision at period t on the whole history of the game up to time t:

$$h(t) = \{x(0), \underline{u}(0), x(1), \underline{u}(1), \ldots, x(t-1), \underline{u}(t-1), x(t)\}.$$

In a more restricted information structure, we could assume that the players do not have a direct access to the actions used in the past by their opponent; they can only observe the *state history*. However, as in the case of single-player Markov decision processes,[5] it can be shown that the use of Markov strategies leads to the same saddle-point solutions for zero-sum Markov games as when history-based memory strategies are allowed. This will not be the case, when we deal with nonzero-sum stochastic games and Nash-equilibrium solution.

9.4 Shapley-Denardo's operator formalism

In this section, we present the formalism of dynamic-programming operators, which has been formally introduced by Denardo in [56] but was already implicit in Shapley's work. The solution of the dynamic-programming equations for the stochastic game is obtained as a fixed point of an operator acting in the space of value functions.

9.4.1 *Dynamic-programming operators*

In reference [219], Shapley proved the existence of optimal (saddle point) stationary Markov strategies using a fixed-point argument, involving a *contracting operator* acting on the space of value functions. We give here a brief account of the method, taking our inspiration from references [56] and [82]. Let $v(\cdot) = (v(s) : s \in S)$ be an arbitrary function

---

[5]A comprehensive treatment of Markov decision processes can be found in [204].

with real values, defined over S. Since we assume that S is finite this function could also be viewed as a vector. Introduce, for any $s \in S$ the so-called *local-reward functions*

$$\mathfrak{h}(v(\cdot), s, u_1, u_2) = r(s, u_1, u_2) + \beta \sum_{s' \in S} p_{s,s'}(u_1, u_2) \, v(s') \quad (u_1, u_2) \in U_1 \times U_2. \quad (9.5)$$

Let us define for each $s \in S$ a *local zero-sum matrix game* with pure strategy sets U_1 and U_2 and payoffs

$$H(v(\cdot), s) = [\mathfrak{h}(v(\cdot), s, u_1^k, u_2^\ell)]_{\substack{k = 1, \ldots, m_1 \\ \ell = 1, \ldots, m_2}}.$$

This matrix game has a value[6] (saddle point solution) denoted $\mathrm{val}[H(v(\cdot), s)]$. Then consider the function that associates with a vector $v(\cdot) \in \mathbf{R}^n$ and a state $s \in S$ the value of the matrix game defined above

$$T(v(\cdot))(s) := \mathrm{val}[H(v(\cdot), s)]. \quad (9.6)$$

Now, let us define a mapping $\mathbf{T} : \mathbf{R}^n \mapsto \mathbf{R}^n$ from the space of n dimension vectors into itself as follows

$$\mathbf{T}(v(\cdot)) = (T(v(\cdot))(s) : s \in S).$$

This mapping will be called the *dynamic-programming operator* of the Markov game.

9.4.2 Existence of sequential saddle points

To prove that a saddle point solution exists for a Markov game, we need to first establish the following auxiliary results:

Lemma 9.1. *Let A and B be two matrices of same dimensions, and $\mathrm{val}[A]$, respectively, $\mathrm{val}[B]$ the values for the associated matrix games. Then*

$$|\mathrm{val}[A] - \mathrm{val}[B]| \leq \max_{k,\ell} |a_{k,\ell} - b_{k,\ell}|. \quad (9.7)$$

The proof of this lemma can be found in reference [82], page 178, and is left as Exercise 9.2.

Lemma 9.2. *If $v(\cdot)$, γ_1 and γ_2 are such that, for all $s \in S$*

$$v(s) \leq (\text{respectively} \geq; \text{ respectively } =) \quad \mathfrak{h}(v(\cdot), s, \gamma_1(s), \gamma_2(s))$$
$$= r(s, \gamma_1(s), \gamma_2(s)) + \beta \sum_{s' \in S} p_{s,s'}(\gamma_1(s), \gamma_2(s)) \, v(s'), \quad (9.8)$$

then

$$v(s) \leq (\text{respectively} \geq; \text{ respectively } =) \quad V(s; \gamma_1, \gamma_2). \quad (9.9)$$

Proof. The proof is relatively straightforward and consists in iterating the inequality (9.8) on the expression (9.3). See [82], page 24, or Exercise 9.3, for details. $\square$

---

[6]Recall that we use $\mathrm{val}[A]$ to denote the operator that associates its saddle point value to a matrix game A.

We can now establish the following result:

Theorem 9.1. *The mapping* $\mathbf{T}$ *is contracting in the "max"-norm*

$$\|v(\cdot)\| = \max_{s \in S} |v(s)|$$

and admits the value function introduced in Definition 9.1 as its unique fixed point

$$v^*(\cdot) = \mathbf{T}(v^*(\cdot)).$$

Furthermore, the saddle-point (optimal for each player) strategies are defined at each state s *as the mixed strategies in the local zero-sum game yielding this value, i.e.,*

$$\mathfrak{h}(v^*(\cdot), s, \gamma_1^*(s), \gamma_2^*(s)) = \text{val}[H(v^*(\cdot), s)], \quad s \in S. \tag{9.10}$$

Proof. We first establish the contraction property. We use Lemma 9.1 and the transition probability properties to establish the following inequalities

$$\|\mathbf{T}(v(\cdot)) - \mathbf{T}(w(\cdot))\| \leq \max_{s \in S} \left\{ \max_{u_1 \in U_1; u_2 \in U_2} \left| r(s, u_1, u_2) + \beta \sum_{s' \in S} p_{s,s'}(u_1, u_2) v(s') \right. \right.$$

$$\left. \left. - r(s, u_1, u_2) - \beta \sum_{s' \in S} p_{s,s'}(u_1, u_2) w(s') \right| \right\}$$

$$= \max_{s \in S; u_1 \in U_1; u_2 \in U_2} \left| \beta \sum_{s' \in S} p_{s,s'}(u_1, u_2) (v(s') - w(s')) \right|$$

$$\leq \max_{s \in S; u_1 \in U_1; u_2 \in U_2} \beta \sum_{s' \in S} p_{s,s'}(u_1, u_2) |v(s') - w(s')|$$

$$\leq \max_{s \in S; u_1 \in U_1; u_2 \in U_2} \beta \sum_{s' \in S} p_{s,s'}(u_1, u_2) \|v(\cdot) - w(\cdot)\|$$

$$= \beta \|v(\cdot) - w(\cdot)\|.$$

Hence $\mathbf{T}$ is a contraction, since $0 \leq \beta < 1$. By the Banach contraction theorem[7] this implies that there exists a unique fixed point $v^*(\cdot)$ to the operator $\mathbf{T}$.

We now show that there exist stationary Markov strategies γ_1^*, γ_2^* for which the saddle-point condition holds

$$V(s; \gamma_1, \gamma_2^*) \leq V(s; \gamma_1^*, \gamma_2^*) \leq V(s; \gamma_1^*, \gamma_2). \tag{9.11}$$

Let $\gamma_1^*(s), \gamma_2^*(s)$ be the saddle-point strategies for the local matrix game with payoffs

$$[\mathfrak{h}(v^*(\cdot), s, u_1, u_2)]_{\substack{u_1 \in U_1 \\ u_2 \in U_2}} = H(v(\cdot), s). \tag{9.12}$$

Consider any strategy γ_2 for Player 2. Then, by definition

$$\mathfrak{h}(v^*(\cdot), s, \gamma_1^*(s), \gamma_2(s)) \geq v^*(s) \quad \forall s \in S. \tag{9.13}$$

---

[7]*Banach contraction theorem*, [22]. Let (X, d) be a non-empty complete metric space. Let $T : X \to X$ be a contraction mapping on X, i.e., there is a nonnegative real number $q < 1$ such that $d(T(x), T(y)) \leq qd(x, y)$ for all $x, y \in X$. Then the map T admits one and only one fixed point x^* in X (this means $T(x^*) = x^*$). Furthermore, this fixed point can be found as follows: start with an arbitrary element x_0 in X and define an iterative sequence by $x_n = T(x_{n-1})$ for $n = 1, 2, 3, \ldots$. This sequence converges, and its limit is x^*.

By Lemma 9.2, the inequality (9.13) implies that for all $s \in S$

$$V(s; \gamma_1^*, \gamma_2) \geq v^*(s). \tag{9.14}$$

Similarly, we would obtain that for any strategy γ_1 and for all $s \in S$

$$V(s; \gamma_1, \gamma_2^*) \leq v^*(s), \tag{9.15}$$

and thus

$$V(s; \gamma_1^*, \gamma_2^*) = v^*(s). \tag{9.16}$$

This establishes the saddle-point property. $\square$

9.4.3 *Numerical illustration*

We use the data of Figure 9.1 to illustrate this contraction property. We start with a value function set to 0 for each state

$$v^0(1) = 0, \quad v^0(2) = 0.$$

We now define the local matrix games, using the extended rewards defined in equation (9.5) as

$$\mathfrak{h}(v(\cdot), s, u_1, u_2) = r(s, u_1, u_2) + \beta \sum_{s' \in S} p_{s,s'}(u_1, u_2) \, v(s'), \quad (u_1, u_2) \in U_1 \times U_2.$$

Since the value function has a 0 value, these matrix games correspond to the transition reward matrices as in Table 9.3. These are matrix games that can be solved using linear programs (see Chapter 3). We obtain the mixed strategies indicated in the margins (called

Table 9.3   Local matrix games: Step 1

Reward matrix			State 1	
J1/ J2	1	2	3	Proba-P1
1	1	4	-3	0.5
2	0	2	4	0.5
Proba-P2	0.875	0	0.125	1

Reward matrix			State 2	
J1/ J2	1	2	3	Proba-P1
1	1	2	-4	0
2	0	1	-2	1
Proba-P2	0	0	1	1

Proba-P1 and Proba-P2, respectively) of the two matrices and the expected rewards

$$v^1(1) = 0.5, \quad v^1(2) = -2.$$

Using these values, we repeat the construction of the local matrix games using equation (9.5). We obtain

Table 9.4   Local matrix games: Step 2

Reward matrix		State 1			
J1/ J2		1	2	3	Proba-P1
1		1	3.81	-3.38	0.48
2		-0.94	1.25	4	0.52
Proba-P2		0.77	0	0.23	1

Reward matrix		State 2			
J1/ J2		1	2	3	Proba-P1
1		0.063	1.25	-5.13	0
2		-0.56	0.16	-3.03	1
Proba-P2		0	0	1	1

These are matrix games that are solved again using linear programs. We obtain the mixed strategies indicated in the margins of the two matrices and the expected rewards

$$v^2(1) = 0.0013, \quad v^2(2) = -3.03.$$

We repeat the procedure one more time and we get the local matrix games

Table 9.5   Local matrix games: Step 3

Reward matrix		State 1			
J1/ J2		1	2	3	Proba-P1
1		0.55	3.32	-3.91	0.49
2		-1.59	0.64	2.75	0.51
Proba-P2		0.76	0	0.24	1

Reward matrix		State 2			
J1/ J2		1	2	3	Proba-P1
1		-0.59	0.64	-5.82	0
2		-1.14	-0.48	-3.70	1
Proba-P2		0	0	1	1

The saddle-point reward vector is now

$$v^3(1) = -0.54, \quad v^3(2) = -3.70.$$

As we can already observe from these three steps, the max norm of the differences of the successive value functions decreases at each step. In Exercise 9.1, we ask the reader to iterate by repeating this operation 10 more times and observing what happens to the max norm of the difference in values.

Table 9.6   Evolution of value functions

State	v0	v1	v2	v3
1	0	0.5	0.0013	-0.5364
2	0	-2	-3.0313	-3.7048

Norm of Differences

$|v^{n+1}(1) - v^n(1)|$		0.5	0.4987	0.5377
$|v^{n+1}(2) - v^n(2)|$		2	1.0313	0.6736
Max norm		2	1.0313	0.6736

9.5   How to compute a solution to a Markov game

In this section, we investigate the numerical algorithms that have been proposed to solve Markov games. We take our inspiration from Refs. [33, 81].

In the single-player case, a Markov game is a Markov-decision process (MDP). In this class of problems, the contraction and monotonicity of the dynamic-programming operator can be exploited to design two types of numerical algorithms for the computation of optimal policies, which are called *value-iteration* and *policy-iteration* procedure, respectively. Both procedures are converging. The policy-iteration procedure converges in a finite number of steps and can be shown to be equivalent to solving a linear program. We refer to [127, 204] for details. The situation is more complicated for Markov games.

9.5.1   *Shapley value iteration algorithm*

The proof of the existence of a saddle-point in a Markov game is based on a contraction property of the dynamic-programming operator. Therefore, the iterated use of this operator must converge to the unique fixed point. This will provide a value-iteration procedure for Markov games, which we shall call the *Shapley algorithm*. The extension of the policy-iteration procedure to Markov games however, has proved more difficult.

9.5.2   *Hoffman and Karp algorithm*

Hoffman and Karp ([126]) proposed the following algorithm to solve Markov games:

step 0   Start with a function (vector) $v^0 \geq v^*$ componentwise.
step n   Find the strategies γ_1^n, γ_2^n, which solve

$$\mathfrak{h}(v^n(\cdot), s, \gamma_1^n(s), \gamma_2^n(s)) = T(v^n(\cdot))(s), \forall s \in S,$$

where $T(v^n(\cdot))(s)$ has been defined in (9.6). Define v^{n+1} as the fixed point of the operator $M_{\gamma_2} : v(\cdot) \mapsto v(\cdot)$ defined by

$$M_{\gamma_2}(v(\cdot))(s) = \max_{\gamma_1} \mathfrak{h}(v(\cdot), s, \gamma_1, \gamma_2^n(s)),$$

that is,

$$v^{n+1}(s) = M_{\gamma_2}(v^{n+1}(\cdot))(s), \forall s \in S.$$

stop Repeat **step n** until $\|v^{n+1} - v^n\| < \varepsilon$.

Rao, Chandrasekaran and Nair [209] showed that, for any $\varepsilon > 0$, this algorithm will converge to an ε-approximation of v^* in a finite number of iterations. At each **step n**, we first compute the values and associated saddle-point strategies for the card(S) local games associated with the function $v^n(s)$, and then we solve the MDP obtained when we take Player 2's strategy as fixed, and we optimize the payoff to Player 1. This consists in solving an MDP by policy iteration. These two operations can be made through the use of linear programs. The convergence is improved, compared to Shapley's algorithm, due to the exploitation of the information contained in Player 2 strategy at each iteration.

9.5.3 *Pollatschek and Avi-Itzhak policy iteration algorithm*

Pollatschek & Avi-Itzhak [202] proposed an algorithm, which can be considered as an extension of the policy iteration procedure of MDPs to the case of Markov games. It proceeds as follows:

step 0 Start with a function $v^0 \geq v^*$ componentwise.
step n Find the strategies γ_1^n, γ_2^n which solve

$$\mathfrak{h}(v^n(\cdot), s, \gamma_1^n(s), \gamma_2^n(s)) = T(v^n(\cdot))(s), \forall s \in S.$$

Then compute the new value function v^{n+1} as the fixed point of the operator $L_{\gamma_1^n, \gamma_2^n}$: $v(\cdot) \mapsto v(\cdot)$ defined by

$$L_{\gamma_1^n, \gamma_2^n}(v(\cdot))(s) = \mathfrak{h}(v(\cdot), s, \gamma_1^n(s), \gamma_2^n(s)),$$

that is:

$$v^{n+1}(s) = L_{\gamma_1^n, \gamma_2^n}(v^{n+1}(\cdot))(s), \forall s \in S.$$

stop Repeat **step n** until $\|v^{n+1} - v^n\| < \varepsilon$.

Pollatschek and Avi-Itzhak showed that this algorithm is guaranteed to converge if the following condition is satisfied:

$$\max_{s \in S} \left\{ \sum_{s' \in S} \left(\max_{\underline{u}} p_{ss'}(\underline{u}) - \min_{\underline{u}} p_{ss'}(\underline{u}) \right) \right\} \leq \frac{1 - \beta}{\beta}. \tag{9.17}$$

Van der Wal [235] showed, via a counterexample, that the algorithm does not converge in general (i.e., when (9.17) is not verified). However, numerical experiments [33] show that it often converges, even for problems where the condition (9.17) is not satisfied. So this is a case of an algorithm that can be useful to compute saddle-point solutions to Markov game even though it is not guaranteed to always converge.

9.5.4 *Newton-type methods for Markov games*

An improved policy-iteration method was proposed by Filar & Tolwinski [81, 228, 229] (see also the book [82]). It is based on the Newton method used to find the 0 of a differentiable function.

The link between the Newton method in unconstrained optimization and the Avi-Itzhak algorithm for stochastic games can be established as follows. Shapley showed that the value $v^*(s)$ of the game at different initial states $s \in S$ is obtained as a fixed point of the operator T defined in (9.6), i.e., $v^* = \mathbf{T}(v^*)$ or

$$0 = T(v^*(\cdot))(s) - v^*(s), \ \forall s \in S. \tag{9.18}$$

Let us denote $\Psi(v) = \mathbf{T}(v) - v$ and $\Psi'(v)$ the Jacobian matrix[8] of the vector valued function $\Psi(v)$. It can be shown that

$$\Psi'(v) = \beta P(v) - I, \tag{9.19}$$

where I is the unit matrix and $P(v)$ is the Markov-chain matrix associated with the saddle-point solutions of the games $H(v(\cdot), s)$ in different states $s \in S$. Since $\beta \in (0, 1)$ and $P(v)$ is a Markov-chain matrix, the matrix $[\beta P(v) - I]^{-1} = [\Psi'(v)]^{-1}$ exists.

For any $v(\cdot) \in \mathbf{R}^n$ let us call $\underline{\gamma}^v = (\gamma_1^v, \gamma_2^v)$ the saddle point strategy pair used in the definition of $\mathbf{T}(v(\cdot))$. This pair of strategies defines, for each $s \in S$, two probability-distribution vectors over actions (controls) of the two players. Let us call $r(s, \underline{\gamma}^v(s))$ the expected transition reward of Player 1 when the state is s and strategies $\underline{\gamma}^v$ are used. Similarly, call $P(\underline{\gamma}^v)$, or more simply as we did above, $P(v)$, the Markov-chain matrix obtained from the use of the stationary strategy pair $\underline{\gamma}^v$. It can easily be shown that the expected payoff over the infinite horizon is given by

$$V(\cdot, \underline{\gamma}^v) = [I - \beta P(\underline{\gamma}^v)]^{-1} r(\cdot, \underline{\gamma}^v(\cdot)) \tag{9.20}$$

or

$$V(\cdot, \underline{\gamma}^v) = [I - \beta P(v)]^{-1} r(\cdot, \underline{\gamma}^v(\cdot)).$$

To solve the game, we need to find v^* such that $\Psi(v^*) = 0$. Consider a candidate v^k. The Taylor expansion of order 1 for $\Psi(v^k)$ is

$$\Psi(v^k + h) = \Psi(v^k) + h\Psi'(v^k) + o(h).$$

A Newton step consists of taking

$$h = -[\Psi'(v^k)]^{-1}\Psi(v^k),$$

that is, a new candidate v^{k+1}, such that

$$\begin{aligned} v^{k+1} &= v^k - [\Psi'(v^k)]^{-1}\Psi(v^k) \\ &= v^k + [I - \beta P(v^k)]^{-1}[T(v^k) - v^k] \\ &= v^k + [I - \beta P(v^k)]^{-1}[r(\cdot, \underline{\gamma}^{v^k}(\cdot)) + \beta P(v^k)v^k - v^k] \\ &= [I - \beta P(v^k)]^{-1}[r(\cdot, \underline{\gamma}^{v^k}(\cdot))]. \end{aligned} \tag{9.21}$$

$$ \tag{9.22}$$

---

[8]Recall that $v = v(\cdot)$ is a vector in $\mathbf{R}^n$.

By (9.20) this last expression is also the expected payoff vector (indexed over the set of possible initial states) $V(\cdot, \underline{\gamma}^v)$. This corresponds to the main step in the Pollatschek & Avi-Itzhak algorithm, which can then be interpreted as a pure Newton method.

The Newton method of unconstrained minimization is very efficient but is not guaranteed to converge for nonquadratic functions. However, modified Newton methods can be devised, where the step size is restricted to a value that makes the norm of the gradient decrease. This is the approach proposed by Filar & Tolwinski.

They consider the function

$$J(v) = \frac{1}{2}[\mathbf{T}(v) - v]^T(\mathbf{T}(v) - v) = \frac{1}{2}\|\mathbf{T}(v) - v\|^2.$$

For all v in $\mathbf{R}^n$, the search for a fixed point v^* is equivalent to the search for the minimum of $J(v)$ which must be 0. The gradient of $J(v)$ is given by

$$J'(v) = [\Psi(v)]^T[\Psi'(v)].$$

They modify iterative step 9.21 according to

$$v^{k+1} = v^k - w^k[\Psi'(v^k)]^{-1}\Psi(v^k),$$

where w^k is selected so as to ensure that $J(v^{k+1}) < J(v^k)$.

The globally convergent modified Newton algorithm is the following:

Step 0 Set $k := 0$. Select two parameter values: $\alpha \in (0, 1)$ and $\mu \in [0.5, 0.8]$. Also select v^0 to be an estimate of the value vector.

Step k Calculate for each $s \in S$, for the matrix game $H(v^k(\cdot, s)$ the saddle-point (mixed) strategies γ^{v^k}, and the value vector $T(v^k)(s)$. Then compute $\Psi(v^k)$ and $J(v^k)$.

Step $k : 1$ If $J(v^k) = 0$, stop; v^k is the value vector of the stochastic game.

Step $k : 2$ Calculate $s^k = -[\Psi'(v^k)]^{-1}\Psi(v^k)$.

Step $k : 3$ Set $w^k := 1$.

Step $k : 4$ Test the inequality

$$J(v^k + w^k s^k) - J(v^k) \leq \alpha w^k J'(v^j)s^k.$$

Step $k : 5$ If the above inequality is satisfied, set $v^k := v^k + w^k s^k$ and return to Step k.

Step $k : 6$ Set $w^k := \mu w^k$ and return to step $k : 4$.

9.5.5 *Multiagent reinforcement learning for Markov games*

Recently the theory of Markov games has been revisited in the context of artificial intelligence. Littman [160, 161] extended the so-called "Q-Learning" algorithm, which had originally been proposed for solving MDPs, to zero-sum stochastic games. The algorithm is provided below. The notion of a Q function ("Q" for "quality") is used to maintain the value of joint set-actions, and the backup operation computes the value of states using the "value" operator. More precisely, we consider a player who does not know anything about the data of the game (i.e., the reward and transition probability functions) and will learn from experience. At stage t he knows the current state s and he has an evaluation function $Q(s, u)$, which summarizes his learning from past experiments. Now he organizes a new random experiment, which will result in observing a pair of actions $\underline{u} = (u_1, u_2)$, a reward r and a transition to a new state s'. The Q-learning procedure is defined as follows:

(1) **Initialize** $Q(s, \underline{u})$, $s \in S, \underline{u} \in U_1 \times U_2$ arbitrarily and set α to be the learning rate.
(2) **Repeat,**

    (a) From state s select action $\underline{\bar{u}}$ that solves the matrix game $\left[Q(s, \underline{u})_{\underline{u} \in U_1 \times U_2}\right]$.
    (b) Observing joint action $\underline{\bar{u}}$, reward r and the realization of next state s' update the quality function as follows:

$$Q(s, \underline{\bar{u}}) \leftarrow (1 - \alpha)Q(s, \underline{\bar{u}}) + \alpha(r + \beta V(s')],$$

    where

$$V(s) = \text{value} \left(\left[Q(s, \underline{u})_{\underline{u} \in U_1 \times U_2}\right] \right).$$

Littman proved that this algorithm defines a sequence of functions $Q(s, \underline{u})$ converging almost surely to the stochastic game's saddle-point solution, assuming that the other agent executes all of its actions infinitely often. This is true even if the other agent does not converge to the equilibrium; thus, it provides an opponent-independent method for learning an equilibrium solution.

9.6 Nonzero-sum stochastic games with finite state and action sets

Nonzero-sum stochastic games provide a paradigm to model competition under uncertainty. However, the mathematical apparatus that is necessary to deal with them is relatively complicated, requiring a mastery of deep results in measure theory, topology and functional analysis. This goes beyond the moderate mathematical level we promised to maintain in this textbook. Therefore, we will often give theorems without entering into a complete and detailed proof, hence provide an overall feeling for this approach, which has been used to deal with uncertainty and infinite horizon in m-player games.

9.6.1 *Stochastic game data*

A nonzero-sum stochastic game with finite state and action sets is defined by the data summarized in the quintuple

$$(M, S, U_j(\cdot), \mathbf{P}(\cdot, \cdot), \mathbf{r}(\cdot, \cdot)),$$

where

- $M = \{1, \ldots, m\}$ is the set of m players;
- $S = \{1, 2, \ldots, n\}$ is the set of possible states;

- for each state $s \in S$ the admissible finite action set is defined by $U_j(s)$ for each Player $j = 1, \ldots, m$;
- $\mathbf{P}(\cdot, \cdot)$ is the transition probability matrix, given s and $\underline{u}$. Its elements are the transition probabilities $p_{s,s'}(\underline{u}) = \mathrm{P}[x(t+1) = s'|x(t) = s, \underline{u}]$, with $s, s' \in S$ and $\underline{u} = (u_1, \ldots, u_m) \in U_1(s) \times \cdots \times U_m(s)$;
- $\mathbf{r}(\cdot, \cdot)$ is the transition-reward vector when the process is in state s and the players take the actions $\underline{u}$. Its components are, for each $j = 1, \ldots, m$ by $r_j(s, \underline{u})$, $s \in S$, $\underline{u} \in U_1(s) \times \cdots \times U_m(s)$.

Remark 9.2. Notice that we allow the different players' action sets to depend on the game's current s state. Shapley's theory still remains valid in that slightly more general setting.

9.6.2 *Stationary-Markov strategies and feedback-Nash equilibria*

Markov-stationary strategies are defined as in the zero-sum case. As for Markov games, we define Markov strategies as mappings from S into the players' *mixed actions*, i.e., $\gamma_j : s \in S \to \mathcal{P}\{U_j(s)\}$ where $\mathcal{P}\{U_j(s)\}$ is the set of probability distributions on the finite action set $U_j(s)$. We denote by $\mu_k^{\gamma_j(s)}$ the probability given to action u_j^k by Player j when he uses strategy γ_j and the current state is s.

Player j's payoff is thus defined as

$$V_j(s_0; \gamma_1, \ldots, \gamma_m) = \mathbb{E}_{\gamma_1, \ldots, \gamma_m} \left[\sum_{t=0}^{\infty} \beta^t \sum_{k=1}^{\lambda_1(x(t))} \cdots \sum_{\ell=1}^{\lambda_m(x(t))} \mu_k^{\gamma_1(x(t))} \cdots \mu_\ell^{\gamma_m(x(t))} \right.$$
$$\left. \times r_j(x(t), u_1^k, \ldots, u_m^\ell)|x(0) = s_0 \right],$$

where $x(t)$ is the random state visited at time t and $\lambda_j(x(t))$ is the number of elements in $U_j(x(t)), j = 1, \ldots, m$.

We call Γ_j the set of all stationary strategies for Player j and $\Gamma = \prod_{j=1}^{m} \Gamma_j$ the product strategy space.

Definition 9.2. An m-tuple of Markov strategies $\underline{\gamma}^* \triangleq (\gamma_1^*, \ldots, \gamma_m^*) \in \Gamma$ is a feedback-Nash equilibrium if, for all $s \in S$ and any stationary strategy $\gamma_j \in \Gamma_j$ of Player $j, j \in M$, the following holds:

$$V_j(s; [\gamma_j, \underline{\gamma}^{*-j}]) \triangleq V_j(s; \gamma_1^*, \ldots, \gamma_j, \ldots, \gamma_m^*)$$
$$\leq V_j(s; \gamma_1^*, \ldots, \gamma_j^*, \ldots, \gamma_m^*) \triangleq V_j(s; \underline{\gamma}^*). \quad (9.23)$$

The payoff value

$$v^*(s, j) = V_j(s; \underline{\gamma}^*)$$

is called the equilibrium value of the game at state s for Player j.

9.6.3   *Denardo-Whitt operator formalism*

The first authors to extend Shapley's work to a nonzero-sum framework were Rogers [211], Sobel [223] and Parthasarathy [195]. We follow the more recent treatment of nonzero-sum sequential games proposed by Whitt [245], who exploited the operator formalism developed by Denardo [56].

Let us introduce the *local-reward functions*

$$\mathfrak{h}_j(s, v(\cdot, j), \underline{u}) = r_j(s, \underline{u}) + \beta \sum_{s' \in S} p_{ss'}(\underline{u}) v(s', j), \quad j \in M \tag{9.24}$$

where the functions $v(\cdot, j) : S \mapsto \mathbf{R}$ are given *reward-to-go* functionals[9] (in this case, they are vectors of dimension $n = \mathrm{card}(S)$) defined for each player j. The local reward (9.24) is the sum of the transition reward for Player j and the discounted expected reward-to-go from the new state reached in the transition. For a given s and a given set of reward-to-go functionals $v(\cdot, j)$, $j \in M$, the local rewards (9.24) define an auxiliary m-matrix game where the pure strategies are the elements of the action sets $U_j(s)$, $j \in M$.

We now define, for any given Markov policy vector $\underline{\gamma} = \{\gamma_j\}_{j \in M}$, an operator $\mathcal{H}_{\underline{\gamma}}$, acting on the space of reward-to-go functionals, defined as[10]

$$\mathcal{H}_{\underline{\gamma}}(v(\cdot))(s, j) = \{\mathbb{E}_{\gamma(s)}[\mathfrak{h}_j(s, v(\cdot, j), \underline{u}]\}. \tag{9.25}$$

It can be shown that this operator, defined for each player j, is contracting and thus admits a unique fixed point, denoted by $v_{\underline{\gamma}}(s, j) : s \in S$, which is the expected discounted payoff to Player j, associated with strategy vector $\underline{\gamma}$, when the initial state is s.

Let $f_{\underline{\gamma}}(s, j) = \sup\{v_{[\gamma_j, \gamma_{-j}]}(s, j) : \gamma_j \in \Gamma_j\}$ denote the maximum expected payoff by Player j when replying to the strategy choice $\gamma_{-j} = \{\gamma_i : i \in M, i \neq j\}$ by the other players, and also introduce the operator $\mathcal{F}_{\underline{\gamma}}$ defined as

$$\mathcal{F}_{\underline{\gamma}}(v(\cdot))(s, j) = \{\sup_{\gamma_j} \mathbb{E}_{[\gamma_j, \gamma_{-j}](s)}[\mathfrak{h}_j(s, v(\cdot, j), \underline{\hat{u}})]\}. \tag{9.26}$$

Here $\underline{\hat{u}}$ is the random action vector generated by $[\gamma_j, \gamma_{-j}]$, which is the Markov strategy vector obtained when only Player j adjusts his policy, while the other players keep their γ-policies fixed. It can be shown that this operator is also contracting and admits a unique fixed point which is $f_{\underline{\gamma}}(s, j)$ for $s \in S$ and $j \in M$.

In other words, equation (9.26) serves to define the optimal reply of each Player j to the Markov strategies chosen by the other players. Therefore, the expression

$$\eta_{\underline{\gamma}}(s, j) \triangleq f_{\underline{\gamma}}(s, j) - v_{\underline{\gamma}}(s, j) \geq 0$$

can be interpreted as a disequilibrium indicator. Hence γ^* is a feedback-Nash equilibrium iff

$$\eta_{\underline{\gamma}^*}(s, j) = 0, \forall s \in S, \forall j = 1, \dots, m. \tag{9.27}$$

It is thus natural to use the point-to-set mappings

$$\psi_j(\underline{\gamma}) = \{\gamma_j \in \Gamma_j : \eta_{\underline{\gamma}}(s, j) = 0 \quad \forall s\}, \tag{9.28}$$

which give the set of optimal reply strategies for Player j when the other players use γ_{-j}.

---

[9]We use the term *functional* to denote a function which maps the state set S into the reals $\mathbf{R}$.

[10]We change slightly our notations here, in order to accommodate the m-player case. $\mathcal{H}(v(\cdot))$ is a function with argument $(s, j) \in S \times M$. $\mathcal{H}(v(\cdot))(s, j)$ is the value of this function at (s, j).

9.6.4 Existence of a feedback-Nash equilibrium

The existence theorem for a feedback-Nash equilibrium in sequential games will be based on the following fixed-point theorem, which is an important result in topology and functional analysis.

Theorem 9.2. *(Kakutani, Glicksberg and Fan). If X is a convex, compact subset of a Hausdorff locally convex topological vector space*[11] *and $\Psi : X \to 2^X$ is convex-valued and upper semi continuous, then $x \in \Psi(x)$ for some $x \in X$.*

Proof. This extension of the Kakutani fixed-point theorem was proved by Glicksberg [94] and Ky Fan [74]. □

We will apply this theorem with $X = \mathbf{\Gamma} = \prod_{j=1}^{m} \Gamma_j$ and

$$\Psi(\underline{\gamma}) = \prod_{j=1}^{m} \psi_j(\underline{\gamma}), \quad \underline{\gamma} \in \mathbf{\Gamma} \tag{9.29}$$

where $\psi_j(\underline{\gamma})$ was defined in (9.28). The reader should compare the construct of this point-to-set mapping $\Psi(\underline{\gamma})$ with the similar development that concerns global reaction functions in concave games, and which is covered in Chapter 3, Section 3.4.2.[12] There are obvious similarities. Here we have also defined a global reaction function that admits feedback-Nash equilibria as fixed points. So the existence proof reduces to proving that $X = \mathbf{\Gamma}$ is compact and convex in a Hausdorff space, while $\Psi(\underline{\gamma})$ is convex-valued and upper-semicontinuous.

This offers a sense of the existence proof for a feedback-Nash equilibrium in a finite state nonzero-sum Markov game. To be complete, below, we provide the different steps used to obtain this result when using the dynamic-programming-operator formalism introduced above.

Theorem 9.3. *Consider the m-player stochastic sequential game defined above, where both the state and the action sets are finite. Then, the following holds:*

(1) The expected payoff vector associated with a vector stationary Markov strategy $\underline{\gamma}$ is given by the unique fixed point $v_{\underline{\gamma}}(\cdot)$ of the contracting operator $H_{\underline{\gamma}}$.

(2) The operator $\mathcal{F}_{\underline{\gamma}}$ is also contracting and thus admits a unique fixed point, which is $f_{\underline{\gamma}}(s, j) : \forall s \in \bar{S}, j = 1, \ldots, m$.

(3) The stationary Markov strategy vector $\underline{\gamma}^*$ is a feedback-Nash equilibrium strategy iff

$$f_{\underline{\gamma}^*}(s, j) = v_{\underline{\gamma}^*}(s, j), \quad \forall s \in S, j = 1, \ldots, m. \tag{9.30}$$

(4) There exists an equilibrium defined by a stationary Markov policy.

---

[11] A Hausdorf locally convex topological vector space is a generalization of normed vector space (like $\mathbf{R}^n$ with the usual norm). The precise definition is out of the scope of this book.

[12] Of course, this is for a case without coupled constraints.

Proof.   The details of the proof can be found in [245]. Results (1)–(3) are derived in a very similar way as in the zero-sum case of Shapley-Markov games. Result (4) is obtained by verifying that the assumption of the Kakutani-Glicksberg-Fan Theorem are satisfied by Γ and $\Psi(\gamma)$ as indicated above.   □

Remark 9.3. Notice that the existence result for a feedback-Nash equilibrium in a nonzero-sum stochastic game is not based on a contraction property of the equilibrium dynamic-programming operator, as it was for the zero-sum Shapley game. Instead, it is obtained as the fixed-point property of a global response function in the space of strategies. This existence proof is topological and not constructive. Consequently, contrary to the zero-sum case, the proof of existence of an equilibrium does not provide an algorithm for computing an equilibrium.

9.7   Stochastic games with continuous state variables

The theory of noncooperative Markov games has been extended to the case where the state and the actions are in continuous sets, by several authors, in particular Whitt [245], Parthasarathy & Shina [196] and Nowak [186–188, 191]. Since we are dealing with stochastic processes taking values in continuous (uncountable) spaces, the apparatus of measure theory and advanced topology becomes essential. In keeping with the philosophy of this textbook, we will only provide a very rapid overview of the main results recently obtained in this area. Notice that the theory of multistage games, which was examined in Chapter 5.7.5, could be considered a subarea of this field dealing with sequential games. Unfortunately the existence theorems that we will be able to prove, will require non-degenerate transition probability distributions and therefore will not apply to deterministic games.

9.7.1   *Description of the game*

An m-player *stochastic game* is defined by the data

$$((M, S, \Sigma), (U_j, U_j(\cdot), r_j(\cdot, \cdot); j \in M), Q(\cdot, \cdot), \beta),$$

where:

(1) (S, Σ) is a measurable state space $S \subset \mathbf{R}^n$ with a countably generated σ-algebra Σ of Borel subsets.[13]
(2) U_j is a compact metric space of actions for Player j.
(3) $U_j(\cdot)$ is a lower measurable map from S into nonempty compact subsets of U_j. For each s, $U_j(s)$ represents the set of admissible actions for Player j.
(4) $r_j(\cdot, \cdot) : S \times \underline{U} \mapsto \mathbf{R}$ is a bounded measurable *transition reward function* for Player j. These functions are assumed to be continuous on $\underline{U} = U_1 \times \cdots \times U_m$, for every $s \in S$.

---

[13] A Borel set is the class of all measurable subsets of S obtained by countable unions, intersections and finite differences of open intervals in S.

(5) $Q(\cdot,\cdot)$ is a product measurable transition probability kernel[14] from $S \times \underline{U}$ to S. It is assumed that $Q(\cdot,\cdot)$ satisfies some regularity conditions, which are too technical to be given here. We refer the reader to [191] for a more precise statement.

(6) $\beta_j \in (0,1)$ is the *discount factor* for Player j.

A *stationary Markov strategy* for Player j is a measurable map $\gamma_j(\cdot)$ from S into the set $\mathcal{P}\{U_j\}$ of probability measure on U_j, such that $\gamma_j(s) \in \mathcal{P}\{U_j(s)\}$ for every $s \in S$.

9.7.2 *Dynamic-programming formalism*

Let $v(s,j)$ denote the reward-to-go function from the initial state s, for Player j. We call $v(\cdot,\cdot)$, or simply v, this function for all $s \in S$ and $j \in M$. The definition of *local reward functions* given in (9.24) for the discrete-state case has to be adapted to the continuous-state format and becomes

$$\mathfrak{h}_j(s, v(\cdot, j), \underline{u}) = r_j(s, \underline{u}) + \beta \int_S v(t, j) Q(dt|s, \underline{u}), \quad j \in M,$$

where $Q(dt|s,\underline{u})$ is the elementary probability of having the next state in the interval $[t, t + dt)$, given that the current state is s and the players' actions are defined by the vector[15] $\underline{u} = (u_j)_{j \in M}$.

9.7.2.1 *Existence of approximate ϵ-equilibria*

The operators $\mathcal{H}_\gamma$ and $\mathcal{F}_\gamma$ are defined as above in (9.25). The existence of equilibria is difficult to establish for this general class of sequential games. In [245], the existence of ϵ-equilibria is proved using an approximation theory in dynamic programming . The existence of equilibria was obtained only for special cases in [186–188] and [196].

We refer here to the approximation theory proposed by Whitt [245] to deal with continuous-state sequential games. A more general result was later proved by Nowak [186] and this is the version that we give below.

Theorem 9.4. *We make the following assumptions:*

(A1) *The σ-algebra[16] $\mathcal{S}$ of measurable subsets of S is countably generated;*

(A2) *The probability kernel $Q(\cdot,\cdot)$ has a density function $z : S \times S \times U_1 \times \cdots \times U_m \to \mathbf{R}$ with respect to some measure μ on $(S, \mathcal{S})$ such that*

   (a) *z is product measurable[17] and $z(\cdot|s, \underline{u})$ is continuous for each $s \in S$ and measurable in s;*

   (b) *$\int_S |z(t|s, \underline{u})| \mu(dt) < \infty$ for each $s \in S$.*

---

[14] In probability theory and statistics, a stochastic kernel is the transition function of a stochastic process.

[15] From now on we will use boldface letter to represent vector actions or vector strategies.

[16] σ-algebra over a set $\mathcal{S}$ is a nonempty collection of subsets of $\mathcal{S}$ (including $\mathcal{S}$ itself) that is closed under complementation and countable unions of its members.

[17] Given two measurable spaces and measures on them, the *product measurable space* and the *product measure* on that space can be defined.

(A3) *The reward functions $r_j(s, \underline{u})$ are continuous in $\underline{u}$ for each $s \in S$ and measurable in s.*

Then, for each $\epsilon > 0$, there exists an ϵ-equilibrium to this game.

Proof.   We refer to [186] for a proof of this theorem.   □

9.7.2.2   *Existence of correlated equilibria*

Let us extend the sets of strategies available to the players by allowing them to correlate their choices. Let $\{\xi_n : n \geq 0\}$ be a sequence of signals drawn independently from $[0, 1]$ according to a uniform distribution. Suppose that, at each period n of the game, the players are informed of the history

$$h_n = (s_0, \xi_0, s_1, \xi_1, \ldots, s_n, \xi_n\}.$$

So the game now has an extended state set, which is $S \times [0, 1]$. The one-stage transition probabilities are easily obtained by taking the product of the original transition kernel and the uniform distribution. For this game, let $V_j(s_0, \xi_0; \underline{\gamma})$ be the expected payoff for Player j when the initial state of the game is (s_0, ξ_0) and the strategy m-tuple $\underline{\gamma}$ is used. Here a strategy γ_j for Player j sends the history h_n into $\mathcal{P}[U_j(s_n)]$.

A strategy m-tuple $\underline{\gamma}^*$ is a *correlated equilibrium with public signal* if the following holds for any initial state s_0 and for each player j:

$$\int_0^1 V_j(s_0, \xi_0; [\gamma_j, \gamma_{-j}^*]) \, \eta(d\xi_0) \leq \int_0^1 V_j(s_0, \xi_0; \underline{\gamma}^*]) \, \eta(d\xi_0), \tag{9.31}$$

where $\eta(\cdot)$ is the uniform distribution elementary probability.

Indeed, the players can exploit the known public signal to introduce some correlation in their choice of action. In the case of stationary strategies, Nowak and Raghavan proved in reference [191] the existence of correlated equilibrium solutions for a general class of sequential games. The *correlated stationary strategy with symmetric information* that they propose consists of $m + 1$ measurable functions

$$\lambda^i(\cdot) : S \mapsto [0, 1], \quad i = 1, \ldots m + 1,$$

satisfying

$$\sum_{i=1}^{m+1} \lambda^i(s) = 1$$

for every $s \in S$, and $m + 1$ measurable maps $\underline{\gamma}^i(\cdot), \ldots, \underline{\gamma}^{m+1}(\cdot)$ where

$$\underline{\gamma}^i(\cdot) : S \mapsto \mathcal{P}(U_1) \times \cdots \times \mathcal{P}(U_m), \quad i = 1, \ldots m + 1,$$

and such that

$$\underline{\gamma}^i(s) \in \mathcal{P}(\Gamma_1(s)) \times \cdots \times \mathcal{P}(\Gamma_m(s))$$

for every $s \in S$. Such a strategy can be implemented as follows: *Let ξ be a random variable with values in $\{1, 2, \ldots, m + 1\}$ with probability distribution $\mathrm{P}[\xi = i] = \lambda^i(s)$*

when the game is in state s. If $\xi = i$, all players are informed of that sample value and are given the suggestion to use $\underline{\gamma}^i(s)$ as their combined strategy.

Definition 9.3. A correlated stationary strategy

$$\left(\lambda^1(\cdot), \ldots, \lambda^{m+1}(\cdot) : \underline{\gamma}^1(\cdot), \ldots, \underline{\gamma}^{m+1}(\cdot)\right)$$

is a **correlated stationary equilibrium** iff no player can unilaterally improve his/her expected payoff by changing his component of any of the $\underline{\gamma}^i(\cdot)$'s.

We can state now an interesting result.

Theorem 9.5. *Consider an m-person nonzero-sum stochastic game for which:*

(i) *S is a nonempty Borel state space.*

(ii) *U_j is a nonempty compact metric space of actions for Player j. We let $\underline{U} = U_1 \times U_2 \times \cdots \times U_m$.*

(iii) *$U_j(s)$ is a nonempty compact subset of U_j and represents the set of actions available to Player j in state s. We assume that $\{(s, u_j) : s \in S \text{ and } u_j \in U_j(s)\}$ is a Borel subset of $S \times U_j$. We put $\underline{U}(s) = U_1(s) \times U_2(s) \times \cdots \times U_m(s)$, $s \in S$.*

(iv) *$r_j : S \times \underline{U} \to R$ is a bounded Borel-measurable reward function for Player j. It is assumed that $r_i(s; \cdot)$ is continuous on X, for every $s \in S$.*

(v) *Q is a Borel-measurable transition probability kernel from $S \times \underline{U} \to S$, called the law of motion among states. If s is a state at some stage of the game and the players select $\underline{u} \in \underline{U}(s)$, then $Q(\cdot|s, \underline{u})$ is the probability distribution of the next state of the game. We assume that the transition probability $Q(\cdot|s, \underline{u})$ has a density function, say $z(\cdot|s, x)$, with respect to a fixed probability measure μ on S, satisfying the following continuity condition:[18] For any sequence of joint actions $\underline{u}^n \to \underline{u}^*$*

$$\int_S |z(t|s, u^n) - z(t, u^*)|\mu(dt) \to 0 \text{ as } n \to \infty.$$

*Then, there exists a **correlated stationary equilibrium** for the sequential game defined in 9.7.1 and satisfying conditions (i)–(v).*

Proof. We refer to [189] and [191] for a detailed proof of this theorem. We limit ourselves to the following hints on the principal steps of the proof.

$M = \{1, \ldots, m\}$ is the set of players. Consider the class $\mathcal{V}$ of functions $v(\cdot, \cdot) : S \times M \mapsto v(s, j) \in \mathbf{R}$, which are measurable and uniformly bounded. This set is compact. For any $v(\cdot, \cdot) \in \mathcal{V}$ and any state $s \in S$, define the local game with set of players M, pure action sets $U_j(s)$ and payoffs

$$w(s, j; \underline{u})(v) = r_j(s, \underline{u}) + \beta \int_S v(t, j)Q(dt|s, \underline{u}) \quad j \in M. \tag{9.32}$$

Given the assumptions of continuity made above, we can establish that this local game admits Nash-equilibria (NEs) in the class of mixed strategies and that the set $N_v(s)$ of all NEs of this local game is compact (see Theorem 3.9, page 69.)

---

[18]called L_1 continuity.

A mixed strategy for Player j is defined by a probability measure $\mu_j(\cdot)$ on $U_j(s)$. The payoff to Player j associated to an m-tuple of mixed strategies $\underline{\mu} = (\mu_1, \ldots, \mu_m)$ is given by

$$w(s, j; \underline{\mu})(v) = \int \cdots \int w(s, j; u_1, \ldots, u_m)\mu_1(du_1) \times \cdots \times \mu_m(du_m) \quad j \in M. \quad (9.33)$$

We characterize the set of NEs $N_v(s)$ as the set of probability distributions $\underline{\mu}^*$ for which a disequilibrium indicator $\tilde{w}(s, \underline{\mu}^*)(v)$ is equal to 0, where

$$\tilde{w}(s, \underline{\mu}^*)(v) = \sum_{j \in M} \left(w(s, j; \underline{\mu})(v) - \max_{\mu_j \in \mathcal{P}[U_j(s)]} w(s, 1; [\mu_j, \boldsymbol{\mu}^*_{-j}])(v) \right). \quad (9.34)$$

so

$$N_v(s) = \left\{ \underline{\mu}^* : \tilde{w}(s, \underline{\mu}^*)(v) = 0 \right\}. \quad (9.35)$$

Now we define $P_v(s)$ as the set of payoff vectors corresponding to all Nash equilibria from the set $N_v(s)$. A fixed point v^* of the set valued mappings $v \mapsto P_v$, i.e., verifying $v^* \in P_{v^*}$, corresponds to a Nash equilibrium solution; unfortunately we don't have the upper-semicontinuity and convex-compact valued properties necessary to invoke the Kakutani-Glicksberg-Fan theorem.

So we consider the point-to-set mapping $v \mapsto \mathrm{co}P_v$, where $\mathrm{co}P_v(s)$ is the convex hull of $P_v(s)$. In the correspondence $s \mapsto \mathrm{co}P_v(s)$, we consider the Borel-measurable selectors[19] and we define M_v as the set of all Borel-measurable selectors, for v fixed.

Now it can be proved, using assumptions (i)–(v) and results in topology and measure theory, that the mapping $v \mapsto M_v$ is convex, compact-valued and upper-semicontinuous. So we can invoke the Kakutani-Glicksberg-Fan theorem and declare that there exists $(v^*(s, j) : s \in S, j \in M)$ such that, for almost all $s \in S$, the following holds:

$$v^*(s) \in M_{v^*}(s) \subset \mathrm{co}P_{v^*}(s).$$

We can invoke Carathéodory's theorem[20] to find, for each $s \in S$, $m + 1$ vectors $v^i \in P_{v^*}(s)$, $i = 1, \ldots, m + 1$ and $m + 1$ weights $\lambda^1(s), \ldots, \lambda^{m+1}(s)$, with $\lambda^i(s) \geq 0$ and $\sum_{i=1}^{m+1} \lambda^i(s) = 1$, such that

$$v^*(s, j) = \sum_{i=1}^{m+1} \lambda^i(s)v^i(s, j). \quad (9.36)$$

Since $v^i \in P_{v^*}(s)$, there exists a Nash-equilibrium stationary mixed strategy m-tuple, $\underline{\mu}^i(\cdot) : s \mapsto \mathcal{P}[U_1(s)] \times \cdots \times \mathcal{P}[U_m(s)])$, such that, $\underline{\mu}^i \in N_{v^*}(s)$ and

$$v^i(s, j) = r_j(s, \underline{\mu}^i(s)) + \beta \int_S v(t, j)Q(dt|s, \underline{\mu}^i(s)). \quad (9.37)$$

---

[19] A Borel measurable selector is a measurable function $f : s \mapsto f(s)$ such that $f(s) \in \mathrm{co}P_v(s)$, for each $s \in S$.

[20] In convex geometry, Carathéodory's theorem states that if a point x of R^m lies in the convex hull of a set P, then there is a subset P' of P consisting of $m + 1$ or fewer points, such that x lies in the convex hull of P'.

Hence if we put

$$\underline{\mu}_{\lambda}(s) = \sum_{i=1}^{m+1} \lambda^i(s)\underline{\mu}^i(s)$$

we get, for almost all $s \in S$, $\underline{\mu}_{\lambda}(s) \in \text{co}N_{v^*}(s)$ and

$$v^*(s,j) = r_j(s,\underline{\mu}_{\lambda}(s)) + \beta \int_S v(t,j)Q(dt|s,\underline{\mu}_{\lambda}(s)). \tag{9.38}$$

The above equation 9.38 shows that

$$v^*(s,j) = V_j(s;\underline{\mu}_{\lambda}). \tag{9.39}$$

and since $\underline{\mu}_{\lambda}(s) \in \text{co}N_{v^*}(s)$, $\underline{\mu}_{\lambda}$ is a correlated equilibrium for the stochastic game under study. $\qquad\square$

Remark 9.4. We introduced correlated equilibria with symmetric information in Section 4.2 of Chapter 4 (see page 87). For m-matrix games, these equilibria were obtained as convex combinations of the existing Nash equilibria. The set of outcomes from symmetric correlated equilibria was the convex hull of the finite set of Nash-equilibria outcomes, which were the extreme points of the outcome set. Also in this new context, if a feedback-Nash equilibrium exists in stationary strategies, it is also a correlated equilibrium, with the probabilities $\lambda^i(s)$ concentrated on a single i. The interesting property of nonzero-sum Markov games is that we can show, by a fixed-point argument, that a correlated equilibrium exists, but we cannot show that "extreme points" consisting of feedback-Nash equilibria exist.

9.7.3 *The case of supermodular stochatic games*

9.7.3.1 *The importance of supermodularity*

We have already noticed the important difference between existence theorems in the zero-sum (Shapley game) and in nonzero-sum stochastic games. In the former case, we managed to prove that the *saddle-point dynamic-programming operator*, acting on the space of value functions, admits a unique fixed point. In the latter, this approach could not be followed since the equilibrium dynamic-programming operator, based on the equilibrium value vector of the local reward game, had a priori no useful property. So we relied on a point-to-set *optimal reply map*, defined on the set of mixed-stationary strategies, to obtain by the Kakutani theorem, the desired fixed-point result in the case of a stochastic game with finite state and action sets. This approach has been difficult to extend to the case of stochastic games with continuous (Borel) state set.

Amir [4] and Curtat [52] have shown that if we consider a class of stochastic games, with state in a Euclidean space, satisfying assumptions leading to supermodularity and strict diagonal dominance for the local reward auxiliary games, then the equilibrium dynamic-programming operator could be used to prove existence of a fixed point defining a feedback-Nash equilibrium in pure feedback strategies. This is an interesting result since supermodularity, as already discussed in Section 4.6 of Chapter 4, is linked to the important and common economic assumption of complementarity. In essence, this feature was responsible for pure strategy equilibria existence in "static" supermodular games, considered in Section 4.6. We give below the essential elements of this approach and we detail the successive steps followed to establish this result.

Nowak [190] has shown how Amir and Curtat results can be related to Theorem 9.5, as they provide conditions under which one can find a fixed point v^* of the set valued mappings $v \mapsto P_v$, i.e., verifying $v^* \in P_{v^*}$ (instead of $v^* \in \mathrm{co} P_{v^*}$,) where P_v is the set of all pure Nash equilibria of the local game associated with the value functional v.

Before treating supermodular stochastic games. let us recall the following definition and properties for supermodular static games:

Definition 9.4. The m-player static game defined by the payoff vector

$$\mathbf{G}(\underline{u}) = [G_j(u_1, \ldots, u_m)]_{j=1,\ldots,m}$$

is called **smooth supermodular** if for every player $j \in M = \{1, \ldots, m\}$,

- $u_j \in U_j$, where U_j is a compact subset of $\mathbf{R}^{m_j}$,
- $\dfrac{\partial^2 G_j}{\partial u_{jk} \partial u_{j\ell}} \geq 0$ for all $1 \leq k < \ell \leq m_j$,
- $\dfrac{\partial^2 G_j}{\partial u_{jk} \partial u_{i\ell}} \geq 0$ for each player $i \neq j$ and for all $1 \leq k \leq m_j, 1 \leq \ell \leq m_j$.

These conditions insure[21] that G_j is a supermodular function w.r.t. u_j, for fixed $\mathbf{u}_{-j}$ and that G_j has increasing differences in u_j and $\mathbf{u}_{-j}$.

The following result is easily obtained

Lemma 9.3. *Any smooth supermodular game* $\mathbf{G}$ *such that the following strict diagonal dominance conditions hold for every* $j \in M$ *and* $1 \leq k \leq m_j$,

$$\frac{\partial^2 G_j}{\partial u_{jk}^2} + \sum_{i \in M-j} \sum_{\ell \neq k} \frac{\partial^2 G_j}{\partial u_{jk} \partial u_{i\ell}} < 0 \tag{9.40}$$

has a unique pure Nash equilibrium.

Proof. Because $\dfrac{\partial^2 G_j}{\partial u_{jk} \partial u_{i\ell}} \geq 0$, condition (9.40) implies

$$\left| \frac{\partial^2 G_j}{\partial u_{jk}^2} \right| > \sum_{i \in M-j} \sum_{\ell \neq k} \left| \frac{\partial^2 G_j}{\partial u_{jk} \partial u_{i\ell}} \right| < 0. \tag{9.41}$$

The uniqueness results from a theorem[22] of Gabay and Moulin [91]. Note that condition (9.40) can also be written, for every $j \in M$,

$$\forall k = 1, \ldots, m_j, \quad \sum_{i \in M} \sum_{\ell=1}^{m_k} \frac{\partial^2 G_j}{\partial u_{jk} \partial u_{i\ell}} < 0. \tag{9.42}$$

$\square$

---

[21] See Theorem 4 in Milgrom and Roberts [176].
[22] The theorem proved in [91] is the following: Let $U_j \subset \mathbf{R}$, for each $j \in M$. Assume the following strict diagonal dominance property

$$\forall j \in M, \quad \left| \frac{\partial^2 G_j}{\partial u_j^2} \right| > \sum_{i \neq j} \frac{\partial^2 G_j}{\partial u_j \partial u_i}$$

9.7.3.2 *The equilibrium dynamic-programming operator*

The data of the game are the same as those given in 9.7.1, with the simplifying assumption[23] that $U_j(s) \equiv U_j$, for all $s \in S$.

We consider the space $\mathcal{V}$ of reward-to-go functionals $v(\cdot, j) : S \to \mathbf{R}, j = 1, \dots, m$, endowed with the sup-norm topology, $\|v\| = \sup\{|v(s, j)| : s \in S, j \in M\}$. We define the local rewards

$$\mathfrak{h}_j(s, v(\cdot, j), \underline{u}) = r_j(s, \underline{u}) + \beta \int_S v(t, j) Q(dt|s, \underline{u}), \quad j \in M,$$

and the local game, associated with initial state s and functional $v(\cdot, \cdot)$, where, for each Player j, the pure strategy set is U_j and the payoff is defined by $\mathfrak{h}_j(s, v(\cdot, j), \underline{u})$.

Assume that, for each $s \in S$, the local game admits a unique Nash equilibrium in pure strategies, and call $\mathcal{H}^*(v(\cdot, \cdot))(s, j)$ the equilibrium local-reward value for Player j associated with this unique equilibrium solution. This defines the *equilibrium operator* $\mathcal{H}^*$ acting on the space $\mathcal{V}$, which associates, with an element $v(\cdot, \cdot) \in \mathcal{V}$, a new element $w(\cdot, \cdot) \in \mathcal{V}$ defined by

$$w(s, j) = \mathcal{H}^*(v(\cdot, \cdot))(s, j), \quad s \in S, j \in M. \tag{9.43}$$

We call this operator the *equilibrium dynamic-programming operator*. A fixed point of $\mathcal{H}^*$ is a reward-to-go functional $v^*(\cdot, \cdot)$ such that $v^*(\cdot, \cdot) \triangleq \mathcal{H}^*(v^*(\cdot, \cdot))$. By the dynamic-programming argument, which is fully detailed in Chapter 5.7.5 and easily extended to the stochastic-game environment, a fixed point of $\mathcal{H}^*$ defines a feedback-Nash equilibrium solution, with pure feedback strategies $\underline{\gamma}^*(s, v) = (\gamma_j^*(s, v))_{j \in M}$ defined by

$$\mathfrak{h}_j(s, v(\cdot, j), \underline{\gamma}^*(s, v)) = \mathcal{H}^*(v(\cdot, \cdot))(s, j), \quad s \in S, j \in M. \tag{9.44}$$

Now if we can show that $\mathcal{H}^*$ is continuous on $\mathcal{V}$ and maps a compact subset $\mathcal{K} \subset \mathcal{V}$ into itself, then, by the Schauder fixed-point theorem, stated below, we can assert that a fixed point of $\mathcal{H}^*$ exists in $\mathcal{K}$.

Theorem 9.6. *Schauder fixed-point theorem*: *If $\mathcal{K}$ is a compact, convex subset of a topological vector space $\mathcal{V}$ and $\mathcal{T}$ is a continuous mapping of $\mathcal{V}$ into itself such that $\mathcal{T}(\mathcal{K}) \subset \mathcal{K}$, then $\mathcal{T}$ has a fixed point in $\mathcal{K}$.*

Proof. The proof of this result of functional analysis can be found in reference [251]. $\square$

Remark 9.5. Notice that we implemented a search for a fixed point for the equilibrium dynamic operator when, in Section 6.6, we looked for a solution to the infinite-horizon dynamic-programming equation in the fisheries game. In that case, the operator was mapping a reward-to-go function with the form $C\sqrt{x}$ into the same class of functions. The parameter-determination procedure was then a way to find the desired fixed point.

---

holds. Then there exists a unique Nash equilibrium. If $U_j \subset \mathbf{R}^{m_j}$, for each $j \in M$, the strict diagonal dominance property becomes: for every $j \in M$ and $1 \leq k \leq m_j$,

$$\left| \frac{\partial^2 G_j}{\partial u_{jk}^2} \right| > \sum_{i \in M-j} \sum_{\ell \neq k} \left| \frac{\partial^2 G_j}{\partial u_{jk} \partial u_{i\ell}} \right| < 0.$$

[23] This assumption is made here for notational convenience. The theory remains valid also with state-dependent action sets.

9.7.3.3   *Uniqueness of pure-strategy equilibrium solution to the local reward games*

We saw, in Theorem 4.6, that a supermodular game having upper-semicontinuous reward functions admits a largest and a smallest equilibrium in pure strategies. If the smallest and the largest equilibrium coincide, then the equilibrium is unique.

So let us find conditions under which the local reward game is supermodular. Recall Definition 4.13, where the game $(M, (U_j, \psi_j; j \in M))$ is said to be *supermodular* if $\psi_j(\underline{u})$ is supermodular on $u_j \in U_j$ for each $\mathbf{u}_{-j} \in \Pi_{k \neq j} U_k$ and, for each $j \in M$, $\psi_j([u_j, \mathbf{u}_{-j}])$, has increasing differences in $(u_j, \mathbf{u}_{-j})$.

In our case, the function $\psi_j(\underline{u})$ is defined by $\mathfrak{h}_j(s, v(\cdot, j), \underline{u})$ for each $s \in S$ and $v \in \mathcal{V}$. It is composed of two parts: (i) the transition reward $r_j(s, \underline{u})$ and (ii) the expected reward-to-go from the next period $\beta \int_S v(t, j) Q(dt|s, \underline{u})$.

Curtat [52] proposes the following set of assumptions:

Assumption 9.1. Assume the following:

Smoothness   The state space S is a compact interval of a Euclidean space $\mathbf{R}^n$. For every $j \in M$, the action (control) set U_j is a compact interval of a Euclidean space $\mathbf{R}^{m_j}$ and $r_j(s, \underline{u})$ is twice continuously differentiable in $(s, \underline{u})$. For every t, the transition probability kernel $Q(dt|s, \underline{u})$ is twice continuously differentiable in $(s, \underline{u})$.

Cardinal complementarity   The transition reward $r_j(s, u_j, \mathbf{u}_{-j})$ satisfies the cardinal complementarity condition in u_j and $(\mathbf{u}_{-j}, s)$, for every $j \in M$. For every t, the transition probability kernel $Q(dt|s, \underline{u})$ satisfies the cardinal complementarity condition (see definition below) in u_j and $(\mathbf{u}_{-j}, s)$ for all j.

Monotonicity   For every $j \in M$, the reward function $r_j(s, [u_j, \mathbf{u}_{-j}])$ is increasing in $(s, \underline{u}_{-j})$. For every t, the transition probability kernel $Q(dt|s, \underline{u})$ is stochastically increasing in $(s, \underline{u})$.

Strict diagonal dominance   For every t, the probability kernel $Q(dt|s, \underline{u})$ has doubly stochastically increasing differences (see definition below) in $\underline{u}$ and s. For every $j \in M$, the reward function $r_j(s, [u_j, \mathbf{u}_{-j}])$ satisfies a strict dominant-diagonal condition in u_j and $\mathbf{u}_{-j}$, namely,

$$\forall j \in M, \forall l \in \{1, \ldots, \ell_j\} \quad \sum_{i=1}^{m} \sum_{\alpha=1}^{\ell_i} \frac{\partial^2 r_j(s, [u_j, \mathbf{u}_{-j}])}{\partial u_j^l \partial u_i^k} < 0, \qquad (9.45)$$

where u_i^k is the k-th component of the vector u_j and ℓ_i is the number of components of the vector u_i.

A stochastic game that satisfies all conditions listed in Assumption 9.1 was called by Curtat a *supermodular stochastic game* . Let us now give precise definitions of the different properties assumed above.

Cardinal complementarity   Milgrom and Shannon [177] have introduced the notion of *cardinal complementarity* : if S is a lattice and T is a poset, a functional f on $S \times T$

satisfies the cardinal complementarity conditions in s and t if and only if it is *super-modular*[24] in s for every fixed t and has increasing differences in s and t, which means that, for $s \geq s'$, the difference $f(s,t) - f(s',t)$ increases with t.

Stochastic cardinal complementarity Supermodularity and increasing differences can be extended to probability distributions on $\mathbf{R}^n$. The transition probability kernel $Q(dt|s,\underline{u})$ satisfies the cardinal complementarity condition in u_j and $(\mathbf{u}_{-j}, s)$ for all j if and only if for every increasing integrable function $v(\cdot, j)$, the expression

$$\int_S v(t,j)Q(dt|s,[u_j,\mathbf{u}_{-j}]) \tag{9.47}$$

considered as a function of u_j and $(\mathbf{u}_{-j}, s)$ is supermodular in u_j and has increasing differences in u_j and $(\mathbf{u}_{-j}, s)$.

Stochastic monotonicity For every t, for every increasing integrable function $v(\cdot, j)$ the expression $\int_S v(t,j)Q(dt|s,\underline{u})$ is increasing in $s, \underline{u}$.

Doubly increasing differences If $X \subset \mathbf{R}^n$ is a lattice and $W \subset \mathbf{R}^p$ is a poset, $f : X \times W \to \mathbf{R}$ has *doubly increasing differences* in x and w iff $f(x,w)$ has increasing differences in x and w and there exists an increasing Lipschitz continuous[25] function $\phi : \mathbf{R}^p \to \mathbf{R}$ such that[26] $f(\phi(w)\mathbf{1}_n - x, w)$ also has increasing differences in x and w. The concept extends to *doubly stochastically increasing differences* in the same manner as above.[27]

Under these assumptions, we can prove that the local reward games have unique Nash-equilibrium solutions in pure strategies. To define these local games, we consider the class

---

[24]Recall (see Definition 4.11, page 114) that $f : X \to \mathbf{R}$, where X is a lattice, is supermodular if, for any pair x_1, x_2 of elements in S, the following holds:

$$f(x_1) + f(x_2) \leq f(x_1 \wedge x_2) + f(x_1 \vee x_2).$$

If f is twice continuously differentiable, this property is implied by the condition

$$\frac{\partial^2 f(x)}{\partial x_i \, \partial x_k} \geq 0 \text{ for all } i \neq k. \tag{9.46}$$

[25]Given two metric spaces (X, d_X) and (Y, d_Y), where d_X denotes the metric on the set X and d_Y is the metric on set Y (for example, Y might be the set of real numbers $\mathbf{R}$ with the metric $d_Y(x,y) = |x - y|$, and X might be a subset of $\mathbf{R}$), a function $f : X \to Y$ is called Lipschitz continuous if there exists a real constant $K \geq 0$ such that, for all x_1 and x_2 in X, $d_Y(f(x_1), f(x_2)) \leq K d_X(x_1, x_2)$.

[26]Here $\mathbf{1}_n$ is the vector of $\mathbf{R}^n$ with all components being equal to 1.

[27]Curtat [52] shows that if f is twice continuously differentiable, the property of doubly increasing differences is closely related to the notion of diagonal dominance. In that case, if f has doubly increasing differences it satisfies a weak dominant-diagonal condition in x, i.e.,

$$\partial^2 f/\partial x_i^2 + \sum_{k \neq i} \partial^2 f/\partial x_i \partial x_k \leq 0,$$

for every $i = 1, \ldots, n$. Conversely, if $f(x,w)$ has increasing differences in x and w and satisfies a strict dominant-diagonal condition in x, i.e.,

$$\partial^2 f/\partial x_i^2 + \sum_{k \neq i} \partial^2 f/\partial x_i \partial x_k < 0,$$

for every $i = 1, \ldots, n$, then it has doubly increasing differences in x and w.

$\mathcal{C}_I(S, \mathbf{R})$ of continuous and increasing functionals $v(\cdot, j) : S \to \mathbf{R}, j \in M$ endowed with the sup norm. These functionals are used as cost-to-go functions to value the next state reached in a transition. Each functional $v(\cdot, j)$ admits a directional derivative almost everywhere. We call ∂v the M-vector function with components $\partial v(\cdot, j) : j \in M$. We say $\partial v \geq \partial \mathbf{w}$ if $\partial v(s) \geq \partial w(s)$ almost everywhere on S.

The local rewards are defined as

$$\mathfrak{h}_j(s, v(\cdot, j), \underline{u}) = r_j(s, \underline{u}) + \beta \int_S v(t, j) Q(dt|s, \underline{u}), \quad j \in M.$$

Note that if v and w are two candidate cost-to-go value functions, and v has a greater slope than w, then $v - w$ is an increasing function since it is continuous and has a.e. (almost everywhere) positive directional derivatives. Then, since $Q(dt|s, \underline{u})$ is stochastically increasing in $\underline{u}$, for every $j \in M$, $\int_S (v(t, j) - w(t, j)) Q(dt|s, \underline{u})$ is increasing in $\underline{u}$. This shows that the local reward function has increasing differences in $\underline{u}$ and ∂v.

Theorem 9.7. *In a supermodular stochastic game, each local reward game has a unique Nash equilibrium in pure strategies. The corresponding Nash strategy profile $\gamma^*(s, v)$ is an increasing function in the state variable s and the slope ∂v of the reward-to-go function $v(\cdot, \cdot)$.*

Proof. The action sets U_j are compact lattices in Euclidean spaces. The local rewards $\mathfrak{h}_j(s, v(\cdot, j), \underline{u})$ satisfy the cardinal complementarity conditions in u_j and $(\mathbf{u}_{-j}, s, \partial v)$ and are continuous in $\mathbf{u}$. Therefore the game admits a smallest and a largest pure Nash-equilibrium solution, both of which are increasing in s and ∂v. The first term $r_j(s, \underline{u})$ in the expression of $\mathfrak{h}_j(s, v(\cdot, j), \underline{u})$ satisfies a strict diagonal-dominance condition, the second term $\beta \int_S v(t, j) Q(dt|s, \underline{u})$ has doubly increasing differences in $\underline{u}$ and s. Therefore, it satisfies a weak diagonal-dominance condition. Therefore, $\mathfrak{h}_j(s, v(\cdot, j), \underline{u})$ satisfies a strict diagonal-dominance condition, namely,

$$\forall j \in M, \forall l \in \{1, \dots, \ell_j\} \quad \sum_{i=1}^{m} \sum_{\alpha=1}^{\ell_i} \frac{\partial^2 h_j}{\partial u_j^l \partial u_i^k} < 0. \tag{9.48}$$

The uniqueness of the Nash equilibrium is then derived by applying a result of Gabay and Moulin [91], which is also implied by Rosen's theorem (Theorem 3.6).[28]   □

Remark 9.6. This theorem is a crucial result, as it allows us to define the equilibrium dynamic programming operator $\mathcal{H}^*$, which will transform a candidate $v \in \mathcal{V}$ into a new candidate $w = \mathcal{H}^*(v) \in \mathcal{V}$.

9.7.3.4 *Curtat existence theorem*

Curtat [52] proved the following result:

Theorem 9.8. Curtat 1996. *A supermodular stochastic game admits a feedback (Markov) Nash equilibrium characterized by reward-to-go value functions and feedback strategies that are Lipschitz continuous and increasing functions of the state variable s.*

---

[28]It can be shown that if $f(x, w)$ is smooth, supermodular in x, and satisfies a strict diagonal-dominance condition in x, then f is strictly concave in x.

Proof. We have already indicated the principle of the proof, which consists of using Schauder fixed-point theorem to show that the equilibrium dynamic-programming operator $\mathcal{H}^*$ has a fixed point in a compact subset of $\mathcal{C}_I(S, \mathbf{R})$. We refer to [52] for the details of this proof, which is very technical for the verification of Lipschitz continuity of the transforms $\mathcal{H}^*(v)$, and for the verification that there exists a compact subset $\mathcal{K}$ of $\mathcal{C}_I(S, \mathbf{R})$, which is transformed into itself by $\mathcal{H}^*$, i.e., $\mathcal{H}^*(\mathcal{K}) \subset \mathcal{K}$. $\square$

9.7.3.5 *An example of a stochastic supermodular game: a search with learning*[29]

At every stage, each one of m agents ("traders") expands his effort, searching for more trading partners. Denoting by $u_j \in [0, 1]$ the effort level exerted by agent j and by $C_j(u_j)$ the corresponding search cost, which is assumed to be strongly convex. In this example we use $C(u_j) = 1.74(1 + u_j)^3$.

The state variable $s \in [1, 2]$ gives the current productivity level of the search process. The j-th player's reward at time t is

$$r_j(x_t, \underline{u}_t) = s u_{j,t} \sum_{i \neq j} u_{i,t} - C_j(u_{j,t}).$$

The state-transition probability kernel is determined by a cumulative distribution function $F(s'|s, \underline{u}) = \mathrm{P}[s_{t+1} \leq s'|s_t = s, \underline{u}_t = \underline{u}]$ that is assumed to satisfy the conditions for stochastic cardinal complementarity and doubly-stochastic increasing differences. Following [52] and [190]), we use

$$F(s'|s, \underline{u}) = \frac{s + u_1 + u_2 + \cdots u_m}{m + 2} G(s')$$
$$+ \left(1 - \frac{s + u_1 + u_2 + \cdots u_m}{m + 2}\right) H(s'), \quad (9.49)$$

where distributions $G(\cdot)$ and $H(\cdot)$ are defined on $[1, 2]$ and are such that $G(\cdot)$ stochastically dominates $H(\cdot)$.[30]

Given the above assumptions, we can conclude that the general existence and uniqueness results presented above apply directly to the dynamic search model. So, there exists a stationary feedback (Markovian) equilibrium in pure strategies, in which the agents' efforts are increasing functions of the productivity level.

9.7.3.6 *A limitation of the existence theorem*

A model that satisfies Assumption 9.1 is claimed to reflect a rather general economic situation, which may occur among competitive agents. However, interestingly, the assumptions concerning stochastic cardinal complementarity exclude *deterministic* state transitions, as illustrated below.

---

[29]We borrow this example from [52].
[30]For example, in reference [190], it is assumed that the densities for $G(\cdot)$ and $H(\cdot)$ are $s/2$ and $1 - s/2$, respectively.

Consider for the purpose of this illustration the special case of one-dimensional state and action spaces. Following [233], we use the integral characterization of stochastic dominance to see that stochastic supermodularity means that $Q(dt|s, [u_j, \mathbf{u}_{-j}])$ in (9.47) satisfies

$$Q(dt|s, [u'_j, \mathbf{u}_{-j}]) < Q(dt|s, [u''_j, \mathbf{u}_{-j}]) \tag{9.50}$$

if $u'_j > u''_j$. I.e., if kernel $Q(dt|s, [u'_j, \mathbf{u}_{-j}])$ *stochastically* dominates $Q(dt|s, [u''_j, \mathbf{u}_{-j}])$ then the relation between the kernels is like in (9.50). We will also assume that

$$Q(dt|s, [u_j, \mathbf{u}_{-j}]) \triangleq F(\cdot|s, [u_j, \mathbf{u}_{-j}])dt. \tag{9.51}$$

We can say that if distribution function $F(\cdot|s, [u_j, \mathbf{u}_{-j}])$ is *stochastically* supermodular then "function" $F(\cdot|s, [u'_j, \mathbf{u}_{-j}])$ is submodular.[31]

For simplicity let there be only two players (j and $-j$) and assume that the state and the action sets are all given by $[0, 1]$. Then, consider a deterministic transition law given by $s(t+1) = f(s(t), [u_j(t), u_{-j}(t)])$, where f is a continuous function. The distribution function of this deterministic transition can be written as

$$F(s'|s, [u_j, u_{-j}]) = \begin{cases} 0 \text{ if } x(t+1) < f(x, u_j, u_{-j}), \\ 1 \text{ if } x(t+1) \geq f(x, [u_j, u_{-j}]). \end{cases} \tag{9.52}$$

Then we will see that $F(s'|s, [u_j, u_{-j}])$, as defined by (9.52), *cannot* be stochastically supermodular in (u_j, u_{-j}) for a fixed s, unless f is actually independent of one of the u's.

To demonstrate this, we can graph F on the (u_j, u_{-j})-unit square and suppose that $F(\cdot|s, [u_j, u_{-j}]) = 1$ for $u_{-j} \geq 1 - u_j$ and $F(\cdot|s, u_j, u_{-j}) = 0$ for $u_{-j} < 1 - u_j$. Let $b = (b_j, b_{-j})$ and $b = (b'_j, b'_{-j})$ where $b'_j < b_j$) and $b'_{-j} > b_{-j}$ so that

$$b \wedge b' = (b'_j, b_{-j}), \qquad b \vee b' = (b_j, b'_{-j}).$$

Hence, $F(b \wedge b') = F(b'_j, b_{-j}) = 0$ and $F(b \vee b') = F(b_j, b'_{-j}) = 1$. Let us write the stochastic supermodularity, or "plain" submodularity, condition

$$\begin{cases} F(b) + F(b') \geq F(b \wedge b') + F(b \vee b') \\ \quad 0 \ + \ \ 0 \ \ \ngeq \ \ \ \ 0 \ \ + \ \ \ 1. \end{cases} \tag{9.53}$$

It is evident that this condition is not satisfied. A similar argument holds for the other pairs of arguments.

We conclude that unless the zero-one discontinuity of F is contained along a vertical or a horizontal line, F will not be submodular in $(u_j, u_{-j}) \in [0, 1] \times [0, 1]$ for fixed s, or stochastically supermodular. This means that Theorem 9.8 cannot be applied to deterministic dynamic games.

9.7.4 *Links with the existence of correlated equilibria*

Nowak [190] has pointed out that for a class of transition probability kernels, similar to that used in the example of Section 9.7.3.5, the existence theorem for supermodular stochastic games could be derived as a corollary to Theorem 9.5.

---

[31] For the definition of submodularity, just invert the sign in (4.17) on page 114.

The transition probability kernels are assumed to be combinations of finitely many measures on the state space. More precisely we consider the following nonzero-sum stochastic game in which:

- S is a Borel state space;
- $U_j \subset \mathbf{R}^{m_j}$ is the action space of Player j, $j \in M = \{1, \dots, m\}$;
- $U_j(s)$ is a nonempty compact convex subset of U_j, which represents the set of admissible actions to Player j in state s;
- $r_j(s, \underline{u})$ is the transition reward to Player j when the state is s and the combined action $\underline{u} = (u_1, \dots, u_m)$ has been selected by the m players. We assume that $r_j(s, \cdot)$ is continuous on $\underline{U}(s) = U_1(s) \times \cdot \times U_m(s)$;
- a transition probability kernel $Q(dt|s, \underline{u})$ defines the elementary probability that the new state reached in the one-stage transition be in the interval $[t, t + dt) \subset S$, given that it is currently in s and the combined action $\underline{u}$ has been taken;
- there exist two transition probability kernels $\mu_1(dt|s)$ and $\mu_2(dt|s)$ and a measurable function $g(s, \underline{u}) : S \times \underline{U} \to [0, 1]$, such that $g(s, \cdot)$ is continuous in $\mathbf{u} \in \underline{U}(s)$ and

$$Q(dt|s, \underline{u}) = g(s, \underline{u})\mu_1(dt|s) + (1 - g(s, \underline{u}))\mu_2(dt|s); \tag{9.54}$$

- the players discount their rewards with a discount factor β.

We make the following assumption

Assumption 9.2. For each state s, the static game with payoffs $r_j(s, \underline{u})$, $j \in M$ and combined actions $\underline{u} \in \underline{U}(s)$ is a smooth supermodular game satisfying the strict diagonal dominance condition (9.40). Moreover, there are Borel measurable functions $g_{jk}(s, u_{jk})$, twice differentiable in u_{jk}, such that

$$g(s, \underline{u}) = \sum_{j \in M} \sum_{k=1}^{m_j} g_{jk}(s, u_{jk}), \tag{9.55}$$

where $u_j = (u_{j1}, \dots, u_{jk}, \dots, u_{jm_j})$ and $g_{jk}(s, u_{jk})$ is continuous in u_{jk} for a given s.

Theorem 9.9. *(Nowak* [190]*) Under Assumption 9.2, the nonzero-sum stochastic game admits a pure stationary feedback (Markov) Nash equilibrium.*

Proof. As in the proof of Theorem 9.5, consider the class $\mathcal{V}$ of functions $v(\cdot, \cdot) : S \times M \mapsto v(s, j) \in \mathbf{R}$, which are measurable and uniformly bounded. This set is compact. For any $v(\cdot, \cdot) \in \mathcal{V}$ and any state $s \in S$, define the local game with the set of players M, pure action sets $U_j(s)$ and payoffs

$$\mathfrak{h}_j(s, v(\cdot, j), \underline{u}) = r_j(s, \underline{u}) + \beta \int_S v(t, j)Q(dt|s, \underline{u}), \quad j \in M. \tag{9.56}$$

Under Assumption 9.2, it is easy to show that that for any $s \in S$ and $v = v(\cdot, \cdot) \in \mathcal{V}$, the local game with payoffs defined by (9.56) is smooth, supermodular and satisfies the diagonal dominance property. Therefore this local game admits a unique Nash-equilibrium in the class of pure strategies and, as a consequence, the set $P_v(s)$ of equilibrium values for this game reduces to a singleton. So, by the same arguments as in Theorem 9.5, we can find $v^* \in V$ such that $v^*(s) = P_{v^*}(s)$. Hence there exists a pure strategy equilibrium m-tuple $\underline{\sigma}^*(s)$ for the local game with payoffs

$$r_j(s, \underline{u}) + \beta \int_S v^*(t, j) Q(dt|s, \underline{u}) \quad j \in M,$$

for which the following equalities holds

$$v^*(s, j) = r_j(s, \underline{\sigma}^*(s)) + \beta \int_S v^*(t, j) Q(dt|s, \underline{\sigma}^*(s)), \quad j \in M, \tag{9.57}$$

whereas the following inequalities hold and for all $j \in M$,

$$v^*(s, j) \geq r_j(s, [u_j, \boldsymbol{\sigma}^*_{-j}(s)]) + \beta \int_S v^*(t, j) Q(dt|s, [u_j, \boldsymbol{\sigma}^*_{-j}(s)]), \quad \forall u_j \in U_j(s).$$

Therefore, by the usual verification theorem, the correspondence $s \mapsto \underline{\sigma}^*(s)$ defines a feedback (Markov) Nash equilibrium for the stochastic game. □

9.7.5 *Another existence theorem for supermodular Markov games*

An existence theorem has been provided in reference [21], under a slightly different set of assumptions. This existence theorem is particularly useful because it is based on a successive approximation argument, which can also be used as a numerical approximation scheme to solve this class of stochastic games. This is demonstrated and illustrated in reference [21]. The proof of a monotonicity property for the operators $\mathfrak{T}^-(v)$ and $\mathfrak{T}^+(v)$, to be introduced below, appears to be an important result, which brings the computation of equilibrium solution in the realm of successive approximation methods, like in the case of (single actor) Markov decision processes, or zero-sum Markov games.

Assumption 9.3. On rewards: Let

- $S = \prod_{i=1}^n [0, \bar{s}_i]$ be an interval in $\mathbf{R}^n$;

    then, for all $j \in M$,
- $r_j(s, \underline{u})$ be continuous in $\underline{u}$ and measurable in s; it is also bounded $r_j(s, \underline{u}) \leq \bar{u}$;
- $\forall \underline{u}, r_j(0, \underline{u}) = 0$ and $r_j(s, [u_j, \mathbf{u}_{-j}])$ be increasing in $\mathbf{u}_{-j}$;
- $r_j(s, \underline{u})$ be supermodular in u_j, for each $(s, \mathbf{u}_{-j})$, and has increasing differences in $[u_j, \mathbf{u}_{-j}]$;
- for all $s \in S$, $u_j \in U_j(s)$, where $U_j(s)$ is an interval in $\mathbf{R}^{m_j}$ and $s \mapsto U_j(s)$ is a measurable correspondence.

Assumption 9.4. On transitions: Let Q be given by

- $Q(\cdot|s, \underline{u}) = g_0(s, \underline{u})\delta_0(\cdot) + \sum_{k=1}^L g_k(s, \underline{u})\lambda_k(\cdot|s)$, where

- for $k = 1, \ldots, L$ the function $g_k : (s, \underline{u}) \mapsto [0, 1]$ is continuous in $\underline{u}$ and measurable in s, increasing in $(s, \mathbf{u})$, supermodular in $\underline{u}$ for fixed s. and $g_k(0, \underline{u}) = 0$; furthermore these functions satisfy the identity

$$\sum_{k=1}^{L} g_k(s, \underline{u}) + g_0(s, \underline{u}) \equiv 1;$$

- $\forall s \in S$, and $\forall k = 1, \ldots, L$, $\lambda_k(\cdot | s)$ is a transition probability, measurable in s;
- For any bounded and measurable functional $v(s)$, for all $k = 1, \ldots, L$, the function $s \mapsto \int_S v(t) \lambda_k(dt|s)$ is measurable and bounded ;
- $\delta_0(\cdot)$ is a probability measure concentrated at point 0.

For a given $s \in S$ and value functions $v(s, j)$, $s \in S$ and $j \in M$, that are measurable in s and bounded, with $v(0, j) \triangleq 0$, define the local game with actions $\underline{u} = (u_j : j \in M)$ and payoffs

$$\pi_j(s, v(\cdot, j), \underline{u}) = (1 - \beta) r_j(s, \underline{u}) + \beta \sum_{k=1}^{L} g_k(s, \underline{u}) \int_S v(t, j) \lambda_k(dt|s), \quad j \in M. \quad (9.58)$$

Notice that we have used above the fact that $v(0, j) \triangleq 0$, so the term g_0 does not appear.

Lemma 9.4. *Under Assumptions 9.3 and 9.4 the local game defined by equation (9.58) admits a maximal Nash equilibrium $\underline{u}^+(s, v)$ and a minimal Nash equilibrium $\underline{u}^-(s, v)$, both in pure strategies; furthermore, both equilibria are increasing in $v(\cdot, \cdot)$.*

Proof. Denote $\tau_{jk} = \int_S v(t, j) \lambda_k(dt|s)$, then (9.58) becomes

$$\pi_j(s, v(\cdot, j), \underline{u}) = (1 - \beta) r_j(s, \underline{u}) + \beta \sum_{k=1}^{L} \tau_{jk} g_k(s, \underline{u}), \quad j \in M. \quad (9.59)$$

It is clear that this game is supermodular. So, by Theorem 4.6 there exists a complete lattice of Nash equilibria with a greatest and a least Nash equilibrium. Furthermore, for each j the payoff to Player j has increasing differences in $\mathbf{u}_{-j}$ and $\tau = [\tau_{jk}]_{j \in M; k=1,\ldots,L}$. Therefore the least and greatest equilibrium are increasing in τ. The definition of τ given above implies that the least and greatest equilibrium are increasing in $v(\cdot, \cdot)$. $\square$

Lemma 9.5. *Under Assumptions 9.3 and 9.4 the equilibrium values associated with the greatest and least Nash equilibrium of the local game defined by (9.58) are monotone in $v(\cdot, \cdot)$.*

Proof. We have denoted $\underline{u}^+(s, v)$ (resp. $\underline{u}^-(s, v)$) the greatest (resp. least) Nash equilibrium of the local game defined by defined by (9.58). Define

$$\pi_j^+(s, v(\cdot, \cdot)) = \pi_j(s, v(\cdot, j), \underline{u}^+(s, v(\cdot, \cdot))) \quad (9.60)$$

and

$$\pi_j^-(s, v(\cdot, \cdot)) = \pi_j(s, v(\cdot, j), \underline{u}^-(s, v(\cdot, \cdot))). \quad (9.61)$$

Consider two functions $v_2 = (v_2(\cdot, j) : j \in M)$ and $v_1 = (v_1(\cdot, j) : j \in M)$ such that $v_1(\cdot, j) \geq v_1(\cdot, j), \forall j \in M$. Since the payoff π_j is increasing in $\mathbf{u}_{-j}$ and $v(\cdot, j)$, , by Lemma 9.4, we have $\underline{u}^-(s, v_2(\cdot, \cdot)) \geq \underline{u}^-(s, v_1(\cdot, \cdot))$. Hence,

$$\pi_j^-(s, v_2(\cdot, \cdot)) \geq \max_{u_j \in U_j(s)} \pi_j(s, v_2(\cdot, j), [u_j, \mathbf{u}_{-j}^-](s, v_2(\cdot, \cdot)))$$

$$\geq \max_{u_j \in U_j(s)} \pi_j(s, v_1(\cdot, j), [u_j, \mathbf{u}_{-j}^-](s, v_2(\cdot, \cdot)))$$

$$\geq \max_{u_j \in U_j(s)} \pi_j(s, v_1(\cdot, j), [u_j, \mathbf{u}_{-j}^-](s, v_1(\cdot, \cdot)))$$

$$= \pi_j^-(s, v_1(\cdot, \cdot)).$$

We have thus proved that $\pi_j^-(s, v(\cdot, \cdot))$ is monotone in v. A similar argument applies for $\pi_j^+(s, v(\cdot, \cdot))$. □

Define the operators

$$\mathfrak{T}^+(v)(s) = \pi^+(s, v(\cdot, \cdot)) = [\pi_j^+(s, v(\cdot, \cdot))]_{j \in M} \qquad (9.62)$$

and

$$\mathfrak{T}^-(v)(s) = \pi^-(s, v(\cdot, \cdot)) = [\pi_j^-(s, v(\cdot, \cdot))]_{j \in M}. \qquad (9.63)$$

acting on the space of functionals $v(s, j) \in \mathbf{R} : s \in S, j \in M$, that are measurable in s, and bounded. It can be proved that $\mathfrak{T}^-(v)(s)$ and $\mathfrak{T}^+(v)(s)$ are measurable on S if v is measurable and bounded on S. It can also be shown that the correspondences $s \mapsto \underline{u}^+(s, v(\cdot, \cdot))$ and $s \mapsto \underline{u}^-(s, v(\cdot, \cdot))$ are measurable for any v measurable and bounded on S. (See [21] for details.)

Theorem 9.10. *Consider the successive approximation scheme defined as follows:*

(1) Take as initial guess $v_0(s, j) \triangleq 0$, (respectively[32] $w_0(s, j) \triangleq \bar{v}$, for $s > 0$, $w_0(0, j) = 0$) for $j \in M$.

(2) Generate recursively a sequence of lower (respectively upper) bounds for equilibrium values $\{v^t(s, j)\}_{t=0}^\infty$ (respectively $\{w^t(s, j)\}_{t=0}^\infty$) where $v^{t+1} = \mathfrak{T}^-(v^t)$ (respectively $w^{t+1} = \mathfrak{T}^+(w^t)$).

(3) Associate sequences of feedback (Markov) Nash equilibrium strategies $\phi^t(s) = \underline{u}^-(s, v^t)$ (respectively $\underline{\psi}^t(s) = \underline{u}^+(s, w^t)$).

Then the following limits exist:

(1) $(\forall s \in S)$, $\lim_{t \to \infty} \phi^t(s) = \phi^*(s)$ and $\lim_{t \to \infty} \psi^t(s) = \psi^*(s)$;
(2) $(\forall s \in S)$, $\lim_{t \to \infty} v^t(s) = v^*(s)$ and $\lim_{t \to \infty} v^t(s) = w^*(s)$;

and $\underline{\phi}^*(s)$ or $\underline{\psi}^*(s)$ are feedback (Markov) Nash-equilibrium m-tuples in the infinite horizon stochastic game, verifying $\underline{\phi}^*(s) \leq \underline{\psi}^*(s)$, whereas $\frac{v^*(s)}{1-\beta}$ or $\frac{w^*(s)}{1-\beta}$ are the associated equilibrium value functions.

---

[32]Here $\bar{v}$ is an upper bound for the values of $v(s, j)$.

Proof. Clearly $\underline{\phi}^1(s) \geq \underline{\phi}^0$ and $v_1 \geq v_0$. Suppose $\underline{\phi}^{t+1}(s) \geq \underline{\phi}^t(s)$ and $v^{t+1} \geq v^t$, by Lemma 9.5 we have $v^{t+2} \geq v^{t+1}$. Then, by Lemma 9.4 , we obtain $\underline{\phi}^{t+2}(s) \geq \underline{\phi}^{t+1}(s)$. Similarly, we can prove monotonicity of $\underline{\psi}^t(s)$ and $w^t(s)$.

Now we can prove that, for all t we have $\forall S, \underline{\phi}^t(s) \leq \underline{\psi}^t(s)$. We proceed by induction. Suppose that the property is satisfied for some t. Since $v^t \leq w^t$, by Lemma 9.5 we obtain

$$v^{t+1} = \pi^-(s, v(\cdot, \cdot)) \leq \pi^-(s, w(\cdot, \cdot)) \leq \pi^+(s, w(\cdot, \cdot)) = w^{t+1}.$$

Then, by Lemma 9.4, we obtain

$$
\begin{aligned}
\underline{\phi}^{t+1}(s) &= \pi^-(s, v^{t+1}(\cdot, \cdot)) \\
&\leq \pi^-(s, w^{t+1}(\cdot, \cdot)) \\
&\leq \pi^+(s, w^{t+1}(\cdot, \cdot)) \\
&= \underline{\psi}^{t+1}(s).
\end{aligned}
$$

By definition of v^t and ϕ^t, we have, for all $s \in S$ and $u_j \in U_j(s)$

$$v^{t+1}(s, j) = (1 - \beta)r_j(s, \underline{\phi}^t(s)) + \beta \sum_{k=1}^{L} g_k(s, \underline{\phi}^t(s)) \int_S v^t(s', j)\lambda_k(ds'|s)$$

$$\geq (1 - \beta)r_j(s, [u_j, \phi^t_{-j}(s)]) + \beta \sum_{k=1}^{L} g_k(s, [u_j, \phi^t_{-j}(s)]) \int_S v^t(s', j)\lambda_k(ds'|s).$$

By the continuity of r_j and g_k, $k = 1, \ldots, L$, and using the Lebesgue dominated convergence theorem[33] we obtain, in the limit when $t \to \infty$

$$v^*(s, j) = (1 - \beta)r_j(s, \underline{\phi}^*(s)) + \beta \sum_{k=1}^{L} g_k(s, \underline{\phi}^*(s)) \int_S v^*(s', j)\lambda_k(ds'|s)$$

$$\geq (1 - \beta)r_j(s, [u_j, \phi^*_{-j}(s)]) + \beta \sum_{k=1}^{L} g_k(s, [u_j, \phi^*_{-j}(s)]) \int_S v^*(s', j)\lambda_k(ds'|s).$$

This implies (by the verification theorem) that $\underline{\phi}^*(s)$ is a pure stationary feedback-Nash equilibrium m-tuple and $\frac{v^*(s)}{1-\beta}$ is the associated equilibrium value functional. Analogously, we can prove that $\underline{\psi}^*(s)$ is a pure stationary feedback Nash equilibrium and $\frac{w^*(s)}{1-\beta}$ is the associated equilibrium value functional. $\qquad \square$

9.8 Memory, trigger and communication strategies

In this section, we present an application of nonzero-sum Markov games to the definition of equilibria in a class of memory strategies with threats in repeated games with uncertainties. In this, we follow the work of Porter [203] and Green & Porter [96].

---

[33]Lebesgue dominated convergence theorem: Let $\{f_n\}$ denote a sequence of real-valued measurable functions on a measure space (S, Σ, μ). Assume that the sequence converges pointwise to a function f and is dominated by some integrable function g in the sense that $|f_n(x)| \leq g(x)$ for all n and all points $x \in S$. Then the limiting function f is integrable and

$$\lim_{n \to \infty} \int_S f_n \, d\mu = \int_S f \, d\mu.$$

9.8.1 *A stochastic repeated duopoly*

Consider the stochastic duopoly model defined by the linear-demand equation

$$p(t+1) = \alpha - \rho[q_1(t) + q_2(t)] + \varepsilon(t),$$

which determines the price $p(t+1)$ of a good at period $t+1$ given the total supply $q_1(t) + q_2(t)$ decided at the end of period t by Players 1 and 2. Assume a unit-production cost equal to c for both firms. The profits at the end of period t for both players (firms) are then determined as

$$\pi_j(t) = (p(t+1) - c)q_j(t).$$

Assume that the two firms have the same discount rate β; then, over an infinite time horizon, the payoff to Player j will be given by

$$V_j = \sum_{t=0}^{\infty} \beta^t \pi(t).$$

This game is repeated and an obvious equilibrium solution consists of repeatedly playing the (static) Cournot solution

$$q_j^c(t) = \frac{\alpha - \delta}{3\rho}, \quad j = 1, 2, \tag{9.64}$$

which generates the payoffs

$$V_j^c = \frac{(\alpha - \rho)^2}{9\rho(1 - \beta)} \quad j = 1, 2. \tag{9.65}$$

A symmetric Pareto (nondominated) solution is given by the repeated actions

$$q_j^P(t) = \frac{\alpha - \delta}{4\rho}, \quad j = 1, 2,$$

and the associated payoffs

$$V_j^P = \frac{(\alpha - \delta)}{8\rho(1 - \beta)} \quad j = 1, 2,$$

where $\delta = \alpha - c$.

The Pareto outcome dominates the Cournot equilibrium but it does not represent an equilibrium. The question is the following: *is it possible to construct a pair of memory strategies that would define an equilibrium with an outcome dominating the repeated Cournot strategy outcome and that would be as close as possible to the Pareto (nondominated) solution?*

9.8.2 A class of trigger strategies based on a monitoring device

The random perturbations affecting the price mechanism do not allow a direct extension of the approach described in the deterministic context, which made it possible for an equilibrium solution to also be Pareto optimal. Since it is assumed that the players' actions are not directly observable, there is a need to proceed to some form of filtering of the sequence of observed states in order to monitor the possible breaches of agreement by one player or the other.

Green and Porter [96] constructed a dominating memory-strategy equilibrium, based on a *one-step memory* scheme. Below, we propose another scheme, using a *multistep memory*, that yields an outcome that lies closer to the Pareto solution.

The basic idea is to extend the state space by introducing a new state variable, denoted v that is used to monitor a *cooperative policy* that all players have agreed to play and which is defined as $\phi : v \mapsto q_j = \phi(v)$. The state equation governing the evolution of this state variable is designed as follows:

$$v(t+1) = \max\{-K, v(t) + p^e - p(t+1)\}, \tag{9.66}$$

where p^e is the expected outcome if both players use the cooperative policy, i.e.,

$$p^e = \alpha - 2\rho\phi(v).$$

It should be clear that the new state variable v provides a cumulative measure of the positive discrepancies between the expected prices p^e and the realized ones $p(t)$. The parameter $-K$ defines a lower bound for v. This is introduced to prevent negative discrepancies from compensating for positive discrepancies. A positive discrepancy could be an indication of *oversupply*, i.e., an indication that at least one player is not respecting the agreement and is maybe trying to take advantage of the other player.

If these discrepancies accumulate too fast, the evidence of cheating is mounting and thus some retaliation should be expected. To model the retaliation process we introduce another state variable, denoted y, which is a binary variable, i.e., $y \in \{0,1\}$ (compare variable y in (6.342), on page 239). This new state variable will be an indicator of the prevailing mode of play. If $y = 1$ then the game is played cooperatively; if $y = 0$, then the the game is played in a noncooperative manner, interpreted as a punitive or retaliatory mode of play.

This state variable is assumed to evolve according to the following state equation

$$y(t+1) = \begin{cases} 1 & \text{if } y(t) = 1 \text{ and } v(t+1) < \theta(v(t)) \\ 0 & \text{otherwise,} \end{cases} \tag{9.67}$$

where the positive valued function $\theta : v \mapsto \theta(v)$ is a design parameter of this monitoring scheme.

According to this state equation, the cooperative mode of play will be maintained, provided the cumulative positive discrepancies do not increase too fast from one period to the next. Also, this state equation tells us that once $y(t) = 0$, then $y(t') \equiv 0$ for all periods $t' > t$, i.e., that the punitive mode of play lasts forever. In the model discussed later on, in the $\mathbb{GE}$ section, we will relax this assumption of everlasting punishment.

When the mode of play is noncooperative, i.e., when $y = 0$, both players use the static Cournot solution as a punishment (or retaliation) forever. This generates the expected payoffs V_j^c, $j = 1, 2$ defined in equation (9.65). Since the two players are identical, we shall not use the subscript j anymore.

When the mode of play is cooperative, i.e., when $y = 1$, both players use an agreed-upon policy that determines their respective controls as a function of the state variable v. This agreement policy is defined by the function $\phi(v)$. The expected payoff is then a function $W(v)$ of this state variable v.

For this agreement to be stable, i.e., for it not to tempt any player to cheat, we impose that it be an equilibrium. Note that the game is now a Markov game with a continuous state space. The dynamic-programming equation characterizing an equilibrium is given below:

$$W(v) = \max_u \{ [\alpha - \delta - \rho(\phi(v) + u)]u$$
$$+ \beta P[v' \geq \theta(v)]V^c$$
$$+ \beta P[v' < \theta(v)]\mathbb{E}[W(v')|v' < \theta(v)]\}, \tag{9.68}$$

where we have denoted by

$$v' = \max\{-K, v + \rho(u - \phi(v)) - \varepsilon\}$$

the random value of the state variable v after the transition generated by the controls $(u, \phi(v))$.

In equation (9.68), we recognize the immediate reward $[\alpha - \delta - \rho(\phi(v) + u)]u$ of Player 1 when he plays u while the opponent sticks to $\phi(v)$. This is added to the conditional expected payoffs after the transition to either the punishment mode of play, corresponding to the values $y = 0$, or to the cooperative mode of play, corresponding to $y = 1$.

A solution of these DP equations can be found by solving an associated fixed-point problem, as indicated in [118]. To summarize the approach, we introduce the operator

$$(T_\phi W)(v, u) = [\alpha - \delta - \rho(u + \phi(v))]\,u + \beta(\alpha - \delta)^2 \frac{F(s - \theta(v))}{9\rho(1 - \beta)}$$
$$+ \beta W(-K)[1 - F(s - K)]$$
$$+ \beta \int_{-K}^{\theta(v)} W(\tau)f(s - \tau)\,d\tau \tag{9.69}$$

where $F(\cdot)$ and $f(\cdot)$ are, respectively, the cumulative distribution function and the density-probability function of the random disturbance ε. We have also used the following notation:

$$s = v + \rho(u - \phi(v)).$$

An equilibrium solution is a pair of functions $(w(\cdot), \phi(\cdot))$ such that

$$W(v) = \max_u T_\phi(W)(v, u), \tag{9.70}$$
$$W(v) = (T_\phi W)(v, \phi(v)). \tag{9.71}$$

reference [118] shows how an adaptation of the Howard *policy-improvement algorithm* [127] makes it possible to compute a solution for this sort of fixed-point problem.

The case treated in [118] corresponds to the use of a quadratic function $\theta(\cdot)$ and a Gaussian distribution law for ε. The numerical experiments reported in [118] enables the definition of a subgame-perfect equilibrium that dominates the repeated Cournot solution.

Porter, in [203], studied this problem in the case where the (inverse) demand law is subject to a multiplicative noise. He obtained a qualitative result that proves the existence of a dominating equilibrium, based on a simple *one-step memory* scheme where the variable v satisfies the following equation:

$$v(t+1) = \frac{p^e - p(t+1)}{p(t)}.$$

This is the case where we do not monitor the cooperative policy through the use of a cumulated discrepancy function, but rather on the basis of repeated identical tests. Also in Porter's approach, the punishment period is finite.

In [118] it is also shown that the approach could be extended to a fully fledged Markov game, i.e., a sequential game rather than a repeated one.[34] A simple model of fisheries management was used in that work to illustrate this type of cooperative equilibrium for sequential games. This model will be presented in the $\mathbb{GE}$ illustration, at the end of this chapter.

9.8.3 *Interpretation as a communication device*

In our approach, by extending the state-space description (i.e., introducing the new variables v and y), we retained a Markov-game formalism for an extended game and this has allowed us to use dynamic programming for the characterization of subgame-perfect equilibria. This, of course, is reminiscent of the concept of *communication device*, which is considered by Forges [83] for repeated games and which is discussed in Part 1. An easy extension of the approach described above would lead to random transitions between the two modes of play, with transition probabilities depending on the monitoring statistic v. Also a a random-duration punishment is possible in this model. In the next section we illustrate these features when we propose a differential game model with random modes of play.

The monitoring scheme is a communication device which receives as input the observation of the state of the system and that outputs a public signal, which suggests that the player play according to two different modes of play.

9.9   What have we learned in this chapter?

- Markov games are obtained as a combination of controlled Markov chains and matrix games. They were introduced by Shapley before the invention of Markov decision processes (MDP) by Howard [127], in 1960.
- We can solve Markov games using a dynamic-programming approach. The solution is obtained as the fixed-point of an operator acting on the space of value functions. This

---

[34]This game is "repeated" because variable $p(t)$ does not cumulate.

operator is contracting, which therefore simultaneously provides proof of the existence of a saddle-point solution and an algorithm that converges numerically to this solution.

- Other algorithms have been proposed to extend, in some manner, the policy iteration algorithm of MDPs. However, the theory of Markov games does not benefit from a class of algorithms as efficient as those used to solve MSPs.
- Nonzero-sum Markov games with a discrete state set have been shown to admit an equilibrium solution in the class of feedback strategies. However the existence of a fixed-point of the dynamic-programming operator is not obtained through a contraction argument, but rather, through the use of a purely topological result based on Kakutani theorem. This does not provide a numerical algorithm. An extension of the policy-iteration algorithm to nonzero-sum games has been observed to work in many examples; however this not generally guaranteed to converge to a feedback-equilibrium solution.
- The theory has been extended to the case where the state space is continuous. Approximate (ϵ) equilibrium solutions have been shown to exist for a subclass of these games. A correlated equilibrium has been shown to exist in a larger class of these games.
- There are assumptions (mainly about the supermodularity of payoffs and state transitions), under which pure strategy-feedback equilibria exist.
- Practically, the games that satisfy these assumptions are stochastic with strategic complementarities among the players.
- Stochastic games can be used to model collusive equilibria based on threats in repeated games with uncertainty and imperfect information.

9.10 Exercises

Exercise 9.1. Program Shapley's value-iteration algorithm, using Excel. Use the Excel solver to solve the associated zero-sum local games.

Exercise 9.2. Prove Lemma 9.1. Hint: refer to Lemma 4.3.4 in [82], page 178.

Exercise 9.3. Prove Lemma 9.2. Hint: refer to [82], page 24.

Exercise 9.4. Consider the state equation

$$x(t+1) = ax(t) - u_1(t) - u_2(t) + p + w(t), \quad x(0) = x_0, \qquad (9.72)$$

and payoff

$$J_j(x(0), u_1, u_2) = E\left\{\sum_{t=0}^{\infty} \beta^t [x(t) + \log(u_j(t))]\right\}, \quad j = 1, 2, \qquad (9.73)$$

where $\beta, 0 < \beta < 1$ is the common discount factor and $\tilde{u}_j = (u_j(t))_{t=0,1,\dots}$.[35] Here $x(t)$ is equal to the logarithm of the quantity of fish in a transboundary fishery at time period

---

[35]Notice that the reward function is also equal to $\log[u_j(t)y(t)]$ where $y(t)$ is the size of the fish population at time t. The product of the fishing effort by the population size can be taken as an indication of the consumption of fish. The logarithm takes care of the satiety effect.

$t, u_j(t)$ is the fishing effort of country j and $w(t)$ is a random disturbance identically and independently distributed with a mean 0 and standard deviation v.[36]

Show that the feedback-Nash equilibrium solution for this stochastic game is the same as in the deterministic case ($v = 0$), which was illustrated in Exercise 6.6, i.e., the unique feedback-Nash equilibrium is defined by

$$\sigma_1(x) = \sigma_2(x) \equiv \frac{1 - a\beta}{\beta}, \tag{9.75}$$

with the corresponding equilibrium expected payoffs

$$V_1^*(x) = V_2^*(x) = V^N(x) = Ax + B, \tag{9.76}$$

where

$$A = \frac{1}{1 - a\beta}, \quad B = \frac{1}{1 - \beta}\left[\log\left(\frac{1 - a\beta}{\beta}\right) + \frac{\beta p}{1 - a\beta} - 2\right].$$

Exercise 9.5. Consider the following stochastic game. Let $j = 1, 2$; the state and action sets are $S = U_1(s) = U_2(s) = [0, 1]$ and the transition function $F(\cdot|x, a)$ is stochastically supermodular. Prove that $g_j(x, [a_j, a_{-j}]) = a_1 + a_2 + a_1 a_2 \sqrt{x} - a_j^2$ is supermodular for any $x \in S$. Is $g_j(\cdots)$ Lipschitz continuous in x?

Exercise 9.6. Consider the following stochastic game. Let $j = 1, 2$; the state and action sets are $S = U_1(s) = U_2(s) = [0, 1]$ and the transition function $F(\cdot|x, a)$ is stochastically supermodular. Prove that $g_j(x, [a_j, a_{-j}]) = \ln(1 + a_j + a_{-j}) - a_j^2$ is not supermodular for any $x \in S$.

9.11 GE illustration: A fishery-management game

We illustrate in this GE how to construct cooperative equilibria in a discounted stochastic sequential game played over an infinite time horizon.[37] The difference with the repeated-duopoly game discussed above is that here we have a system described by a state variable that evolves over time.

9.11.1 A cumulative state variable

Our setting of the management of a fishery by n countries is a typical example of exploitation of a common resource by economic agents (see, e.g., the survey in Long [163]).

---

[36] Let $y(t)$ be the size of the fish population at time t. Notice that the state equation, in the absence of exploitation and random chock is given by $\log[y(t+1)] = a\log[y(t)] + p$, which yields $\log[\frac{y(t+1)}{y(t)}] = (a-1)\log[y(t)] + p$. One then obtains the biological law for this species

$$y(t + 1) = y(t)e^{(a-1)\log[y(t)] + p}. \tag{9.74}$$

This is an equation similar to the one adopted in the Great Fish War model [159]. If $a < 1$ and $p > 0$, the steady state for the unexploited resource is $\log[\bar{y}] = \frac{p}{1-a}$. Now, using control variables defined as fishing efforts, we assume that these efforts act directly on the growth rate of the population of fish. Similarly the uncertainty is about the growth rate.

[37] This GE Illustration is based on Haurie and Tolwinski (1990).

The literature has established that, in the absence of any regulatory mechanism, e.g., fishing quotas, these agents would maximize their individual payoffs, which would result in a higher harvest than is socially desirable. This behavior would have two negative impacts for the players. As the market price is negatively related to the total harvest, the first impact is a lower utility for each fisherman. This effect occurs in any game where strategies, here fishing efforts, are strategic complements. The second effect, which is a long-term one, would be the (risk of) extinction of the stock of fish. See (VI.6.4.) for a full analysis of a multistage fishery game, where the long-term behavior of the biomass is characterized in terms of the model's parameters.

We consider a simple setting where two countries exploit the same fishery. Denote by $x(t)$ the logarithm of fish in the fishery at time t, $t = 0, 1, 2, \ldots$. The evolution of this state variable is described by the following difference equation:

$$x(t+1) = ax(t) - u_1(t) - u_2(t) + p + w(t), \quad x(0) = x_0, \qquad (9.77)$$

where $u_j(t)$ is the fishing effort undertaken by country j, $j = 1, 2$, at time t; the parameter $a, 0 < a < 1$ represents the regeneration (or net growth) of the fish stock, and p is a positive parameter; and $w(t)$ is a random disturbance identically independently distributed for all t, having a normal p.d.f. with zero mean and standard deviation σ. We suppose that the parameters p and a are common knowledge.

Remark 9.7. In the fishery-games literature, especially since the seminal paper by Levhari and Mirman ([159]) on Fish Wars, the one-period growth function of the fish stock is typically taken as

$$x(t+1) = b(x(t))^\alpha,$$

where α is a biological parameter and b a scaling constant, with $0 < \alpha < 1 \leq b$. The restriction on α is meant to capture the saturation effect in reproduction. By letting $x(t)$ be the logarithm of the stock of fish, the state equation (9.77) accounts for this effect.

Let country j's one-period utility depend on the stock of fish $x(t)$ and on the effort level $u_j(t)$ taken as equal to consumption. We adopt the following specification:[38]

$$x(t) + \log(u_j(t)).$$

The logarithm is to account for marginal decreasing utility of consumption (compare with the example in (VI.6.4.)). Assuming that the players maximize the expected discounted sum of utilities, then their payoff functionals are

$$J_j(x(0), u_1, u_2) = E\left\{\sum_{t=0}^{\infty} \beta^t [x(t) + \log(u_j(t))]\right\}, \quad j = 1, 2, \qquad (9.78)$$

where $\beta, 0 < \beta < 1$, is the common discount factor and $u_j = (u_j(t))_{t=0,1,\ldots}$. The state space is $X \subset R^+$.

---

[38] We could write the per-period utility as

$$\beta x(t) + \delta \log(u_i(t)),$$

where β and δ are positive scaling parameters. To save on notation, we set both of them as equal to one.

Remark 9.8. The per-period utility of a player, and hence his total payoff, does not depend on the other player's control. The interdependence between the two players is indirect, i.e., through the state variable. This feature, which is common to many economic models, is not a necessary requirement for the design of a cooperative equilibrium.

Before formally introducing the stochastic sequential-game formulation and the solutions of interest, let us give an overview of what is at stake.

In the absence of any cooperation between the two countries, their fishing fleets will rationally play a noncooperative game and implement a feedback-Nash equilibrium. This equilibrium can be seen as a benchmark to assess the benefits of cooperation.

If the two countries can agree to play the game cooperatively, then they will seek a Pareto-optimal solution, and share the total resulting payoff according to a certain rule. In a cooperative game, we typically start by determining the optimally collective solution, and next address the issue of allocating, through, e.g., side payments, this total cooperative payoff. Indeed, in our fully symmetric context, both players face the same parameters, and we can assume that, under cooperation, the players jointly maximize the sum of their payoffs. In this case, the allocation problem of the total cooperative outcome is automatically solved, that is, there is no need to introduce a side payment. This cooperative solution provides an upper bound on what the players can achieve.

As we saw in different contexts, notably in games with a prisoner's dilemma payoff structure, cooperation is not self-supporting. This means that when an agreement is not an equilibrium, there is no guarantee that the players will abide by the agreement's terms and implement the optimal harvest policy throughout the game. Trigger strategies have proven to be effective tools to endow the cooperative solution with an equilibrium property. This is the line followed in this GE. Implementing such an approach requires that the players agree on what will trigger a switch from a cooperative mode of play to a noncooperative mode of play. This implies that some preplay communication between the players is necessary to establish the trigger rule and the duration of the non-cooperative mode of play before reverting to cooperation. Note that, notwithstanding this preplay agreement, the game itself is still played noncooperatively, in the sense that there is no need to assume that the agreement between the two countries will be binding. Finally, it is intuitively clear that in a noisy environment, we cannot expect to reach the cooperative solution exactly, through the use of trigger strategies.

9.11.2 *Stochastic sequential-game formulation*

9.11.2.1 *Information structure*

Let us assume that each player can observe the state variable directly, but can observe neither his opponent's actions nor the realizations of the random variable $w(t)$. This assumption is typical in the literature on cartel stability. For instance in a model of the OPEC oil cartel, we would assume that each player can observe the market price of the commodity, but not the extraction or export levels of the other cartel members. The methodological implication is that any coordination mechanism that the players would put in place, including the trigger strategies to be used here, has to rely on the observed state values rather than on the decisions made by the players. Consequently, we assume that Player i selects his control $u_j(t) \in U_j \subset R^+$ on the basis of the information represented by the random sequence

$$\varsigma(t) = \{x(0), x(1), \ldots, x(t-1), x(t)\}, \quad t = 0, 1, 2, \ldots.$$

Note that, consistently with the observation made above, $\varsigma(t)$ does not include any private information on Player j, e.g., his previous controls.

A strategy of Player j is defined as a sequence of mappings

$$\tilde{\gamma}_j = \{\gamma_{jt} : t = 0, 1, 2, \ldots\}, \quad j = 1, 2,$$

where γ_{jt} associates a harvest level $u_j(t) \in U_j$ with every $\varsigma(t)$. The strategy space of Player j is denoted by Γ_j.

Although this formulation has the merit of being general, it is easy to realize that a strategy requiring that a player recall the whole sequence $\varsigma(t)$ for every t, may be too demanding in terms of information storage, and of little practical value. In many instances, as here, it fully makes sense to restrict the strategy space to Markovian strategies where the history of the play up to time t is summarized by the state value at that time.

9.11.2.2   *The extended state*

When the fishery game is played noncooperatively, as in Section (VI.6.4.), or cooperatively, the state of the game is defined straightforwardly as the stock of fish. When we use memory (trigger) strategies, we must extend the state vector to include all information on the history of the game that is relevant to both players to monitor their preplay arrangement.

Denote by $z(t)$ the state vector at period t, with

$$z(t) = (x(t), v(t), y(t), c(t)), \tag{9.79}$$

where $v(t)$ is a statistic utilized to monitor the cooperative policies; $y(t)$ is a binary variable indicating the mode of play (cooperative or noncooperative) at period t; and, $c(t)$ is an auxiliary variable denoting the number of periods since the last switch to a noncooperative mode of play. The evolution over time of this state vector is described by the following state equations:

$$x(t+1) = ax(t) - u_1(t) - u_2(t) + p + w(t), \tag{9.80}$$

$$v(t+1) = f_1(x(t), u_1(t), u_2(t), w(t), v(t)), \tag{9.81}$$

$$y(t+1) = f_2(x(t), v(t), y(t), c(t)), \tag{9.82}$$

$$c(t+1) = \begin{cases} 0, & \text{if } y(t) = 1, \\ c(t) + 1, & \text{if } y(t) = 0, \end{cases} \tag{9.83}$$

where the function f_2 takes values in the set $\{0, 1\}$. Three remarks are in order:

(1) The choice of the variable $v(t)$ and of the state-transition functions f_1 and f_2 are a matter of design, and as such, they are an integral part of the definition of a cooperative equilibrium. These items will be defined rigorously below. As a hint, $v(t)$ will be related, quite intuitively, to the difference between the expected size of the stock of fish at t under a cooperative mode of play and the observed fish stock. For instance, if this measurement shows a high enough difference, then the players would conclude that some cheating has occurred.

(2) The model (9.79)–(9.83) has to be known and accepted by both players. This is what we meant by preplay communication and agreement on the rules. The players must also have access to information about the variable $v(t)$, and consequently, about the variables $y(t)$ and $c(t)$.

(3) The model assumes that the two players are using the same monitoring statistics. This must not necessarily always be the case. The model can be easily extended to have $v(t) = (v_1(t), v_2(t))$, i.e., the two players use different monitoring statistics, for as long the evolution and values of these statistics are common knowledge.

9.11.2.3 *Defining cooperative and noncooperative policies*

We restrict our interest to stationary Markovian strategies, that is, strategies that do not explicitly depend on time. This assumption is often made in dynamic games of infinite duration. Denote by $\gamma_j \in \Gamma_j$ a stationary Markovian strategy of Player j. Note that the symbol γ_j will be used to denote both γ_{jt} (that is, Player j's decision rule at period t) as well as his strategy (that is, the whole infinite sequence of those rules). To complete the definition of the game in normal form, we express the payoff functionals in (9.78) in terms of the strategies:

$$J_j\left(x(0), \tilde{\underline{\gamma}}\right) = \mathbb{E}_{\tilde{\underline{\gamma}}}\left\{\sum_{t=0}^{\infty} \beta^t \left[x(t) + \log\left(\gamma_{jt}\left(z(t)\right)\right)\right]\right\}, \quad j = 1, 2.$$

The expectation is taken with respect to the probability measure induced by a strategy pair γ, where

$$\tilde{\underline{\gamma}} = (\tilde{\gamma}_1, \tilde{\gamma}_2) = \{\gamma_t = (\gamma_{1t}, \gamma_{2t}) : t = 0, 1, 2, \ldots\}.$$

Definition 9.5. An admissible strategy pair $\tilde{\underline{\gamma}}^* = (\tilde{\gamma}_1^*, \tilde{\gamma}_2^*)$ is an equilibrium at $z(0)$ if

$$J_1\left[z(0), \tilde{\underline{\gamma}}^*\right] \geq J_1\left[z(0), \tilde{\gamma}_1, \tilde{\gamma}_2^*\right], \text{ for all } \gamma_1 \in \Gamma_1 \text{ s.t. } (\gamma_1, \gamma_2^*) \text{ is admissible,}$$

$$J_2\left[z(0), \tilde{\underline{\gamma}}^*\right] \geq J_2\left[z(0), \tilde{\gamma}_1^*, \tilde{\gamma}_2\right], \text{ for all } \gamma_2 \in \Gamma_2 \text{ s.t. } (\gamma_1^*, \gamma_2) \text{ is admissible.}$$

Definition 9.6. A strategy pair $\tilde{\underline{\gamma}}^* = (\tilde{\gamma}_1^*, \tilde{\gamma}_2^*)$ is a subgame-perfect equilibrium if it is an equilibrium at every $z(0) \in Z = X \times V \times \{0, 1\} \times D$, where X is the original state space, while V and D represent the sets of all values taken by the variables $v(t)$ and $c(t)$, respectively.

If the game is played noncooperatively, then the players will implement the (noncooperative) stationary policies

$$\mu(x, c) = (\mu_1(x, c), \mu_2(x, c)),$$

and collect the outcomes

$$J_j^N\left[x\left(0\right),c\left(0\right)\right] = \mathbb{E}_\mu\left\{\sum_{t=0}^{\infty}\beta^t\left[x\left(t\right)+\log\left(\mu_j\left(x,c\right)\right)\right]\right\}, \quad j=1,2.$$

As there is nothing to be monitored when the players behave selfishly, the above policies are clearly independent of the monitoring statistics v.

If the game is played cooperatively, then the players will adopt the stationary policies

$$\eta\left(x,v\right) = \left(\eta_1\left(x,v\right),\eta_2\left(x,v\right)\right),$$

and realize the outcomes

$$J_j^C\left[x\left(0\right),v\left(0\right)\right] = \mathbb{E}_\eta\left\{\sum_{t=0}^{\infty}\beta^t\left[x\left(t\right)+\log\left(\eta_j\left(x,v\right)\right)\right]\right\}, \quad j=1,2.$$

The above policies and resulting outcomes do not depend on variable c, which counts the number of noncooperative periods since the last switch to that mode of play. The reason is that c will always be equal to zero under a cooperative mode of play.

We assume, and later verify, that the cooperative outcomes strictly dominate their non-cooperative counterparts, that is,

$$J_j^N\left[x\left(0\right),c\left(0\right)\right] < J_j^C\left[x\left(0\right),v\left(0\right)\right], \text{ for } j=1,2 \text{ and every } \left(x,v,c\right) \in X \times V \times D.$$

9.11.2.4 *Trigger strategies*

The policy pairs μ and η, together with the state equations (9.79)–(9.83), determine the trigger strategies

$$\gamma_j\left(x\left(t\right),v\left(t\right),y\left(t\right),c\left(t\right)\right) = \begin{cases} \eta_j\left(x\left(t\right),v\left(t\right)\right), \text{ if } y\left(t\right)=1, \\ \mu_j\left(x\left(t\right),c\left(t\right)\right) \text{ if } y\left(t\right)=0, \end{cases} \tag{9.84}$$

for $j=1,2$. In other words, a trigger strategy tells Player j to implement his noncooperative policy when the indicator variable y has a value of one, and his cooperative policy when y is equal to zero.

The strategies defined in (9.84) are self-enforcing if they part of a Nash equilibrium. This leads to the following definition.

Definition 9.7. A quintuple $\{v, f_1, f_2, (\eta_1, \xi_1), (\eta_2, \xi_2)\}$ is said to constitute a cooperative equilibrium if the trigger strategy pair $\underline{\gamma} = (\gamma_1, \gamma_2)$ defined in (9.84) is a subgame-perfect equilibrium.

Denote by $W_j\left(z\right), j = 1,2$, the Bellman value function generated by an equilibrium strategy pair γ, defined by (9.84). The functions $W_1\left(z\right), W_2\left(z\right)$ and the strategies $\gamma_1\left(z\right), \gamma_2\left(z\right)$ must satisfy the following two equations:

$$W_j\left(z\right) = \max_{u_j \in U_j}\left\{x\left(t\right)+\log\left(u_j\left(t\right)\right)+\beta\mathbb{E}\left[W_j\left(z^j\right)\right]\right\}, \quad j=1,2 \tag{9.85}$$

where

$$z = (x, v, y, c), \quad z^1 = \left(x^1, v^1, y^1, c^1\right), \quad z^2 = \left(x^2, v^2, y^2, c^2\right),$$
$$x^1 = f_0 \left[x, u, \gamma_2 \left(z\right), w\right], \quad v^1 = f_0 \left[x, u, \gamma_2 \left(z\right), w\right],$$
$$y^1 = f_2 \left[x, u, \gamma_2 \left(z\right), w, v, y, c'\right],$$
$$x^2 = f_0 \left[x, \gamma_1 \left(z\right), u, w\right], \quad v^2 = f_0 \left[x, \gamma_1 \left(z\right), u, w\right],$$
$$y^2 = f_2 \left[x, \gamma_1 \left(z\right), u, w, v, y, c'\right],$$
$$c' = \begin{cases} c, & \text{if } y = 1, \\ c+1, & \text{if } y = 0. \end{cases}$$

Put differently, z^1 (respectively z^2) is the state to which the dynamic system moves from z if the players' actions are u and $\gamma_2 \left(z\right)$ (respectively $\gamma_1 \left(z\right)$ and u) and the realization of the random variable is w.

The remaining pending issue is when y changes its value from 1 to 0 and from 0 to 1. Following the trigger-strategies literature, we assume that the players start the game by cooperating, i.e., $y\left(0\right) = 1$. They will switch to noncooperative policies if a certain clearly defined event occurs, e.g., the observed stock of fish is below a certain threshold. As stated before, the definition of that event must be part of the preplay communication. We shall come back to this item later on. Also part of the preplay agreement is the duration of the noncooperative mode of play triggered by this same event. This duration can be endogenously determined, that is, computed as a function of, e.g., the deviation with respect to cooperation, or defined as a given number of periods that is independent of the deviation.

Assuming that the number of time intervals during which the play can be noncooperative is fixed in advance at M, the state equation (9.82) can be written as follows:

$$y\left(t+1\right) = \begin{cases} 1, & \text{if } y\left(t\right) = 1, \text{ and } v\left(t+1\right) \in \Theta \left(v\left(t\right)\right), \\ & \text{or if } y\left(t\right) = 0, \text{ and } c\left(t+1\right) = M, \\ 0, & \text{otherwise,} \end{cases} \quad (9.86)$$

where Θ is a set-valued mapping associating to each v a subset of V. The above state equation implies that the cooperative mode of play is maintained as long as the monitoring variable $v\left(t+1\right)$ stays in the set $\Theta \left(v\left(t\right)\right)$, while a period of noncooperative play is initiated each time $v\left(t+1\right)$ fails to be in $\Theta \left(v\left(t\right)\right)$. After such an event, the noncooperative mode of play lasts for M periods, after which the players resume their cooperation. Note that the mapping Θ and the integer M are additional design parameters in the definition of the cooperative equilibrium.

Define

$$W_j^1 \left(x, c\right) = W_j \left(x, v, 0, c\right), \quad c = 0, 1, 2, \ldots, M - 1,$$
$$W_j^2 \left(x, v\right) = W_j \left(x, v, 1, c\right),$$

where W_j^1 and W_j^2 denote the Bellman functions describing Player j's payoffs at periods of noncooperative and cooperative play, respectively. Notice that W_j^1 does not depend on the monitoring variable v (this variable is undefined under a noncooperative mode of play) and that W_j^2 is independent of c, which is always zero when the mode of play is cooperative. To keep the computations simple, we make the assumption that M is infinite, that is, that the mode of play remains noncooperative forever, after the first time the monitoring variable $v(t+1)$ is not contained in $\Theta(v(t))$. (In Exercise 5.4, consistent with the literature on repeated games, we use the term "grim-trigger strategy" for a strategy instructing players never to return to cooperation once cheating occurs.) Under this assumption, it is no longer necessary to consider the state variable $c(t)$, and moreover, the Bellman functions W_j^1 can be calculated independently of W_j^2.

The dynamic-programming equations in (9.85) lead to the following equations:

$$W_j^1(x) = \max_{u_j \in U_j} \left\{ x(t) + \log(u_j(t)) + \beta \mathbb{E}\left[W_j^1(x^j)\right] \right\},$$

$$W_j^2(x, v) = \max_{u_j \in U_j} \left\{ x(t) + \log(u_j(t)) \right.$$

$$+ \beta P\left[v^j \in \Theta(v)\right] \mathbb{E}\left[W_j^2(x^j, v^j) \,|v^j \in \Theta(v)\right] \left.\right\}$$

$$+ \beta P\left[v^j \in \Theta(v)\right] \mathbb{E}\left[W_j^1(x^j) \,|v^j \in \Theta(v)\right] \text{ for } j = 1, 2. \quad (9.87)$$

9.11.2.5 *Defining the monitoring variable*

As previously stated, the monitoring variable involves a comparison of the expected stock of fish under cooperative and noncooperative modes of play. To define this variable, we therefore need to compute the solutions under these two modes of play.

Noncooperative equilibrium.   Under a noncooperative mode of play, the dynamic programming equations in (9.85) can be written as

$$V_j(x(t)) = \max_{u_j \in U_j} \left\{ x(t) + \log(u_j(t)) + \beta \mathbb{E}\left[V_j(x(t+1))\right] \right\}, \quad j = 1, 2.$$

Note that the only relevant state variable here is the stock of fish. It can be shown[39] that the unique (symmetric) feedback-Nash equilibrium policies are constant and given by

$$\mu_1 = \mu_2 = \left(\frac{1 - a\beta}{\beta}\right), \quad (9.88)$$

with the corresponding payoffs

$$V_1^N(x) = V_2^N(x) = V^N(x) = Ax + B, \quad (9.89)$$

where

$$A = \frac{1}{1 - a\beta}, \quad B = \frac{1}{1 - \beta}\left[\log\left(\frac{1 - a\beta}{\beta}\right) + \frac{\beta p}{1 - a\beta} - 2\right].$$

---

[39] See Exercise 9.4. It is a good exercise to obtain this solution yourself. Simply apply the method of undetermined coefficients explained in Chapter VI.

Cooperative solution. Assuming that the two countries maximize the sum of their objectives, the dynamic programming equation of this optimization problem is

$$V\left(x\left(t\right)\right) = \max_{u_1, u_2} \left\{2x\left(t\right) + \log\left(u_1\left(t\right)\right) + \log\left(u_2\left(t\right)\right) + \beta\mathbb{E}\left[V\left(x\left(t+1\right)\right)\right]\right\}, \; j = 1, 2.$$

It can be shown that the optimal controls are constant and given by

$$u_1^* = u_2^* = u^* = \left(\frac{1 - a\beta}{2\beta}\right), \tag{9.90}$$

and the corresponding payoffs by

$$V_1^*\left(x\right) = V_2^*\left(x\right) = V^*\left(x\right) = Ax + D, \tag{9.91}$$

where

$$D = B + \frac{1 - \log 2}{1 - \beta} \approx B + \frac{0.3069}{1 - \beta}.$$

Note that the cooperative and noncooperative value functions have the same coefficient A for x. Observe that each country's fishing effort is lower under cooperation than under noncooperation, and the cooperative individual total payoff is higher than its noncooperative counterpart. Indeed,

$$u_j^* - \mu_j = -\left(\frac{1 - a\beta}{2\beta}\right) < 0,$$

$$V^*\left(x\right) - V^N\left(x\right) = D - B = \frac{0.3069}{1 - \beta} > 0, \quad \forall x \in X.$$

The value $V^*\left(x\right)$ can be seen as an upper bound for individual collusive payoffs, which are achievable under noisy information.

The monitoring variable. In view of the model's symmetric structure and the fact that the cooperative controls are constant, we shall seek cooperative policies $\eta_j\left(x, v\right)$ in the form of simpler mappings depending only on v, say $\phi\left(v\right)$, i.e., we shall assume that

$$\eta_j\left(x, v\right) = \phi\left(v\right), \; \text{for all } x \text{ and } j = 1, 2.$$

Suppose that in the preplay arrangement, the two countries agree to set their cooperative harvest policy at a lower level than under noncooperation, i.e., $\phi\left(v\right) < \mu_j$, for every v. We define the monitoring variable by the expression

$$v\left(0\right) = 0,$$

$$v\left(t+1\right) = \max\left\{-K, v\left(t\right) + x^e - x\left(t+1\right)\right\}, \quad t = 0, 1, 2, \ldots, \tag{9.92}$$

where x^e is the expected value of $x\left(t+1\right)$ if each country's fishing effort is $\phi\left(v\left(t\right)\right)$, i.e.,

$$x^e = ax\left(t\right) + p - 2\phi\left(v\left(t\right)\right).$$

Substituting for $x\left(t+1\right)$ for its value from (9.77) in (9.92) leads to

$$v\left(0\right) = 0,$$

$$v\left(t+1\right) = \max\left\{-K, v\left(t\right) - 2\phi\left(v\left(t\right)\right) + u_1\left(t\right) + u_2\left(t\right) - w\left(t\right)\right\}, \; t = 0, 1, 2, \ldots.$$

The interpretation of the monitoring variable is similar to the one offered in the repeated-duopoly game. Indeed, the variable v provides a cumulative measure of the discrepancies between the expected stock x^e and the realized ones $x(t+1)$, with $-K$ being a lower bound to $v\left(t\right)$.

9.11.3 *Collusive equilibrium*

9.11.3.1 *Analytical solution*

Let $\theta(v)$ be a positive-valued function, and define

$$\Theta(v) = \{s < \theta(v)\}.$$

The state equation for the mode-of-play indicator can now be written as

$$y(0) = 1, \quad y(t+1) = \begin{cases} 1, & \text{if } v(t+1) < \theta(v) \text{ and } y(t) = 1, \\ 0, & \text{otherwise.} \end{cases}$$

The dynamic programming equation (9.87) becomes

$$W(x,v) = \Big\{ \max_u \{x + \log u + \beta \mathrm{P}\left[v^1 \geq \theta(v)\right] \mathbb{E}\left[Ax^1 + B | v^1 \geq \theta(v)\right]\}$$
$$+ \beta \mathrm{P}\left[v^1 < \theta(v)\right] \mathbb{E}\left[W^1(x^1, v^1) | v^1 < \theta(v)\right] \Big\}, \tag{9.93}$$

where

$$W(x,v) = W_j^2(x,v), \text{ for } j = 1, 2,$$
$$x^1 = ax + p - \phi(v) - u + w,$$
$$v^1 = \max\{-K, v - w + u - \phi(v)\}.$$

It can be verified that the Bellman function satisfying (9.93) has the form

$$W(x,v) = Ax + V(v).$$

Substituting $Ax + V(v)$ for $W(x,v)$ in (9.93) and noticing that the event $v^1 \geq \theta(v)$ holds if and only if

$$w \leq v - \theta(v) + u - \phi(v),$$

we obtain

$$Ax + V(v) = \max_u \Big\{ x + \log u + \beta \int_{-\infty}^{s} \left[Ax^1 + B\right] f(w)\, dw$$
$$+ \beta \int_{s}^{\infty} \left[Ax^1 + V(v^1)\right] f(w)\, dw \Big\}$$
$$= \max_u \Big\{ x + \log u + \beta A \left[ax + p - \phi(v) - u\right] + \beta B \int_{-\infty}^{s} f(w)\, dw$$
$$+ \beta \int_{s}^{\infty} V(v^1) f(w)\, dw \Big\}$$
$$= x(1 + \beta a A) + \beta A \left[p - \phi(v)\right]$$
$$+ \max \Big\{ \log u - \beta A u + \beta \left[BF(s) + \int_{s}^{\infty} V(v^1) f(w)\, dw\right] \Big\} \tag{9.94}$$

where

$$s = v - \theta(v) + u - \phi(u).$$

By identification, from (9.94) we get

$$A = 1 + \beta a A,$$

which is the same as before. The equation $V(v)$ can be further simplified by noting that

$$\int_s^\infty V(v^1) f(w) \, dw = V(-K)[1 - F(v + u - \phi(v) + K)]$$

$$+ \int_{v+u-\phi(v)-\theta(v)}^{v+u-\phi(v)+K} V(v - w + u - \phi(v)) f(w) \, dw$$

$$= V(-K)[1 - F(v + u - \phi(v) + K)]$$

$$+ \int_{-K}^{\theta(v)} V(w) f(v - w + u - \phi(v)) \, dw.$$

Define the operator

$$T_\phi(V)(v, u) = \beta A[p - \phi(v)] + \log u - \beta A u + \beta A[ax + p - \phi(v) - u]$$

$$+ \beta V(-K)[1 - F(v + u - \phi(v) + K)] + \int_{-K}^{\theta(v)} V(w) f(v - w + u - \phi(v)) \, dw.$$

The problem of obtaining cooperative equilibria is now reduced to selecting a number K, a positive-valued function $\theta(v)$, and then determining mappings $V(\cdot)$ and $\phi(\cdot)$ that satisfy

$$V(v) = \max_u T_\phi(V)(v, u), \text{ for all } v,$$

$$V(v) = T_\phi(V)(v, \phi(v)), \text{ for all } v.$$

9.11.3.2   *Computational results*

Using a policy-improvement algorithm (as in the duopoly-repeated game of the previous section), we obtain convergent sequences $\{V_n(\cdot)\}$ and $\{\phi_n(\cdot)\}$ for various configurations of the problem parameters. To illustrate, consider the following parameter values:

$$a = 0.4, \quad p = 4.0, \quad \beta = 0.9.$$

Then the feedback equilibrium and the Pareto-optimal solutions become

$$\mu_j = 0.7111, \quad V_j^N(x) = 1.5625x + 32.8407, \quad j = 1, 2,$$

$$u_j^* = 0.3556, \quad V_j^*(x) = 1.5625x + 35.9092, \quad j = 1, 2.$$

Recall that the cooperative policies $\phi(v)$ generate the payoffs given by

$$W(x, v) = Ax + V(v) = 1.5625x + V(v).$$

Let us assume that the mapping $\theta(v)$ has the following quadratic form

$$\theta(v) = -rv^2 + v + g,$$

where r and g are positive parameters.

Table 1 gives the difference between the payoffs in a cooperative equilibrium and under noncooperation, as measured by $V(0) - B$, under near optimal choice of $\theta(\cdot)$ for various values of the standard deviation σ in the random noise. Note that the higher the value of

Table 9.7   Gain from cooperation for different values of σ

σ	$V(0) - B$	$\dfrac{V(0)-B}{V^*(x)-V^N(x)}$	r	g
0.1	2.5495	0.8309	0.0525	0.21
0.2	2.0063	0.6538	0.0875	0.35
0.3	1.4927	0.4865	0.1125	0.45
0.4	1.0189	0.3321	0.1300	0.52
0.5	0.5832	0.1901	0.1425	0.57
0.6	0.1886	0.0615	0.1525	0.61
0.7	0.0192	0.0063	0.1550	0.62
0.8	0.0000	0.0000	–	–

σ, the lower the registered gain. To appreciate these numbers, ranging from 0.0000 (for $\sigma = 0.8$) to 0.25495 (for $\sigma = 0.1$), we need to compare them to the difference between Pareto-optimal and noncooperative outcomes, i.e.,

$$V^*(x) - V^N(x) = 1.5625x + 35.9092 - (1.5625x + 32.8407) = 3.0685.$$

This is done in the third column of Table 9.7.

Table 9.8 presents mappings $\phi(\cdot)$ and $V(\cdot)$ for two values of σ. The numbers in the two $V(v)$ columns fall in the interval $[B, D] = [32.8407, 35.9092]$. These numbers are closer to D, the constant in the cooperative value function of Player j, $j = 1, 2$, for a lower σ.

Table 9.8   Mappings $\phi(\cdot)$ and $V(\cdot)$

| v | $\sigma = 0.2$ | | $\sigma = 0.4$ | |
	$\phi(v)$	$V(v)$	$\phi(v)$	$V(v)$
0.00	0.4516	34.8477	0.6146	33.4239
0.20	0.4455	34.8412	0.6134	33.4235
0.40	0.4227	34.8283	0,6100	33.4227
0.60	0.3848	34.7864	0.6042	33.4209
0.80	0.3455	34.6840	0.5963	33.4169
1.00	0.3148	34.5235	0.5869	33.4090
1.20	0.2956	34.3206	0.5768	33.3948
1.40	0.2873	34.0988	0.5675	33.3724
1.60	0.2899	33.8799	0.5607	33.3402
1.80	0.3043	33.6806	0.5580	33.2977
2.00	0.3336	33.5091	0.5610	33.2455

Chapter 10

Piecewise Deterministic Differential Games

10.1 Introduction and definitions

In this chapter, we consider a class of stochastic games involving the control of a hybrid system by several players. The system is called *hybrid* because its state $s = (\xi, \mathbf{x})$ is composed of a discrete part ξ that takes its value from a discrete set I, e.g., a set of integer numbers, and a continuous part $\mathbf{x}$ that takes its value from a subset X of $\mathbf{R}^n$. This game format was proposed in reference [105] under the name of *Piecewise Deterministic Differential Games* (PDDG). A first exploitation of this paradigm in the study of stochastic oligopoly model was presented in [116]. A motivation for these games can be found in processes that can "jump," e.g., due to discoveries. For instance, in a game of R&D investment between "Pharma" firms, the players need to account for a possibility that, with a certainly probability, a new "blockbuster" drug will become available, some time in the future.

The dynamics of the system is described by

- A set of state equations

$$\dot{\mathbf{x}}(t) = f^i(\mathbf{x}(t), \mathbf{u}_1(t), \ldots, \mathbf{u}_m(t), t) \tag{10.1}$$

where, as usual $\mathbf{u}_j(t) \in U_j \subset \mathbf{R}^{m_j}$ is the control variable of Player j, and where the velocity function $f^i(\cdot, \cdot, \ldots, \cdot, \cdot)$ is indexed over the discrete set I, i.e., a different function is associated with each $i \in I$; these functions are assumed to be continuously differentiable in $\mathbf{x}, t$ and continuous in each $\mathbf{u}_j$.

- A stochastic jump process $\{\xi(t) : t \geq 0\}$, taking its values from the discrete set I, with jump rates

$$P\left[\xi(t + dt) = \ell | \xi(t) = k, \mathbf{x}(t) = \mathbf{x}, \mathbf{u}_1(t) = \mathbf{u}_1, \ldots, \mathbf{u}_m(t) = \mathbf{u}_m, t\right]$$
$$= q_{k,\ell}(\mathbf{x}, \mathbf{u}_1, \ldots, \mathbf{u}_m, t) + o(dt) \tag{10.2}$$

where each function $q_{k,\ell}(\cdot, \cdot, \ldots, \cdot, \cdot)$ is also assumed to be continuously differentiable in x and t, and continuous in each $\mathbf{u}_j$ and $\lim_{dt \to 0} \frac{o(dt)}{dt} = 0$.

Player j receives a reward defined by the rate functions $\mathcal{L}_j^i(\mathbf{x}(t), \mathbf{u}_1(t), \ldots, \mathbf{u}_m(t), t)$, also indexed over the discrete set I.

The players control this system through the use of *piecewise open-loop strategies*. They are defined as follows:

- At jump time τ, the state observed is $s^\tau = (\xi^\tau, \mathbf{x}^\tau)$. Each player j selects a function $\alpha_j : [0, \infty) \to U_j$. Until the next jump occurs, at time τ', the control of Player j is defined as $\mathbf{u}_j(t) = \alpha(t - \tau)$. For short, we will call $\tilde{\mathbf{u}}_j^\tau$ the open-loop control thus defined over the time interval $[\tau, \infty)$ and $\mathcal{U}$ the set of all these admissible controls.
- When the m players have selected their controls, a trajectory $\tilde{\mathbf{x}}(\cdot) : [\tau, \infty) \to X$ is generated and a probability measure on the next jump time and next discrete state visited τ' is determined, through the jump rate functions defined in (10.2), where $k = \xi(\tau)$.
- A transition reward is obtained by each player between the two successive jump times. For Player j, this reward is given by

$$r_j(\tau, \xi^\tau, \mathbf{x}^\tau; \tilde{\mathbf{u}}_1, \dots, \tilde{\mathbf{u}}_m) = \int_\tau^{\tau'} e^{-\rho(t-\tau)} \mathcal{L}_j^i(\mathbf{x}(t), \mathbf{u}_1(t), \dots, \mathbf{u}_m(t), t)\, dt, \quad (10.3)$$

  which is a random variable, since jump time τ' is random. Note that, in expression (10.3) one assumes implicitly that the state trajectory $\mathbf{x}(t) : \tau \leq t \leq \tau'$ is defined as the solution of (10.1), where $i = \xi^\tau$, with initial condition $\mathbf{x}(\tau) = \mathbf{x}^\tau$ and controls $\underline{\mathbf{u}}(t) : \tau \leq t \leq \tau'$.
- The system is observed from time $\tau^0 = 0$ with initial state $s^0 = (\xi^0, x^0)$, and the players select piecewise open-loop strategies $\gamma_j : [0, \infty) \times I \times X \to \mathcal{U}_j$. This means that they use open-loop controls, which will be adapted to the state of the system at each random jump time of the $\xi(\cdot)$ process. Let τ^n, $n = 0, \dots, \infty$ denote the n-th random jump time. At this time the system's state is $s^n = (\xi^n, x^n)$. The payoff to Player j is the expected discounted sum of the transition rewards

$$J_j(0, \xi^0, \mathbf{x}^0; \underline{\gamma})$$

$$= \mathbb{E}_{\underline{\gamma}}\left[\sum_{n=0}^\infty e^{-\rho\tau^n} r_j(\tau, \xi^{\tau^n}, \mathbf{x}^{\tau^n}; \tilde{\mathbf{u}}_1, \dots, \tilde{\mathbf{u}}_m) | \mathbf{x}(0) = \mathbf{x}^0, \xi(0) = \xi^0 \right], \quad (10.4)$$

  which, according to (10.3) can also be written as the expected value of a stochastic integral

$$J_j(0, \xi^0, \mathbf{x}^0; \underline{\gamma})$$

$$= \mathbb{E}_{\underline{\gamma}}\left[\int_0^\infty e^{-\rho t} \mathcal{L}_j^i(\mathbf{x}(t), \mathbf{u}_1(t), \dots, \mathbf{u}_m(t), t)\, dt | \mathbf{x}(0) = \mathbf{x}^0, \xi(0) = \xi^0 \right], \quad (10.5)$$

  where $\underline{\gamma} = (\gamma_1, \dots, \gamma_m)$ designates the m player strategy vector; $\mathbf{x}(\cdot), \dots, \mathbf{u}_j(\cdot), \dots$ are stochastic processes generated by the dynamics (10.1)-(10.2) associated with the piecewise open-loop strategies. The expected values are taken with respect to the probability measure generated by the strategy vector $\underline{\gamma}$.

Definition 10.1. A piecewise open-loop m-strategy vector $\underline{\gamma}^*$ is a **piecewise open-loop Nash equilibrium** if the following holds for each player j

$$J_j(t, i, \mathbf{x}; \underline{\gamma}^*) \geq J_j(t, i, \mathbf{x}; [\gamma_j, \underline{\gamma}^*_{-j}]), \quad \forall t \geq 0, i \in I, \mathbf{x} \in X, \quad (10.6)$$

where t, i and x represent any initial condition defining first jump time at t, with initial state $s^0 = (i, x)$ and as usual $[\gamma_j, \underline{\gamma}^*_{-j}]$ designates the strategy vector obtained when Player j uses strategy γ_j but all other players stick to the equilibrium strategies.

Remark 10.1. Notice that the Nash equilibrium property is supposed to hold for any possible initial data. This definition[1] implies that the equilibrium will be subgame perfect at the jump times. It also implies that we need to implement some form of dynamic programming to characterize these equilibrium solutions.

The game that we have just defined combines features of differential games and sequential stochastic games. It provides interesting opportunities to model competition under uncertainty. This will be illustrated in the forthcoming section through a detailed example of a stochastic duopoly game and in the $\mathbb{GE}$ illustration by a stochastic model of climate change, where two groups of countries compete to define the proper timing of abatement and R&D policies. Through these examples, we will show how to formulate and solve the dynamic-programming equations characterizing the equilibrium solution.

10.2 A stochastic duopoly model

10.2.1 *System's dynamics and payoffs*

In this first example, we present the stochastic oligopoly model introduced by Haurie & Roche in [116] and the numerical method they proposed to solve this PDDG.

Consider m firms competing in a market where the demand varies randomly. More precisely, let E be a finite set of possible market conditions. A homogeneous continuous-time Markov chain $\xi(t) : t \geq 0$ with value in E and transition kernel matrix $Q = [q_{k\ell}]$ describes the random changes in market conditions for $q_{k\ell} \geq 0$, if $\ell \neq k$ and $q_{kk} = -\sum_{\ell \neq k} q_{k\ell}$. Recall that, in a continuous-time Markov chain, the transition rate $q_{k\ell}$ is defined by

$$q_{k\ell} = \lim_{dt \to 0} \frac{P[\xi(t + dt) = \ell | \xi(t) = k]}{dt}. \tag{10.7}$$

With each element $i \in E$, a specific inverse-demand function $\bar{x} \mapsto D^i(\bar{x})$ is associated, where $\bar{x}$ is the total supply on the market and $p = D^i(\bar{x})$ is the market-clearing price, i.e., the price that equates supply and demand.

Let $x_j(t)$ denote the production capacity of firm j at time t, and $u_j(t)$ the investment (capacity increase) rate at time t. We assume a capital-depreciation rate for firm j equal to $\mu_j \geq 0$. Assuming that the firms always produce at capacity, the total supply at time t is given by

$$\bar{x}(t) = \sum_{j \in M} x_j(t), \tag{10.8}$$

where $M = \{1, \dots, m\}$.

The state equation describing the dynamics of capacity is

$$\dot{x}_j(t) = u_j(t) - \mu_j x_j(t), \quad x_j(0) = x_j^0, \tag{10.9}$$

---

[1] *Piecewise open-loop* Nash equilibrium is the traditional name, see [105]. Sometimes *pointwise feedback* Nash equilibrium is also used to describe this kind of stategry.

where x_j^0 is the initial production capacity of firm j. The adjustment of capacity (or investment) cost of firm j at time t is given by $\phi_j(x_j(t), u_j(t))$. The resulting instantaneous reward, when the market is in "mode" $i \in E$, is given by

$$g^i(\bar{x}(t), x_j(t), u_j(t)) = x_j(t) D^i(\bar{x}(t)) - \phi_j(x_j(t), u_j(t)). \qquad (10.10)$$

We assume that each firm j uses a piecewise open-loop policy. This means that it uses open-loop controls that are adapted at every jump time of the process. We describe an action a_j of firm j as a function $a_j : [0, \infty) \to \mathbf{R}^+$ and we call $\mathcal{A}_j$ the set of all possible actions of firm j.

Let us denote by $t^0 = 0, t^1, \ldots, t^n, \ldots$ the successive jump times of the Markov chain $\xi(t) : t \geq 0$. At jump time t^n, if firm j chooses action a_j^n, then, during the random interval $[t^n, t^{n+1})$, the control used by firm j is defined as $u_j(t) = a_j^n(t - t^n)$. We assume that each firm knows when a jump occurs and has then access to a complete observation of the state $s^n = (\mathbf{x}^n, \xi^n)$, where $\mathbf{x}^n = \mathbf{x}(t^n) = (x_j(t^n))_{j \in M}$ and $\xi^n = \xi(t^n)$. Therefore, on the random time interval $[t^n, t^{n+1})$, the profit accrued to firm j is given by the random variable (whose probability law depends on both state s^n and action vector $\underline{a}^n$)

$$\pi_j^n(s^n, \underline{a}^n) = \int_{t^n}^{t^{n+1}} e^{-\rho(t-\tau^n)} \left(x_j(t) D^{\xi^n}(\bar{x}(t)) - \phi_j(x_j(t), u_j(t)) \right) dt. \qquad (10.11)$$

A policy for firm j is defined as a mapping γ_j, which associates a specific open-loop control with the observed state s of the system at a given jump time. We assume that the class of admissible policies is such that the vector policy $\underline{\gamma} = (\gamma)_{j \in M}$ defines a Markov process $\{s^n : n = 0, 1, \ldots, n, \ldots\}$.

10.2.2 *Markov-game representation*

The oligopoly model described above defines a (stochastic) sequential game with state transitions at random times t^n. A stationary strategy dictates a choice of action to Player j for all random times in the sequential game. We associate the following expected infinite-horizon payoff from initial state s^0, called the value function, with the policy vector $\underline{\gamma}$ as follows:

$$V_j(s^0; \underline{\gamma}) = \mathbb{E}_{\underline{\gamma}} \left[\sum_{n=0}^{\infty} e^{-\rho t^n} \pi_j^n(s^n, \gamma_j(s^n)) \right]. \qquad (10.12)$$

The above equation fits the format of nonzero-sum Markov games with continuous state and action sets, as discussed in Chapter 9. Therefore, we can here apply the formalism developed for the treatment of that class of games.

So, let us consider the space $\mathcal{V}$ of bounded continuous functions $v : S \times M \to \mathbf{R}$, endowed with the sup norm and define, for given state $s = (\mathbf{x}, i)$, the reward-to-go function $v(\cdot, j), j \in M$ and action $\underline{a} = (a_j)_{j \in M}$ and the local reward functions

$$h_j((\mathbf{x}, i), v(\cdot), \underline{a}) = \mathbb{E} \left[\int_0^T e^{-\rho t} \left(x_j(t) D^i(\bar{x}(t)) - \phi_j(x_j(t), u_j(t)) \right) dt \right.$$

$$\left. + e^{-\rho T} v(\mathbf{x}(T), \xi(T), j) | \mathbf{x}(0) = \mathbf{x}, \xi(0) = i \right], \qquad (10.13)$$

where T is the random time at which the first jump of the Markov chain $\xi(\cdot)$ occurs. Notice that here, since we consider the game at initial time, the action a_j and the control u_j are the same function for each $j \in M$.

As in equation (9.25) of Chapter 9, we define, for any given Markov policy vector $\underline{\gamma} = \{\gamma_j\}_{j \in M}$, an operator $\mathcal{T}_{\underline{\gamma}}$, acting on the space of the reward-to-go functionals, defined as

$$\mathcal{T}_{\underline{\gamma}}(v(\cdot))(s, j) = \{\mathbb{E}_{\underline{\gamma}(s)} h_j(s, v(\cdot, j), \tilde{a})\}.$$

Let $f_{\underline{\gamma}}(s, j) = \sup\{v_{[\gamma_j, \underline{\gamma}_{-j}]}(s, j) : \gamma_j \in \Gamma_j\}$ denote the maximum expected payoff by Player j when replying to the strategy choice of the other players, i.e., $\underline{\gamma}_{-j} = (\gamma_i)_{i \neq j}$, and also introduce the operator $\mathcal{T}_{\underline{\gamma}}^*$ defined as

$$\mathcal{T}_{\underline{\gamma}}^+(v(\cdot))(s, j) = \{\sup_{\gamma_j} E_{[\gamma_j, \underline{\gamma}_{-j}](s)}[h_j(s, v(\cdot, j), \underline{\tilde{a}})]\}. \tag{10.14}$$

Here $\underline{\tilde{a}}$ is the action vector generated by $[\gamma_j, \underline{\gamma}_{-j}](s)$, which is the Markov strategy vector obtained when only Player j adjusts his policy, but the other players maintain fixed their γ policies. We showed in Chapter 9 that the operator $\mathcal{T}_{\underline{\gamma}}^+$ is contracting and admits a unique fixed point, which is $f_{\underline{\gamma}}(s, j)$ for $s \in S$ and $j \in M$.

Definition 10.2. A stationary strategy vector $\underline{\gamma}$ is an ϵ-equilibrium ($\epsilon > 0$) if

$$f_{\underline{\gamma}}(s, j) - v_{\underline{\gamma}}(s, j) \leq \epsilon, \forall s \in M, j \in M. \tag{10.15}$$

A 0-equilibrium is a Nash equilibrium.

10.2.3 *The associated infinite-horizon open-loop differential games*

The RHS of the definition of the local reward function $h(s, v(\cdot), a)$ in (10.13), after writing the elementary probability distribution of the Markov stopping time T explicitly, can also be written as follows:

$$h_j((\mathbf{x}, i), v(\cdot), a) = \int_0^\infty q^i e^{-q^i t} \left[\int_0^t e^{-\rho s} \left(x_j(s) D^i((\bar{x}(t)) - f_j(x_j(t), u_j(t))) \right) dt \right.$$

$$\left. + e^{-\rho t} \sum_{k \neq i} \frac{q_{ik}}{q^i} v(\mathbf{x}(t), k, j) \right], \tag{10.16}$$

where $q^i = \sum_{k \neq i} q_{ik}$ and $q^i e^{-q^i t} dt = P[T \in (t, t + dt)] + o(dt)$, $\lim_{dt \to 0} \frac{o(dt)}{dt} \to 0$. Also the ratio $\frac{q_{ik}}{q^i}$ is the conditional probability of a transition from i to k knowing that the current discrete state is i and that a transition takes place at time t.

We now let

$$A(t) = \int_0^t e^{-\rho s} \left(x_j(s) D^i((\bar{x}(s)) - \phi_j(x_j(s), u_j(s))) \right) ds$$

$$B(t) = e^{-q^i t}$$

and we notice that the integral in the RHS of equation (10.16) can be written as $-\int_0^\infty A(t)dB(t)dt$; hence, it can be integrated by parts. Assuming that

$$\lim_{t\to\infty} A(t)B(t) = 0,$$

we obtain

$$h_j((\mathbf{x}, i), v(\cdot), \tilde{a}) = \int_0^\infty B(t)dA(t) - [A(t)B(t)]_0^\infty + e^{-(\rho+q^i)t}\sum_{k\neq i} q_{ik}v(\mathbf{x}(t), k, j)$$

$$= \int_0^\infty e^{-(\rho+q^i)t}[x_j(t)D^i(\bar{x}(t)) - \phi_j(x_j(t), u_j(t))$$

$$+ \sum_{k\neq i} q_{ik}v(\mathbf{x}(t), k, j)] dt. \tag{10.17}$$

We notice that the expression in the RHS of equation (10.17) is an infinite-horizon integral payoff, for an open-loop differential game with state equations

$$\dot{x}_j(t) = u_j(t) - \mu_j x_j(t), \quad x_j(0) = x_j, \tag{10.18}$$

where x_j is the j-th component of $\mathbf{x}$.

We have just shown that, given a candidate reward-to-go function $v(\cdot)$, we can define at any initial position $s = (\mathbf{x}, i)$ an infinite-horizon open-loop differential game defined by the data (10.17)-(10.18). If this differential game admits a unique open-loop Nash-equilibrium solution, at each initial position, it then defines, at each initial position, a new *equilibrium operator* $\mathcal{E} : \mathcal{V} \to \mathcal{V}$, where

$$\mathcal{E}v(\cdot)(s, j) = h_j((\mathbf{x}, i), v(\cdot), \underline{a}^*), \tag{10.19}$$

where $\underline{a}^*$ is the equilibrium action (or open-loop strategy) in the open-loop differential game (10.17)-(10.18).

Lemma 10.1. *Suppose the operator $\mathcal{E}$ defined in (10.19), verifies the following inequality for a given reward function:*

$$\|v^{\text{new}}(\cdot) - v(\cdot)\| = \|\mathcal{E}v(\cdot) - v(\cdot)\| < \epsilon.$$

Then the policy associated with the solution of (10.17)-(10.18) obtained for each initial position, defines an ϵ-Nash equilibrium stationary strategy vector for the stochastic game, where A is a given number that depends on the contraction ratios of the two operators $\mathcal{T}, \mathcal{T}^*$.

Proof. We know that the operators $\mathcal{T}$ and $\mathcal{T}^*$ are contracting. Let $\alpha < 1$ be the largest of the contraction ratios for the two operators $\mathcal{T}, \mathcal{T}^*$. This means that, for each operator, e.g., $\mathcal{T}$, we have

$$\|\mathcal{T}v^1 - \mathcal{T}v^2\| \leq \alpha\|v^1 - v^2\|,$$

for any pair of candidate reward-to-go functions, v^1 and v^2.

Now, call γ^* the policy m-tuple defined by the equilibrium solution to the infinite-horizon open-loop differential game (10.17)–(10.18) at any possible initial position. By construction, we have

$$T_{\underline{\gamma}^*} v(\cdot) = \mathcal{E}v(\cdot);$$

therefore, the following holds:

$$\|\mathcal{E}v(\cdot) - v(\cdot)\| = \|T_{\underline{\gamma}^*} v(\cdot) - v(\cdot)\| < \epsilon.$$

Iterating with the operator $T_{\underline{\gamma}^*}$, we obtain at the limit

$$\|v_{\underline{\gamma}^*}(\cdot) - v(\cdot)\| \le \frac{1}{1-\alpha}\epsilon,$$

where $v_{\underline{\gamma}^*}(\cdot)$ is the unique fixed-point of $T_{\underline{\gamma}^*}$.

Consider now the operator $T_{\underline{\gamma}^*}^*$ defined in (10.14). Again, by construction of the equilibrium solution, we have

$$T_{\underline{\gamma}^*}^* v(\cdot) = \mathcal{E}v(\cdot);$$

therefore, for the same reason as above,

$$\|f_{\underline{\gamma}^*}(\cdot) - v(\cdot)\| \le \frac{1}{1-\alpha}\epsilon,$$

where $f_{\underline{\gamma}^*}(\cdot)$ is the unique fixed-point of $T_{\underline{\gamma}^*}^*$.

Now it is easy to obtain

$$\|f_{\underline{\gamma}^*}(\cdot) - v_{\underline{\gamma}^*}(\cdot)\| \le \|f_{\underline{\gamma}^*}(\cdot) - v(\cdot)\| + \|v_{\underline{\gamma}^*}(\cdot) - v(\cdot)\| \le \frac{2}{1-\alpha}\epsilon. \tag{10.20}$$

This establishes the desired property, according to Definition 10.2, with $A = \frac{2}{1-\alpha}$. $\quad\square$

10.2.4 *Numerical illustration*

Consider a market with an inverse demand function defined by

$$D^i(\bar{x}) = \frac{a^i}{x + b^i} - c^i, \tag{10.21}$$

where

$$b^i = 20, \quad i = 1, 2, 3,$$
$$a^1 = 120, \quad a^2 = 100, \quad a^3 = 80,$$
$$c^1 = 3, \quad c^2 = 2.5, \quad c^3 = 2.$$

The cost functions are defined as

$$\phi_j(x_j, u_j) = x_j^2 + u_j^2,$$

and the continuous-time Markov chain is defined by the following matrix of transition probability rates:

$$\begin{matrix} -0.2 & 0.2 & 0 \\ 0.01 & -0.05 & 0.04 \\ 0.1 & 0 & -0.1 \end{matrix} \tag{10.22}$$

Below, we give the Hamiltonian for Player j in the infinite-horizon differential game associated with the local game, when the discrete state is i. (The variable p_j^i is the costate variable associated with the state equation $\dot{x}_j = u_j - \mu_j x_j$, when $\xi = i$.)

$$H_j^i \triangleq x_j D^i(\bar{x}) - \phi_j(x_j, u_j) + \sum_{k \neq i} q_{ik} v(x_1, x_2, k; j) + p_j(u_j - \mu_j x_j),$$

$$= x_j \frac{a^i}{x_1 + x_2 + b^i} - x_j^2 - u_j^2 + \sum_{k \neq i} q_{ik} v(x_1, x_2, k; j) + p_j^i(u_j - \mu_j x_j).$$

The maximization w.r.t. to u_j yields

$$\frac{\partial}{\partial u_j} H_j^i = -2u_j + p_j^i = 0, \tag{10.23}$$

$$u_j = p_j^i/2.$$

Therefore, we can write the maximized Hamiltonian as follows:

$$\mathcal{H}_j^i \triangleq x_j \frac{a^i}{x_1 + x_2 + b^i} - x_j^2 + \frac{(p_j^i)^2}{4} + \sum_{k \neq i} q_{ik} v(x_1, x_2, k; j) - p_j^i \mu_j x_j.$$

Taking the partial derivatives with respect to x_j

$$\frac{\partial \mathcal{H}_j^i}{\partial x_j} = \frac{a^i}{x_1 + x_2 + b^i} - \frac{a^i x_j}{(x_1 + x_2 + b^i)^2} - 2x_j + \sum_{k \neq i} q_{ik} \frac{\partial v(x_1, x_2, k; j)}{\partial x_j} - p_j^i \mu_j$$

$$= \frac{a^i x_{-j} + a_i b_i}{(x_1 + x_2 + b^i)^2} - 2x_j + \sum_{k \neq i} q_{ik} \frac{\partial v(x_1, x_2, k; j)}{\partial x_j} - p_j^i \mu_j$$

$$= \frac{a^i}{(x_1 + x_2 + b^i)} - \frac{a^i x_j}{(x_1 + x_2 + b^i)^2} - 2x_j + \sum_{k \neq i} q_{ik} \frac{\partial v(x_1, x_2, k; j)}{\partial x_j} - p_j^i \mu_j$$

and a second time

$$\frac{\partial^2 \mathcal{H}_j^i}{\partial x_j^2} = -\frac{2a^i}{(x_1 + x_2 + b^i)^2} + \frac{2a^i x_j}{(x_1 + x_2 + b^i)^3} - 2x_j + \sum_{k \neq i} q_{ik} \frac{\partial^2 v(x_1, x_2, k; j)}{\partial x_j^2}$$

$$= -\frac{2a^i(x_{-j} + b^i)}{(x_1 + x_2 + b^i)^3} - 2x_j + \sum_{k \neq i} q_{ik} \frac{\partial^2 v(x_1, x_2, k; j)}{\partial x_j^2},$$

we observe that the maximized Hamiltonian is concave w.r.t. x_j if the functions $v(\mathbf{x}, k; j)$ are concave w.r.t. x_j. Also, computing the second derivative w.r.t. p_j^i, we observe, as expected, that the Hamiltonian is convex in p_j^i. Indeed,

$$\frac{\partial^2 \mathcal{H}_j^i}{\partial (p_j^i)^2} = \frac{1}{2}.$$

Hence $\mathcal{H}_j^i$ is convex in p_j^i ($\frac{\partial^2 \mathcal{H}_j^i}{\partial (p_j^i)^2} > 0$) and concave in x_j ($\frac{\partial^2 \mathcal{H}_j^i}{\partial x_j^2} < 0$), if each $v(x_1, x_2, k; j)$ is concave in x_j.

The Hamiltonian system

$$\dot{x}_j = \frac{\partial \mathcal{H}_j^i}{\partial p_j^i},$$

$$\dot{p}_j^i - (\rho + q^i)p_j(t) = -\frac{\partial \mathcal{H}_j^i}{\partial x_j},$$

which is concave in x_j and convex in p_j, can admit a global attractor (or *turnpike*[2]) steady-state solution, which is given by the solution of the algebraic equations

$$0 = \frac{\partial \mathcal{H}_j^i}{\partial p_j^i}, \tag{10.24}$$

$$-(\rho + q^i)p_j(t) = -\frac{\partial \mathcal{H}_j^i}{\partial x_j}. \tag{10.25}$$

Denote by $(\bar{x}_j^i, \bar{p}_j^i)$ the solution of equations (10.24) and (10.25), which is also called the "turnpike" associated with the discrete state i. When the Markov chain ξ is in state i, all the equilibrium trajectories for $(\tilde{x}_1, \tilde{x}_2)$ and costate trajectories for $(\tilde{p}_1^i, \tilde{p}_2^i)$ are attracted by the same turnpike $(\bar{x}_j^i, \bar{p}_j^i)$. The turnpike values for the x_j variables can be computed through the use of a solver for nonlinear equations.[3]

Table 10.1 gives the results of these calculations, for successive iterations of the operator $\mathcal{E}$.

Table 10.1   Turnpike values at equilibrium

Iteration	$\bar{x}_1^1$	$\bar{x}_2^1$	$\bar{x}_1^2$	$\bar{x}_2^2$	$\bar{x}_1^3$	$\bar{x}_2^3$
1	1.053	1.061	0.926	0.931	0.799	0.803
2	1.047	1.051	0.925	0.929	0.8	0.804
3	1.046	1.051	0.925	0.929	0.8	0.804
4	1.046	1.051	0.925	0.929	0.8	0.804
5	1.046	1.051	0.925	0.929	0.8	0.804

Knowing the asymptotic values for the $\tilde{x}_j^{i*}$ trajectories at equilibrium, we can compute these trajectories by solving the two-point boundary value problem (TPBVP)

$$\dot{x}_j = \frac{\partial \mathcal{H}_j^i}{\partial p_j^i},$$

$$\dot{p}_j^i - (\rho + q^i)p_j(t) = -\frac{\partial \mathcal{H}_j^i}{\partial x_j},$$

$$x_j(0) = x,$$

$$x_j(\infty) = \bar{x}_j^i, \quad j \in M,$$

---

[2] See [40, 41] for sufficient conditions ensuring the existence of turnpikes for nonzero-sum differential games.
[3] Mathematica or Maple may be used to perform these calculations. Another possibility would be to implement a Newton method to solve these equations.

for all initial positions (x_1, x_2, i). In fact, we use a grid of initial values for (x_1, x_2, i) and for each $i \in \{1, 2, 3\}$. We solve the TPBVP for each initial point in the grid (the asymptotic value is always the same for a given i) and, after that calculation, we adjust a smooth polynomial function, using least-square approximation, to obtain a representation of the functions $v(\mathbf{x}, i, j)$ for all $\mathbf{x}$.

In this way, we can produce an estimation of the approximation of the $v(\cdot, i, j)$ functions when the operator $\mathcal{E}$ is applied recurrently. Table 10.2 shows the evolution of the error estimate. We observe a monotone decrease in all Δ_j^i.

Table 10.2   Value-function error estimate: $\Delta_j^i = \|v^{\text{new}}(\cdot, i, j) - v(\cdot, i, j)\|$

error estimate Iteration	Δ_1^1	Δ_2^1	$\Delta 1^2$	Δ_2^2	Δ_1^3	Δ_2^3
1	3.9	4.1	3.4	3.6	5	5.2
2	3.4	3.4	1.4	1.4	2.1	2.1
3	1.1	1.1	0.6	0.6	0.5	0.5
4	0.4	0.4	0.16	0.16	0.2	0.2
5	0.1	0.1	0.06	0.06	0.06	0.06

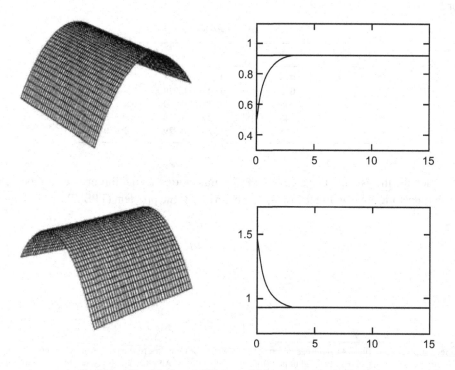

Fig. 10.1   Value functions and Nash-equilibrium trajectories for $\xi = 2$

The graphs of the value functions, given in Figure 10.1, show that the concavity with respect to x_j is very strong. The trajectories on the right side of the figure illustrate the turnpike property of the equilibrium trajectories of the infinite-horizon differential game associated with these value functions, when the discrete state is $\xi = 2$. The two trajectories start from different points ($x_j = 0.5$ or $x_j = 1.5$) and they both converge toward the attractor (turnpike), which is $\bar{x}_j = 1.05$.

The theory of turnpikes for open-loop differential games on infinite-time horizons is fully developed in reference [40].

Remark 10.2. In this example, the Markov chain that defines the randomness in the market dynamics is unaffected by the players' decisions. We could therefore approach the problem from a different perspective and look for an S-adapted equilibrium solution, instead of a piecewise open-loop equilibrium characterized by a dynamic-programming equation. We refer the reader to reference [121] for an exploration of this type of solution for games with uncontrolled Markov disturbances.

10.3   What have we learned in this chapter?

- Piecewise deterministic games constitute a class of stochastic games where the players control a hybrid system where the discrete state component evolves as a controlled Markov process.
- This class of games can be viewed as particular instances of Markov games with state and control variables in measurable (Borel) spaces. As such, they are amenable to the use of dynamic programming to characterize and compute the equilibrium solutions.
- In the dynamic-programming approach, we defined a local game that was used to define an operator acting on the value-function space. This local game can be reformulated as an infinite-horizon open-loop differential game. When the controlled system is stationary, these infinite-horizon differential games are likely to benefit from a "turnpike" property, which means that a global attractor exists for the equilibrium trajectories, regardless of their initial state. This simplifies the numerical solution of these games.

10.4   Exercises

Exercise 10.1. Let $x(t)$ denote the amount of a firm's R&D capital available at time t. The elementary probability that a technological breakthrough will happen at a time T, included in the time interval $[t, t + dt)$, knowing that it did not happen before, is given by

$$P[T \in [t, t + dt)|T \geq t] = q((x(t))\, dt + o(dt), \text{ where } \lim_{dt \to 0} \frac{o(dt)}{dt} = 0.$$

Show that the following holds:

$$P[T \in [t, t+dt)] = q(x(t))e^{-\int_0^t q(x(s))ds} dt + o(dt),$$

$$P[T \le \tau)] = \int_0^\tau q(x(t))e^{-\int_0^t q(x(s))ds} dt.$$

Exercise 10.2. Two firms compete in quantity in a market and invest in R&D with the aim of achieving a technological breakthrough. We index by i the four possible modes of R&D investment:

$i = 0$ The two firms are investing in R&D;
$i = 1$ Firm 1 has made the technological change; Firm 2 continues to invest in R&D;
$i = 2$ Firm 2 has made the technological change; Firm 1 continues to invest in R&D;
$i = 3$ Both firms have made the technological change and no longer invest.

Denote by y_j^i the output of firm $j, j = 1, 2$, in status $i, i = 0, 1, 2, 3$. We assume the following linear-demand law:

$$p = A - B(y_1^i + y_2^i), \quad i = 0, 1, 2, 3,$$

where A and B are positive constants.
The two firms' production costs are given by

$$C_j^i(y_j) = \alpha^i y_j + \beta_j^i(y_j)^2,$$

where α^i and β_j^i are positive constants.
Technological progress is triggered by an investment in R&D capital described by the state equations

$$\dot{x}_j(t) = u_j(t) - \mu x_j(t),$$
$$x_j(0) = x_j^0, \quad j = 1, 2,$$

where $x_j(t)$ is the stock of R&D capital available to Firm j at time t, $u_j(t)$ is the investment in R&D at time t and μ is the depreciation rate of R&D capital. The R&D investment and maintenance costs are defined by a function $L_j^i(x_j, u_j)$.

The mode $\xi(t)$ changes at random times according to a continuous-time controlled Markov chain with transition rates $q_{01}(x_1)$, $q_{02}(x_2)$, $q_{13}(x_2)$, $q_{23}(x_1)$. These functions describe the influence of R&D capital on the occurrence of a technological change (breakthrough) in one firm or the other. The other transition rates are null ($q_{kl} \equiv 0, \ k \ne l$).

When a mode switches from i to a new value k, there is an adjustment cost for firm j, given by $\Phi_j^{ik}(x_j)$. We assume that $\Phi_j^{ik}(x_j) \equiv 0$ when $ik = 01$ and $j = 2$, or $ik = 02$ and $j = 1$ or $ik = 13$ and $j = 1$ or $ik = 23$ and $j = 2$, as these cases correspond to situations where the innovation is done by the competitor.

(1) Formulate a Markov game, with hybrid state variables, to describe the competition through supply and R&D, by these two firms. Assume a positive discount rate $\rho > 0$.

(2) Show that decisions about supply (y_j) are decoupled from decisions about R&D (u_j) and that the two competing firms' profits, before R&D expenditures, are given by

$$\Pi_j^i = (A - By^i)y_j^i - \alpha_j^i y_j^i - \beta_j^i (y_j^i)^2,$$

with the equilibrium supply defined by

$$y_j^i = \frac{2(B + \beta_{-j}^i)(A - \alpha_j^i) - B(A - \alpha_{-j}^i)}{4(B + \beta_j^i)(B + \beta_{-j}^i) - B^2},$$

where, as usual, $-j$ designates the player who is not j.
(3) Formulate the local game when $\xi(t) = 3$. Show that the equilibrium value function for each firm j, when $\xi(t) = 3$ is given by

$$v(\cdot, 3; j) = \frac{\Pi_j^i}{\rho}.$$

(4) Formulate the local games for the 3 other possible discrete modes.
(5) Propose a method to numerically solve this game and find the equilibrium R&D investment policy for the two firms.

10.5 GE illustration: Competitive timing of climate policies

10.5.1 *Introduction*

In this GE, we present a stochastic climate-game model due to Bahn et al. [19], where two groups of countries, namely, developing countries and industrialized ones (essentially OECD countries), strategically interact on climate policies.[4] The countries are sure to suffer from climate change triggered by an increase in atmospheric GHG concentrations. The countries schematically have two options to control for the environmental damage. The first option consists of reducing economic activities to reduce the use of fossil energy, and hence of pollutants emissions. The second option, which is more attractive in terms of the standard of living, is to adopt a carbon-free production technology. The problem, however, with this second option is that this clean technology (e.g., nuclear fusion, or efficient CCS[5]) is not yet available and the countries need to invest in R&D to increase their probability of inventing or accessing such a technology. An additional inherent difficulty is that the severity of the damage caused by climate change cannot be precisely predicted at the outset; the players will experience the true extent of this damage as time goes by.

To model this situation, [19] proposes a stochastic non-cooperative game model that takes into account the inherent uncertainty of a technological breakthrough in clean technologies, the existence of a spillover effect that distributes the technological know-how to all players, and the possibility of finding out that more extensive damage than is currently being predicted occur.

The essence of the model is given below.

---

[4]See, e.g., the COP-15 and COP-16 conferences organized by the United Nations Framework Convention on Climate Change. http://unfccc.int.
[5]Carbon capture and sequestration.

10.5.1.1 *Variables*

Let us define:

$j = 1, 2$:   index for the two groups (or coalitions) of countries;

t: model running time;

$C(j, t)$:   total consumption in coalition j at time t, in trillions (10^{12}) of dollars;

$E_1(j, t)$:   yearly emissions of GHG (in GtC[6] equivalent) in the carbon economy by coalition j at time t;

$E_2(j, t)$:   yearly emissions of GHG in the clean economy by coalition j at time t, in GtC;

$\mathrm{ELF}^{\omega(t)}(t)$:   economic loss factor due to climate change, under condition of severity $\omega(t)$ at time t, in %;

$I_i(j, t)$:   investment in capital $i = 1, 2$ by coalition j at time t, in trillions of dollars;

$K_1(j, t)$:   physical stock of productive capital in the carbon economy of coalition j at time t, in trillions of dollars;

$K_2(j, t)$:   physical stock of R&D or productive capital in the clean economy of coalition j at time t, in trillions of dollars;

$L_1(j, t)$:   part of the (exogenously defined) labor force $L(j, t)$ of coalition j allocated at time t to the carbon economy, in millions (10^6) of persons;

$L_2(j, t)$:   part of the labor force of coalition j allocated at time t to the clean economy, in millions of persons;

$M(t)$:   atmospheric concentrations of GHG at time t, in GtC equivalent;

$\mathrm{WRG}(j)$:   discounted welfare of coalition j;

W: total discounted welfare;

$Y(j, t)$:   economic output of coalition j at time t, in trillions of dollars;

$\xi(t) \in \{0, 1\}$:   indicator of access to the clean technology;

$\omega(t) \in \{0, 1\}$:   indicator of intensity of climate change damage.

10.5.1.2 *Equations*

For each coalition j, social welfare (WRG) is given by the accumulation of a discounted utility obtained from per capita consumption, with discount rate $\rho > 0$:

$$\mathrm{WRG}(j) = \int_0^\infty e^{-\rho t} L(j, t) \log\left[\frac{C(j, t)}{L(j, t)}\right] dt. \tag{10.26}$$

Total labor (L) is divided between labor allocated to the carbon economy (L_1) and labor allocated to the clean economy (L_2):

$$L(j, t) = L_1(j, t) + L_2(j, t). \tag{10.27}$$

Capital stock $K_i(j, \cdot)$ evolves according to the choice of investment $I_i(j, \cdot)$ and a depreciation rate δ_K, as follows:

$$\dot{K}_i(j, t) = I_i(j, t) - \delta_K K_i(j, t) \quad i = 1, 2. \tag{10.28}$$

Economic output $Y(j, \cdot)$ occurs in the two economies according to an extended Cobb-Douglas production function in three inputs, capital $K(j, \cdot)$, labor $L(j, \cdot)$ and energy (its use is measured by the associated emission level $E(j, \cdot)$):

$$Y(j, t) = A_1(j, t) K_1(j, t)^{\alpha_1(j)} (\phi_1(j, t) E_1(j, t))^{\theta_1(j, t)} L_1(j, t)^{1-\alpha_1(j)-\theta_1(j, t)}$$
$$+ \xi(t) A_2(j, t) K_2(j, t)^{\alpha_2(j)} (\phi_2(j, t) E_2(j, t))^{\theta_2(j, t)} L_2(j, t)^{1-\alpha_2(j)-\theta_2(j, t)}, \quad (10.29)$$

where $A_i(j, t)$ is the total factor productivity in the carbon (resp. clean) economy (when $i = 1$, resp. $i = 2$) of coalition j at time t; $\alpha_i(j)$ is the elasticity of output with respect to capital $K_i(j, t)$; $\phi_i(j, t)$ is the energy-conversion factor for emissions $E_i(j, t)$; and θ_i is the elasticity of output with respect to emissions E_i. Notice that in equation (10.29) production from the carbon-free economy occurs only when $\xi(t) = 1$, namely when a technological breakthrough has occurred. Economic output, net of damages caused by climate change, is used for consumption (C), investment (I) and the payment of energy costs:

$$\text{ELF}^{\omega(t)}(t) Y(j, t) = C(j, t) + I_1(j, t) + I_2(j, t) + p_{E_1}(j, t) \phi_1(j, t) E_1(j, t)$$
$$+ p_{E_2}(j, t) \phi_2(j, t) E_2(j, t), \quad (10.30)$$

where $p_{E_i}(j, t)$ is the energy price in the carbon (resp. clean) economy (when $i = 1$, resp. $i = 2$) for coalition j at time t. Concentration $M(t)$ of GHG in the atmosphere evolves according to

$$\dot{M}(t) = \beta \sum_{j=1}^{2} (E_1(j, t) + E_2(j, t)) - \delta_M (M(t) - M_p), \quad (10.31)$$

where β is the marginal atmospheric retention rate, δ_M the natural atmospheric elimination rate and M_p is the preindustrial level of atmospheric concentrations. Finally, increased atmospheric GHG concentrations yield economic losses (due to climate change) that reduce available consumption.[7]

We assume that the economic loss factor is defined by

$$\text{ELF}^{\omega(t)}(t) = 1 - \left(\frac{M(t) - M_d}{cat_M^{\omega(t)} - M_d} \right)^2, \quad (10.32)$$

where M_d is the concentration level at which damages start to occur and $cat_M^{\omega(t)}$ is a "catastrophic" concentration level, which depends on the uncertain intensity of climate damage, which is represented by the indicator $\omega(t) \in \{0, 1\}$.

Finally, we introduce a constraint on the rate of investment in the clean technology for each group of countries:

$$I_2(j, t) \leq 0.05 \, Y(j, t). \quad (10.33)$$

---

[7]Obviously, we abstract in this model from the new British vineyards, ice-free travel across the Arctics, etc., which could, locally, increase consumption. For developing countries in particular, global warming means a loss in agriculture production and, therefore in consumption.

10.5.1.3 *Uncertainty about a technological breakthrough*

Access to a cleaner economy at time t is indicated by a binary variable $\xi(t) \in \{0,1\}$. The clean technology becomes available only when ξ switches from an initial value of 0 to 1. The initial value $\xi(0) = 0$ indicates that there is no access to the clean technology at the initial time. The switch to the value 1 occurs at a random time, which is influenced by the global accumulation of R&D capital $\tilde{K}_2$, where

$$\tilde{K}_2(t) = \sum_{j=1}^{2} K_2(j,t). \tag{10.34}$$

More precisely, we introduce a jump-rate function $q_b(t, \tilde{K}_2(t))$, which determines the elementary probability of a switch around time t, given that no switch has occurred up to that time:

$$P[\xi(t+dt) = 1 | \xi(t) = 0, \tilde{K}_2(t)] = q_b(t, \tilde{K}_2(t))\, dt + o(dt). \tag{10.35}$$

We choose the following affine form for $q_b(t, K_2(t))$:

$$q_b(t, \tilde{K}_2(t)) = \omega_b + \upsilon_b \tilde{K}_2(t), \tag{10.36}$$

where ω_b is the initial probability rate of discovery and υ_b is the slope with respect to $\tilde{K}_2(t)$ of the probability rate of discovery.

Notice that this representation of the breakthrough dynamics implies a full spillover effect whereby each coalition has immediate access to the new technology when it becomes available.

10.5.1.4 *Uncertainty about intensity of damage*

Assume that the two possible damage intensities are uncertain with a priori probabilities $\pi_0 = \pi_1 = 0.5$. Consider an indicator function $\omega(t)$ with two possible values:

$\omega(t) = 0$ low damage intensity with $cat_M^0 = 2500$ GtC;
$\omega(t) = 1$ high damage intensity with $cat_M^1 = 2000$ GtC.

Assume that $\omega(t) = 0$ at the initial time ($t = 0$), i.e., expect climate-change damage to be low. At a random (Markov) time θ, we will discover the true extent of the damage. This random time has an intensity that depends on GHG concentrations:

$$P[\theta \in (t, t+dt) | \theta \geq t, M(t)] = q_c(t, M(t))dt + o(dt), \tag{10.37}$$

where the jump-rate function $q_c(t, M(t))$ is supposed to be an affine function

$$q_c(t, M(t)) = \varpi_c + \upsilon_c M(t). \tag{10.38}$$

In the above equation, the parameters (ϖ_c, υ_c) are defined as follows:

ϖ_c initial probability rate of finding the real extent of damages;
υ_c slope w.r.t. $M(t)$ of the probability rate of of finding the real extent of damages.

10.5.1.5 *Stochastic-game formulation*

Define the five state-variable process:

$$s(\cdot) = ((\mathbf{K}(j,\cdot) : j = 1, 2), M(\cdot), \xi(\cdot), \omega(\cdot)), \tag{10.39}$$

where $\mathbf{K}(j,\cdot) \in \mathbf{R}^{+^2}$ represents the evolution of the two capital stocks in coalition j; $M(\cdot) \in \mathbf{R}^+$ is the evolution of atmospheric GHG concentrations; $\xi(\cdot)$, which takes its value from $\{0, 1\}$, indicates the contingent or possible availability of the advanced (clean) technology; and $\omega(\cdot)$, which takes its value from $\{0, 1\}$, indicates the contingent or possible switch to a high level of climate-change damage. At initial time $t^0 = 0$, the state $s(0) = s^0 = (\mathbf{K}^0, M^0, \xi^0, \omega^0)$ is such that $K_2^0 = 0$ and $\xi^0 = \omega^0 = 0$ since the advanced technology is not yet available and climate-change damage is initially low. For ease of notation, we have denoted by $\underline{\mathbf{K}}(\cdot) := ((\mathbf{K}(j,\cdot) : j = 1, 2)$, the vector of capital stocks for all the coalitions.

It will be convenient to introduce a special notation for the continuous state variable $\underline{\mathbf{x}} = (\underline{\mathbf{K}}, M)$ and for the discrete variables $\zeta = (\xi, \omega)$. Control variables are the emission rates $E_1(j, t), E_2(j, t)$, the investment rates $I_1(j, t), I_2(j, t)$ in the different types of capital and the labor allocations $L_1(j, t), L_2(j, t) = L(j, t) - L_1(j, t)$ for $j = 1, 2$. Control variables are denoted by $\underline{\mathbf{u}} = (\underline{\mathbf{E}}, \underline{\mathbf{I}}, \underline{\mathbf{L}})$ where $\underline{\mathbf{E}}(t) = (E_i(j, t))_{i,j=1,2}, \underline{\mathbf{I}}(t) = (I_i(j, t))_{i,j=1,2}$, $\underline{\mathbf{L}}(j, t) = (L_i(j, t))_{i,j=1,2}$. The dynamics of the state variable $\underline{\mathbf{x}}$ is a piecewise deterministic process[8] denoted by

$$\underline{\dot{\mathbf{x}}}(t) = f^{\zeta(t)}(t, \underline{\mathbf{x}}(t), \underline{\mathbf{u}}(t)). \tag{10.40}$$

Given a state $s = (\underline{\mathbf{x}}, \zeta)$ and a control variable $\underline{\mathbf{u}}$, we determine each coalition's instantaneous consumption utility; therefore, we can introduce the reward function:

$$\mathcal{L}^{\zeta(t)}(j, t, \underline{\mathbf{x}}(t), \underline{\mathbf{u}}(t)) = L(j, t) \log \left(\frac{1}{L(j, t)} \left[\mathrm{ELF}^{\omega(t)}(t) \, F^{\xi(t)}(j, t, \mathbf{E}(j, t), \mathbf{K}(j, t), \mathbf{L}(j, t)) \right. \right.$$
$$\left. \left. - \sum_{i=1}^{2} (I_i(j, t) + p_{E_i}(j, t) \phi_i(j, t) E_i(j, t)) \right] \right), \tag{10.41}$$

where $F^{\xi(t)}(j, t, \mathbf{E}(j, t), \mathbf{K}(j, t), \mathbf{L}(j, t)) = Y(j, t)$, as in Equation 10.30. The controls are subject to the constraints

$$E_i(j, t) \geq 0, \ I_i(j, t) \geq 0, \ L_i(j, t) \geq 0, \ i = 1, 2,$$
$$L_1(j, t) + L_2(j, t) \leq L(j, t), \ j = 1, 2.$$

This is summarized in general notations by $\underline{\mathbf{u}}(t) \in U(t)$.

---

[8]We adopt here the general notation of a velocity for the state variable $\mathbf{x}$ that is depending on the discrete state $\zeta(t)$, even though in this particular example the discrete state does not enter in the definition of velocities. This will help to identify the change of control regimes at each jump time of the $\zeta(\cdot)$ process.

10.5.1.6 *Piecewise open-loop strategies*

Let $\tau^0 = 0$ be the initial time and τ^1, τ^2 be the jump times of the $\zeta(\cdot)$ process indicating the time at which a technological breakthrough occurs and/or the time when the true extent of the damage is learned. Denote by s^0, s^1, s^2 the state observed at times τ^0, τ^1, τ^2. A strategy γ_j for Player j is a mapping that associates with τ^0 and s^0 a control $\mathbf{u}_{j,\tau^0,s^0}(\cdot)$: $(\tau^0, \infty \mapsto U(j, \cdot)$ that will be used by Player j until the system jumps at some (random) time τ^1. At that point, strategy γ_j selects a new control for Player j, denoted by $\mathbf{u}_{j,\tau^1,s^1}(\cdot)$[9] according to the time τ^1 and the state s^1 reached at that time. Similarly at time τ^2, the controls of the two players are adapted through their strategies γ_j, $j = 1, 2$ to $\mathbf{u}_{j,\tau^2,s^2}(\cdot)$, knowing the state s^2 was reached at jump time τ^2.

This corresponds to the concept of piecewise deterministic control. We obtain an expected reward (welfare gain) associated with a policy vector $\underline{\gamma} = \{\gamma_j : j = 1, 2\}$ and an initial state s^0, for Player j, defined by

$$J(j, \underline{\gamma}, s^0) = \mathbb{E}_{\underline{\gamma}} \left[\int_0^\infty e^{-\rho(j)t} \mathcal{L}^{\zeta(t)}(j, t, \mathbf{x}(t), \underline{\mathbf{u}}(t)) \, dt \right], \tag{10.42}$$

where the expectation $\mathbb{E}_{\underline{\gamma}}$ is taken with respect to the probability measure induced by the strategy vector $\underline{\gamma}$. Such a strategy is admissible if, for each player, it generates a control that almost surely[10] satisfies all constraints.

10.5.1.7 *Nash equilibrium*

A strategy m-tuple $\underline{\gamma}^*$ is a Nash equilibrium if the following inequality holds:

$$J(j, \underline{\gamma}^*, s^0) \geq J(j, [\gamma_j, \underline{\gamma}^*_{-j}], s^0) \tag{10.43}$$

for all initial state s^0 and Player j, where $[\gamma_j, \underline{\gamma}^*_{-j}]$ stands for the policy vector where Player $\ell \neq j$ uses equilibrium strategy γ_ℓ^* whereas Player j uses any admissible strategy γ_j.

10.5.2 *Characterization of Nash-equilibrium solutions*

10.5.2.1 *Dynamic programming*

In a piecewise open-loop information structure, players adapt their controls at the jump times, observing the system state right after the jump. The Nash-equilibrium solution can thus be characterized by using a dynamic-programming argument for the discrete-event system, which is obtained by observing this control system and deciding on the subsequent control policy only at jump times. In this case, there are three jump times: the initial time τ^0 followed by two jump times τ^1 and τ^2 when the breakthrough occurs or the uncertainty about whether the damage intensity is high or low, is settled.

---

[9]For simplicity, in the remainder of the chapter, we drop the indices τ and s and write $u(j, \cdot)$ instead of $\mathbf{u}_{j,\tau,s}(\cdot)$.
[10]We recall that "almost surely" means "with probability one".

10.5.2.2 *After the last jump*

Assume that the last jump occurs at time τ^2. From time τ^2 onwards, the carbon-free technology is available and the intensity of damages is known to be low or high.

At time τ^2, for a given state $s^2 = (\mathbf{x}^2, 1, \iota) = (\mathbf{K}^2, M^2, 1, \iota)$, where $\iota = 0, 1$ indicates the level of damage intensity that has been revealed, the value function $V_\iota^2(j, \tau^2, x^2)$ is defined as the payoff to Player j in the equilibrium solution to an open-loop differential game. In this equilibrium solution, each player solves the following optimal response problem:

$$
V_\iota^2(j, \tau^2, x^2) = \max_{\mathbf{u}(j,\cdot)} \left[e^{\rho(j)\tau^2} \int_{\tau^2}^{\infty} e^{-\rho(j)t} \mathcal{L}^{(1,\iota)}(j, t, \underline{\mathbf{x}}(t), [\mathbf{u}_j(t), \mathbf{u}^*_{-j}(t)]) \, dt \right],
$$

where the control $\mathbf{u}^*_{-j}(t); -j \neq j$ is given, subject to the state equations

$$
\dot{\underline{\mathbf{x}}}(t) = f^{(1,\iota)}(t, \underline{\mathbf{x}}(t), [\mathbf{u}_j(t), \mathbf{u}^*_{-j}(t)]);
$$
$$
[\mathbf{u}_j(t), \mathbf{u}^*_{-j}(t)] \in U(t); \quad t \geq \tau^2; \quad x(\tau^2) = x^2, \tag{10.44}
$$

where $\mathbf{u}^*_{-j}(t)$ is the equilibrium control of the other player.

10.5.2.3 *At the first jump after the initial time*

The first jump, occurs at time τ^1. At this jump time, two scenarios are possible: either the discrete state has switched from $(0,0)$ to $(1,0)$, which means that the cleaner technology becomes available while the damage intensity is still low; or it has switched from $(0,0)$ to $(0, \iota)$, where $\iota = 0, 1$, which means that the true extent of climate-change damage is discovered. It can remain low or switch to high, before the cleaner technology becomes available.

10.5.2.4 *Earlier availability of clean technology*

In the first type of transition, let $s^1 = (\mathbf{K}^1, M^1, 1, 0)$ be the system state right after the jump time. The Nash-equilibrium value function for Player j is $V_{1,0}^1(j, \tau^1, x^1)$, defined as the solution to the following optimization problem:

$$
V_{1,0}^1(j, \tau^1, x^1) = \max_{\mathbf{u}(j,\cdot)} \mathbb{E}_{M(\cdot)} e^{\rho\tau^1} \left[\int_{\tau^1}^{\tau^2} e^{-\rho t} \mathcal{L}^{(1,0)}(j, t, \underline{\mathbf{x}}(t), [\mathbf{u}_j(t), \mathbf{u}^*_{-j}(t)]) \, dt \right.
$$
$$
\left. + e^{-\rho\tau^2} V_{1,\omega(\tau^2)}^2(\tau^2, x(\tau^2)) \right] \tag{10.45}
$$
$$
= \max_{\mathbf{u}(j,\cdot)} e^{\rho\tau^1} \left[\int_{\tau^1}^{\infty} \left[e^{-\int_{\tau^1}^{s} q_c(\alpha, \underline{\mathbf{x}}(\alpha)) \, d\alpha} \right. \right.
$$

$$q_c(s, x(s)) \int_{\tau^1}^{s} e^{-\rho t} \mathcal{L}^{(1,0)}(j, t, \underline{x}(t), [\mathbf{u}_j(t), \mathbf{u}^*_{-j}(t)]) \, dt \bigg] ds$$

$$+ \int_{\tau^1}^{\infty} \left[e^{-\int_{\tau^1}^{s} q_c(\alpha, \underline{x}(\alpha)) \, d\alpha} q_c(s, \underline{x}(s)) e^{-\rho s} \right] \times$$

$$\left[\sum_{\iota=0,1} \pi_\iota V_\iota^2(j, s, x(s)) \right] ds \bigg] \tag{10.46}$$

s.t.

$$\dot{\underline{x}}(t) = f^{(1,0)}(t, \underline{x}(t), [\mathbf{u}_j(t), \mathbf{u}^*_{-j}(t)]); \quad \mathbf{u}_j(t) \in U(j, t);$$
$$t \geq \tau^1; \quad \underline{x}(\tau^1) = \underline{x}^1,$$

where $\mathbf{u}^*_{-j}(t)$ is the equilibrium control of the other player.

In equation (10.45), the second jump time τ^2 is stochastic. The associated jump rate is $q_c(t, M(t))$ at any time $t \geq \tau^1$. Thus the expectation $\mathbb{E}_\gamma$ is replaced here by $\mathbb{E}_M$ to stress the fact that the effect of γ on τ^2 is mediated by $M(\cdot)$. We denote by $s^2(\tau^2)$ the random state reached after the second jump time. In equation (10.45) we make explicit the calculation of the expected value. After integration by parts of the first term in equation (10.45), and if we assume that the following constraint has to be satisfied by all admissible controls:

$$\lim_{s \to \infty} \left[e^{-\int_{\tau^1}^{s} q_c(\alpha, \underline{x}(\alpha)) \, d\alpha} \int_{\tau^1}^{s} e^{-\rho t} \mathcal{L}^{(1,0)}(t, \underline{x}(t), \underline{u}(t)) \, dt \right] = 0,$$

we obtain the following equivalent infinite-horizon deterministic-control problem:

$$V_{1,0}^1(j, \tau^1, x^1) = \max_{\mathbf{u}(j, \cdot)} e^{\rho \tau^1} \int_{\tau^1}^{\infty} e^{-(\rho t + \int_{\tau^1}^{t} q_c(s, M(s)) \, ds)}$$

$$\left[\mathcal{L}^{(1,0)}(j, t, \underline{x}(t), [\mathbf{u}_j(t), \mathbf{u}^*_{-j}(t)]) \right.$$

$$\left. + q_c(t, M(t)) \sum_{\iota=0,1} \pi_\iota V_\iota^2(j, t, \underline{x}(t)) \right] dt \tag{10.47}$$

s.t.

$$\dot{\underline{x}}(t) = f^{(1,0)}(t, \underline{x}(t), [\mathbf{u}_j(t), \mathbf{u}^*_{-j}(t)]); \quad \mathbf{u}_j(t) \in U(j, t);$$
$$t \geq \tau^1; \quad \underline{x}(\tau^1) = \underline{x}^1,$$

where $\mathbf{u}^*_{-j}(t)$ is the equilibrium control of the other player.

10.5.2.5 *Earlier knowledge of damage severity*

In the second type of transition, let $s^1 = (\mathbf{K}^1, M^1, 0, \iota)$ be the system state right after the jump time when the true extent of climate-change damage ($\iota = 0, 1$) has been revealed.

The equilibrium problem that must be solved by Player j can be described as follows:

$$V_{0,\iota}^1(j, \tau^1, x^1) = \max_{u(\cdot)} \mathbb{E}_{K_2(\cdot)} e^{\rho \tau^1} \left[\int_{\tau^1}^{\tau^2} e^{-\rho t} \mathcal{L}^{(0,\iota)}(j, t, \underline{x}(t), [u(j, t), u_{-j}^*(t)]) \, dt \right.$$

$$\left. + e^{-\rho \tau^2} V_\iota^2(j, \tau^2, x^2(\tau^2)) \right]$$

s.t.

$$\dot{\underline{x}}(t) = f^{(0,\iota)}(t, \underline{x}(t), [u_j(t), u_{-j}^*(t)]); \quad u_j(t) \in U(j, t);$$
$$t \geq \tau^1; x(\tau^1) = x^1.$$

The second jump time τ^2 is still stochastic. The associated jump rate is $q_b(t, K_2(t))$ at any time $t \geq \tau^1$. Thus the expectation $\mathbb{E}_\gamma$ is replaced here by $\mathbb{E}_{K_2}$ to stress the fact that the effect of γ on τ^2 is mediated by K_2. We again denote by $s^2(\tau^2)$ the random state reached after the second jump time. Once more, as for equation (10.45), by using integration by parts, we obtain the following equivalent infinite-horizon deterministic-control problem:

$$V_{0,\iota}^1(j, \tau^1, x^1) = \max_{u(j, \cdot)} e^{\rho \tau^1} \int_{\tau^1}^\infty e^{-(\rho t + \int_{\tau^1}^t q_b(s, \tilde{K}_2(s)) \, ds)} \left[\mathcal{L}^{(0,\iota)}(j, t, \underline{x}(t), \right.$$

$$\left. [u_j(t), u_{-j}^*(t)]) + q_b(t, \tilde{K}_2(s)]) V_\iota^2(j, t, \underline{x}(t)) \right] dt \qquad (10.48)$$

s.t.

$$\dot{\underline{x}}(t) = f^{(0,\iota)}(t, \underline{x}(t), [u(j, t), u_{-j}^*(t)]); \quad u_j(t) \in U(j, t); t \geq \tau^1; x(\tau^1) = x^1,$$

where $u_{-j}^*(t)$ is the equilibrium control (open-loop strategy) of the other player.

10.5.2.6 *At the initial time*

At the initial time, the discrete state is $(0, 0)$, meaning that there is no access to the clean technology and the damage intensity is still low. The stochastic-optimization problem describing the equilibrium response of Player j to the equilibrium control of the other player $-j$ can be written as follows:

$$V^0(j, x^0) = \max_{u_j(\cdot)} \mathbb{E}_{K_2(\cdot), M(\cdot)} \left[\int_0^{\tau^1} e^{-\rho(j)t} \mathcal{L}^{(0,0)}(j, t, \underline{x}(t), [u_j(t), u_{-j}^*(t)]) \, dt \right.$$

$$\left. + e^{-\rho(j)\tau^1} V_{\xi(\tau^1), \omega(\tau^1)}^1(j, \tau^1, x^1(\tau^1)) : \text{ given } u_{-j}^*(\cdot); -j \neq j \right]$$

s.t.

$$\dot{\underline{x}}(t) = f^{(0,0)}(t, \underline{x}(t), [u_j(t), u_{-j}^*(t)]);$$
$$u_j(t) \in U(j, t); \quad t \geq 0; \quad x(0) = x^0,$$

where $u_{-j}^*(t)$ is the equilibrium control of the other player.

The deterministic equivalent infinite-horizon differential game problem is given by

$$V^0(\underline{x}^0) = \max_{u(j,\cdot)} \int_0^\infty e^{-\left(\rho(j)t + \int_0^t q_b(s,\tilde{K}_2(s)) + q_c(s,M(s))\right)ds}$$

$$\left[\mathcal{L}^{(0,0)}(j,t,\underline{x}(t),[\mathbf{u}_j(t),\mathbf{u}^*_{-j}(t)]) + q_b(t,\tilde{K}_2(t))V^1_{1,0}(j,t,\underline{x}(t))\right.$$

$$\left. + q_c(t,M(t))\sum_{\iota=0,1}\pi_\iota V^1_{0,\iota}(j,t,\mathbf{x}(t))\right] dt \tag{10.49}$$

s.t.

$$\underline{\dot{x}}(t) = f^{(0,0)}(t,\underline{x}(t),[\mathbf{u}_j(t),\mathbf{u}^*_{-j}(t)]); \quad \mathbf{u}_j(t) \in U(j,t); \quad t \geq 0; \quad x(0) = x^0.$$

10.5.3   *Solving the deterministic equivalent dynamic games*

We have seen how, in a piecewise open-loop information structure, the characterization of equilibrium solutions can be obtained through a sequence of solutions to deterministic differential games. Finding an equilibrium solution in a differential game is not an easy task. It can be done, first by discretizing the time domain (and obtaining an approximating multistage game) and then by implementing a nonlinear complementarity algorithm or the heuristic cobweb search of a fixed point in the two players' optimal response function.

Consider, for example the differential game played from jump time τ^2 onward. For simplicity of notation, denote:

$$\psi(j,\tau^2,s^2,u(\cdot)) = e^{\rho(j)\tau^2}\int_{\tau^2}^\infty e^{-\rho(j)t}\mathcal{L}^{(1,\iota)}(j,t,\underline{x}(t),[\mathbf{u}_j(t),\mathbf{u}_{-j}(t)])\,dt$$

where $\underline{x}(\cdot)$ is the state trajectory resulting from the choice of control

$$\underline{u}(\cdot) = (\mathbf{u}(1,\cdot),\mathbf{u}(2,\cdot))$$

and initial state $s^2 = (\underline{x}^2,1,\iota)$. Let us take any weighting $r = (r_j > 0 : j = 1,2)$ and define the following "reply function":

$$\theta(\tau^2,s^2,\underline{u}(\cdot),\underline{v}(\cdot);r) = \sum_{j=1}^2 r_j\psi(j,\tau^2,s^2,[\mathbf{u}(-j,\cdot),\mathbf{v}(j,\cdot)]), \tag{10.50}$$

where, as before, $-j = 2$ if $j = 1$ and $-j = 1$ if $j = 2$ and where $v(\cdot) \in U(\cdot)$. Now define the generalized best-reply mapping as the point-to-set map:

$$\Theta(\underline{u}(\cdot)) = \left\{\tilde{\underline{v}}(\cdot) : \theta(\tau^2,s^2,\underline{u}(\cdot),\tilde{\underline{v}}(\cdot);r) = \max_{\underline{v}(\cdot)\in U(\cdot)} \theta(\tau^2,s^2,\underline{u}(\cdot),\underline{v}(\cdot);r)\right\}. \tag{10.51}$$

A fixed point $\underline{u}^*(\cdot) \in \Theta(\underline{u}^*(\cdot))$ is a Nash equilibrium for this game. To compute such a fixed point, we can implement the following cobweb method:

Let $G \subset \mathbf{R}^5$ be a sample of initial values $s^2 = (K_1^2(j),K_2^2(j),M^2)$ and let $\mathcal{T}$ be a set of initial times.

Step 0 For each $\tau^2 \in \mathcal{T}$ and each $s^2 \in G$

Step 1 Find $\underline{\mathbf{u}}^*(\cdot) = \text{argmax}_{\underline{\mathbf{u}}(\cdot)} \left[\sum_{j=1}^2 r_j \psi(j, \tau^2, s^2, \underline{\mathbf{u}}(\cdot)) \right]$

Step 2 Take $\underline{\mathbf{u}}(\cdot) = \underline{\mathbf{u}}^*(\cdot)$; find

$$\underline{\mathbf{v}}^*(\cdot) = \text{argmax}_{\underline{\mathbf{v}}(\cdot)} \left[\sum_{j=1}^2 r_j \psi(j, \tau^2, s^2, [\mathbf{u}_{-j}(\cdot), \mathbf{v}_j(\cdot)]) \right]$$

Step 3 Take $\underline{\mathbf{u}}(\cdot) = \mathbf{v}^*(\cdot)$; find new

$$\underline{\mathbf{v}}^*(\cdot) = \text{argmax}_{\underline{\mathbf{v}}(\cdot)} \left[\sum_{j=1}^2 r_j \psi(j, \tau^2, s^2, [\mathbf{u}_{-j}(\cdot), \mathbf{v}_j(\cdot)]) \right]$$

Step 4 Repeat step-3 until $\text{max}_{\underline{\mathbf{v}}(\cdot)} \left[\sum_{j=1}^2 r_j \psi(j, \tau^2, s^2, [\mathbf{u}_{-j}(\cdot), \mathbf{v}_j(\cdot)]) \right]$ does not change significantly.

Step 5 Record for each $j = 1, 2$ the value $V_\iota^2(j, \tau^2 x^2) = \psi(j, s, \underline{\mathbf{u}}^*(\cdot))$ where $s = (\underline{\mathbf{x}}^2, 1, 1)$, and $\underline{\mathbf{x}}^2 \in G$.

Step 6 Adjust (e.g., by using least squares) an analytical form for the function $V_\iota^2(j, \tau^2, \underline{\mathbf{x}}), \underline{\mathbf{x}} \in \mathbf{R}^5$.

It is well known that a cobweb approach is not sure to converge; but when it does, the limit is an equilibrium solution. In practice the cobweb approach has converged for almost all the cases we have tried to solve. An approach with a greater guarantee of convergence would consist of solving the Nash-equilibrium search by using a variational-inequality or nonlinear-complementarity method (see, e.g., references [77, 80, 79]).

A similar method is used to compute the Nash equilibrium of the equivalent deterministic differential games at the first jump time τ^1 and initial time τ^0.

Chapter 11

Stochastic-Diffusion Games

11.1 Introduction

In this chapter, we consider a class of dynamic games called stochastic-diffusion games, where the system is subject to a continuous flow of random shocks. To account for this randomness, the state dynamics are described by *stochastic* differential equations. These games are relevant in many areas, e.g., in finance where the evolution of portfolio's wealth is naturally stochastic and, also, in renewable resources (forests, fisheries) where, again, the way the stock of different species fluctuates is fully stochastic.

11.2 Stochastic diffusions and stochastic calculus

Stochastic-diffusion processes require a specific calculus method to be correctly handled. In this section, we recall some fundamental features of stochastic calculus. In particular, we establish the Ito formula, which plays a central role in the development of dynamic-programming reasoning for controlled stochastic-diffusion processes.

11.2.1 *Stochastic-differential equation and stochastic integral*

Consider a stochastic process defined by the following expression:

$$x(t) = x(0) + \int_0^t b(x(s))\, ds + \int_0^t \varsigma(x(s)) dw(s), \qquad (11.1)$$

where $x(\cdot)$ is an **R**-valued process with continuous sample paths; $w(\cdot)$ is a standard *Wiener process* also called *Brownian motion*.[1] Symbols $b(\cdot)$ and $\varsigma(\cdot)$ denote given functions of x. Recall also the definition of a standard Wiener process.

Definition 11.1. A **Wiener process** is a stochastic process with value in **R**, such that

---

[1] Another popular term, relevant to stochastic processes, is a Gaussian *white noise* whose integral is the Wiener process.

(1) $w(0) = 0$ with probability 1;
(2) the random variables defined by the increments $(w(s) - w(t) : s \geq t)$ are independent of the history of $w(\cdot)$ up to t;
(3) the increments $(w(s) - w(t) : s \geq t)$ are normally distributed with the mean equal to 0 and the variance equal to $s - t$;
(4) the sample paths of $w(\cdot)$ are continuous functions of t.

The term $\int_0^t \varsigma(x(s))dw(s)$ in (11.1), where $\varsigma(x(s))$ is a positively valued function, is a stochastic integral, a concept we clarify below.

Formally, we can differentiate both sides of (11.1) and obtain the stochastic-differential equation (SDE) as follows[2]:

$$dx(t) = b(x(t))dt + \varsigma(x(t))dw(s), \quad t \geq 0, \quad x(0) \text{ given.} \tag{11.3}$$

We explain now how the stochastic integral $\int_0^t y(s)dw(s)$ is defined. Let $y(s)$ be a process adapted[3] to the history of $w(\cdot)$ up to time s and also a *simple function process*. The latter implies that there exists a sequence of deterministic times $0 = t_0 < t_1 < \cdots < t_i \to \infty$, such that $\mathbf{y}(t) \equiv y(t_i)$ for $t \in [t_i, t_{i+1})$. For such functions, because of this piecewise-constant structure, we can set, for any $t \in [t_n, t_{n+1})$

$$\int_0^t y(s)dw(s) = \sum_{i=0}^{n-1} y(t_i)[w(t_{i+1}) - w(t_i)] + y(t_n)[w(t) - w(t_n)]. \tag{11.4}$$

For more general processes $y(s)$, but still adapted to the history of $w(\cdot)$, it can be shown that there exists a sequence of simple function processes $\{y_n(\cdot), n \in \mathbf{N}\}$ such that, for each $T \in [0, \infty)$,

$$\int_0^T \mathbb{E}[|y_n(s) - y(s)|^2]\, ds \to 0, \tag{11.5}$$

as $n \to \infty$. It is then natural to define $\int_0^t y(s)dw(s)$ as the limit (in some appropriate sense) of the processes $\int_0^t y_n(s)dw(s)$.

For the class of simple function processes, it is easy to show, using the definition of the integral given in (11.4), that the following properties are satisfied, when $\mathbb{E}_{\mathcal{F}_s}$ refers to the

---

[2]Some authors tend to write (11.3) as

$$\frac{dx(t)}{dt} = b(x(t)) + \varsigma(x(t))\frac{dw(t)}{dt}, \quad t \geq 0, \quad x(0) \text{ given;} \tag{11.2}$$

however, this is a notation abuse because the derivative $\frac{dw(t)}{dt}$ of the Wiener process does **not** exist, and, as a consequence, neither does $\frac{dx(t)}{dt}$.

[3]The term "adapted" has a special meaning in measure theory; it refers to the measurability of the $y(\cdot)$ process in the probability space generated by the possible histories of $w(\cdot)$. In other words, we say that $y(s)$ is a process adapted to the history of $w(\cdot)$ up to time s when we gather information on $y(s)$ out of the observations of $w(\tau)$, $\tau \leq s$.

conditional expectation, given the history of the $w(\cdot)$ process up to time $s \le t$:

$$\mathbb{E}_{\mathcal{F}_s}\left[\int_0^t y(u)dw(u)\right] = \int_0^s y(u)dw(s), \tag{11.6}$$

$$\mathbb{E}_{\mathcal{F}_s}\left[\int_s^t y(u)dw(u)\right]^2 = \int_s^t \mathbb{E}_{\mathcal{F}_s}[y(u)]^2 \, ds, \tag{11.7}$$

$$\int_0^t y_1(s)dw(s) + \int_0^t y_2(s)dw(s) = \int_0^t (y_1(s) + y_2(s)) \, dw(s). \tag{11.8}$$

These properties extend by approximation to the more general class of processes $y(s)$ and allow us to define the stochastic integral in (11.1). We also notice that $\varsigma(x(s))$, the positive valued function of process $x(s)$ appearing in (11.1), can correspond to process $y(s)$ considered in the general definition of a stochastic integral given above.

11.2.2 *Ito's formula*

This is the stochastic calculus counterpart of the chain rule in ordinary calculus.

In ordinary calculus, if $f(x(t), t)$ is a function of $x(t)$ and t, then we can write its total differential as

$$df = \frac{\partial f(x(t), t)}{\partial x} \, dx + \frac{\partial f(x(t), t)}{\partial t} \, dt,$$

and, dividing by dt, we obtain the total derivative

$$\frac{df(x(t), t)}{dt} = \frac{\partial f(x(t), t)}{\partial x} \frac{dx(t)}{dt} + \frac{\partial f(x(t), t)}{\partial t}.$$

In stochastic calculus, we have to allow for the fact that the stochastic integral uses $dw(t)$ rather than dt; so, the integrating variable is not time t, but the standard Wiener process $w(t)$, for which the following holds:

$$\mathbb{E}[dw(t)]^2 = dt, \tag{11.9}$$

(see item (3) in Definition 11.1.)

Consider a function $f(x(t), t)$, which is twice continuously differentiable in x; assume that $x(\cdot)$ is a stochastic process satisfying the SDE

$$dx = \mu \, dt + \varsigma \, dw,$$

where, to simplify the notation, we omit the arguments of functions μ and ς, which could be x or t or both. Expanding $f(x, t)$ into the Taylor series yields

$$df = \frac{\partial f}{\partial x} \, dx + \frac{\partial f}{\partial t} \, dt + \frac{1}{2} \frac{\partial^2 f}{\partial x^2} \, dx^2 + \cdots \tag{11.10}$$

and substituting $\mu \, dt + \varsigma \, dt$ for dx gives

$$df = \frac{\partial f}{\partial x} (\mu \, dt + \varsigma \, dt) + \frac{\partial f}{\partial t} \, dt + \frac{1}{2} \frac{\partial^2 f}{\partial x^2} (\mu^2 dt^2 + 2\mu\varsigma dt \, dw + \varsigma^2 dw^2) + \cdots \tag{11.11}$$

In the limit, as $dt \to 0$, the dt^2 and $dtdw$ terms become negligible but the dw^2 term tends to dt. This is a consequence of (11.7) and (11.9) and property (3) of the Wiener process. Therefore, we obtain the stochastic differential of $f(x(t), t)$ as

$$df = \left(\mu \frac{\partial f}{\partial x} + \frac{\partial f}{\partial t} + \frac{1}{2} \varsigma^2 \frac{\partial^2 f}{\partial x^2} \right) dt + \varsigma \frac{\partial f}{\partial x} dw. \tag{11.12}$$

equation (11.12) above is *Ito's differentiation formula.*

Remark 11.1. The derivations (11.10)–(11.12) of the Ito formula are informal. A rigorous proof would involve a stochastic approximation procedure. For a complete presentation of stochastic calculus, we refer the reader to Karatzas and Shreve's book [136].

If the process takes values in $\mathbf{R}^n$, then the SDE can be written as follows:

$$d\mathbf{x} = \boldsymbol{\mu}(\mathbf{x})dt + \varsigma(\mathbf{x})d\mathbf{w}, \tag{11.13}$$

where $\mathbf{w}(\cdot)$ is a vector of n independent standard Wiener processes, $\boldsymbol{\mu}$ is a vector in $\mathbf{R}^n$ and ς is an $n \times n$ covariance matrix. Then, the Ito formula takes the following form[4]:

$$df = \left(\frac{\partial f}{\partial \mathbf{x}} \boldsymbol{\mu}(\mathbf{x}) + \frac{\partial f}{\partial t} + \frac{1}{2} \text{trace} \left[\frac{\partial^2 \mathrm{f}}{\partial \mathbf{x}^2} \varsigma(\mathbf{x}) \varsigma(\mathbf{x})' \right] \right) dt + \frac{\partial f}{\partial \mathbf{x}} \varsigma(\mathbf{x})dw. \tag{11.14}$$

11.2.3 Application of Ito's formula to option valuation

In finance, a derivative is a contract whose payoff depends on the behavior of an underlying asset price S. The most common derivatives have market values and are traded on exchanges.

Definition 11.2. Options are contracts that give the owner the right, but not the obligation, to buy (in the case of a **call option**) or sell (in the case of a **put option**) an asset. The price at which the sale takes place is known as the **strike price.** It is specified at the time the parties enter into the contract (option). The option contract also specifies a maturity date. In the case of a European option, the owner has the right to require that the sale takes place on (but not before) the maturity date; in the case of an American option, the owner can require that the sale takes place at any time up to the maturity date. If the owner of the contract exercises this right, the counterpart in the contract has the obligation to carry out the transaction.

Options are among the most common derivatives.

An asset price, $S(t)$, is supposed to follow a (stochastic) process called a *geometric Brownian motion*

$$dS = \mu S dt + \varsigma S dw, \tag{11.15}$$

where μ is a real number called the "drift" and $\varsigma \geq 0$ is the volatility, both assumed constant; $w(\cdot)$ is a standard Wiener process. We notice that if $\varsigma = 0$, then the price grows exponentially for $\mu > 0$.

---

[4]The trace of a square matrix $Q = [q_{ij}]_{i,j=1,\dots,n}$ is $\text{trace}(Q) = \sum_i q_{ii}$.

Let $V = V(S, t)$ be the market value of an option (contingent claim), at time t, when the underlying price is S. We expect $V(\cdot, \cdot)$ to be a smooth function of its arguments; in particular we assume V is twice continuously differentiable in S (so of class $\mathcal{C}^2$) and $\mathcal{C}^1$ in t.

When the price of the underlying asset fluctuates, so does the value $V(S(t), t)$. Applying Ito's formula (11.12), we obtain the following SDE, satisfied by the process $V(S(\cdot), \cdot)$;

$$dV = \left(\mu S \frac{\partial V}{\partial S} + \frac{\partial V}{\partial t} + \frac{1}{2} \varsigma^2 S^2 \frac{\partial^2 V}{\partial S^2} \right) dt + \varsigma S \frac{\partial V}{\partial S} dw. \tag{11.16}$$

Let Π be the value of a portfolio consisting of one short position[5] of value V and α units of the underlying asset. The value of the portfolio at time t is thus

$$\Pi = -V + \alpha S. \tag{11.17}$$

Differentiating (11.17), we get

$$\tag{11.18}$$

$$\begin{aligned} d\Pi &= -dV + \alpha \, dS \\ &= -\left(\mu S \left(\frac{\partial V}{\partial S} - \alpha \right) + \frac{\partial V}{\partial t} + \frac{1}{2} \varsigma^2 S^2 \frac{\partial^2 V}{\partial S^2} \right) dt + \left(\alpha - \frac{\partial V}{\partial S} \right) \varsigma S \, dw. \end{aligned}$$

It is clear from the above expression that if we constantly modify the quantity α of the stock held, by maintaining the equality $\alpha = \frac{\partial V}{\partial S}$, then the differential (11.18) becomes

$$d\Pi = -\left(\frac{\partial V}{\partial t} + \frac{1}{2} \varsigma^2 S^2 \frac{\partial^2 V}{\partial S^2} \right) dt. \tag{11.19}$$

This is a deterministic expression, since the term in dw has a zero coefficient. Furthermore, the drift μ has also disappeared from the expression of the differential $d\Pi$.

Imposing the no-arbitrage principle[6] means that the risk-free return r is equal to the return obtained from (11.19), which is also riskless. Investing the portfolio in the risk-free asset gives a variational change of wealth:

$$d\Pi = r \, \Pi = r(-V + \alpha S)dt = \left(-rV + r S \frac{\partial V}{\partial S} \right) dt. \tag{11.20}$$

The RHS's in Eqs. (11.20) and (11.19) must be equal. From this equality we derive the famous *Black-Scholes equation*,

$$\frac{\partial V}{\partial t} + \frac{1}{2} \varsigma^2 S^2 \frac{\partial^2 V}{\partial S^2} + r S \frac{\partial V}{\partial S} - rV = 0. \tag{11.21}$$

---

[5] Short position means: (i) the sale of a borrowed security, commodity or currency with the expectation that the asset will fall in value; (ii) in the context of options, it is the sale (also known as "writing") of an option's contract – mathematically, it is equivalent to buying a "negative" amount of the assets.

[6] "In economics and finance, *arbitrage* is the practice of taking advantage of a price difference between two or more markets: striking a combination of matching deals that capitalize upon the imbalance, the profit being the difference between the market prices. [...] an arbitrage is a transaction that involves no negative cash flow at any probabilistic or temporal state and a positive cash flow in at least one state; in simple terms, it is the possibility of a risk-free profit at zero cost" (quoted from *Wikipedia*).

This is a partial differential equation that describes the evolution of the derivative's market value. This equation is completed by boundary conditions that depend on the type of option we are dealing with. E.g., for a European call option with maturity T and strike price K, these boundary conditions are

$$V(0, t) = 0, \forall t \geq 0, \tag{11.22}$$

$$V(S, t) \to S \text{ as } S \to \infty, \tag{11.23}$$

$$V(S, T) = \max\{S - K, 0\}. \tag{11.24}$$

The last condition gives the value of the option at maturity. In European options the time of maturity is the only time the option can be exercised.

For an American call option, it can be proved that it is never optimal to exercise the option before maturity. Therefore, the value of an American call is the same as a European one, if the underlying asset pays no dividends.

Remark 11.2. It is remarkable that, in the option's market valuation, the drift term μ disappears. Therefore, the expected return on the underlying asset, and hence, the expected return of the derivative itself, does not influence the valuation. It could be shown that the valuation based on the "no-arbitrage" principle is also given by the expected value of returns, but under a modified probability measure called a **risk-neutral** measure.

11.2.4   *Ito's formula and calculation of a value function as the expected discounted sum of rewards*

In the previous section, we showed how Ito's formula can be used to value an option contract when using a no-arbitrage principle. In this section, we show how Ito's formula allows for the computation of the value function in dynamic programming , which is generally obtained as the expected discounted sum of a reward function.

Consider a diffusion process defined as in (11.1) or by the SDE

$$dx(t) = b(x(t))dt + \varsigma(x(t))dw(t) \quad t \geq 0, \quad x(0) \text{ given,}$$

and assume that a reward is obtained by an agent at a rate $L(x(t))$, so that the total discounted sum of rewards is given by

$$\int_0^\infty e^{-\rho t} L(x(t)) \, dt. \tag{11.25}$$

We are interested in identifying the value function

$$V(x) = \mathbb{E}\left\{ \int_0^\infty e^{-\rho t} L(x(t)) \, dt \,\Big|\, x(0) = x \right\}. \tag{11.26}$$

We will use Ito's formula to derive a (partial) differential equation, to characterize $V(x)$.

Consider a time increment δ, which will eventually tend to 0, and decompose (11.26) in the following way:

$$V(x) = \mathbb{E}\left\{\int_0^\delta e^{-\rho t} L(x(t)\,dt\right\} + \mathbb{E}\left\{\int_\delta^\infty e^{-\rho t} L(x(t)\,dt\right\}$$

$$= \mathbb{E}\left\{\int_0^\delta e^{-\rho t} L(x(t)\,dt\right\} + e^{-\rho\delta}\mathbb{E}\left\{\int_\delta^\infty e^{-\rho(t-\delta)} L(x(t)\,dt\right\}$$

$$= \mathbb{E}\left\{\int_0^\delta e^{-\rho t} L(x(t)\,dt\right\} + e^{-\rho\delta}V(x(\delta)). \tag{11.27}$$

From this expression, we get

$$\frac{1}{\delta}\mathbb{E}\left\{e^{-\rho\delta}V(x(\delta)) - V(x) + \int_0^\delta e^{-\rho t} L(x(t)\,dt\right\} = 0, \tag{11.28}$$

or

$$0 = \mathbb{E}\left\{\frac{V(x(\delta)) - V(x)}{\delta}\right\} + \frac{(e^{-\rho\delta} - 1)}{\delta}\mathbb{E}\left\{V(x(\delta)\right\}$$

$$+ \mathbb{E}\left\{\frac{1}{\delta}\int_0^\delta e^{-\rho t} L(x(t)\,dt\right\}. \tag{11.29}$$

Applying Ito's formula to the term $\Delta V = V(x(\delta)) - V(x)$ yields

$$\Delta V = \left(b(x)\frac{\partial V}{\partial x} + \frac{1}{2}\varsigma(x)^2\frac{\partial^2 V}{\partial x^2}\right)\delta + \varsigma(x)\frac{\partial V}{\partial x}\,dw + o(\|\delta\|), \tag{11.30}$$

where $\frac{o(\|\delta\|)}{\delta} \to 0$ when $\delta \to 0$.

Substitute (11.30) in (11.29) and let $\delta \to 0$. The term $\frac{(e^{-\rho\delta}-1)}{\delta}$ tends to the derivative of $e^{-\rho t}$ at $t = 0$, which is equal to $-\rho$. The term $\mathbb{E}\left\{V(x(\delta)\right\}$ tends to $V(x)$. The term $\mathbb{E}\left\{\frac{1}{\delta}\int_0^\delta e^{-\rho t} L(x(t)\,dt\right\}$ tends to $L(x)$. Finally the term $\mathbb{E}\left\{\varsigma\frac{\partial V}{\partial x}dw\right\}$ is equal to zero (recall that $w(t)$ is a standard Wiener process). Collecting all these limits, we obtain

$$\rho V(x) = b(x)\frac{\partial V(x)}{\partial x} + \frac{1}{2}\varsigma(x)^2\frac{\partial^2 V(x)}{\partial x^2} + L(x). \tag{11.31}$$

If $x \in \mathbf{R}$, then this is a differential equation satisfied by the value function. We keep the notation with partial derivatives because this formula is formally the same for $\mathbf{x} \in \mathbf{R}^n$. It is then written as follows:

$$\rho V = \frac{\partial V}{\partial \mathbf{x}}b(\mathbf{x}) + \frac{1}{2}\frac{\partial \mathbf{x}^2 V}{\partial \mathbf{x}^2}\varsigma(\mathbf{x})^2 + k(\mathbf{x}). \tag{11.32}$$

Remark 11.3. Equation (11.32) is a partial differential equation satisfied by the current-valued "reward-to-go" functional $V(x)$ defined in (11.25). This will provide an important building block for the construction of the dynamic programming equations in stochastic diffusion games.

11.3   Dynamic-programming equations for feedback-Nash equilibrium

Consider a controlled diffusion process defined by the SDE

$$dx(t) = b(x(t), u(t))dt + \varsigma(x(t))dw(t) \quad t \geq 0, \quad x(0) = x^0 \text{ given},$$

and assume that a reward is obtained by an agent at a rate $L(x(t), u(t))$, so that the total discounted sum of rewards is given by

$$\int_0^\infty e^{-\rho t} L(x(t), u(t)) \, dt. \tag{11.33}$$

We want to characterize a feedback control $u(t) = \sigma^*(x(t))$ which solves

$$\max_{\sigma(\cdot)} V^\sigma(x^0) = \mathbb{E}\left[\int_0^\infty e^{-\rho t} L(x(t), \sigma(x(t)) \, dt \,|\, x(0) = x^0\right].$$

Theorem 11.1. *Let a function $V^*(x)$ satisfy the functional equation*[7]

$$\rho V^*(x) = \max_u \left\{ b(x, u)\frac{\partial V^*(x)}{\partial x} + \frac{1}{2}\varsigma(x)^2\frac{\partial^2 V^*(x)}{\partial x^2} + L(x, u) \right\}. \tag{11.34}$$

Then

(1) the feedback law defined by

$$\sigma^*(x) = \operatorname{argmax}_u \left\{ b(x, u)\frac{\partial V^*(x)}{\partial x} + L(x, u) \right\} \tag{11.35}$$

is an optimal feedback control;
(2) $V^*(x)$ is the optimal value function.

Proof.   Consider $\sigma^*(x)$ defined as in (11.35), then, as shown in the previous section

$$V^*(x^0) = \mathbb{E}\left[\int_0^\infty e^{-\rho t} L(x(t), \sigma^*(x(t)) \, dt \,|\, x(0) = x^0\right].$$

Now take a time $T > 0$, then we can write

$$V^*(x^0) = \mathbb{E}\left[\int_0^T e^{-\rho t} L(x(t), \sigma^*(x(t)) \, dt + e^{-\rho T} V^*(x(T)) \,|\, x(0) = x^0\right].$$

Using equation (11.34) with $u = \sigma^*(x)$ we can write

$$V^*(x^0) = \mathbb{E}\Bigg[\int_0^T e^{-\rho t} \left\{ \rho V^*(x(t)) - b(x(t), \sigma^*(x(t)))\frac{\partial V^*(x(t))}{\partial x} \right.$$
$$\left. - \frac{1}{2}\varsigma(x)^2\frac{\partial^2 V^*(x(t))}{\partial x^2} \right\} dt + e^{-\rho T} V^*(x(T)) \,|\, x(0) = x^0\Bigg].$$

According to the Ito formula this expression boils down to

$$V^*(x^0) = \mathbb{E}\left[V^*(x(0)) - e^{-\rho T} V^*(x(T)) + e^{-\rho T} V^*(x(T))\right].$$

---

[7]We say *functional equation* because it is a combination of a (partial) differential equation and a max operator w.r.t. u. See equation (7.17), on page 252.

Consider any feedback control $\sigma(x)$. According to equation (11.34) we have, for all x,

$$0 \leq \rho V^*(x) - b(x, \sigma(x))\frac{\partial V^*(x)}{\partial x} - \frac{1}{2}\varsigma(x)^2\frac{\partial^2 V^*(x)}{\partial x^2} - L(x, \sigma(x))$$

or

$$L(x, \sigma(x)) \leq \rho V^*(x) - b(x, \sigma(x))\frac{\partial V^*(x)}{\partial x} - \frac{1}{2}\varsigma(x)^2\frac{\partial^2 V^*(x)}{\partial x^2}. \tag{11.36}$$

Assume we use feedback control $\sigma(x)$ up to time T and from T to ∞ we use feedback control $\sigma^*(x)$. Then the expected payoff is

$$V^T(x^0) = \mathbb{E}\left[\int_0^T e^{-\rho t} L(x(t), \sigma(x(t)))\, dt + e^{-\rho T}V^*(x(T)) \mid x(0) = x^0\right].$$

According to (11.36) we have

$$V^T(x^0) \leq \mathbb{E}\left[\int_0^T e^{-\rho t}\left\{\rho V^*(x) - b(x, \sigma(x))\frac{\partial V^*(x)}{\partial x} - \frac{1}{2}\varsigma(x)^2\frac{\partial^2 V^*(x)}{\partial x^2}\right\} dt\right.$$
$$\left. +e^{-\rho T}V^*(x(T)) \mid x(0) = x^0\right].$$

Again, applying Ito formula to the RHS we get

$$V^T(x^0) \leq V^*(x^0).$$

This is true for all $T > 0$. Letting $T \to \infty$ and noticing that $\mathbb{E}[e^{-\rho T}V^*(x(T))] \to 0$, we obtain

$$V^\infty(x^0) = \mathbb{E}\left[\int_0^\infty e^{-\rho t} L(x(t), \sigma(x(t)))\, dt \mid x(0) = x^0\right] \leq V^*(x^0).$$

This establishes the optimality of $\sigma^*(x)$. $\qquad\square$

Equation (11.34) is the Hamilton-Jacobi-Bellman equation for stochastic control.[8] This equation is useful to identify feedback (Markov) Nash equilibria in diffusion games, i.e., in games where the state equations are stochastic-diffusion differential equations.

Corollary 11.1. *Consider a diffusion process controlled by two agents $j = 1, 2$ defined by the SDE*

$$dx(t) = b(x(t), u_1(t), u_2(t))dt + \varsigma(x(t))dw(t) \quad t \geq 0, \quad x(0) = x^0 \text{ given.}$$

Assume each player uses a stationary feedback strategy σ_j and defines his control as $u_j(t) = \sigma_j(x(t))$. A reward is obtained by each agent at a rate $L_j(x(t), u_1(t), u_2(t))$, so that the payoff to player j is given by the total discounted sum of rewards

$$V_j(x^0; \underline{\sigma}) = \mathbb{E}\left[\int_0^\infty e^{-\rho t} L_j(x(t), \underline{\sigma}(x(t)))\, dt \mid x(0) = ?x^0\right] \quad j = 1, 2, \tag{11.37}$$

where, as usual $\underline{\sigma} = (sigma_1, \sigma_2)$.

---

[8]See with equation (7.17), on page 252, for deterministic-control problems.

Let the functions $V_j^*(x)$ satisfy the functional equations

$$\rho V_1^*(x) = \max_{u_1} \left\{ b(x, u_1, \sigma_2^*(x)) \frac{\partial V_1^*(x)}{\partial x} \right.$$
$$\left. + \frac{1}{2}\varsigma(x)^2 \frac{\partial^2 V_1^*(x)}{\partial x^2} + L_1(x, u_1, \sigma_2^*(x)) \right\}. \tag{11.38}$$

and

$$\rho V_2^*(x) = \max_{u_2} \left\{ b(x, \sigma_1^*(x), u_2) \frac{\partial V_2^*(x)}{\partial x} \right.$$
$$\left. + \frac{1}{2}\varsigma(x)^2 \frac{\partial^2 V_2^*(x)}{\partial x^2} + L_2(x, \sigma_1^*(x), u_2) \right\}. \tag{11.39}$$

With

$$\sigma_1^*(x) = \operatorname{argmax}_{u_1} \left\{ b(x, \sigma_1^*(x), u_2) \frac{\partial V_1^*(x)}{\partial x} + L_1(x, u_1, \sigma_2^*(x)) \right\}, \tag{11.40}$$

$$\sigma_2^*(x) = \operatorname{argmax}_{u_2} \left\{ b(x, \sigma_1^*(x), u_2) \frac{\partial V_2^*(x)}{\partial x} + L_2(x, \sigma_1^*(x), u_2) \right\}. \tag{11.41}$$

Then the feedback laws $(\sigma_1^*(x), \sigma_2^*(x))$ constitute a feedback-Nash equilibrium for this diffusion game and

$$V_j^*(x) = V_j(x; \underline{\sigma}^*), \quad j = 1, 2.$$

Proof.   Use the preceding theorem for each player.   □

11.4   The Grenadier model for the equilibrium investment strategies of firms

The theory of continuous-time finance offers many examples of stochastic-diffusion games. In this section, we briefly present the model proposed by Grenadier [97], as an archetypal example of stochastic games where the solution is obtained by implementing the Black-Scholes method of option valuation.

Consider an oligopolistic market, on which m identical firms compete with a single homogenous good. At time t, firm j produces $q_j(t)$ and $Q(t) = \sum_{j=1}^{m} q_j(t)$ is the total market supply. The price of the good that clears the market fluctuates stochastically over time according to the following inverse demand function:

$$P(t) = D(x(t), Q(t)), \tag{11.42}$$

where $x(\cdot)$ is an exogenous random-shock process. We assume that $D(\cdot, \cdot)$ is of class C^2 in x and Q, strictly increasing in x and strictly decreasing in Q. The shock process $x(\cdot)$ is a stochastic diffusion with value in **R** satisfying the following SDE:

$$dx = \mu x dt + \varsigma x dw, \tag{11.43}$$

where μ and ς are assumed to be constant, which defines $x(\cdot)$ as a geometric Brownian motion. So, at time t, the state of this system is $s(t) = (x(t), q_1(t), \ldots, q_m(t))$. Let us introduce the notation

$$Q_{-j} = \sum_{i \neq j} q_i.$$

At any time t, each firm can invest in additional capacity to increase its output by an infinitesimal amount dq_j with a linear cost $K dq_j$. Grenadier proposes that we view this opportunity to invest as an American option. More precisely, for each output level q_j, Player j holds an option on the marginal flow of profits by increasing its output to $q_j + dq_j$, with an exercise price K. Similarly, the other firms (players) hold options to increase their output, which directly impacts their profits.

Player j controls the process $\tilde{q}_j(\cdot)$ for firm j, which is adapted to the processes $x(\cdot)$ and $\tilde{Q}_{-j}(\cdot)$. This means that the decision to increase $q_j(t)$ by dq_j is contingent on the observed values for $x(t)$ and $Q_{-j}(t)$. To be consistent with the definition of feedback strategies, we write

$$dq_j = \sigma_j(x, q_j, Q_{-j}) dt, \tag{11.44}$$

where $\sigma : \mathbf{R}^3 \to \mathbf{R}$ is a given feedback law. The value for firm j, for given strategies $\sigma_j, \boldsymbol{\sigma}_{-j}$ at initial values (x^0, q_j^0, Q_{-j}^0) is defined by

$$V_j \left(x^0, q_j^0, Q_{-j}^0; [\sigma_j, \boldsymbol{\sigma}_{-j}] \right) = \mathbb{E} \left\{ \int_0^\infty e^{-\rho t} q_j(t) D(x(t), q_j(t) + Q_{-j}(t)) dt \right.$$

$$\left. - \int_0^\infty e^{-\rho t} K \sigma_j(x(t), q_j(t), Q_{-j}(t)) dt \right\}, \tag{11.45}$$

where ρ is a common discount rate, which corresponds to the risk-free rate of return, since here (recall Remark 11.2), we assume that the expectation is taken with respect to the risk-neutral measure. So we use the Black-Scholes equation to represent the value's evolution.

Definition 11.3. The strategy m-tuple $\boldsymbol{\sigma}^* = (\sigma_1^*, \ldots, \sigma_m^*)$ is a feedback (Markov) Nash equilibrium if, for each initial state (x^0, q_j^0, Q_{-j}^0), the following holds:

$$\tag{11.46}$$

$$V_j \left(x^0, q_j^0, Q_{-j}^0; \boldsymbol{\sigma}^* \right) = \max_{\sigma_j} V_j \left(x^0, q_j^0, Q_{-j}^0; [\sigma_j, \boldsymbol{\sigma}_{-j}^*] \right).$$

Now, we use the option-valuation procedure to characterize an equilibrium solution. For all states for which no option is exercised, the value of firm j evolves according to the Black-Scholes equation with dividends $q_j(t) D(x(t), q_j(t) + Q_{-j}(t))$. This yields[9]

$$0 = q_j D(x, q_j + Q_{-j}) + r x \frac{\partial V_j^*}{\partial x} + \frac{1}{2} \varsigma^2 x^2 \frac{\partial^2 V_j^*}{\partial x^2} - r V_j^*, \quad j = 1, \ldots, m. \tag{11.47}$$

---

[9]The attentive reader will notice the difference with Black-Scholes formula for a simple European option given in (11.21). The dividend term appears now as a forcing term in the equation. We refer the reader to [57] for a derivation of BS formula with dividends.

We know the boundary conditions for this partial differential equation. The first one corresponds to the states where firm j exercises the option. The exercise occurs on a manifold that is called the trigger and that is defined by a function $\mathcal{X}^j(q_j, Q_{-j})$. On this manifold, the value-matching condition imposes that

$$\frac{\partial V_j^*}{\partial q_j}\left(\mathcal{X}^j(q_j, Q_{-j}), q_j, Q_{-j}\right) = K, \quad j = 1, \ldots, m. \tag{11.48}$$

The second boundary condition is obtained by requesting that the trigger define an optimal decision. This yields

$$\frac{\partial^2 V_j^*}{\partial q_j \partial x}\left(\mathcal{X}^j(q_j, Q_{-j}), q_j, Q_{-j}\right) = 0, \quad j = 1, \ldots, m. \tag{11.49}$$

The final boundary condition is obtained by imposing the value-matching condition on the manifold corresponding to the other firms' (competitors') trigger (denoted by $\mathcal{X}^{-j}(q_j, Q_{-j})$). This means that

$$\frac{\partial V_j^*}{\partial Q_{-j}}\left(\mathcal{X}^{-j}(q_j, Q_{-j}), q_j, Q_{-j}\right) = 0, \quad j = 1, \ldots, m. \tag{11.50}$$

We summarize these results in the following theorem.

Theorem 11.2. *The equilibrium value of each firm j, denoted by $V_j^*(x, q_j, Q_{-j})$, satisfies the partial differential equations (11.47) with boundary conditions (11.48), (11.49) and (11.50). Each firm's equilibrium investment strategy is characterized by increasing its output incrementally (by dq_j) whenever $x(t)$ reaches the trigger function $\mathcal{X}^j(q_j, Q_{-j})$.*

Remark 11.4. We obtained the equilibrium conditions through a market-value reasoning, based on the no-arbitrage principle. However, the same results could have been obtained by the techniques of stochastic control. This example shows that, for some stochastic-control problems, the control can be exercised only on boundary manifolds of the state space. The precise treatment of these control problems is well beyond the mathematical scope of this textbook.[10]

Before ending this presentation of a stochastic game of competitive investment, note that Grenadier showed we could exploit the fact that all of the firms are identical, by reducing the state space to (x, Q), each firm producing $\frac{Q}{m}$. The value of each firm would then be determined by the following equations:

$$D(x, Q) + \frac{Q}{n}\frac{\partial D}{\partial Q}(x, Q) + r x \frac{\partial v}{\partial x} + \frac{1}{2}\varsigma^2 x^2 \frac{\partial^2 v}{\partial x^2} - rv = 0, \tag{11.51}$$

$$v(X^*(Q), Q) = K, \tag{11.52}$$

$$\frac{\partial}{\partial x}v(X^*(Q), Q) = 0. \tag{11.53}$$

---

[10]We refer the reader to Kushner & Dupuis [149] for a comprehensive treatment of solving stochastic optimal-control problems. In the next section, we apply some methods proposed by Kushner and Dupuis to numerically solve a stochastic switching-diffusion game, where the players' control is distributed over the whole state space.

See [97] for more details about solving these equations, which are much simpler than the previous ones.

Remark 11.5. As is often the case in finance, modelers are mainly interested in characterizing the value dynamics of the asset under consideration. Equations (11.51)–(11.52) thus provide a relatively simple characterization of the value of each firm in a symmetric oligopoly. The characterization of the equilibrium strategy is much more complicated, as it involves an infinite number of infinitesimal adjustments each time the random state trajectory hits the trigger surface.

11.5 Subgame perfect collusive equilibria in infinite-horizon stochastic games

For infinite-horizon deterministic-dynamic games,[11] we showed that it may be possible to design *collusive* equilibria. These equilibria are supported by the threat that any deviation by any player from a mutually agreed-upon strategy will trigger a (credible) retaliation from all of the other players in the form of playing a low-payoff antagonistic Nash game. This mechanism was introduced under the name of *memory-based trigger strategies* because the players could observe and *remember* their opponents' actions and base their own strategies on what they recalled.

The construction of memory-based trigger strategies was possible in deterministic games because every deviation from a collusive path, or "cheating," is observable with no error by all players. Knowing that any deviation is fully detectable, the players naturally choose to play cooperatively. Thus, Definition 5.2 (on page 144) says that collusion promises Pareto-efficiency. In the stochastic context considered in this chapter, we cannot expect that collusion will generate *non-dominated* payoffs. However, to be attractive, the collusive payoffs should dominate the antagonistic Nash game outcomes.

In a stochastic game, knowing the current state and the players' actions, there is a probability of reaching a given state in the future. So, an observation that the state does not conform to an expected outcome makes sense only in probabilistic terms. A consequence of this is that a slight instance of cheating might pass for a stochastic deviation. Significant cheating will be detected when we know the stochastic features of the process but, even then, only probabilistically. Then, the natural question then arises whether we can expect economic agents to play collusive equilibria in infinite-horizon stochastic games.

Here we show, in concrete rather than abstract terms, that the answer to this question is yes. We do so by engineering a collusive equilibrium for a two-player stochastic fishery-management game. We say that we *engineer* this equilibrium because we propose which quantity can be a monitoring variable and construct a specific retaliation trigger.

Focusing on a particular stochastic-differential game model enables us to propose a tractable solution method. However, the method is numerical; therefore, it delivers parameter-specific results. This means that, for a different game, we would have to ap-

---

[11] Such games were discussed in Chapters 5, 5.7.5 and 7. In particular, see Sections 5.2.3, 5.4.1, 6.9 and the GE illustration on page 231.

ply this method to that particular game before being able to formulate a claim about the equilibrium.

We call this game a switching-diffusion game because the state equation is a stochastic differential equation, which describes a diffusion process. The jump refers to the discrete changes in the monitoring variable that triggers a retaliatory action, which then modifies the state process.

We solve the game at hand using the sufficient conditions for a feedback equilibrium, which are given by a set of coupled Hamilton-Jacobi-Bellman (*HJB*) equations (see equation (11.32)). The resulting equilibria make use of a particular kind of memory strategy that exploits the nonuniqueness of the solution to an associated *HJB* equation. A numerical analysis, approximating the solution of the *HJB* equations through an associated Markov game, enables us to show that there exist collusive ϵ-equilibria, which dominate the antagonistic feedback-Nash equilibrium of the original diffusion game, *provided that* the triggering mechanism has been adequately calibrated.

Evidence that it is possible to build equilibria with trigger strategies in a discrete-time stochastic game, repeated over an infinite number of periods, through the use of an ad hoc dynamic-programming scheme, was first given in [203] and [96]. The approach of [203] and [96] has been extended to a fully dynamic game setting in [118] where a multistage game was studied. A numerical method was used to show the dominance of the collusive equilibria over the feedback-Nash non-cooperative equilibrium. Here, following [110], we show that the same type of result may be obtained in the realm of continuous-time stochastic-diffusion games. To deliver this result, we will exploit ideas first presented in [106] and [107] and model the triggering mechanism as a stochastic Markov-jump process and the retaliation duration as an exponential random variable. We will thus associate the original stochastic-diffusion game with a family of switching-diffusion games, each having a specific monitoring-and-retaliation scheme. Using a numerical technique adapted from [147] and [2], we will demonstrate that *collusive* equilibria can dominate the non-cooperative feedback-Nash equilibria in a fishery-management game.

11.5.1   *A stochastic game of a fishery exploited by two companies*

Differential-game models have been proposed by several authors (e.g., [45, 178, 99, 100]) to study the competitive behavior of (mainly) two fishing companies exploiting the same stock of biomass. In all these studies, a deterministic evolution equation was assumed for the biomass. We will study a stochastic version of these models.

11.5.1.1   *State equations*

Let $x(t) \in \mathbf{R}^+$ be the stock of biomass at time t. Two fishing companies, $j = 1, 2$, exploit this stock. Let $u_j(t) \geq 0$ be the fishing effort of company j. Let $\pi_j(t)$ be the cumulative profit of company j between time 0 and t. We assume the following system dynamics:

$$dx(t) = (A - Bx(t) - u_1(t) - u_2(t))x(t)\,dt + \varsigma\,dw(t), \tag{11.54}$$

$$x(0) = x_0, \tag{11.55}$$

$$d\pi_j(t) = \left\{ [a - bx(t)(u_1(t) + u_2(t))]x(t)u_j(t) \right.$$
$$\left. - (\alpha_j u_j(t)^2 + \beta_j u_j(t)) \right\} dt,$$
$$\pi_j(0) = 0, \quad j = 1, 2.$$
(11.56)

The above equations are based on usual assumptions in fishery-management models. In particular, equation (11.54) represents a nonlinear growth model in which A and B are positive constants. The biomass $x(t)$ has a natural yield, defined as a quadratic function of the current stock $(A - Bx(t))x(t)$. The initial condition on the biomass x_0 is deterministic. The fishing catch rate $u_j(t)$ is proportional to the amount of fishing effort and the current stock level $x(t)$.

By $w(t)$ we denote a Wiener process, or a one-dimensional standard Brownian motion and parameter $\varsigma > 0$ is the standard deviation of the process that perturbs the deterministic growth (could be a "noise" caused by measurement error or a "shock" such as weather conditions). Although the presence of the disturbance term $\varsigma w(t)$ can drive the state variable toward negative values, we assume that this event has a very low probability; hence, we consider that equation (11.54) will be a valid representation of the true biomass dynamics for the modeled species.

Equation (11.54) gives the instantaneous profit of company j. The term $[a - bx(t)(u_1(t) + u_2(t))]$ is the inverse demand giving the price of fish $(a > 0, b > 0)$ and $x(t)u_j(t)$ the catch by firm j. Therefore, the product of these two terms is the instantaneous revenue. The fishing-effort cost is assumed to be quadratic and given by $\alpha_j u_j(t)^2 + \beta_j u_j(t)$, with $\alpha_j > 0, \beta_j > 0$. Finally, the instantaneous cost at time zero is 0.

11.5.1.2 *Equilibria and cooperative solutions*

To fully define the differential game played between the fishing companies, we need to specify the information structure, the associated strategies – or policies – as well as the types of solutions that the game can admit.

We assume that each player (company) can observe the stock level $x(t)$ at any time t but is unable to observe the other player's fishing effort. A strategy[12] for Player j will be a mapping γ_j, which associates a fishing effort $u_j(t) = \gamma_j(\tilde{x}_t) \geq 0$ to a current time t and available observations $\tilde{x}_t = \{x(s) : s \leq t\}$ of the biomass stock. A pair of strategies $\gamma = (\gamma_1, \gamma_2)$ defines an *admissible policy* if it generates a well-defined stochastic process[13] $(x(\cdot), u_1(\cdot), u_2(\cdot))$.

---

[12]We use the notation γ for a more general strategy pair that exploits the history of the game and we keep the notation σ for feedback or Markov strategies.

[13]Here, we mean that the mean and covariance matrices of the three-dimensional stochastic process $(x(\cdot), u_1(\cdot), u_2(\cdot))$ are defined.

Given the initial stock level x_0 and an admissible strategy vector $\underline{\gamma}$, Player j expects the following discounted profit:

$$V_j^\tau(x_0, \underline{\gamma}) = \mathbb{E}_{x^\tau, \underline{\gamma}} \left[\int_\tau^\infty e^{-\rho t} \, d\pi_j(t) \, | \, x(\tau) = x^\tau \right], \quad j = 1, 2, \tag{11.57}$$

where ρ is a positive discount rate. For ease of notation, we shall write $V_j(x_0, \underline{\gamma})$ when $\tau = 0$, instead of $V_j^0(x_0, \underline{\gamma})$.

Definition 11.4. A pair $\underline{\gamma}^* = (\gamma_1^*, \gamma_2^*)$ is an **equilibrium** at x_0 if (i) it is admissible and (ii) for any admissible pairs $\underline{\gamma}^{*^{-1}} = (\gamma_1, \gamma_2^*)$ or $\underline{\gamma}^{*^{-2}} = (\gamma_1^*, \gamma_2)$, the following inequalities hold:

$$V_j^0(x_0, \underline{\gamma}^*) \geq V_j^0(x_0, \underline{\gamma}^{*^{-j}}), \quad j = 1, 2.$$

Since the controlled system is stochastic, the equilibrium property should hold everywhere in the state space. This corresponds to the property, called *subgame perfectness*, which makes an equilibrium credible.[14]

Definition 11.5. A pair $\underline{\gamma}^* = (\gamma_1^*, \gamma_2^*)$ is a **subgame perfect equilibrium** on $X \subset \mathbf{R}$ if (i) it is admissible at any $x_0 \in X$ and (ii) for any history $\tilde{x}^t$ where $x(t) = x_i$ and for any admissible pairs $\underline{\gamma}^{*^{-1}} = (\gamma_1, \gamma_2^*)$ or $\underline{\gamma}^{*^{-2}} = (\gamma_1^*, \gamma_2)$, the following inequalities hold:

$$V_j^t(x_i, \underline{\gamma}^*) \geq V_j^t(x_i, \underline{\gamma}^{*^{-j}}), \quad j = 1, 2.$$

Implicitly, $X \subset \mathbf{R}$ is the set of biomass values, from which a transition to a negative value is highly unlikely.

11.5.2   *Feedback versus memory strategies*

We suppose that the players consider implementing a different strategy that could improve their feedback-Nash equilibrium payoffs. This "meta" strategy will be composed of a "cooperative" feedback strategy pair $\underline{\sigma}^C$ that will be used as long as there is confidence in the other player behavior and a retaliation feedback strategy pair $\underline{\sigma}^R$ that will be used as a threat if the confidence is broken. The mechanism of confidence requires some form of memory in the meta-strategy. For that purpose, we will modify the game and introduce an additional (discrete) state with a dynamics reflecting a monitoring of the cooperative behavior of the two players.

We will refer to the game at hand as the "original" and will call "new" the game that should generate higher payoffs.

11.5.2.1   *Pure feedback policies in the original game*

Among the admissible policies, we specifically consider the stationary feedbacks, which could also be called *homogeneous nonrandomized Markov policies*.

---

[14] See Section 4.4 in Chapter 4 and other chapters' subsections on subgame perfectness; also see Definition 11.3.

Definition 11.6. A measurable[15] map $\sigma_j : \mathbf{R} \mapsto \mathbf{R}^+$ defines a pure stationary-feedback strategy for Player j. The control at time t of Player j is thus defined as $u_j(t) = \sigma_j(x(t))$, $\forall t \geq 0$.

Let us introduce the notations

$$b(x, \underline{\sigma}(x)) = (A - Bx(t) - \sigma_1(x(t)) - \sigma_2(x(t)))x(t), \tag{11.58}$$

$$L_j(x, \underline{\sigma}) = [a - bx(t)(\sigma_1(x(t)) + \sigma_2(x(t)))]x(t)\sigma_j(x(t)) \tag{11.59}$$
$$- \left(\alpha_j \sigma_j(x(t))^2 + \beta_j \sigma_j(x(t))\right), \quad j = 1, 2.$$

The "reward-to-go" (11.57) is now expressed as

$$V_j^\tau(x^\tau; \underline{\sigma}) = \mathbb{E}_{x_0, \underline{\sigma}} \left[\int_\tau^\infty e^{-\rho t} L_j(x, \underline{\sigma}(x)) \, dt | x(\tau) = x^\tau \right], \quad j = 1, 2,$$

and the current valued reward-to-go functional $V_j(x, \underline{\sigma})$ defined by

$$V_j(x; \underline{\sigma}) = e^{-\rho\tau} V_j^\tau(x; \underline{\sigma})$$

satisfies the partial differential equation

$$\rho V_j(x; \underline{\sigma}) = \frac{\partial V(x; \underline{\sigma})}{dx} b(x, \underline{\sigma}(x)) + \frac{1}{2}\varsigma(x)\frac{\partial^2 V(x; \underline{\sigma}))}{\partial x^2} + L_j(x, \underline{\sigma}(x)), \tag{11.60}$$

according to Ito's formula given in equation (11.32).

We will assume that this game has a unique pure feedback-Nash equilibrium strategy pair $\underline{\sigma}^*$.

An equilibrium is the prevalent solution concept when the players are competing; however, they may choose to cooperate. If so, the game will be cooperative, which may be solved under the concept of Pareto optimality (or efficiency).

Definition 11.7. A pair of feedback strategies $\underline{\sigma}^\mathcal{P} = (\sigma_1^\mathcal{P}, \sigma_2^\mathcal{P})$ is a **Pareto-optimal solution** if (i) it is admissible and (ii) it maximizes a convex sum of payoffs to the two players, i.e., for any other admissible pair $\underline{\sigma}$ the following inequality holds:

$$\sum_{j=1,2} \alpha_j V_j(x^0; \underline{\sigma}^\mathcal{P}) \geq \sum_{j=1,2} \alpha_j V_j(x^0; \underline{\sigma}) \quad \forall x^0.$$

with $\alpha_j > 0, \sum_{j=1,2} \alpha_j = 1$.

Different Pareto solutions are obtained, by changing the players' weights. These Pareto solutions are natural candidates for a particular cooperative solution. By a proper choice of weights one may find feedback-Pareto solution which generates a pair of outcomes (i.e., expected profits) that dominate the Nash-feedback equilibrium. However, as is well known, Pareto solutions are very seldom equilibria. Therefore, in a Pareto-efficient solution, both players will be tempted to deviate from the feedback strategy $\underline{\sigma}^\mathcal{P}$.

In the deterministic, multistage game case, we have seen in section 6.9, how one could construct a subgame-perfect collusive equilibrium by cpmbining the dominating Pareto-optimal pair $\underline{\sigma}^\mathcal{P}$ and the feedback-Nash equilibrium pair $\underline{\sigma}^*$, which is used as a retaliation

---

[15]Function-measurability is a an advanced mathematical notion. Broadly speaking, if a function is integrable in the usual sense of Riemann, then it is measurable (but not necessarily the other way around.)

threat, as soon as a deviation from the cooperative behavior is detected. Similar developments have been made in the realm of differential games, in section 7.12. In the present
case, these approaches would not work, because the state trajectory being random, there
is no way to be sure, 100 %, that a deviation has occurred due to a change of bahavior of
one player. To detect cheating, i.e., purposeful deviation by at least one player, one must
develop a monitoring system and make some sequential statistical testing to decide that,
with some high confidence, cheating has occurred.

In the $\mathbb{GE}$ Section 9.11 we have proposed a method to explore, numerically, the possibility of designing a "dominating" memory strategy for discrete time stochastic games.
This is the method that we will extend to the present framework of diffusion games.

11.5.2.2   *Memory strategies based on monitoring an agreement*

The question we address here is the following:

> *Is it possible to design a collusive policy that retains all the properties of a subgame-
> perfect equilibrium and generates a pair of outcomes (i.e., expected profits) that dominate
> the feedback-Nash equilibrium defined above?*

The basic idea of constructing – or engineering – such an equilibrium is to design
a new game with an extended state space and to construct a feedback-Nash equilibrium
for this new game. This equilibrium will be called collusive, since some form of collusion
among the players is necessary to design this new game out of the data defining the original
game. As the equilibrium will be of the feedback type, it will satisfy the subgame perfectness condition. If the new game is properly designed, the outcome of this extended-game
feedback-Nash equilibrium can dominate the outcome of the original game's feedback-
Nash equilibrium. We proceed below to the design of such an extended game.

To define a new game, whose equilibrium will possibly dominate the pure feedback-
Nash equilibrium of the game at hand, we need to discuss the following notions:

> *extended state, monitoring system, switching between mode of play, punishment strategy,
> resetting the monitoring variable.*

Extended state: The original game is stochastic and the players do not know each other's
actions. Therefore, if the actions are supposed to result from an agreement, knowing state
$x(t)$ is not sufficient for Player j to decide whether the opponent has complied with what
was consented to. We contend that by observing an extended state, the players will be able
to infer the opponents' compliance.

The extended state will be $s = (x, z, \xi) \in \mathbf{R} \times \mathbf{R} \times \{0, 1\}$. In addition to the stock
variable x, the players will also consider the value of a state variable z, which will be a
monitoring variable, as well as the discrete state variable ξ, which will define two possible
modes of play for the game. When $\xi = 0$ the mode of play is retaliatory or punitive; when
$\xi = 1$, the mode of play is conciliatory or cooperative.

Monitoring system: We assume that both players know the fundamental biomass dynamics given by equation (11.54). Let $\tilde{\sigma}_j^1(\cdot, \cdot)$, $j = 1, 2$ be two measurable functions

mapping $\mathbf{R}^2$ into $\mathbf{R}^+$. Let a policy pair

$$\tilde{\underline{\sigma}}^1(x(t), z(t)) = (\tilde{\sigma}_1^1(x(t), z(t)), \tilde{\sigma}_2^1(x(t), z(t)))$$

represent a cooperative mode of play that has to be monitored. (Presumably, $\tilde{\underline{\sigma}}^1$ can generate a better payoff to each player than $\underline{\sigma}^*$ for the original game.) This policy is a pure feedback in the extended state space (x, z). We define a monitoring system via an auxiliary-state equation

$$dz(t) = dx(t) - b(x(t), \tilde{\underline{\sigma}}^1(x(t), z(t))) \, dt, \tag{11.61}$$

$$z(0) = 0, \tag{11.62}$$

where we recall that $b(.,.)$ is the drift term in the dynamics of the fish stock in (11.54).

By this design, $z(t)$ will behave as a pure Wiener process as long as the players actually use the policy pair $\tilde{\underline{\sigma}}^1(t)$. If one player does not behave in accordance with this agreed-upon policy, then process $z(\cdot)$ will exhibit a drift. A detection of the drift could then trigger a retaliation against the other player who is suspected of cheating.

Switching between modes of play: We introduce a new discrete state variable ξ with values in $\{0, 1\}$. If $\xi(t) = 1$, then the current mode of play at time t is cooperative; if $\xi(t) = 0$, then the mode of play at time t is retaliatory. The retaliation triggering mechanism changes the game's mode of play. In the scheme proposed here, the mode of play evolves according to a stochastic-jump process characterized by jump rates

$$q_{01} = \lim_{dt \to 0} \frac{P[\xi(t + dt) = 1 \mid \xi(t) = 0]}{dt}, \tag{11.63}$$

$$q_{10}(z) = \lim_{dt \to 0} \frac{P[\xi(t + dt) = 0 \mid \xi(t) = 1, \ z(t) = z]}{dt}. \tag{11.64}$$

We assume in equation (11.63) that the duration of a punitive mode of play is an exponential random variable with mean $\frac{1}{q_{01}}$. In (11.64), we assume that the switch from a cooperative to a punitive mode of play is triggered with a probability that is a function of the observed value of the monitoring variable z. This functional dependence implements the cheating-detection scheme discussed above.

Punishment strategy: We notice that probability (11.63) does not depend on the monitoring variable. This is because there will be nothing to monitor in the punitive mode of the game. In this mode, the players implement a pair of feedback policies $\sigma_j^0(x)$, $j = 1, 2$ defined over the original (biomass) x-space.

Resetting the monitoring variable: Once a punitive period is terminated, i.e., when $\xi(t)$ jumps to 1, then the z variable is reset to zero.

Remark 11.6. The extended game is now characterized by a switching diffusion dynamics, i.e., a dynamics composed of coupled stochastic diffusions and jump processes. This extended game, including the switch in the modes of play, the monitoring system and the jump rates, is entirely built from the original game data. The monitoring state $z(t)$ is a function of the past history $\tilde{x}^t$; hence the policies $\sigma_j^1(x, z)$ and $\sigma_j^0(x)$ are in accordance with the information structure of the original game.

11.5.2.3 *Feedback policies and Ito equations for a switching-diffusion game*

The system defined by equations (11.54)–(11.55), (11.61)–(11.62), (11.63)–(11.64) is a particular instance of a switching-diffusion control system. For an analysis of such systems, see [111] and [95]. In these references, only the single-player case of optimal control was considered. A general theory of noncooperative switching-diffusion games is developed in [148]. For the present study, we will use some results obtained in [148] for optimal control of switching-diffusion systems and their game extensions.

In a general formulation, a switching (or jump) diffusion system is characterized by a hybrid state $(\mathbf{y}, \xi) \in \mathbf{R}^n \times E$, where E is a finite set; and, by a control $\underline{\mathbf{u}} \in \mathbf{R}^p$. The continuous state variable has an evolution driven by the Ito equations

$$d\mathbf{y}(t) = f^k(\mathbf{y}(t), \mathbf{u}(t))\, dt + \varsigma\, d\mathbf{w}(t), \tag{11.65}$$

$$\underline{\mathbf{u}}(t) = (\mathbf{u}_j(t))_{j \in M} \in \prod_{j \in M} \mathbf{U}_j^k \subset \mathbf{R}^m, \tag{11.66}$$

$$\mathbf{y}(0) = \mathbf{y}_0, \tag{11.67}$$

where $k \in E$, ς is an $n \times n$ matrix and $\mathbf{w}(t)$ a vector Wiener process. The discrete state variable ξ evolves according to a jump process with state dependent jump rates

$$q_{k\ell} = \lim_{dt \to 0} \frac{P[\xi(t + dt) = \ell \,|\, \xi(t) = k\,,\, \mathbf{y}(t) = \mathbf{y}]}{dt}, \quad k, \ell \in E,\ k \neq \ell.$$

A reward rate $L_j^{\xi(t)}(x, \tilde{\underline{\sigma}}(\xi(t), \mathbf{y}(t)))$, for each player $j \in M$, is associated with current state $(\xi(t), \mathbf{y}(t))$.

A pure-feedback policy for such a system is defined as a measurable mapping $\tilde{\sigma}_j$: $(\mathbf{y}, \xi) \mapsto U_j^\xi$.

Consider the "reward-to-go" functional associated with an initial state $(k, \mathbf{y})$ and feedback strategy pair $\tilde{\underline{\sigma}}$

$$V_j(k, \mathbf{y}; \tilde{\sigma}) = \mathbb{E}_{\mathbf{y};\tilde{\sigma}} \left[\int_0^\infty L_j^{\xi(t)}(\mathbf{y}, \tilde{\underline{\sigma}}(\xi(t), \mathbf{y}(t)))\, dt \,|\, \mathbf{y}(0) = \mathbf{y} \right]. \tag{11.68}$$

Then the functions $V_j(\cdot, \mathbf{y}; \tilde{\sigma})$, which are a.e. C^2 in $\mathbf{y}$ satisfy the following coupled[16] partial differential equations

$$\rho V_j(k, \mathbf{y}) = L_j^k(\mathbf{y}, \tilde{\underline{\sigma}}(k, \mathbf{y})) + \frac{1}{2} \sum_{ii'=1}^n a_{i,i'} \frac{\partial^2 V_j(k, \mathbf{y})}{\partial y_i \partial y_{i'}}$$

$$+ \sum_{i=1}^n f_i^k(\mathbf{y}, \tilde{\underline{\sigma}}(k, \mathbf{y})) \frac{\partial V_j(k, \mathbf{y})}{\partial y_i}, + \sum_{\ell \in E} q_{k\ell}(\mathbf{y}) V_j(\ell, \mathbf{y}). \tag{11.69}$$

Here, $\mathbf{a} = [a_{ii'}]_{i,i'=1,\dots n}$ stands for trace$[\varsigma'\varsigma]$.

This result, which extends the Ito's formula to switching diffusion processes is established, in particular, in Ref. [95].

---

[16] We call these equations coupled because they are written for the different modes of ξ and each equation involves all the functionals corresponding to different values of ξ in its RHS.

11.5.2.4 *Collusive equilibria with monitoring*

Consider a pair of memory strategies monitoring an agreement as described in Section 11.5.2.2. These strategies are pure feedback policies for the associated switching diffusion game. Now assume that these policies define a feedback-Nash equilibrium for this switching diffusion game. Then, we have defined a monitoring scheme and a retaliation scheme that satisfy the conditions for being a subgame perfect equilibrium.

Remark 11.7. There is an implicit fixed-point argument in the definition of a cooperative equilibrium since the cooperative policy, which has to satisfy the feedback-equilibrium condition, is also the policy we want to monitor. Since an equilibrium is also a concept based on a fixed-point argument (each strategy is the best reply to the other players' strategies), we have a double fixed-point property in the definition of this or collusive equilibrium.

11.6 **HJB** equations for equilibrium policies

11.6.1 **HJB** *equations for the original diffusion game*

[17]

As indicated in the Introduction, our ultimate goal is to demonstrate that a properly designed monitoring-and-retaliation ϵ-equilibrium[18] generate higher payoffs for all players than those achieved under the self-enforcing feedback-Nash equilibrium strategy obtained for the original diffusion-stochastic game.

11.6.2 **HJB** *equations for the associated switching-diffusion game*

The extended state $((x, z), \xi)$ is hybrid (it has a continuous and a discrete part). The discrete-state variable evolves according to a random-jump Markov process, while x and z are stochastic diffusions. A feedback-Nash equilibrium can be determined in this extended formulation if we find a solution to the following *coupled HJB* dynamic-programming equations[19]

$$
\rho \tilde{V}_j^*(1, (x, z)) = \frac{1}{2} \varsigma^2 \left\{ \frac{\partial^2}{\partial x^2} \tilde{V}_j^*(1, (x, z)) + \frac{\partial^2}{\partial z^2} \tilde{V}_j^*(1, (x, z)) \right.
$$

$$
+ 2 \frac{\partial^2}{\partial x \partial z} \tilde{V}_j^*(1, (x, z)) \bigg\} + q_{10}(z) \left(\tilde{V}_j^*(0, x) - \tilde{V}_j^*(1, (x, z)) \right)
$$

$$
+ \max_{u_j} \left\{ L_j(x, [\tilde{\sigma}^{1(-j)}(x, z), u_j]) + b(x, [\tilde{\sigma}^{1(-j)}(x, z), u_j]) \frac{\partial}{\partial x} \tilde{V}_j^*(1, (x, z)) \right.
$$

$$
\left. + \left[b(x, [\tilde{\sigma}^{1(-j)}(x, z), u_j]) - b(x, \tilde{\sigma}^1(x, z)) \right] \frac{\partial}{\partial z} \tilde{V}_j^*(1, (x, z)) \right\}, \tag{11.70}
$$

---

[17]The diffusion term represents the stochastic nature of the game.

[18]We anticipate the game approximations and a numerical solution that will be an ϵ-equilibrium; see Section 11.7.

[19]We use here, as the attentive reader will have noticed, an extended version of the sufficient dynamic-programming conditions, for switching diffusion games, using the Ito formula (11.69).

$$\rho \tilde{V}_j^*(0, x) = \frac{1}{2} \varsigma^2 \frac{\partial^2}{\partial x^2} \tilde{V}_j^*(0, x) + q_{01} \left(\tilde{V}_j^*(1, (x, 0)) - \tilde{V}_j^*(0, x) \right)$$

$$+ \max_{u_j} \left[L_j(x, [\tilde{\sigma}^{0(-j)}(x), u_j]) + b(x, [\tilde{\sigma}^{0(-j)}(x), u_j]) \frac{\partial}{\partial x} \tilde{V}_j^*(0, x) \right] \quad (11.71)$$

where $[\tilde{\sigma}^{1(-j)}(x, z), u_j]$ is the control obtained when the other player uses the collusive policy $\tilde{\sigma}^1(x, z)$, while Player j uses his control u_j. Similarly $[\tilde{\sigma}^{0(-j)}(x), u_j]$ is the control obtained when all the other players use the punishment policy $\tilde{\sigma}^0$ and Player j uses his control u_j.

If we manage to solve the above coupled equations, and if the monitoring and switching schemes are properly defined, then the resulting equilibrium not necessarily unique, could dominate the feedback-Nash equilibrium of the original game obtained in Section 11.6.1. Given the complexity of the equations, only a numerical approach can attempt to verify this statement. This is the objective of the next section.

11.7 Numerical approximations to equilibria in stochastic-differential games

11.7.1 *A general formulation of HJB equations*

In a numerical approach to the equilibrium determination, only approximate ϵ−equilibria, rather strict equilibria, can be identified. In this section, we address the problem of how to compute feedback ϵ−equilibria for the class of differential games described in the previous section. We propose to adapt the method in [147], which has been successfully applied to stochastic-control problems.

Suppose we want to solve the infinite-horizon *HJB* equations, where $x \in \mathbf{R}^n$,

$$\rho V_j(x, k) = \frac{1}{2} \sum_{i, i'=1}^{n} a_{i, i'}^k(x) \frac{\partial^2 V_j(x, k)}{\partial x_i \partial x_{i'}} + \max_{u_j \in U_j^k(x)} \left\{ L_j(x, [\sigma^{*(-j)}(x, k), u_j]) \right.$$

$$+ f^k(x, [\sigma^{*(-j)}(x, k), u_j])' \frac{\partial V_j(x, k)}{\partial x} + \sum_{\ell \in E} q_{k\ell}(x, [\sigma^{*(-j)}(x, k), u_j]) V_j(x, \ell) \left. \right\},$$

$$j = 1, 2. \quad (11.72)$$

Here, we consider the general formulation where the noise variance-covariance matrix is state dependent and the jump rates are state and control dependent.

11.7.2 *Discretization scheme*

In broad terms, the discretization scheme consists of the state space being replaced by a finite grid, of substituting finite differences for the partial derivatives, and of approximating the stochastic state process through a Markov chain whose transition probabilities are derived from the characteristics of the state process.

Let $\mathbf{R}_h^n$ denote the $h-$grid on $\mathbf{R}^n$ defined by

$$\mathbf{R}_h^n = \{x : x = \sum_{i=1}^n r_i e_i h, \ r_i \text{ integers}\}$$

where e_i denotes the unit vector in the i-th direction. According to the scheme, we use the following rules for approximating the partial derivatives:

$$v_{x_i}(x) \to [v(x + e_i h) - v(x)]/h \text{ if } f_i^k(x, \mathbf{u}) \geq 0,$$
$$v_{x_i}(x) \to [v(x) - v(x - e_i h)]/h \text{ if } f_i^k(x, \mathbf{u}) < 0,$$
$$v_{x_i x_i}(x) \to [v(x + e_i h) + v(x - e_i h) - 2v(x)]/h^2.$$

For $i \neq j$ and $a_{ij}^k(x) \geq 0$

$$v_{x_i x_j}(x) \to [2v(x) + v(x + e_i h + e_j h) + v(x - e_i h - e_j h)]/2h^2$$
$$-[v(x + e_i h) + v(x - e_i h) + v(x + e_j h) + v(x - e_j h)]/2h^2$$

where $\mathbf{u}$ are the control realizations when the players are using the strategy σ. For $i \neq j$ and $a_{ij}^k(x) < 0$

$$v_{x_i x_j}(x) \to -[2v(x) + v(x + e_i h - e_j h) + v(x - e_i h + e_j h)]/2h^2$$
$$+[v(x + e_i h) + v(x - e_i h) + v(x + e_j h) + v(x - e_j h)]/2h^2.$$

Then, we define the interpolation interval

$$\Delta t^h(x, \mathbf{u}) = h^2/Q_h^k(x, \mathbf{u})$$

where

$$Q_h^k(x, \mathbf{u}) = -q_{kk}(x, \mathbf{u})h^2 + \sum_{i=1}^n a_{ii}^k(x) - \sum_{j \neq i} \frac{|a_{ij}^k(x)|}{2} + h \sum_{i=1}^n |f_i^k(x, \mathbf{u})|.$$

Now we define the transition probabilities

$$p^h((x, k), (x \pm e_i h, k)|\mathbf{u}) = \left[\frac{a_{ii}^k(x)}{2} - \sum_{j \neq i} \frac{|a_{ij}^k(x)|}{2} + h f_i^{k\pm}(x, \mathbf{u})\right] \Big/ Q_h^k(x, \mathbf{u}),$$

$$p^h((x, k), (x + e_i h + e_j h, k)|\mathbf{u}) = p^h((x, k), (x - e_i h - e_j h, k)|\mathbf{u})$$
$$= a_{ij}^{k+}(x)/2Q_h^k(x, \mathbf{u}),$$

$$p^h((x, k), (x + e_i h - e_j h, k)|\mathbf{u}), = p^h((x, k), (x - e_i h + e_j h, k)|\mathbf{u})$$
$$= a_{ij}^{k-}(x)/2Q_h^k(x, \mathbf{u}),$$

$$p^h((x, k), (x, \ell)|\mathbf{u}) = q_{k\ell}(x, \mathbf{u})h^2/Q_h^k(x, \mathbf{u}),$$

where a_{ij}^{k+} and a_{ij}^{k-} represent the positive and negative parts of $a_{ij}^k(x)$ respectively.

After substituting and rearranging terms, we get the following difference equations:

$$v_j(x, k) = \frac{1}{(1 + \rho \Delta t^h(x, \mathbf{u}))} \left\{ \sum_{x'} p^h((x, k), (x', k)|\mathbf{u}) v_j(x', k) + \right.$$

$$\sum_{l \in E, l \neq k} p^h((x, k), (x, l)|\mathbf{u}) v_j(x, l)$$

$$\left. + \Delta t^h(x, \mathbf{u}) g(x, \mathbf{u}) \right\}, x \in \mathbf{R}_h^n, \ j \in \{1, 2\}. \tag{11.73}$$

This set of equations can be rewritten as the fixed-point of a discrete-state dynamic-programming operator

$$v_j(x, k) = (T_{j,\mathbf{u}} v_j(\cdot, \cdot))(x, k), \quad x \in \mathbf{R}_h^n, \quad j \in M. \tag{11.74}$$

The operator $T_{j,\mathbf{u}}$ is contracting. Therefore, if a feedback law $\mathbf{u}(\cdot)$ has been defined, the solution of the above fixed-point problem gives an approximation of the value functions of the players $j \in M$ defined on the grid-points of $\mathbf{R}_h^n$.

Kushner has shown that the Markov chain built according to the above scheme gives a converging[20] approximation to the solution of the *HJB* equations associated with a given policy $\sigma(\cdot)$. When the feedback policy of the other players $-j$ is kept fixed, the scheme can generate the optimal control of Player j (see [148]).

To establish a fixed point of (11.74), we will use a policy-improvement algorithm [127] in conjunction with a multi-grid method.[21]

11.7.3   *Policy-improvement algorithm*

The approximation scheme assigns a multistage Markov game to the jump-diffusion differential game (which we have created out of the original non-monitored diffusion-differential game). We will solve the Markov game using an adaptation of Howard's policy-improvement method [127], which is an efficient method for numerically solving Markov decision problems.

Our important modification of the Howard algorithm is that instead of establishing, at each step, the players' payoff-maximization policies and improving them in the next step, we compute the static-Nash equilibria (i.e., the optimal responses that maximize the square brackets in (11.70)) and attempt to improve them in the next iteration.

We say "we attempt" because the procedure of arriving at equilibrium is not necessarily monotonous and therefore contraction may fail to occur. However, our game with a logistic-growth function and concave payoffs converged satisfactorily to an equilibrium, which we can interpret economically.

11.7.4   *Grid selection techniques*

The discretization scheme introduced in Section 11.7.2 depends on the grid step size h and so does the equilibrium delivered by the policy-improvement algorithm. Presumably, the finer the grid, the more closely the numerical solution approximates the "true" solution. However, we cannot use an arbitrarily fine grid because of the demand this places on computer memory and time, which may become unrealistic given the dimensionality of the problem.

---

[20]In the weak convergence topology.

[21]The latter means that we first solve the problem on a coarse grid, then make the grid finer and compare the solutions. If the difference between them is greater than an assumed tolerance, we make the grid finer and compare the solutions again, etc.

The multigrid technique provides a means for a suitable grid-size selection. We will not enter into further details of the numerical technique, which are clealy out of scope for this textbook.

11.8 Calibration of the game

11.8.1 *Parameter specification*

We propose the following choice of parameters for the model of the fishery-management game described in Section 11.5.1:

$$A = 1 \quad , \quad B = 1$$
$$\rho = 0.1 \quad , \quad \varsigma = 0.1$$
$$a = 1 \quad , \quad b = 0.1$$
$$\alpha_j = 1 \quad , \quad \beta_j = 0.1$$
$$j = 1, 2 .$$

The monitoring variable z satisfies equations (11.61)–(11.62). The jump rates defining the probabilities of switching from one mode to the other are defined as follows:

$$q_{10}(z) = q_{10}z^2, \quad q_{10} > 0, \tag{11.75}$$
$$q_{01} = \text{constant} \geq 0. \tag{11.76}$$

To model the players' impatience (or "trigger happiness") and the probability of restoring cooperation, we will assume different values of the above jump rates. We will specify these values when we comment on the results in Section 11.9.

11.8.2 *Boundary conditions and grid management*

When the mode is noncooperative ($\xi = 0$, i.e., retaliation), the value function $v_j(x, 0)$ does not depend on the monitoring variable z. The grid is thus of dimension one (one state), i.e., contained in the interval $[0, \frac{A}{B}]$.

We need to define what the algorithm should do if the state were pushed outside the grid by the optimization routine. We impose the following boundary conditions:

$$p[(0, 0), (-h, 0)] = 0, \tag{11.77}$$
$$p\left[\left(\frac{A}{B}, 0\right), \left(\frac{A}{B} + h, 0\right)\right] = 0. \tag{11.78}$$

When the mode of play is cooperative ($\xi = 1$), then the value function $v_j((x, z), 1)$ depends on two continuous state variables (x, z). The grid is then of dimension two, i.e., contained in the rectangle $[0, \frac{A}{B}] \times [z_{\min}, z_{\max}]$. Here, the following boundary conditions have been imposed:

$$p[((0, z), 1), ((-h, z), 1)] = 0 \quad \forall z, \tag{11.79}$$

$$p\left[\left(\left(\frac{A}{B}, z\right), 1\right), \left(\left(\frac{A}{B} + h, z\right), 1\right)\right] = 0 \quad \forall z, \tag{11.80}$$

$$p[((x, z_{\min}), 1), (x, 0)] = 1 \quad \forall x, \tag{11.81}$$

$$p[((x, z_{\max}), 1), (x, 0)] = 1 \quad \forall x. \tag{11.82}$$

Selecting values for z_{min} and z_{max} depends on the design parameters defined in Section 11.8.1, and in particular, on q_{10}. The basic idea is to choose z_{min} and z_{max} such that the probability function $p(z) := p^h[((x, z), 1), (x, 0)]$ is smooth enough at the interval ends $z = z_{min}$ and $z = z_{max}$, for a fixed x. Since

$$p(z) = \frac{q_{10} z^2 h^2}{q_{10} z^2 h^2 + h \sum_{i=1}^{2} |f_i^1(x, z, \mathbf{u})|}, \tag{11.83}$$

a possible choice is $z_{max} = -z_{min} = 1/h$ and h, such that for $z = z_{min}$ and $z = z_{max}$, the expression $h \sum_{i=1}^{2} |f_i^1(x, z, \mathbf{u})|$ is negligible compared to q_{10}, for all $x \in [0; A/B]$. We would thus obtain

$$p(z_{max}) = \frac{q_{10}}{q_{10} + h \sum_{i=1}^{2} |f_i^1(x, z, \mathbf{u})|} \approx 1, \tag{11.84}$$

and similarly

$$p(z_{min}) \approx 1. \tag{11.85}$$

Note that this choice for z_{min}, z_{max} and h implies that N_z, the number of grid points in the z-direction, is $N_z = 2/h^2$, since $N_z \cdot h = z_{max} - z_{min}/h$. If q_{10} is too small, then h has to be chosen to be very small, resulting in a very large number of grid points.

In the numerical experiments we usually worked with $N_z = 64$ grid points in the z-direction, and $N_x = 32$ or 64 in the x-direction. Sparse matrix techniques were used to handle the linear systems obtained on these grids.

11.9   Numerical results

We applied the numerical method introduced in Section 11.7.4 to the fishery-exploitation model, with the choice of parameters given in Section 11.8.1 and the boundary conditions specified by equations (11.77)–(11.82). We considered the case of symmetric players, to simplify the presentation of the results, but intensive numerical tests indicate that symmetry is not a crucial assumption.

The numerical results are displayed in Figures 11.1 and 11.2. We are using the following line legend:

(1) For the Pareto-optimal solution (the two players fully cooperate), see the dashed-dotted $(- \cdot - \cdot -)$ line;
(2) For the Nash-equilibrium solution without monitoring (i.e., the classical noncooperative Nash equilibrium), see the solid line;

(3) For the Nash-equilibrium solution with monitoring (collusive) when the game is in state 1 (cooperative solution with monitoring system), see the dotted line.

(4) For the Nash-equilibrium solution with monitoring (collusive) when the game is in state 0 (noncooperative mode of play with monitoring system), see the spaced dotted (. . . .) line.

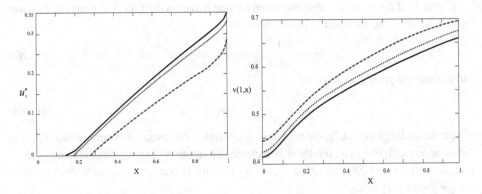

Fig. 11.1 Fishing strategies and value functions; $q_{01} = 0, q_{10}(z) = \frac{z^2}{100}$

In the left panel of Figure 11.1, we plotted the optimal policy $u_1^* = \sigma_1^*(x)$ for Player 1 (note that $u_1^* = u_2^*$ since the model is symmetric).

In the right panel of this figure, $v(1, x)$ – the associated value functions of Player 1 are plotted versus the stock of biomass x.

Both figures correspond to the same choice of parameters used for the monitoring situation (3), i.e.,

$$q_{01} = 0, \quad q_{10}(z) = \frac{1}{100}z^2. \tag{11.86}$$

This means that both players cooperate at the beginning of play when $z = 0$, but the probability of a jump from the cooperative mode ($\xi = 1$) to the non-cooperative one ($\xi = 0$) increases with z, very slowly at first (for a small z, z^2 is obviously very small) but then growing faster. Once $\xi = 0$ (in the retaliatory mode), there is no possibility of returning to a cooperative mode, since $q_{01} = 0$.

The value functions displayed in the right panel of Figure 11.1 clearly show that the Nash-equilibrium solution with monitoring (collusive) dominates the feedback equilibrium of the original game, whereas, as usual, the Pareto-optimal solution dominates both Nash-equilibrium solutions. This is different from what we saw in the deterministic capitalists-versus-workers game discussed in the GE illustration of Chapter 5.7.5 (see page 231), where the collusive equilibrium could fully support a Pareto-efficient solution.

The collusive equilibrium solution for when the game is in state 0 was not plotted in the above figure, since, in this case, there is no possibility of jumping from the noncooperative mode of play to the cooperative mode because $q_{01} = 0$. Thus, the Nash-equilibrium solution obtained in this situation is identical to the Nash-equilibrium solution of the original game (situation (2) above).

Furthermore, the left panel of Figure 11.1 shows that, in all situations, no fishing occurs if the stock of biomass x is below a certain limit, and that the collusive-equilibrium solution implies less fishing effort than in the Nash equilibrium of the original game. And, not surprisingly, even less fishing effort is needed under the Pareto-optimal solution.

Figure 11.2 displays the value functions for one player in the four situations cited above. In the left panel, the monitoring parameters are

$$q_{01} = \frac{1}{100}, \quad q_{10}(z) = \frac{1}{100} z^2, \tag{11.87}$$

and in the right panel, we selected

$$q_{01} = \frac{1}{100}, \quad q_{10}(z) = z^2. \tag{11.88}$$

So, according to (11.87), the probability of retaliation given an observed value $z(t)$ is the same as in Figure 11.1, but there is a chance of resuming cooperation, since $q_{01} > 0$. The jump rates in (11.88) describe a situation where the players are "trigger happy" because the q_{10} grows very fast.

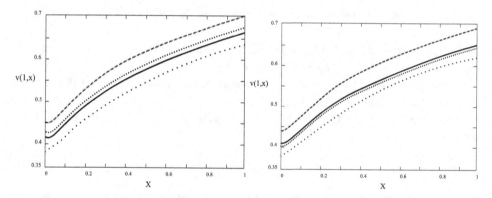

Fig. 11.2   Value functions for $q_{01} = 10^{-2}, q_{10}(z) = \frac{z^2}{100}$ and –trigger happy– $q_{01} = 10^{-2}, q_{10}(z) = z^2$

The solutions plotted in the left panel of Figure 11.2 are very similar to those of the right panel in Figure 11.1 for situations (1)–(3). However, this figure also has a fourth line that corresponds to situation (4), which is also the collusive equilibrium; but here, the game starts from (or transits through) $\xi = 0$, i.e., the non-cooperative (punitive) mode of play. This solution is dominated by the three other solutions. This can be attributed to $q_{01} > 0$, i.e., to the possibility of resuming cooperation. It lessens the threat of retaliation and the players engage in "cheating" that is detected and punished. As a result, the mode of the game is retaliatory for most of the time, and payoffs remain low.

Finally, the right panel of the current figure shows that the (collusive) equilibrium *can* be dominated by the feedback-Nash (non-cooperative) equilibrium of the original game. This is related to the trigger-happy nature of the game in this particular case. Indeed, since q_{10} is large, the players are very sensitive to cheating. Small indications that the other player may be cheating – while in reality $z(t) > 0$ can be caused by a stochastic shock – can cause the other player to switch, very quickly, from a cooperative mode of play to the noncooperative one. As a result, payoffs are low.

To summarize, the numerical results obtained by the implementation of the method described in Section 11.7 show that

(1) The algorithm converges to an ϵ-equilibrium for a variety of values of the parameters q_{10} and q_{01}, which determine the mode-switching jump process.
(2) For some jump rates (e.g., $q_{01} = \frac{1}{100}$ and $q_{10} = \frac{z^2}{100}$), the collusive equilibrium dominates the feedback equilibrium of the original game. This demonstrates that the construction of dominating collusive equilibria is possible in the context of stochastic-differential games.
(3) If the parameter q_{10} is too big, then we are in a situation where the players are trigger happy; this means that they switch to the retaliation mode "too quickly," possibly mistaking a random fluctuation of the stock for cheating by the other player. In that case, the procedure can converge toward a collusive equilibrium that is dominated by the feedback equilibrium of the original game. This tells us that a collusive equilibrium, even if it exists, may be an inappropriate solution concept for some games (e.g., for games whose players have a large q_{10}).

11.10 What have we learned in this chapter?

- We defined stochastic games where m players competitively control a stochastic-diffusion process.
- A stochastic-diffusion process is susceptible to a particular form of calculus leading to the Ito formula, which replaces the fundamental chain rule in ordinary calculus. This formula plays an important role in the derivation of the dynamic-programming equations characterizing equilibrium solutions to stochastic games.
- Continuous-time theoretical finance provides examples of stochastic-diffusion games.
- Options or derivatives can be evaluated using the Ito formula and principle of no-arbitrage in perfect markets. This leads to the Black-Scholes equation, satisfied by the derivative value. We can solve competitive investment games using this approach.
- The theory of collusive equilibria, where each equilibrium is supported by a threat, can be developed in a fully stochastic environment, where both the state equations and the decisions to retaliate are stochastic.
- It is possible to compute ϵ-equilibria for some analytically nontractable games, including stochastic ones.
- These equilibria can be feedback (Markov)-Nash (non-cooperative) equilibria as well as collusive equilibria, i.e., feedback (Markov)-Nash with monitoring.
- The collusive equilibria can dominate feedback (Markov)-Nash (non-cooperative) equilibria, depending on the appropriateness of the design of the monitoring-and-retaliation mechanism.

11.11 Exercises

Exercise 11.1. Consider the SDE

$$dx(t) = \mathbf{A}\mathbf{x}(t) + \mathbf{B}_1\mathbf{u}_1(t) + \mathbf{B}_2\mathbf{u}_2(t) + \varsigma d\mathbf{w}(t)$$

with $\mathbf{x} \in \mathbf{R}^n, u_j \in \mathbf{R}^{m_j}$, describing a linear diffusion controlled by two players. Here $\mathbf{w}(t)$ is a vector of independent Wiener processes taking value in $\mathbf{R}^n$. The respective payoffs of the players are

$$J_j = \mathbb{E}\left\{ \int_0^\infty \left(\frac{1}{2}\mathbf{x}(t)'\mathbf{Q}_j\mathbf{x}(t) + \frac{1}{2}\mathbf{u}_j(t)'\mathbf{R}_j\mathbf{u}_j(t) \right) dt \right\} \quad j = 1, 2.$$

Here the matrices are supposed to have the right dimensions. Find a feedback Nash equilibrium for this diffusion game.

Exercise 11.2. Consider the SDE

$$dx(t) = ax(t)(u_1(t) + u_2(t) + \varsigma dw(t)$$

describing a diffusion controlled by two players with respective payoffs

$$J_j = \mathbb{E}\left\{ \int_0^\infty b_j\, x(t)(u_j(t) - \frac{1}{2}u_j(t)^2) dt \right\} \quad j = 1, 2.$$

Find a feedback Nash equilibrium for this diffusion game.

Exercise 11.3. Propose the functional form of the jump-rate parameters if the players react strongly to small values of z but believe that a large z must be due to a stochastic shock and does not warrant triggering a penalty.

Exercise 11.4. The choice of the monitoring variable z in equation (11.61) is by far non unique. Propose another variable, on which the fishermen could base their decisions regarding the game mode.

11.12 GE illustration: A differential game of debt-contract valuation

11.12.1 *Introduction*

In [7] an interesting dynamic-game model of debt contracts was proposed and used to explain some observed discrepancies on the yield spread of risky debts. The model is cast in a discrete time setting, with a simplifying assumption on the information structure, allowing for a relatively easy sequential formulation of the equilibrium conditions as a sequence of Stackelberg solutions, where the firm owner is the leader and the lender is the follower. Reformulating the debt-contract valuation model as a dynamic game played in continuous time allows us[22] to explore the possible mix of the *Black & Scholes valuation* principles [29] with the stochastic-differential game concepts. Indeed, the debt-contract valuation problem provides a very natural framework where antagonistic parties act strategically to manage risk in an uncertain environment.

---

[22]This GE illustration is based on reference [113].

11.12.2 *The firm and the debt contract*

A firm has a project which is characterized by a stochastic state equation in the form of a *geometric Brownian motion*

$$dx(t) = \alpha x(t)\, dt + \varsigma x(t)\, dw(t) \tag{11.89}$$

where

$w(\cdot)$ is a standard Wiener process,
$\alpha x(t)$ is the instantaneous growth rate,
$(\varsigma x(t))^2$ is the instantaneous variance of process x.

The state could represent, for example, the price of a project output.

The firm expects a stream of cash flows $\pi(x(t))$ defined as a function of the state of the project. Therefore, if the firm has a discount rate ρ, then, when it is debt free (we say the *unlevered*), the equity evaluated as the net present value of expected cash flows is given by

$$\mathcal{E}(x) = \mathbb{E}\left[\int_0^\infty e^{-\rho t}\pi(x(t))dt\,\big|\,x(0) = x\right]. \tag{11.90}$$

Using a standard technique of stochastic calculus, we can characterize the function $\mathcal{E}(x)$ as the solution of the following differential equation (see equation(11.32)):

$$\rho\mathcal{E}(x) = \pi(x) + \mathcal{E}'(x)\alpha v + \frac{1}{2}\mathcal{E}''(x)(\varsigma x)^2, \tag{11.91}$$

$$\mathcal{E}(0) = 0. \tag{11.92}$$

The boundary condition (11.92) comes from the fact that a project with a zero value will remain with a zero value and thus, will generate no cash flow. A tractable and interesting case is when the cash flow is proportional to state x, i.e., $\pi(x) = \beta x$, $\beta > 0$; if so, we obtain

$$\rho\mathcal{E}(x) = \beta x + \mathcal{E}'(x)\alpha x + \frac{1}{2}\mathcal{E}''(x)(\varsigma x)^2, \tag{11.93}$$

$$\mathcal{E}(0) = 0. \tag{11.94}$$

A linear function $\mathcal{E}(x) = Ax$ is tried. It satisfies the boundary condition (11.94) and it satisfies (11.93) if A is such that $\rho Ax = \beta x + A\alpha x$ for all $x > 0$. This defines $A = \frac{\beta}{\rho - \alpha}$ and therefore $\mathcal{E}(x) = \frac{\beta x}{\rho - \alpha}$.

Suppose that the firm needs to borrow an amount C to finance its project. The financing is arranged through the issuance of a debt contract with a lender. The contract is defined by the following parameters, which are also called the *contract terms*:

- outstanding principal amount P
- maturity T
- grace period[23] g
- coupon rate[24] c.

---

[23] A grace period is a time past the deadline for an obligation during which a late penalty that would have been imposed is waived.

[24] A coupon rate is the interest rate stated on a bond when it's issued. The coupon is typically paid semiannually.

The contract terms define a *debt service*, which we represent by a function $t \mapsto \hat{q}(t)$. It gives the cumulative contractual payments that have to be made-up to time $t \in [0, T]$. The contract may also include an interest rate ρ^- in case of delayed payment and an interest rate ρ^+ in case of advance payments. Assuming a linear amortization, the function $\hat{q}(t)$ is then the solution to

$$\hat{q}(0) = 0,$$
$$\hat{q}(t) = cPt \quad \text{when } 0 \le t \le g,$$
$$\hat{q}(t) = cPg + \frac{P}{T-g}(t-g) + \frac{cP}{2}\left(\frac{(T-t)^2}{T-g} - (T-g)\right)$$
$$\text{when } g \le t \le T.$$

This function is illustrated in Figure 11.3. The strategic structure of the problem is due to

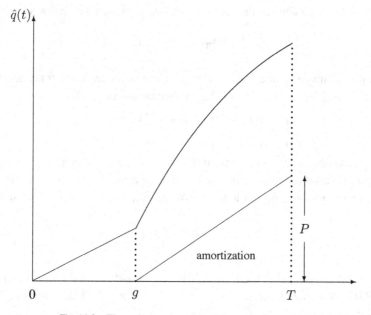

Fig. 11.3   The representation of the debt service in continuous time

the fact that the firm controls its payments and may decide not to respect the contract at some time.

At time t, let the variable $q(t)$ describe the state of the debt service, which is the cumulated payments made by the firm up to time t. This state variable evolves according to the following differential equation:

$$\dot{q}(t) = u(t) + \rho^+(q(t) - \hat{q}(t))^+ - \rho^-(q(t) - \hat{q}(t))^-, \qquad (11.95)$$
$$q_0 = 0, \qquad (11.96)$$

where $u(t)$ is the payment at time t, $(q(t) - \hat{q}(t))^+ = \max\{0, q(t) - \hat{q}(t)\}$ and $(q(t) - \hat{q}(t))^- = \max\{0, -q(t) + \hat{q}(t)\}$. We assume that the payment rate $u(t)$ is upper bounded by the cash flows generated by the project

$$u(t) \in [0, \pi(x(t))].$$

The strategic structure of the problem is also due to the fact that the lender is authorized to take control of the firm when the owner is late in his payments, i.e., at any time $\theta \in \Theta(q(\cdot))$ where $\Theta(q(\cdot))$ is the set

$$\Theta(q(\cdot)) = \{t \in [0, T] : q(t) < \hat{q}(t)\}.$$

The lender's action is an impulse control. If at time $\theta \in \Theta(q(\cdot))$, the lender liquidates the firm, he receives the minimum between the debt balance and the liquidation value of the firm, i.e.,

$$L(\theta, q(\theta), x(\theta)) = \min\{d(\theta, q(\theta)), l(\theta, x(\theta))\}.$$

A possible form for the liquidation value is

$$l(\theta, x(\theta)) = \max\{0, \mathcal{E}(x(\theta)) - K\}. \tag{11.97}$$

This assumes that the lender will find another firm ready to buy the project at equity value $\mathcal{E}(x(\theta))$ and that K is a constant liquidation cost. For the above example of a debt contract, the debt balance is

$$d(\theta, q(\theta)) = \begin{cases} P + \hat{q}(\theta) - q(\theta) & \text{if } \theta \le g \\ P\dfrac{T - \theta}{T - g} + \hat{q}(\theta) - q(\theta) & \text{if } \theta > g. \end{cases}$$

The point made by Anderson & Sundaresan, the authors of the discrete-time model of reference [7], is that, if the firm is in financial distress and its liquidation value is low, the lender may prefer not to liquidate, even if condition $\theta \in \Theta(q(\cdot))$ is satisfied, in order to wait for a possible increase in the project's value.

We will thus model the strategic interaction between the firm owner and the lender, as a noncooperative differential game where the payoffs are linked to the equity and debt value, respectively. These values will depend on time t, the state of the project x, and the state of debt service q, and it will be determined either by optimizing the expected utility of the agents or, if equity and debt can be considered as derivatives traded on a market, through the so-called *equivalent risk-neutral valuation*.

11.12.3 *Stochastic-game formulation*

Let us assume that the risk involved in developing the project, or in financing it through the agreed-upon debt contract, cannot be spanned by assets traded on a market.[25] So, we represent the strategic interaction between two individuals, the firm owner who strives to

---

[25]The *risk is spanned* by assets traded on a market if there is a possibility of replicating this risk through a portfolio of these assets.

maximize the expected present value of the net cash flow, using a discount rate ρ_1, and a lender who maximizes the present value of the debt, using a discount rate ρ_2. The state of the system is the pair $s = (q, V) \in S \dot{=} \mathbf{R} \times \mathbf{R}^+$. An admissible strategy $\sigma_1 \in \Gamma_1$ for the firm owner is given by a measurable function[26] $\tilde{u}(t, s) : [0, T] \times S \mapsto u$. A strategy $\sigma_2 \in \Gamma_2$ for the lender is a stopping time θ defined by $\theta = \inf\{t : (t, s(t)) \in B\}$ where B is a Borel set $\subseteq [0, T] \times S$. Associated with a strategy pair $\sigma = (\sigma_1, \sigma_2)$ we define the payoffs of the two players:

For the firm's owner

$$\Pi_1(\sigma; t, s) = \mathbb{E}_\sigma \left[\int_t^{t_f} e^{-\rho_1(\tau-t)} (\pi(x(\tau)) - u(\tau)) \, d\tau \right.$$

$$\left. + e^{-r(t_f-t)} \Phi(t_f, \tilde{x}(t_f)) \Big| s(t) = s \right], \tag{11.98}$$

$$t_f = \min\{\theta, T\},$$

$$\Phi(t_f, x(t_f)) = \{l(t_f, q(t_f), x(t_f))$$

$$- L(t_f, q(t_f), x(t_f)), 0\}$$

$$\text{if } t_f = \theta \le T, \ q(t_f) < \hat{q}(t_f), \tag{11.99}$$

$$\Phi(T, x(T)) = \mathcal{E}[x(T)] + q(T) - \hat{q}(T) \text{ if } q(T) - \hat{q}(T) \ge 0. \tag{11.100}$$

For the lender

$$\Pi_2(\sigma; t, s) = \mathbb{E}_\sigma \left[\int_t^{t_f} e^{-\rho_2(\tau-t)} u(\tau) \, d\tau \right.$$

$$\left. + e^{-r(t_f-t)} \Psi(t_f, x(t_f)) \Big| s(t) = s \right], \tag{11.101}$$

$$t_f = \min\{\theta, T\},$$

$$\Psi(t_f, x(t_f)) = L(t_f, q(t_f), x(t_f))$$

$$\text{if } t_f = \theta \le T \quad q(t_f) < \hat{q}(t_f), \tag{11.102}$$

$$\Psi(T, x(T)) = \hat{q}(T) - q(T) \text{ if } q(T) - \hat{q}(T) \ge 0. \tag{11.103}$$

Remark 11.8. It is in the treatment of the debt service that this model differs significantly from [7]. The creditor does not forget the late payments as is implicitly assumed in [7]. The creditor can also take control at maturity if condition $q(T) < \hat{q}(T)$ holds. The firm is allowed to "overpay" and it is thus possible for $q(T) - \hat{q}(T)$ to be positive at maturity.[27] It is then normal that the firm get this amount back at time T as indicated in (11.100).

Definition 11.8. A strategy pair $\sigma^* = (\sigma_1^*, \sigma_2^*))$ is a subgame perfect Nash equilibrium for this stochastic game if, for any $(t, s) \in [0, T] \times S$, the following holds:

$$\Pi_1(\sigma^*; t, s) \ge \Pi_1(\sigma_1, \sigma_2^*; t, s) \quad \forall \sigma_1 \in \Gamma_1, \tag{11.104}$$

$$\Pi_2(\sigma^*; t, s) \ge \Pi_2(\sigma_1^*, \sigma_2; t, s) \quad \forall \sigma_1 \in \Gamma_2. \tag{11.105}$$

---

[26] In all generality, the control of the firm owner could be described as a process that is adapted to $x(\cdot)$ and $q(\cdot)$. For our purposes here, its definition as a feedback law will suffice.

[27] Notice that overpayment is a form of investment by the firm. We could allow it to invest a part of its cash flows in another type of asset, and this would be more realistic from a financial point of view, but it would further complicate the model.

We shall not pursue here the characterization of the equilibrium strategies. We recognize that they are affected by several assumptions regarding the agents' attitudes to risk,[28] (i.e., their utility functions) and their relative discount rates. What is problematic is that, in practice, these data are not readily observable and it is difficult to assume that they are common knowledge for the agents. In the next section, we shall use another approach that is valid when the equity and debt can be considered as derivatives obtained from assets traded or spanned by an efficient market.[29] We will see that using the *real option*[30] or the equivalent risk-neutral valuation methods eliminates the need to know these parameters for an equilibrium solution.

11.12.4 *Equivalent risk-neutral valuation*

Here we develop the risk-neutral valuation of this derivative product. There will be, inevitably, some use of a financial jargon, which should not impair understanding for the lay-person.

11.12.4.1 *Assumptions and system dynamics*

Throughout the rest of this section, we make the following assumptions about the financial market where the debt contract takes place.

Assumption 11.1. We assume that the following conditions hold:

A1 no transaction costs

A2 trading takes place continuously

A3 assets are assumed to be perfectly divisible

A4 unlimited short selling is allowed

A5 borrowing and lending is possible at the risk-free rate, i.e., the risk-free asset can be short sold

A6 the firm has no impact on the risk structure of the whole market

A7 no arbitrage

A8 there exists a risk-free security B, for example, a zero-coupon bond, paying the risk-free interest rate r. The dynamics of B is given by

$$dB(t) = rB(t)\,dt. \tag{11.106}$$

To simplify the model, we assume that the firm's project has a value V that is perfectly correlated with the asset x being traded. We assume that this asset pays no dividend, so its

---

[28] Note that we have assumed here that the agents are optimizing discounted cash flows, not the utility of them. If risk aversion has to be incorporated in the model, then the equilibrium characterization will be harder.

[29] An *efficient market* is one in which security prices reflect all available information. This means that every security traded on the market is correctly valued given the available information.

[30] A *real option* itself, is the right, but not the obligation, to undertake some business decision; typically the option to make, abandon, expand, or contract a capital investment. For example, the opportunity to invest in the expansion of a firm's factory, or alternatively to sell the factory, is a real call or put option, respectively.

return stems from capital gains only. Then, x evolves according to

$$dx(t) = \mu x(t)\,dt + \varsigma x(t)\,dw(t).$$

According to the CAPM theory (see [65][31]), the drift rate μ should be equal to the expected rate of return from holding an asset with this risk characteristic, that is

$$\mu = r + \zeta \rho_{xm}\varsigma, \tag{11.107}$$

where r is the risk-free rate, ζ is the market price of risk[32] and ρ_{xm} is the correlation of x with the market portfolio. We also call μ the *risk-adjusted expected rate of return* that investors would require to own the project.

The project is assumed to pay dividends $f(t) = \beta V(t)$, where β is the payout ratio. Then, the value V evolves according to the following geometric Brownian motion:[33]

$$dV(t) = \alpha V(t)\,dt + \varsigma V(t)\,dw(t),$$

where $\alpha = \mu - \beta$. If we repeat the developments of (11.91)–(11.94), but with $\rho = \mu$ and $\alpha = \mu - \beta$, we will prove that V is indeed the equity value for the unlevered firm when the risk-adjusted rate of return is used as the discount rate.

Again we assume that the firm needs an amount C to launch the project. The firm, which is not default free,[34] cannot sell the risk-free security short, since this would be equivalent to borrowing at the risk-free rate. Therefore, assumption A5 applies only to investors who may buy the firm's equity or debt. The firm's owner and the lender are now interested in maximizing the levered firm's equity and debt value, respectively. For that they act strategically, playing a dynamic Nash-equilibrium based on the evolutions of equity and debt.

11.12.4.2 *Debt and equity valuations when bankruptcy is not considered*

At time $t \in [0, T]$ the firm pays $u(t)$ to contribute (it could be partial payment) to the debt service, and the remaining cash, given by $\beta V(t) - u(t)$, is a dividend paid to the equity holder. Assume that the lender's strategy is never to ask for bankruptcy, while the borrower's strategy is given by a feedback law $\sigma(q, V, t)$. We are interested in defining the value of equity under these strategies. The equity is a function of time t, the value of the hypothetical unlevered firm V and the state of debt service q, and is denoted by $\mathcal{E}(t, V, q)$. To simplify the notation, we omit the arguments and simply refer to $\mathcal{E}$. We construct a portfolio composed of risk-free bonds B and derivative E, replicating the unlevered firm. To avoid arbitrage opportunities, this portfolio must provide the same return αV as the

---

[31]In finance, the capital-asset pricing model (CAPM) is used to determine a theoretically appropriate required rate of return for an asset, if that asset is to be added to an already well-diversified portfolio (called the market portfolio), given that asset's non-diversifiable risk.

[32]It is defined by $\zeta = \dfrac{r_m - r}{\varsigma_m}$ where r_m and ς_m are the market's portfolio expected return and the standard deviation respectively.

[33]This can be verified by constructing a portfolio where all dividends are immediately reinvested in the project. Such a portfolio is a replication of $x(t)$ and must therefore follow the same dynamics as $x(t)$.

[34]Default may occur if the debtor is either unwilling or unable to pay their debt.

unlevered firm, since it has the same risk.[35] Applying Ito's lemma, we obtain the following stochastic equation for the equity-value dynamics:

$$d\mathcal{E} = \left[\alpha V \frac{\partial \mathcal{E}}{\partial V} + \frac{\partial \mathcal{E}}{\partial t} + \frac{1}{2}\varsigma^2 V^2 \frac{\partial^2 \mathcal{E}}{\partial V^2} + [u + \rho(q - \hat{q})]\frac{\partial \mathcal{E}}{\partial q} \right] dt + \varsigma V \frac{\partial \mathcal{E}}{\partial V} dw.$$

The dynamics of $\mathcal{E}$ and V are perfectly correlated. Therefore, it is possible to construct a self-financing portfolio Π consisting of $\psi(t)$ shares of the risk-free asset B and $\phi(t)$ shares of the equity $\mathcal{E}$ (all dividends paid by the equity are immediately reinvested in the portfolio), which replicates the risky term of V and pays the same dividend as V. The portfolio value at time t is

$$\Pi(t) = B(t)\psi(t) + \mathcal{E}(t)\phi(t).$$

Keeping in mind that the equity pays a dividend $\beta V - u$ and the portfolio pays a dividend βV, the strategy is self-financing if[36]

$$B(t)d\psi(t) + \mathcal{E}(t)d\phi(t) = -\beta V dt + \phi(t)(\beta V - u)dt, \tag{11.108}$$

which leads to

$$d\Pi(t) = \psi(t)dB(t) + \phi(t)d\mathcal{E}(t) - \beta V dt + \phi(t)(\beta V - u)dt. \tag{11.109}$$

In order to get a risk replication, the weight $\phi(t)$ must be given by

$$\phi(t) = \frac{1}{\frac{\partial \mathcal{E}}{\partial V}}. \tag{11.110}$$

Then the weight $\psi(t)$ must verify

$$\psi(t) = \frac{1}{\frac{\partial \mathcal{E}}{\partial V}} \frac{1}{B} \left[\Pi \frac{\partial \mathcal{E}}{\partial V} - \mathcal{E} \right]. \tag{11.111}$$

We can now write the stochastic equation satisfied by the portfolio value

$$d\Pi = \psi dB + \phi d\mathcal{E} - \beta V dt + \phi(t)(\beta V - u)dt$$

$$= \frac{1}{\frac{\partial \mathcal{E}}{\partial V}} r \left[\Pi \frac{\partial \mathcal{E}}{\partial V} - \mathcal{E} \right] dt + \frac{1}{\frac{\partial \mathcal{E}}{\partial V}} \left[\alpha V \frac{\partial \mathcal{E}}{\partial V} + \frac{\partial \mathcal{E}}{\partial t} + \frac{1}{2}\varsigma^2 V^2 \frac{\partial^2 \mathcal{E}}{\partial V^2} \right.$$

$$\left. + [u + \rho(q - \hat{q})]\frac{\partial \mathcal{E}}{\partial q} \right] dt + \varsigma V dw - \beta V dt + \frac{1}{\frac{\partial \mathcal{E}}{\partial V}}(\beta V - u)dt, \tag{11.112}$$

$$= \frac{1}{\frac{\partial \mathcal{E}}{\partial V}} \left[r\Pi \frac{\partial \mathcal{E}}{\partial V} - r\mathcal{E} + (\alpha - \beta)V \frac{\partial \mathcal{E}}{\partial V} + \frac{\partial \mathcal{E}}{\partial t} \right.$$

$$\left. + \frac{1}{2}\varsigma^2 V^2 \frac{\partial^2 \mathcal{E}}{\partial V^2} + [u + \rho(q - \hat{q})]\frac{\partial \mathcal{E}}{\partial q} + \beta V - u \right] dt + \varsigma V dw. \tag{11.113}$$

---

[35] Notice that the usual way of obtaining the Black-Scholes equation would have been to construct a self-financing portfolio composed of the risk-free asset B and the underlying asset V in proportions that would replicate the derivative (either E or D). Then according to the last two assumptions, we know that this portfolio has to give the same return as the underlying asset. However in our case, the unlevered firm is not traded; so we proceed in a symmetric way and construct a self-financing portfolio composed of the risk-free asset B and the derivative E that will replicate V.

[36] Such a dynamic adaptation of the portfolio composition is feasible, as Π is measurable with respect to the filtration generated by $W(t)$.

We observe that, as intended, portfolio Π replicates the risk of V. Due to our sixth assumption, Π must have the same return αV as V; otherwise, there would be arbitrage opportunities. Moreover, since at time zero, the value of the portfolio is equal to the value of V, we must have $\Pi(t) = V(t)$ for all t. Matching the drift terms and multiplying by $\frac{\partial E}{\partial V}$, we obtain the following partial differential equation, which has to be satisfied by the equity value on the domain $[0, T] \times S$

$$(r - \beta)V\frac{\partial \mathcal{E}}{\partial V} - r\mathcal{E} + \frac{\partial \mathcal{E}}{\partial t} + \frac{1}{2}\varsigma^2 V^2 \frac{\partial^2 \mathcal{E}}{\partial V^2} + [u + \rho(q - \hat{q})]\frac{\partial \mathcal{E}}{\partial q} + \beta V - u = 0. \quad (11.114)$$

We should have at maturity

$$\mathcal{E}(T, s) = \max\{0, V + q - \hat{q}(T)\} \quad \forall s \in S. \quad (11.115)$$

A similar equation could be derived for the other derivative (debt) value D

$$(r - \beta)V\frac{\partial D}{\partial V} - rD + \frac{\partial D}{\partial t} + \frac{1}{2}\varsigma^2 V^2 \frac{\partial^2 D}{\partial V^2} + [u + \rho(q - \hat{q})]\frac{\partial D}{\partial q} + u = 0. \quad (11.116)$$

Again, at maturity, we should have

$$D(T, s) = \min\{V, \hat{q}(T) - q\} \quad \forall s \in S. \quad (11.117)$$

Remark 11.9. As we assume that no bankruptcy occurs, we have to suppose that, at maturity, the balance of the debt service will be cleared, if the value of the project permits it. This is represented in the terminal conditions (11.115) and (11.117).

This formulation is interesting in that the model then reduces to a single controller framework. The firm's owner controls the $q(\cdot)$ trajectory but runs the risk of having a terminal value that is down to 0. Since the payments are upper bounded by the cash-flow value $u(t) \leq f(t)$, it might be optimal for the borrower to anticipate the payments, for some configurations of (t, s). We will not pursue that development, but rather consider the more realistic case where the lender can liquidate the firm when it is late in its payments.

11.12.4.3 *Debt and equity valuations when liquidation may occur*

Now suppose that both players have chosen a strategy, with the lender having a *stopping-time strategy*, which defines the conditions under which liquidation occurs. After liquidation, the equity $\mathcal{E}$ is not available to construct portfolio Π. Therefore, this portfolio has a return that is smaller than or equal to the return of V. We obtain the following relations for $\mathcal{E}$ and D, respectively;

$$(r - \beta)V\frac{\partial \mathcal{E}}{\partial V} - r\mathcal{E} + \frac{\partial \mathcal{E}}{\partial t} + \frac{1}{2}\varsigma^2 V^2 \frac{\partial^2 \mathcal{E}}{\partial V^2} + [u + \rho(q - \hat{q})]\frac{\partial \mathcal{E}}{\partial q} + \beta V - u \leq 0,$$

$$(r - \beta)V\frac{\partial D}{\partial V} - rD + \frac{\partial D}{\partial t} + \frac{1}{2}\varsigma^2 V^2 \frac{\partial^2 D}{\partial V^2} + [u + \rho(q - \hat{q})]\frac{\partial D}{\partial q} + u \leq 0.$$

Equality holds at points where there is no liquidation. At a liquidation time θ, the following boundary conditions[37] hold:

$$q < \hat{q}(\theta),$$
$$\mathcal{E}(\theta, q, V) = \max\{V - L(\theta, q, V), 0\},$$
$$D(\theta, q, V) = L(\theta, q, V)$$

---

[37] We use the notation θ^- for $\theta - \varepsilon$ with $\varepsilon > 0$ arbitrarily small.

where

$$L(\theta, q(\theta), x(\theta)) = \min\{d(\theta, q(\theta)), l(\theta, x(\theta))\}$$

and

$$l(\theta, x(\theta)) = \max\{0, V(\theta) - K\}. \tag{11.118}$$

At maturity T when $q \geq \hat{q}(T)$, the boundary conditions are

$$\mathcal{E}(T, q, V) = V + q - \hat{q}(T),$$
$$D(T, q, V) = \hat{q}(T) - q.$$

Consider a strategy pair $\sigma = (\sigma_1, \sigma_2)$. We can rewrite the debt value as

$$D(t, q, V) = \mathbb{E}_\sigma \left[\int_t^{t_f} e^{-r(\tau-t)} u(\tau) \, d\tau + e^{-r(t_f-t)} \Psi(t_f, \tilde{V}(t_f)) \right],$$
$$t_f = \min\{\theta, T\},$$
$$\Psi(t_f, \tilde{V}(t_f)) = L(t_f, q(t_f), \tilde{V}(t_f)) \quad \text{if } t_f = \theta \leq T \text{ and } q(\theta) < \hat{q}(\theta),$$
$$\Psi(T, \tilde{V}(T)) = \hat{q}(T) - q(T) \text{ if } q(T) - \hat{q}(T) \geq 0,$$

and the equity value as

$$\mathcal{E}(t, q, V) = \mathbb{E}_\sigma \left[\int_t^{t_f} e^{-r(\tau-t)} \right.$$
$$\left. (\beta \tilde{V}(\tau) - u(\tau)) \tau + e^{-r(t_f-t)} \Phi(t_f, \tilde{V}(t_f)) \right],$$
$$t_f = \min\{\theta, T\},$$
$$\Phi(t_f, \tilde{V}(t_f)) = \{l(t_f, q(t_f), \tilde{V}(t_f))$$
$$\qquad -L(t_f, q(t_f), \tilde{V}(t_f)), 0\} \quad \text{if } t_f = \theta \leq T \text{ and } q(\theta) < \hat{q}(\theta),$$
$$\Phi(T, \tilde{V}(T)) = \tilde{V}(T) + q(T) - \hat{q}(T) \text{ if } q(T) - \hat{q}(T) \geq 0,$$

where $\tilde{V}(t)$ is the auxiliary process defined by

$$d\tilde{V}(t) = (r - \beta)\tilde{V}(t) \, dt + \varsigma \tilde{V}(t) \, dw(t)$$

and $\mathbb{E}_\sigma[\cdot]$ denotes the expected value w.r.t. the probability measure induced by the strategies σ. (Here $u(t)$ is the stochastic process induced by the feedback law $\sigma(\cdot)$). This is the usual change of measure occurring in Black-Scholes evaluations.

11.12.5 *Debt and equity valuations for Nash-equilibrium strategies*

Now let us reformulate the noncooperative dynamic game sketched in Section 11.12.3, using the real-option (or risk-neutral) approach to value debt and equity. The players' strategies are now chosen in such a way that a Nash equilibrium is reached at all points

$(t, s) \in [0, T] \times S$. Under the appropriate regularity conditions, a feedback-Nash equilibrium will be characterized by the HJB equations satisfied by equity and debt values

$$-\frac{\partial}{\partial t}\mathcal{E}(t, q, V) = \max_{u \in [0, \beta V]} \left[(\beta V - u) + \frac{\partial \mathcal{E}(t, q, V)}{\partial q}(u - \rho(\hat{q}(t) - q)^+) \right.$$
$$\left. - rE(t, q, V) + \frac{\partial \mathcal{E}(t, q, V)}{\partial V} rV + \frac{1}{2}\varsigma^2 V^2 \frac{\partial^2 \mathcal{E}(t, q, V)}{\partial V^2} \right], \quad (11.119)$$

$$-\frac{\partial}{\partial t}D(t, q, V) = \left[u + \frac{\partial D(t, q, V)}{\partial q}(u - \rho(\hat{q}(t) - q)^+) \right.$$
$$\left. - rD(t, q, V) + \frac{\partial D(t, q, V)}{\partial V} rV + \frac{1}{2}\varsigma^2 V^2 \frac{\partial^2 D(t, q, V)}{\partial V^2} \right]. \quad (11.120)$$

Notice that equation (11.120) does not involve a max operator, because the control of the lender is impulsive (he stops the game when he wishes). The boundary conditions at maturity T when $q \geq \hat{q}(T)$ are

$$\mathcal{E}(T, q, V) = V + q - \hat{q}(T),$$
$$D(T, q, V) = \hat{q}(T) - q.$$

The boundary conditions at θ are[38]

$$q < \hat{q}(\theta),$$
$$E(\theta, q, V) = \max\{V - L(\theta, q, V), 0\},$$
$$D(\theta, q, V) = L(\theta, q, V),$$
$$D(\theta^-, q, V) > L(\theta^-, q, V).$$

The optimal strategy for the firm is termed "bang-bang" and defined by the switching manifold

$$\mathcal{C}(t, q, V) = \text{sign} \left[\frac{\partial \mathcal{E}(t, q, V)}{\partial q} - 1 \right].$$

The equilibrium strategy is given by

$$\mathcal{C}(t, q, V) > 1 \quad \Rightarrow \quad u = \beta V, \quad (11.121)$$
$$\mathcal{C}(t, q, V) < 1 \quad \Rightarrow \quad u = 0, \quad (11.122)$$
$$\mathcal{C}(t, q, V) = 1 \quad \Rightarrow \quad u \in [0, \beta V]. \quad (11.123)$$

The manifold having equation $\mathcal{C}(t, q, V) = 1$ determines the firm's behavior. If the equation can be solved as $q = \Gamma(t, V)$, then

$$q < \Gamma(t, V) \quad \Rightarrow \quad u = \beta V,$$
$$q > \Gamma(t, V) \quad \Rightarrow \quad u = 0,$$
$$q = \Gamma(t, V) \quad \Rightarrow \quad u \in [0, \beta V].$$

The lender takes control as soon as the following conditions are satisfied:

$$\theta \in \Theta(q(\cdot)),$$
$$D(\theta, q(\theta), V(\theta)) \leq V(\theta) - K.$$

---

[38] We use the notation θ^- for $t - \varepsilon$ with $\varepsilon > 0$ arbitrarily small.

11.12.6 *Conclusion*

The design of the "best" debt contract would combine the design parameters T, c, g, P that maximize the Nash-equilibrium value of equity $E(0; (0, V))$, while the Nash-equilibrium debt value $D(0; (0, V))$ would be at least equal to the needed amount C. This is a very complicated problem that can be addressed by some direct search methods where the Nash-equilibria will be computed for a variety of design parameter values. In this section, we have concentrated on the evaluation of the equity and debt values when, given some contract terms, the firm owner and the lender act strategically and seek a feedback (Markov) Nash equilibrium. The interesting aspect of this $\mathbb{GE}$ illustration lies in the use of the equivalent risk-neutral or real option valuation technique in a stochastic-diffusion game, in the spirit of Black-Scholes economics.

Bibliography

[1] Abreu, D., Pearce, D. G. and Stacchetti, E. (1986). Optimal cartel equilibria with imperfect monitoring, *Journal of Economic Theory* **39**, 1, pp. 251–269.

[2] Akian, M., Chancellier, J. P. and Quadrat, J. (1988). Dynamic programming complexity and applications, in *Proceedings 27th IEEE CDC, Austin*, pp. 1551 – 1558.

[3] Amir, R. (1980). Stochastic games in economics and related fields: An overview, in A. Neyman and S. Sorin (eds.), *Stochastic Games and Applications* (Springer), pp. 455–470.

[4] Amir, R. (1989). A lattice-theoretic approach to a class of dynamic games, *Computers & Mathematics with Applications* **17**, 8-9, pp. 1345–1349.

[5] Amir, R. (1996). Cournot oligopoly and the theory of supermodular games, *Games and Economic Behavior* **15**, 2, pp. 132–148.

[6] Amrouche, N., Martín-Herrán, G. and Zaccour, G. (2008). Feedback stackelberg equilibrium strategies when the private label competes with the national brand, *Annals of Operations Research* **164**, pp. 79–95.

[7] Anderson, R. W. and Sundaresan, S. (1996). Design and valuation of debt contracts, *The Review of Financial Studies* **9**, 1, pp. 37–68.

[8] Aubin, J.-P. (1991). *Viability Theory, Systems and Control: Fondations and Applications* (Birkhäuser).

[9] Aubin, J.-P., Bayen, A. and Saint-Pierre, P. (2011). *Viability Theory: New Directions* (Springer).

[10] Audet, C. and Hansen, P. (2002). Enumeration of all extreme equilibria of bimatrix games, *SIAM journal on scientific computing* **23**, no1, pp. 323–338.

[11] Aumann, R. J. (1974). Subjectivity and correlation in randomized strategies, *Journal of Mathematical Economics* **1**, 1, pp. 67–96.

[12] Aumann, R. J. (1989). *Lectures on Game Theory* (Westview Press, Boulder).

[13] Aumann, R. J. (2008). Game engineering, in S. K. Neogy, R. B. Bapat, A. K. Das and T. Parthasarathy (eds.), *Mathematical Programming and Game Theory for Decision Making* (World Scientific), pp. 279–285.

[14] Aumann, R. J. and Maschler, M. (1995). *Repeated Games with Incomplete Information* (MIT Press).

[15] Babiker, M., Criqui, P., Ellerman, D., Reilly, J. and Viguier, L. (2003). Assessing the impact of carbon tax differentiation in the European Union, *Environmental Modeling & Assessment* **8**, 3, pp. 187–197.

[16] Babiker, M., Reilly, J. and Viguier, L. (2004). Is international emission trading always beneficial? *The Energy Journal* **25**, 2, pp. 33–56.

[17] Başar, T. and Bernhard, P. (1991). *H-Infinity Optimal Control and Related Minimax Design Problems: A Dynamic Game Approach* (Birkhäuser, Boston).

[18] Başar, T. and Olsder, G. K. (1982). *Dynamic Noncooperative Game Theory* (Academic Press, New York).

[19] Bahn, O., Haurie, A. and Malhamé, R. (2009). A stochastic control/game approach to the optimal timing of climate policies, in J. Filar and A. Haurie (eds.), *Uncertainty and Environmental Decision Making: A Handbook of Research and Best Practice*, *Springer International Series in Operations Research & Management Science*, Vol. 138, chap. A Stochastic control/game approach to the optimal timing of climate policies (Spinger), pp. 211–237.

[20] Balaguer, J., Orts, V. and Uriel, E. (2007). Testing price-fixing agreements in a multimarket context: The European case of Vitamin C, *International Review of Law and Economics* **27**, 2, pp. 245–257.

[21] Balbus, L., Reffett, K. L. and Wozny, L. P. (2010). A constructive study of Markov equilibria in stochastic games with strategic complementarities, Tech. rep., Available at SSRN: http://ssrn.com/abstract=1723038.

[22] Banach, S. (1922). Sur les opérations dans les ensembles abstraits et leur applications aux équations intégrales, *Fundamenta Mathematicae* **3**, pp. 133–181.

[23] Bellman, R. (1957). *Dynamic Programming* (Princeton Univ. Press).

[24] Bellman, R. (1961). *Adaptive Control Processes, A Guided Tour* (Princeton University Press).

[25] Bernard, A. and Vielle, M. (2003). Measuring the welfare cost of climate change policies: A comparative assessment based on the computable general equilibrium model GEMINI-E3, *Environmental Modeling & Assessment* **8**, 3, pp. 99–217.

[26] Bernard, A. and Vielle, M. (2008). GEMINI-E3, a general equilibrium model of international and national interactions between economy, energy and the environment, *Computational Management Science* **5**, 1-2, pp. 173–206.

[27] Bernheim, D. (1984). Rationalizable strategic behavior, *Econometrica* **52**, pp. 1007–1028.

[28] Birge, J. R. and Louveaux, F. (1997). *Introduction to Stochastic Programming*, Springer Series in Operations Research (Springer-Verlag, New York).

[29] Black, F. and Scholes, M. (1973). The pricing of options and corporate liabilities, *The Journal of Political Economy* **81**, 3, pp. 637–654.

[30] Blaquière, A., Gérard, F. and Leitmann, G. (1969). *Quantitative and Qualitative Games* (Academic Press, New York/London).

[31] Brander, J. A. and Spencer, B. J. (1983). Strategic commitment with R&D: The symmetric case, *The Bell Journal of Economics* **14**, 1, pp. 225–235.

[32] Breitner, M. (2005). The genesis of differential games in light of Issacs' contributions, *Journal of Optimization Theory and Applications* **124**, 3, pp. 523–559.

[33] Breton, M., Filar, J. A., Haurie, A. and Schultz, T. A. (1986). On the computation of equilibria in discounted stochastic dynamic games, in T. Baçar (ed.), *Dynamic Games and Applications in Economics*, *Lecture Notes in Economics and Mathematical Systems*, Vol. 265, pp. 64–87.

[34] Breton, M., Turki, A. and Zaccour, G. (2004). Dynamic model of R&D, spillovers and efficiency of Bertrand and Cournot equilibria, *Journal of Optimization Theory and Applications* **123**, 1, pp. 1–25.

[35] Brooke, A., Kendrick, D. and Meeraus, A. (1992). *GAMS. A User's Guide, Release 2.25* (Scientific Press/Duxbury Press).

[36] Bryson Jr., A. E. and Ho, Y. C. (1975). *Applied Optimal Control* (Taylor & Francis, Boca Raton).

[37] Buckdahn, R., Cardaliaguet, P. and Quincampoix, M. (2011). Some recent aspects of differential game theory, *Dynamic Games and Applications* **1**, 1, pp. 74–114.

[38] Buratto, A. and Zaccour, G. (2009). Coordination of advertising strategies in a fashion licensing contract, *Journal of Optimization Theory and Applications* **142**, 1, pp. 31–53.

[39] Calzolari, G. and Lambertini, L. (2007). Export restraints in a model of trade with capital accumulation, *Journal of Economic Dynamics & Control* **31**, 12, pp. 3822–3842.

[40] Carlson, D. A. and Haurie, A. (1995). A turnpike theory for infinite horizon open-loop differential games with decoupled dynamics, in G. J. Olsder (ed.), *New Trends in Dynamic Games and Applications*, *Annals of the International Society of Dynamic Games*, Vol. 3 (Birkhäuser), pp. 353–376.

[41] Carlson, D. A. and Haurie, A. (2000). Infinite horizon dynamic games with coupled state constraints, in J. A. filar, V. Gaitsgory and K. Misukami (eds.), *Advances in Dynamic Games and Applications*, *Annals of the International Society of Dynamic Games*, Vol. 5 (Birkhäuser, Boston), pp. 195–212.

[42] Carlson, D. A. and Haurie, A. (May 1987). *Infinite Horizon Optimal Control: Theory and Applications*, *Lecture Notes in Economics and Mathematicl Systems*, Vol. 290 (Springer Verlag).

[43] Carlson, D. A., Haurie, A. and Leizarowitz, A. (1991). *Infinite Horizon Optimal Control: Deterministic and Stochastic Systems*, Vol. 332 (Springer Verlag).

[44] Clark, C. W. (1976). *Mathematical Bioeconomics* (Wiley-Interscience, New York).

[45] Clark, C. W. (1980). Restricted access to common-property resources: A game theoretic analysis, in P. Liu (ed.), *Dynamic Optimization and Mathematical Economics* (Plenum), pp. 117–132.

[46] Clemhout, S. and Wan, H. Y. (1974). A class of trilinear differential games, *Journal of Optimization Theory and Applications* **14**, 4, pp. 419–424.

[47] Coase, R. (1960). The problem of social cost, *Journal of Law and Economics* **3**, pp. 1–44.

[48] Conner, J. M. (2006). The great global vitamins conspiracy: Sanctions and deterrence, Working Paper 06-02, American Antitrust Institute.

[49] Cournot, A. (1838). *Recherches sur les principes mathématiques de la théorie des richesses* (Librairie des sciences politiques et sociales, Paris).

[50] Cressman, R. (2003). *Evolutionary Dynamics and Extensive Form Games* (MIT Press).

[51] Criqui, P., Mima, S. and Viguier, L. (1999). Marginal abatement costs of CO_2 emission reductions, geographical flexibility and concrete ceilings: An assessment using the POLES model, *Energy Policy* **27**, 10, pp. 585–601.

[52] Curtat, L. O. (1996). Markov equilibria of stochastic games with complementarities, *Games and Economic Behavior* **17**, 2, pp. 177–199.

[53] Dantzig, G. B. (1963 re-edited 1998). *Linear Programming and Extensions* (Princeton University press).

[54] de Roos, N. (2001). Collusion with a competitive fringe: An application to Vitamin C, Working paper, Yale University.

[55] DeFreitas, G. and Marshall, A. (1998). Labour surplus, worker rights and productivity growth: A comparative analysis of Asia and Latin America, *Labour* **12**, 3, pp. 515–539.

[56] Denardo, E. V. (1967). Contractions mappings in the theory underlying dynamic programming, *SIAM Review* **9**, 2, pp. 165–177.

[57] Dixit, A. K. and Pindyck, R. S. (1994). *Investment under Uncertainty* (Princeton University Press).

[58] Dockner, E., Feichtinger, G. and Jørgensen, S. (1985). Tractable classes of nonzero-sum open-loop nash differential games, *Journal of Optimization Theory and Applications* **45**, pp. 179–198.

[59] Dockner, E., Jørgensen, S., Long, N. V. and Sorger, G. (2000). *Differential Games in Economics and Management Science* (Cambridge University Press, Cambridge).

[60] Dockner, E. J., Gaunersdorfer, A. and Jørgensen, S. (1996). Government price subsidies to promote fast diffusion of a new consumer durable, in S. Jørgensen and G. Zaccour (Eds.), *Dynamic Competitive Analysis in Marketing* (Springer-Verlag, Berlin), pp. 101–110.

[61] Dockner, E. J. and Long, N. V. (1993). International pollution control: Cooperative versus noncooperative strategies, *Journal of Environmental Economics and Management* **25**, 1, pp. 13–29.

[62] Doležal, J. (1976). Necessary optimality conditions for n-player nonzero-sum multistage games, *Kybernetika* **12**, 4, pp. 264–295.

[63] Dorfman, R., Samuelson, P. A. and Solow, R. (1958). *Linear Programming and Economic Analysis* (Dover re-edited 1987).

[64] Drouet, L., Haurie, A., Moresino, F., Vial, J.-P., Vielle, M. and Viguier, L. (2008). An oracle based method to compute a coupled equilibrium in a model of international climate policy, *Computational Management Science* **5**, 1, pp. 119–140.

[65] Duffie, D. (1992). *Dynamic Asset Pricing Theory* (Princeton University Press).

[66] Dutta, P. K. and Radner, R. (2004). Self-enforcing climate-change treaties, *Proceedings of the National Academy of Sciences of the United States of America* **101**, 14, pp. 5174–5179.

[67] Dutta, P. K. and Radner, R. (2009). A strategic analysis of global warming: Theory and some numbers, *Journal of Economic Behavior & Organization* **71**, 2, pp. 187–209.

[68] Engwerda, J. (2005). *Linear-Quadratic Dynamic Optimization and Differential Games* (Wiley, New York).

[69] European Commission (2003). *The EU Emissions Trading Scheme: How to develop a National Allocation Plan* (European Commission, Brussels), non-Paper 2nd meeting of Working 3 Monitoring Mechanism Committee, April 1.

[70] F. Murphy, H. S. and Soyster, A. (1982). A mathematical programming approach for determining oligopolistic market equilibrium, *Mathematical Programming* **24**, 1, pp. 92–106.

[71] Facchinei, F., Fischer, A. and Piccialli, V. (2007). On generalized Nash games and variational inequalities, *Operations Research Letters* **35**, 2, pp. 159–164.

[72] Facchinei, F., Fischer, A. and Piccialli, V. (2009). Generalized Nash equilibrium problems and Newton methods, *Mathematical Programming, Ser. B* **117**, 1-2, pp. 163–194.

[73] Facchinei, F. and Kanzow, C. (2007). Generalized Nash equilibrium problems, *4OR: A Quartely Journal of Operations Research* **5**, 3, pp. 173–210.

[74] Fan, K. (1952). Fixed-point and minimax theorems in locally convex topological linear spaces, *Proc. Nat. Acad. Sci. U.S.A.* **38**, pp. 121–126.

[75] Fan, L. T. and Wang, C. S. (1964). *The Discrete Maximum Principle* (John Wiley and Sons, New York).

[76] Feichtinger, G. and Dockner, E. J. (1985). Optimal pricing in a duopoly: A noncooperative differential games solution, *Journal of Optimization Theory and Applications* **45**, 2, pp. 199–218.

[77] Ferris, M. C. and Munson, T. S. (1999). Interfaces to PATH 3.0, *Computational Optimization and Applications* **12**, 1-3, pp. 207–227.

[78] Ferris, M. C. and Munson, T. S. (2000). Complementarity problems in GAMS and the PATH solver, *Journal of Economic Dynamics and Control* **24**, 2, pp. 165–188.

[79] Ferris, M. C. and Pang, J. S. (1997a). *Complementarity and Variational Problems: State of the Art* (SIAM Publications, Philadelphia, Pennsylvania).

[80] Ferris, M. C. and Pang, J.-S. (1997b). Engineering and economic applications of complementarity problems, *SIAM Review* **39**, pp. 669–713.

[81] Filar, J. A. and Tolwinski, B. (1991). On the algorithm of Pollatschek and Avi-Itzhak, in T. E. S. Raghavan et al. (ed.), *Stochastic Games and Related Topics* (Kluwer Academic Publishers), pp. 59–70.

[82] Filar, J. A. and Vrieze, K. (1997). *Competitive Markov Decision Processes* (Springer-Verlag, New York).

[83] Forges, F. (1986). An approach to communication equilibria, *Econometrica* **54**, 6, pp. 1375–1385.

[84] Fourer, R., Gay, D. M. and Kernighan, B. W. (1993). *AMPL: A Modeling Language for Mathematical Programming* (Scientific Press/Duxbury Press).

[85] Friedman, A. (1971). *Differential Games* (Wiley-Interscience, New York).

[86] Friedman, J. W. (1977). *Oligopoly and the Theory of Games* (North-Holland, Amsterdam).

[87] Friedman, J. W. (1986). *Game Theory with Economic Applications* (Oxford University Press, Oxford).

[88] Friesz, T. L. (2010). *Dynamic Optimization and Differential Games*, International Series in Operations Research and Management Science (Springer).

[89] Fudenberg, D. and Tirole, J. (1991). *Game Theory* (The MIT Press, Cambridge, Massachusetts, London, England).

[90] Fukushima, M. (2009). Restricted generalized Nash equilibria and controlled penalty algorithm, *Computational Management Science* **31**, 5, pp. 1–18.

[91] Gabay, D. and Moulin, H. (1980). On the uniqueness and stability of nash equilibria in noncooperative games, in A. Bensoussan, P. Kleindorfer and C. Tapiero (eds.), *Applied Stochastic Control in Econometrics and Management Science* (Amsterdam: North-Holland), pp. 271–293.

[92] Genc, T., Reynolds, S. S. and Sen, S. (2007). Dynamic oligopolistic games under uncertainty: A stochastic programming approach, *Journal of Economic Dynamics & Control* **31**, 1, pp. 55–80.

[93] Genc, T. and Sen, S. (2008). An analysis of capacity and price trajectories for the Ontario electricity market using dynamic nash equilibrium under uncertainty, *Energy Economics* **30**, 1, pp. 173–191.

[94] Glicksberg, I. (1952). A further generalization of the Kakutani fixed point theorem with application to Nash equilibrium points, *Proceedings of the American Mathematical Society* **3**, pp. 170–174.

[95] Gosh, M. R., Arapostathis, A. and Marcus, S. (1992). Optimal control of switching diffusions with application to flexible manufacturing systems, *SIAM Journal on Control and Optimization* **30**, 6, pp. 1–23.

[96] Green, E. and Porter, R. H. (1984). Noncooperative collusion under imperfect price information, *Econometrica* **52**, 1, pp. 87–100.

[97] Grenadier, S. R. (2002). Option exercise games: An application to the equilibrium investment strategies of firms, *The Review of Financial Studies* **15**, 3, pp. 691–721.

[98] Gürkan, G., Özge, A. Y. and Robinson, S. M. (1999). Sample-path solution of stochastic variational inequalities, *Mathematical Programming* **84**, 2, pp. 313–333.

[99] Hämäläinen, R. P., A.Haurie and Kaitala, V. (1985). Equilibria and threats in a fishery management game, *Optimal Control Applications and Methods* **6**, 4, pp. 315–333.

[100] Hämäläinen, R. P., Kaitala, V. and Haurie, A. (1984). Bargaining on whales: A differential game model with Pareto optimal equilibria, *Operations Research Letters* **3**, 1, pp. 5–11.

[101] Harker, P. T. (1991). Generalized Nash games and quasivariational inequalities, *European Journal of Operational Research* **4**, 1, pp. 81–94.

[102] Harsany, J. (1967-68). Games with incomplete information played by Bayesian players, *Management Science* **14**, pp. 159–182, 320–334, 486–502.

[103] Haurie, A. (1975). On some properties of the characteristic function and the core of a multistage game of coalitions, *IEEE Transactions on Automatic Control* **AC-20**, pp. 238–241.

[104] Haurie, A. (1976). A note on nonzero-sum differential games with bargaining solution, *Journal of Optimization Theory and Applications* **18**, pp. 31–39.

[105] Haurie, A. (1989). Piecewise deterministic differential games, in T. Başar and P. Bernhard (eds.), *Differential Games and Applications*, Lecture Notes in Control and Information Sciences, Vol. 119 (Springer-Verlag), pp. 114–127.

[106] Haurie, A. (1991). Piecewise deterministic and piecewise diffusion differential games, in G. Ricci (ed.), *Lecture Notes in Economics and Mathematical Systems*, Vol. 353 (Springer).

[107] Haurie, A. (1993). From repeated to differential games: How time and uncertainty pervade the theory of games, in K. Binmore, A. Kirman and P. Tani (eds.), *Frontiers of Game Theory* (MIT Press, Cambridge0), pp. 165–193.

[108] Haurie, A. (1995). Environmental coordination in dynamic oligopolistic markets, *Group Decision and Negotiation* **4**, 1, pp. 39–57.

[109] Haurie, A. and Krawczyk, J. B. (1997). Optimal charges on river effluent from lumped and distributed sources, *Environmental Modelling and Assessment* **2**, 3, pp. 177–199.

[110] Haurie, A., Krawczyk, J. B. and Roche, M. (1994). Monitoring cooperative equilibria in a stochastic differential game, *Journal of Optimization Theory and Applications* **81**, 1, pp. 73–95.

[111] Haurie, A. and Leizarowitz, A. (1992). Overtaking optimal regulation and tracking of piecewise diffusion linear systems, *SIAM Journal on Control and Optimization* **3**, 4, pp. 816–837.

[112] Haurie, A. and Moresino, F. (2002a). Computation of S-adapted equilibria in piecewise deterministic games via stochastic programming methods, in E. Altman and O. Pourtallier (eds.), *Advances in Dynamic Games and Applications*, Annals of the International Society of Dynamic Games, Vol. 6, chap. 13 (Birkhäuser), pp. 225–252.

[113] Haurie, A. and Moresino, F. (2002b). A differential game of debt contract valuation, in M. Dror, P. L'Ecuyer and F. Szidarovsky (eds.), *Modeling Uncertainty: An Examination of Stochastic Theory, Methods and Applications*, International Series in Operations Research and Management Science (Kluwer), pp. 267–282.

[114] Haurie, A. and Moresino, F. (2002c). S-adapted oligopoly equilibria and approximations in stochastic variational inequalities, *Annals of Operations Research* **114**, 1-4, pp. 183–201.

[115] Haurie, A. and Pohjola, M. (1987). Efficient equilibria in a differential game of capitalism, *Journal of Economic Dynamics and Control* **11**, 1, pp. 65–78.

[116] Haurie, A. and Roche, M. (1994). Turnpikes and computation of piecewsie open-loop equilibria in stochastic differential games, *Journal of Economic Dynamics and Control* **18**, pp. 317–344.

[117] Haurie, A. and Tolwinski, B. (1985). Definition and properties of cooperative equilibria in a two-player game of infinite duration, *Journal of Optimization Theory and Applications* **46**, 4, pp. 525–534.

[118] Haurie, A. and Tolwinski, B. (1990). Cooperative equilibria in discounted stochastic sequential games, *Journal of Optimization Theory and Applications* **64**, 3, pp. 511–535.

[119] Haurie, A. and Vielle, M. (2007). Le cartel du pétrole et les politiques climatiques mondiales: Une analyse par la théorie de jeux, in P. Burger and R. Kaufmann-Hayoz (eds.), *Nachhaltige Entwicklung: Recherche dans le domaine du développement durable – perspective des sciences sociales et humaines* (Académie suisse des sciences humaines et sociales, Bern), pp. 167–205.

[120] Haurie, A. and Zaccour, G. (1995). Differential game models of global environmental management, in C. Carraro and J. A. Filar (eds.), *Control and Game-Theoretic Models of the Environment*, Annals of the International Society of Dynamic Games, Vol. 2 (Birkhäuser, Boston), pp. 3–23.

[121] Haurie, A. and Zaccour, G. (2005). S-adapted equilibria in games played over event trees: An overview, *Annals of the International Society of Dynamic Games* **7**, pp. 417–444.

[122] Haurie, A., Zaccour, G., Legrand, J. and Smeers, Y. (1987). A stochastic dynamic Nash-Cournot model for the European gas market, Tech. Rep. G-87-24, Les Cahiers du GERAD.

[123] Haurie, A., Zaccour, G., Legrand, J. and Smeers, Y. (1988). Un modèle de Nash-Cournot stochastique et dynamique pour le marché européen du gaz, in *Actes du colloque Modélisation et analyse des marchés du gaz naturel* (HEC, Montréal).

[124] Haurie, A., Zaccour, G. and Smeers, Y. (1990). Stochastic equilibrium programming for dynamic oligopolistic markets, *Journal of Optimization Theory and Applications* **66**, 2, pp. 243–253.

[125] Hofbauer, J. and Sigmund, K. (1998). *Evolutionary Games and Population Dynamics* (Cambridge University Press).

[126] Hofman, A. and Karp, R. (1966). On non-terminating stochastic games, *Management Science* **12**, 5, pp. 359–370.

[127] Howard, R. (1960). *Dynamic Programming and Markov Processes* (MIT Press, Cambridge Mass).

[128] Isaacs, R. (1965). *Differential Games* (Wiley, New York).

[129] Jørgensen, S. and Zaccour, G. (1999). Price subsidies and guaranteed buys of a new technology, *European Journal of Operational Research* **114**, 2, pp. 338–345.

[130] Jørgensen, S. and Zaccour, G. (2000). Optimal output strategies in a two-stage game with entry, learning by doing and spillover, in L. A. Petrosjan and V. V. Mazalov (eds.), *Game Theory and Applications* (Science Publishers Inc.), pp. 65–72.

[131] Jørgensen, S. and Zaccour, G. (2004). *Differential Games in Marketing* (Kluwer, Boston).

[132] Jørgensen, S. and Zaccour, G. (2007). Developments in differential game theory and numerical methods: Economic and management applications, *Computational Management Science* **4**, 2, pp. 159–182.

[133] Judd, K. (1998). *Numerical Methods in Economics* (MIT Press).

[134] Kaldor, N. (1961). Capital accumulation and economic growth, in F. Lutz and D. C. Hague (eds.), *Proceedings of a Conference Held by the International Economics Association* (McMillan, London), pp. 177–222.

[135] Kalish, S. and Lilien, G. L. (1983). Optimal price subsidy for accelerating the diffusion of innovations, *Marketing Science* **2**, 2, pp. 407–420.

[136] Karatzas, I. and Shreve, S. E. (1988). *Brownian Motion and Stochastic Calculus* (Springer-Verlag, New York).

[137] Konnov, I. (2007). *Equilibrium Models and Variational Inequalities, Mathematics in Science and Engineering*, Vol. 210 (Elsevier).

[138] Krassovski, N. N. and Subbotin, A. I. (1977). *Jeux différentiels* (Mir, Moscow).

[139] Krawczyk, J. B. (2005). Coupled constraint Nash equilibria in environmental games, *Resource and Energy Economics* **27**, 2, pp. 157–181.

[140] Krawczyk, J. B. (2007). Numerical solutions to coupled-constraint (or generalised) Nash equilibrium problems, *Computational Management Science* **4**, 2, pp. 183–204.

[141] Krawczyk, J. B. and Shimomura, K. (2003). Why countries with the same technology and preferences can have different growth rates, *Journal of Economic Dynamics and Control* **27**, 10, pp. 1899–1916.

[142] Krawczyk, J. B. and Tidball, M. (2006). A discrete-time dynamic game of seasonal water allocation, *Journal of Optimization Theory and Applications* **128**, 2, pp. 411–429.

[143] Krawczyk, J. B. and Tidball, M. (2009). How to use Rosen's normalised equilibrium to enforce a socially desirable Pareto efficient solution, Working Papers 09-20, LAMETA, Universtiy of Montpellier, URL http://ideas.repec.org/p/lam/wpaper/09-20.html.

[144] Krawczyk, J. B. and Uryasev, S. (2000). Relaxation algorithms to find Nash equilibria with economic applications, *Environmental Modelling and Assessment* **5**, 1, pp. 63–73.

[145] Krawczyk, J. B. and Zuccollo, J. (2006). NIRA-3: A MATLAB package for finding Nash equilibria in infinite games, Working Paper, School of Economics and Finance, VUW.

[146] Kuhn, H. W. (1953). Extensive games and the problem of information, in H. W. Kuhn and A. W. Tucker (eds.), *Contributions to the Theory of Games, Annals of Mathematical Studies*, Vol. 2 (Princeton University press, Princeton new Jersey,), pp. 193–216.

[147] Kushner, H. J. (1977). *Probability Methods for Approximation in Stochastic Control and for Elliptic Equations* (Academic Press).

[148] Kushner, H. J. (2008). Numerical methods for non-zero-sum stochastic differential games: Convergence of the Markov chain approximation method, in P.-L. Chow, G. Yin and B. Mor-

dukhovich (eds.), *Topics in Stochastic Analysis and Nonparametric Estimation*, *The IMA Volumes in Mathematics and its Applications*, Vol. 145 (Springer, New York), pp. 51–84, 10.1007/978-0-387-75111-5_4.

[149] Kushner, H. K. and Dupuis, P. G. (1991). *Numerical Methods for Stochastic Control Problems in Continuous Time* (Springer-Verlag, New-York).

[150] Lee, E. B. and Markus, L. (1967). *Foundations of Optimal Control Theory* (J. Wiley, New York).

[151] Lehrer, E. and Sorin, S. (1992). A uniform Tauberian theorem in dynamic programming, *Mathematics of Operations Research* **17**, 2, pp. 303–307.

[152] Leitmann, G. (1974). *Cooperative and Non-cooperative Many Players Differential Games* (Springer, New York).

[153] Leitmann, G. and Schmitendorf, W. E. (1978). Profit maximization through advertising: A nonzero sum differential game approach, *IEEE Transactions on Automatic Control* **AC-23**, 4, pp. 645–650.

[154] Lemke, C. E. (1962). A method of solution for quadratic programs, *Management Science* **8**, 4, pp. 442–453.

[155] Lemke, C. E. (1965). Bimatrix equilibrium points and mathematical programming, *Management Science* **11**, 7, pp. 681–689.

[156] Lemke, C. E. and Howson, J. T. (1964). Equilibrium points of bimatrix games, *Journal of the Society for Industrial and applied Mathematics* **12**, pp. 413–423.

[157] Léonard, D. and van Long, N. (1992). *Optimal Control Theory and Static Optimization in Economics* (Cambridge University Press, Cambridge.).

[158] Levenstein, M. and Suslov, V. (2006). What determines cartel success, *Journal of Economic Literature* **44**, 1, pp. 43–95.

[159] Levhari, D. and Mirman, L. J. (1980). The great fish war: An example using a dynamic Cournot-Nash solution, *The Bell Journal of Economics* **11**, 1, pp. 322–334.

[160] Littman, M. L. (1994). Markov games as a framework for multi-agent reinforcement learning, in *Proceedings of the Eleventh International Conference on Machine Learning* (Morgan Kaufman, New Brunswick, NJ), pp. 157–163.

[161] Littman, M. L. (2001). Value-function reinforcement learning in Markov games, *Cognitive Systems Research* **2**, 1, pp. 55–66.

[162] Long, N. V. (1977). Optimal exploitation and replenishment of a natural resource, in J. Pitchford and S. Turnovsky (eds.), *Applications of Control Theory in Economic Analysis* (North-Holland, Amsterdam), pp. 81–106.

[163] Long, N. V. (2011). Dynamic games in the economics of natural resources: A survey, *Dynamic Games and Applications* **1**, 1, pp. 115–148.

[164] Luenberger, D. G. (1969). *Optimization by Vector Space Methods* (J. Wiley & Sons, New York).

[165] Luenberger, D. G. (1979). *Introduction to Dynamic Systems: Theory, Models & Applications* (J. Wiley & Sons, New York).

[166] MacCracken, C. N., Edmonds, J. A., Kim, S. H. and Sands, R. D. (1999). The economics of the Kyoto Protocol, *The Energy Journal* **201**, pp. 25–71.

[167] Mangasarian, O. L. (1969). *Nonlinear Programming* (McGraw-Hill, New York).

[168] Mangasarian, O. L. and Stone, H. (1964). Two-person nonzerosum games and quadratic progamming, *Journal of Mathematical Analysis and Applications* **9**, pp. 348–355.

[169] Marx, K. (1967). *Capital* (International Publishers, New York).

[170] McKelvey, R. (1992). A Lyapunov function for Nash equilibria, Social Science Working Paper.

[171] McKelvey, R. D. and Palfrey, T. R. (1995). Quantal response equilibria for normal form games, *Games and Economic Behavior* **10**, 1, pp. 6–38.

[172] McKelvey, R. D. and Palfrey, T. R. (1996). A statistical theory of equilibrium in games, *The Japanese Economic Review* **47**, 2, pp. 187–209.

[173] McKenzie, L. (1976). Turnpike theory, *Econometrica* **44**, 5, pp. 841–865.

[174] McKibbin, W. J., Shackelton, R. and Wilcoxen, P. J. (1999). What to expect from an international system of tradable permits for carbon emissions? *Resource and Energy Economics* **21**, 3-4, pp. 319–346.

[175] Mehlmann, A. (1988). *Applied Differential Games* (Plenum Press, New York).

[176] Milgrom, P. and Roberts, J. (1990). Rationalizability, learning, and equilibrium in games with strategic complementarities, *Econometrica* **58**, 6, pp. 1255–1277.

[177] Milgrom, P. and Shannon, C. (1994). Monotone comparative statics, *Econometrica* **62**, 1, pp. 157–180.

[178] Munro, G. R. (1979). The optimal management of transboundary renewable resources, *Canadian Journal of Economics* **12**, 3, pp. 355–376.

[179] Myerson, R. B. (1997). *Game Theory: Analysis of Conflict* (Harvard University Press, Cambridge Mass.).

[180] Nash, J. F. (1951). Non-cooperative games, *Annals of Mathematics* **54**, pp. 286–295.

[181] Neumann, J. V. (1928). Zur theorie der gesellshaftsspiele, *Math. Annalen* **100**, pp. 295–320.

[182] Neumann, J. V. and Morgenstern, O. (1944). *Theory of Games and Economic Behavior* (Princeton University Press, Princeton).

[183] Neyman, A. and Sorin, S. (2003). *Stochastic Games: Classification and Basic Tools* (NATO ASI, Kluwer Academic Publishers).

[184] Nordhaus, W. D. (1994). *Managing the Global Commons: The Economics of Climate Change* (MIT Press, Cambridge, Mass.).

[185] Nordhaus, W. D. and Yang, Z. (1996). A regional dynamic general-equilibrium model of alternative climate change strategies, *American Economic Review* **86**, pp. 741–765.

[186] Nowak, A. S. (1985). Existence of equilibrium stationary strategies in discounted noncooperative stochastic games with uncountable state space, *Journal of Optimization Theory and Applications* **45**, pp. 591–602.

[187] Nowak, A. S. (1987). Nonrandomized strategy equilibria in noncooperative stochastic games with additive transition and reward structures, *Journal of Optimization Theory and Applications* **52**, 3, pp. 429–441.

[188] Nowak, A. S. (1993). Stationary equilibria for nonzero-sum average payoff ergodic stochastic games with general state space, in T. Baçar and A. Haurie (eds.), *Advances in Dynamic Games and Applications*, no. 1 in Annals of the International Society of Dynamic Games (Birkhäuser), pp. 231–246.

[189] Nowak, A. S. (2003). *N*-person stochastic games: Extensions of the finite state space case and correlation, in A. Neyman and S. Sorin (eds.), *Stochastic Games and Applications*, NATO science series, series C: Mathematical and physical sciences, Vol. 570 (Kluwer, Dordrecht), pp. 93–106.

[190] Nowak, A. S. (2007). On stochastic games in economics, *Mathematical Methods of Operations Research* **66**, 3, pp. 513–530.

[191] Nowak, A. S. and Raghavan, T. E. S. (1992). Existence of stationary correlated equilibria with symmetric information for discounted stochastic games, *Mathematics of Operations Research* **17**, 3, pp. 519–526.

[192] Osborne, M. J. and Rubinstein, A. (1994). *A Course in Game Theory* (MIT Press, Cambridge Mass.).

[193] Owen, G. (1982). *Game Theory* (Academic Press,, New York, London).

[194] Pang, J.-S. and Fukushima, M. (2005). Quasi-variational inequalities, generalized Nash equilibria and multi-leader-follower games, *Computational Management Science* **1**, 1, pp. 21–56.

[195] Parthasarathy, T. (1973). Discounted, positive and non-cooperative stochastic games, *Interna-*

tional Journal of Game Theory **2**, 1, pp. 25–37.

[196] Parthasarathy, T. and Shina, S. (1989). Existence of stationary equilibrium strategies in nonzero-sum discounted stochastic games with uncountable state space and state independent transitions, *International Journal of Game Theory* **18**, 2, pp. 189–194.

[197] Pearce, D. (1984). Rationalizable strategic behavior and the problem of perfection, *Econometrica* **52**, 4, pp. 1029–1050.

[198] Petrosjan, L. (1993). *Differential Games of Pursuit* (World Scientific, Singapore).

[199] Petrosjan, L. and Zenkevich, N. A. (1996). *Game Theory* (World Scientific, Singapore).

[200] Pineau, P.-O. and Murto, P. (2003). An oligopolistic investment model of the Finnish electricity market, *Annals of Operations Research* **121**, pp. 123–148.

[201] Pineau, P.-O., Rasata, H. and Zaccour, G. (2011). Impact of some parameters on investments in oligopolistic electricity markets, *European Journal of Operational Research* **213**, 1, pp. 180–195.

[202] Pollatschek, M. A. and Avi-Itzhak, B. (1969). Algorithm for stochastic games with geometrical interpretation, *Management Science* **15**, 7, pp. 399–415.

[203] Porter, R. H. (1983). Optimal cartel trigger price strategies, *Journal of Economic Theory* **29**, 2, pp. 313–338.

[204] Puterman, M. L. (1994). *Markov Decision Processes* (Wiley).

[205] Quint, T. and Shubik, M. (1997). A theorem on the number of Nash equilibria in a bimatrix game, *International Journal of Game Theory* **26**, 3, pp. 1432–1470.

[206] Radner, R. (1980). Collusive behavior in noncooperative ε-equilibria of oligopolies with long but finite lives, *Journal of Economic Theory* **22**, 2, pp. 136–154.

[207] Radner, R. (1981). Monitoring cooperative agreement in a repeated principal-agent relationship, *Econometrica* **49**, 5, pp. 1127–1148.

[208] Ramsey, F. P. (1928). A mathematical theory of saving, *Economic Journal* **38**, 152, pp. 543–559.

[209] Rao, S., Chandrasekaran, R. and Nair, K. (1973). Algorithms for discounted stochastic games, *Journal of Optimization Theory and Applications* **11**, 6, pp. 627–637.

[210] Rasata, H. and Zaccour, G. (2010). An empirical investigation of open-loop and closed-loop equilibrium investment strategies in an electricity oligopoly market, Tech. Rep. G-2010-54, Cahiers du GERAD.

[211] Rogers, P. D. (1969). *Non-zero sum stochastic games*, Ph.D. thesis, University of California, Berkeley.

[212] Romer, P. M. (1986). Increasing returns and long-run growth, *Journal of Political Economy* **94**, 5, pp. 1002–1037.

[213] Rosen, J. B. (1965). Existence and uniqueness of equilibrium points for concave n-person games, *Econometrica* **33**, 3, pp. 520–534.

[214] Rosenthal, R. W. (1981). Games of perfect information, predatory pricing, and the chain store, *Journal of Economic Theory* **25**, 1, pp. 92–100.

[215] Rowat, C. (2007). Non-linear strategies in a linear quadratic differential game, *Journal of Economic Dynamics and Control* **31**, 10, pp. 3179–3202.

[216] Rutherford, T. F. (1995). Extensions of GAMS for complementarity problems arising in applied economic analysis, *Journal of Economic Dynamics and Control* **19**, 8, pp. 1299–1324.

[217] Sandholm, W. (2010). *Population Games and Evolutionary Dynamics* (MIT Press).

[218] Selten, R. (1975). Rexaminition of the perfectness concept for equilibrium points in extensive games, *International Journal of Game Theory* **4**, 1, pp. 25–55.

[219] Shapley, L. (1953). Stochastic games, *Proceedings of the National Academy of Science* **39**, pp. 1095–1100.

[220] Shor, N. Z. (1985). *Minimization Methods for Non-Differentiable Functions* (Springer-Verlag).

[221] Shubik, M. (1975a). *Games for Society, Business and War* (Elsevier, New York).

[222] Shubik, M. (1975b). *The Uses and Methods of Gaming* (Elsevier, New York).

[223] Sobel, M. J. (1971). Noncooperative stochastic games, *Annals of Mathematical Statistics* **42**, pp. 1930–1935.

[224] Sorin, S. (2002). *A first course on zero-sum repeated games*, *Mathématiques & Applications*, Vol. 37 (Springer-Verlag, Berlin).

[225] Spencer, B. J. and Brander, J. A. (1983). International R&D rivalry and industrial strategy, Working Paper 1192, National Bureau of Economic Research, URL http://www.nber. org/papers/w1192.

[226] Stigler, G. J. (1964). A theory of oligopoly, *Journal of Political Economy* **72**, 1, pp. 44–61.

[227] Tarski, A. (1955). A lattice-theoretical fixpoint theorem and its applications, *Pacific Journal of Mathematics* **5**, 2, pp. 285–309.

[228] Tolwinski, B. (1989). Newton-type methods for stochastic games, in T. Baçar and P. Bernhard (eds.), *Differential Games and Applications*, Vol. 119, pp. 128–143.

[229] Tolwinski, B. (1991). Solving dynamic games via Markov game approximations, in R. Hämäläinen and H. Ehtamo (eds.), *Differential Games – Developments in Modelling and Computation*, Numerical Methods For Dynamic And Hierarchical Games, pp. 265–274.

[230] Tolwinski, B., Haurie, A. and Leitmann, G. (1986). Cooperative equilibria in differential games, *Journal of Mathematical Analysis and Applications* **119**, pp. 182–202.

[231] Topkis, D. M. (1978). Minimizing a submodular function on a lattice, *Operations Research* **26**, pp. 305–321.

[232] Topkis, D. M. (1979). Equilibrium points in non-zero sum n-person submodular games, *SIAM Journal on Control and Optimization* **17**, 6, pp. 773–787.

[233] Topkis, D. M. (1998). *Supermodularity and Complementarity* (Princeton University Press).

[234] Tsutsui, S. and Mino, K. (1990). Nonlinear strategies in dynamic duopolistic competition with sticky prices, *Journal of Economic Theory* **52**, 1, pp. 136–161.

[235] van der Wal, J. (1978). Discounted Markov games: Generalized policy iteration method, *Journal of Optimization Theory and Applications* **25**, 1, pp. 125–138.

[236] Varayia, P. and Lin, J. (1963). Existence of saddle points in differential games, *SIAM Journal on Control and Optimization* **7**, 2, pp. 142–157.

[237] Viguier, L., Vielle, M., Haurie, A. and Bernard, A. (2006). A two-level computable equilibrium model to assess the strategic allocation of emission allowances within the European union, *Computers & Operations Research* **33**, 2, pp. 369–385, game Theory: Numerical Methods and Applications.

[238] Vives, X. (1990). Nash equilibrium with strategic complementarities, *Journal of Mathematical Economics* **19**, pp. 305–321.

[239] Vives, X. (2001). *Oligopoly Pricing: Old Ideas and New Tools* (Cambridge: MIT Press).

[240] von Heusingen, C. and Kanzow, A. (2006). Optimization reformulations of the generalized Nash equilibrium problem using Nikaido-Isoda-type functions, Preprint 269, University of Würzburg, Institute of Mathematics.

[241] von Stackelberg, H. (1934). *Marktform und Gleichgewicht* (J. Springer).

[242] von Stackelberg, H. (2011). *Market Structure and Equilibrium* (Springer, Heidelberg Dordrect London New York).

[243] Weibull, J. (1995). *Evolutionary Game Theory* (MIT Press).

[244] Weyant, J. P. (1999). *The Costs of the Kyoto Protocol: A Multi-Model Evaluation*, special issue (Energy Journal).

[245] Whitt, W. (1980). Representation and approximation of noncooperative sequential games, *SIAM Journal on Control and Optimization* **18**, 1, pp. 33–48.

[246] Wilson, R. (1971). Equilibria of n-person games, *SIAM Journal on Applied Mathematics* **21**, 1, pp. 80–87.

[247] Yeung, D. W. K. and Petrosjan, L. (2005). *Cooperative Stochastic Differential Games* (Springer, New York, NY).

[248] Zaccour, G. (1987). *Théorie des jeux et Marchés énergétiques: Marché Européen de gaz naturel et échanges d'électricité*, Ph.D. thesis, HEC, Montréal.

[249] Zaccour, G. (1996). A differential game model for optimal price subsidy of new technologies, *Game Theory and Applications* **2**, pp. 103–114.

[250] Zadeh, L. A. and Desoer, C. A. (1963). *Linear System Theory: The State Space Approach*, McGraw-Hill Series in System Science (McGraw-Hill, New York).

[251] Zeidler, E. (1995). *Applied Functional Analysis: Applications to Mathematical Physics* (Springer-Verlag, New York).

[252] Zelikin, M. I. (1994). *Theory of Chattering Control with Applications to Astronautics, Robotics, Economics, and Engineering* (Birkhäuser).

Index